AF327505

# STOCK ASSESSMENT IN
# INLAND FISHERIES

# STOCK ASSESSMENT IN INLAND FISHERIES

EDITED BY

## I.G. COWX

*Hull International Fisheries Institute*
*University of Hull, UK*

FISHING NEWS BOOKS

# Contents

# Preface

Following the second international symposium organized and hosted by the University of Hull International Fisheries Institute, on Catch Effort Sampling Strategies: their Application in Freshwater Fisheries Management (Cowx 1991) it was recognized that the subject of stock assessment in inland fisheries was extremely important to the concept of fisheries management. However, there was general concern about the status and development of stock assessment techniques and how they were used by scientists and managers for management purposes. In view of the importance of these techniques in the management of inland fisheries, it was felt that they warranted further discussion and dissemination.

To this end, the fourth biennial, international workshop and symposium to be organized and hosted by the International Fisheries Institute, at the University of Hull, UK, from 11–15 April 1994 focused on the subject of *Stock Assessment in Inland Fisheries*. The objectives of the symposium were to advance the scientific and management basis of fisheries stock assessment, and to provide a medium for the dissemination and exchange of ideas. Preliminary discussions considered the subject material to be extremely broad and it was decided that the appropriate breakdown for discussions should focus on five key areas: Survey Methods; Population Modelling; Assessment of Riverine Stocks; Assessment of Lake and Reservoir Stocks; and Fisheries Management. These proceedings contain selected papers from the symposium and highlight many of the key issues in these areas, both in temperate and tropical inland fisheries. The proceedings are organized into sections based on the working parties, and the key sections are fronted by a chapter summarizing the discussions of the working group. It is hoped they will stimulate fisheries scientists, managers and academics to collaborate in further research to improve our understanding of Stock Assessment in Inland Fisheries.

The production of these proceedings has involved considerable effort by a number of people. In particular, thanks must go to members of the Working Groups on each subject area for the fruitful discussions and synthesis of the information which is found in the introductory chapters to each section. The membership of the Working Groups was as follows. *Survey Methods*: P. Hickley (UK – Chairman), E. Baras (Belgium), J. Coeck (Belgium), M. Cryer (New Zealand), I. Davidson (UK), D. Evans (UK), J. Kubeka (Czech Republic), R. Laughton (UK), H. Löffler (Germany), K. McCarthy (Ireland), I. Winfield (UK). *Population Modelling*: W. Dekker (Netherlands – Chairman), S. Alimoso (Malawi), D. Hoggarth (UK), H. Kristiansen (Denmark), K. Lorenzen (UK), J-1 Noiset (Belgium), C. Pitts (UK), T.K. Shrestha (Nepal). *Lake and Reservoir Fisheries*: E.H. Allison (Malawi – Chairman), B.E. Marshall (Zimbabwe) A. Ali (Malaysia), J. Backx (Netherlands), M. Banda (Malawi), J.J. Mandima (Zimbabwe), R. Ogutu-Othwayo (Uganda), M. Staras (Romania), F. Orach-Meza (Uganda), D. Tweddle (Malawi). *Riverine Fisheries*: D.C. Jackson (USA – Chairman), B.R. Ward (Canada), G. Smith (UK), G. Jones (UK),

P.B. Bayley (USA), I Small (UK), I Navodaru (Romania), A.P. Strevens (UK).

I would also like to thank Julia Cowx, Emma Burnett and Debbie Cutsforth for their considerable assistance in the running of the symposium and production of these proceedings. Finally, I would like to thank the many international funding agencies and organizations for their financial support, thus ensuring truly international coverage of the issues and the success of the symposium.

Ian G. Cowx
*University of Hull*
*International Fisheries Institute*

# I SURVEY METHODS

# Chapter 1
# Fish population survey methods: a synthesis

P. HICKLEY *National Rivers Authority, Severn-Trent Region, 550 Streetsbrook Road, Solihull B91 1QT, UK*

**Abstract**   A synthesis of workshop discussions on the topic of fish population survey methods is presented. For the principal methods, status and future are assessed. It is recommended that efforts be made to further develop hydroacoustics, to optimize the use of fishery catch statistics, to combine complementary methods and to improve interpretation of results.

## 1.1   Introduction

First catch your fish! – The obvious key input to stock assessment is the efficient capture, count or observation of all or a representative proportion of individuals in the fish population under study. In turn, accurate fish stock assessment is often the critical first step in the implementation of many fishery management programmes. Fishing methods range from those which require only simple gear, such as spears and rakes, to the use of ocean-going vessels with enormous trawls or encircling nets (Brandt 1985). In inland fisheries, gill netting, seine netting, electric fishing and angling are probably the most common techniques used for scientific purpose and all involve fish capture. However, observation using divers, electronic fish counters or echo-sounding equipment is an option. Many techniques are well tried and established (Bagenal 1978; Nielsen & Johnson 1983) but, although to some degree dictated by circumstance, the choice of method should be made with the survey objective kept firmly in mind. Factors for consideration could include the importance of quantitative assessment versus the identification of trends and species biology versus population dynamics.

The sampling of fish populations can be very efficient in small rivers and pools, and most freshwater research has been undertaken in such habitats. Nonetheless, the fishery resources of larger aquatic ecosystems are invariably of great importance and these have received more attention in recent years (Hickley & Starkie 1985). The aim of this synthesis of workshop discussions is to provide an overview of the survey techniques available, to assess current status, to examine recent advances and to identify scope for development in the future. A summary of the principal methods under consideration is given in Table 1.1.

**Table 1.1**  Matrix showing the principal methods used for stock assessment survey work in freshwater systems and their most common applications.

| Method | Fish capture | Rivers | Lakes | Migratory species | Non-migratory species |
|---|---|---|---|---|---|
| Counters | No | • | | • | |
| Hydroacoustics | No | • | • | | • |
| Observation | No | • | • | • | • |
| Electric fishing | Yes | • | | | • |
| Netting | Yes | | • | | • |
| Trapping | Yes | • | • | • | • |
| Fishermen's catch | Yes | • | • | • | • |

## 1.2    Non-capture methods

### 1.2.1    *Counters*

Electronic fish counters are usually used for counting migratory salmonids in rivers, for example, as described by Aprahamian *et al.* (Chapter 3). The technology is more or less understood (Bussell 1978) and, in general, if the civil engineering of the structure is correct then the counter is likely to work efficiently, with up to 90% of the fish passing over the electrodes being successfully counted.

The principal constraint is the cost of the associated structure, such as a weir or dam, but the opportunistic installation in existing structures can prove relatively inexpensive. It is difficult to use the counter to differentiate between sizes of fish, such as large sea-trout and small salmon, but the complementary use of photographic equipment can be helpful where conditions permit (Fewings 1994).

The design criteria for combined fish passes and counting units is established, whereas devices for counting across wide weir crests require further research. There is also a need for development of counters in situations where no structure exists. Some progress is being made with weir-less counters but, although upstream counts can be quite good, downstream ones remain poor. In terms of data capture and transfer, recent advances in telemetry systems have proved invaluable.

### 1.2.2    *Hydroacoustics*

Recent developments in the use of hydroacoustics for freshwater survey work have been significant. Quality of output has progressed from not much more than marks on a recorder chart to the computation of biomass based on target strengths. This can, however, introduce the danger of inappropriate confidence in the results obtained whereas it is crucial to recognize the complex nature of the data analysis, and the assumptions that have to be made. The biggest problem is truthing in relation to fish behaviour and the calibration of target strengths. Also the equipment is very expensive.

The advantage of sonar is that it provides access to some fish populations that cannot be conveniently sampled in other ways, providing high precision and good spatial distribution of effort. For example, previous knowledge of the offshore waters of Lake Malawi was confined to limited areas whereas echo-sounding enabled the fishery potential of the entire pelagic zone to be assessed (Allison, Chapter 17). Hydroacoustic equipment counts fish successfully, although it works better in deep, open water with few species present, e.g. in Lock Ness (Bean, Winfield & Fletcher, Chapter 16).

Further research is needed to better provide species and size separation, although considerable progress is being made following the introduction of multiple beam systems. Also, the effect of the sampling vessel itself is not clearly understood. Notwithstanding the constraints, the method will undoubtedly be used more and more as technology continues to improve. In particular, the application of horizontal sonar in shallow freshwater systems, especially rivers, shows considerable potential for the future (Kubecka, Chapter 13).

### 1.2.3    *Observation*

Observational methods can be used both for survey work itself and for the calibration of gear. The human observer can either watch and count the fish directly or deploy remote control camera systems. Fish shoals in shallow water have been successfully counted from a boat whilst in other cases diving (using snorkel, scuba or mini-submarine) has been used. The obvious constraint is the need for good water clarity but other problems include the cost, training and safety considerations when divers are used.

Observation is, perhaps, an under-utilized technique and the scope for its wider application in the future ought to be investigated. Observation is likely to best suit situations where traditional sampling methods are either inefficient or result in unacceptable disturbance of the fish population, especially where repeat estimates of biomass are sought (Bird, Lightfoot & Strevens, Chapter 4).

### 1.3    **Capture methods**

### 1.3.1    *Electric fishing*

To scientists attempting stock assessment of rivers and streams, electric fishing is a method of immense value. In England and Wales alone, about 2000 fishery surveys are carried out annually (National Rivers Authority 1994). Although best suited to smaller watercourses, the method has been adapted for use in all types of inland waters including the littoral zones of large rivers and lakes. Efficiency can vary enormously according to a host of different factors and truly quantitative population estimates for large water bodies will probably remain elusive. Seemingly undaunted, however, field workers continue to modify equipment and techniques in order to maximize the efficiency with which they can sample any freshwater environment (Harvey & Cowx, Chapter 2).

It should be noted that a significant international symposium was held in Hull, UK, in 1988 at which much information on all aspects of electric fishing was discussed and consolidated (Cowx 1990; Cowx & Lamarque 1990). In addition to the many symposium papers, Burridge *et al.* (1990) provided a bibliography of 931 relevant works.

In general terms, therefore, the methodologies for fishing with electricity are now well established with little need for continued development, although some attempt at standardization could be beneficial. Also, further research should be considered in relation to the acute and chronic effects on fish of the different types of electric current in use (Sharber & Carothers 1990).

### 1.3.2    *Netting*

The use of nets for fish capture employs long-established equipment and techniques which have changed little in the general principles applied. The seine net, for instance, was introduced to many countries by the Romans, although the gear was known to the early Greeks, Phoenicians and Egyptians (Brandt 1985). However, the replacement of natural fibres with synthetic materials in the manufacture of nets has not only extended their useful life but has had an impact on efficiency of capture and exploitation rates.

The main types of nets used for survey work are seine, trawl and gill nets. In addition, much netting in freshwater lakes is carried out for commercial purpose and many scientific programmes link into such fisheries.

Seine nets operated from the shore have the principal limitation of needing a good bed profile over which to set and haul the net. Purse seines, whilst not limited by the substratum, have less application in fresh water rather than the marine environment because of the relative shallowness of many lakes and the nature of the target species. Nonetheless, seine netting, especially with micro-mesh nets, is often the only way to catch the smaller representatives of populations and, as stated by Pygott *et al.* (1990), should not be neglected in the repertoire of techniques.

Although more usually associated with food fish harvest at sea, trawling can be a successful method for surveying larger freshwater systems (Banda, Tommasson & Tweedle, Chapter 5; Buijse & Dekker, Chapter 19). The major constraints are the cost implications and the difficulty in deriving quantitative assessments of the fish population under study. Probably, however, there is scope for the increased use of trawl nets for survey work, particularly the smaller, faster mid-water trawls now available.

Gill nets, popular worldwide in inland artisanal fisheries (e.g. Ali, Chapter 38), are notoriously difficult to use for quantitative stock assessment. Not only do they show inherent selectivity for size of fish but mesh shape and type of materials used to make the net can influence efficiency of capture (Hamley 1980). Although this selectivity can aid the setting and enforcement of net specifications in the management of commercial fisheries, difficulties arise with scientific survey work unless attempts are made to take all factors into account. Progress continues to be made in quantifying selectivity and efficiency (Jensen 1986) but, perhaps, the development of a standard gill net for each key target species should be considered.

### 1.3.3     *Trapping*

Entrapment devices essentially fall into two categories, those such as baskets, pots and fyke nets, baited or otherwise, used for non-migratory fish and those traps used to intercept upstream or downstream migrants. Whilst traps are used by many artisanal and commercial fishermen worldwide, with the gear often manufactured from locally available materials, their application to stock assessment surveys is more limited.

Non-interceptory fish traps, like any other passive capture gear, are absolutely dependent upon the behaviour of the fish in relation to the type of trap used. Thus, trapping for non-migratory fish can generally be considered unsuitable as a method of quantitative assessment of populations. However, in other work relating to individual target species, such as estimating age-structure, growth or condition of the fish, traps may prove convenient. For example, the Windermere perch trap, first described by Worthington in 1950, has been a valuable aid to the long-term study of Lake Windermere (LeCren *et al.*, 1977). However, notwithstanding the difficulties with quantitative appraisal, traps remain popular for studies on juvenile fish which may be below the size that can be taken by other methods such as netting or electric fishing (e.g. Kubecka, Chapter 6).

For those species that migrate, either to and from the marine environment or within inland systems, interceptory traps can be a convenient way of catching fish for analysis or tag and release. In particular, in-river trapping has been regularly used during research on both salmon, *Salmo salar* L. (Davidson, Purvis, Milner & Cove, Chapter 21) and eels, *Anguilla anguilla* L. (Knights & White, Chapter 34). A classic situation might be a weir trap to take downstream migrating salmon smolts with a trap chamber in a fish ladder to reclaim the returning adults.

### 1.3.4     *Fishermen's catch*

The collection and use of catch information from fishermen gives access to a vast amount of data, albeit the primary reason for the actual fishing activity being commercial gain, subsistence or leisure. Many surveys based on catch returns have run over a much longer timescale than might be possible with custom designed scientific sampling programmes. For example, Gee and Milner (1980) reported on salmon catch statistics over a 70-year period and, similarly, Schneider and Leach (1977) analysed long-term records for walleye in the Great Lakes of North America.

Problems associated with the use of catch data include the difficulties of estimating abundance, the ability and willingness of the fishermen to make accurate returns, species bias and the quantification of effort. Such disadvantages have to be balanced against the cost-effectiveness of data collection and because observation at fish landings is the only way to sample many of the fisheries under study. Also, there is a potential public relations benefit in direct liaison with the fishermen (Ali, Chapter 38; Amarasinghe, Chapter 25).

It is without doubt, that catch statistics will continue to be used in the assessment and management of commercial fisheries, because they are as crucial to such fisheries as the Great Lakes of Africa (Turner, Chapter 37). Recently, however, the socio-economic importance of recreational fishing is being increasingly recognised (FAO 1994), with the monitoring of catches being a valuable aid to management.

A variety of methods can be used to collect angler catch data. Roving creel census (Malvestuto *et al.*, 1978) is likely to be the most accurate, as it involves face to face contact with the fisherman, but is relatively labour intensive. Methods relying on third party reporting include the use of log books (Evans, Chapter 9), postal questionnaire (North & Hickley 1989) and newspaper reports (Minotti & Malvarez 1991).

Although both commercial and recreational catch data collection can be straightforward, analysis and interpretation of results can be problematical, usually with regard to the appraisal of effort. Löffler (Chapter 29), for instance, states that results for Lake Constance show a high degree of uncertainty even though trials date back to the early 1930s. However, it appears that shortfalls are being recognised and attempts being made to improve the usefulness and quality of catch return information. Example fisheries include Lake Kariba (Mandima, Chapter 27), Lake Tanganyika (Marshall, Chapter 26), salmon of the River Spey (Smith, Laughton & Dora, Chapter 11) and coarse fish contest waters (Steel, O'Hara & Smith, Chapter 8).


### 1.4    Conclusion and recommendations

In general terms, the range of survey methods available means that some level of stock assessment can be carried out, whatever the nature of the water body being studied. The degree to which any appraisal is quantitative, however, often leaves much to be desired. This is especially so where the fishery is large and difficult to sample. Hence there has been significant effort invested in development of methodologies and recent improvements are evident in many fields, especially fish counters, hydroacoustics and electric fishing.

More development is still required for fish counters and hydroacoustics, plus observational techniques and the handling of fishermen's catch data. As development progresses it is likely that the last three methods, especially sonar, will become increasingly used in the future. In addition, there will always be scope for fishery scientists to develop unique site-specific methodologies (e.g. Jackson, pers. comm.) or to exploit theoretical estimates such as those based on secondary production (e.g. Cryer, Chapter 28) A summary of the status and future of the principal fish survey methods is given in Table 1.2 .

**Table 1.2**  Matrix summarizing the status and potential of the principal methods used for stock assessment survey work in freshwater systems.

| Method | Recent advances | Established and usable | Development required | Increased future use |
|---|---|---|---|---|
| Counters | ● | ● | ● | |
| Hydroacoustics | ● | | ● | ● |
| Observation | ● | | ● | ● |
| Electric fishing | ● | ● | | |
| Netting | | ● | | |
| Trapping | | ● | | |
| Fishermen's catch | | ● | ● | ● |

Many workers have used more than one sampling technique. Sometimes this is for the purpose of gear calibration (Bayley 1985), usually to calibrate the primary method against a more efficient but costly and difficult secondary method. Often, however, there is a tendency to adopt a confrontational approach, as if it is necessary to determine an outright winner (e.g. Allison, Chapter 24; Pygott *et al.* 1990). A better strategy would be to use the different methods in a complementary way so as to enhance the accuracy of stock assessment overall. For example, hydroacoustics and boom-boat electric fishing or salmon anglers' catch and fish counters could be successfully used in combination.

Field survey work is, of course, only one component of any fisheries management activity. Data processing and analysis of results play their part but, perhaps, one of the most critical stages is the translation of findings into hard management decisions. Scientists and managers need to be encouraged to further consider mechanisms for interpretation of results (Cowx, Chapter 39), hopefully with the same enthusiasm that has previously been shown towards fish capture techniques.

In summary, it is recommended that:

- the development of hydroacoustics be continued (but beware false accuracy and high cost);
- the capture and handling of fishermen's catch data be improved;
- a move away from the single method approach be made and complementary techniques be used in combination;
- emphasis be redirected from gear development towards result interpretation methodologies in order to aid management decision making.


**Acknowledgements**

The author thanks all those who helped to generate the content of this chapter by contributing to the workshop discussions or by presenting papers to the symposium, in particular M. Aprahamian (UK), E. Baras (Belgium), J. Coeck (Belgium), M.

Cryer (New Zealand), I. Davidson (UK), D Evans (UK), J. Kubecka (UK), R. Laughton (UK), H. Löffler (Germany), K. McCarthy (Ireland) and I. Winfield (UK).

## References


Bagenal T. (ed.) (1978) *Methods for Assessment of Fish Production in Fresh Waters*, 3rd Edition. Oxford: Fishing News Books. 365 pp.

Bayley P.B. (1985) Sampling problems in freshwater fisheries. *Proceedings of the 4th British Freshwater Fisheries Conference*. Liverpool, 3–11.

Brandt A. von (1985) *Fish Catching Methods of the World*. Oxford: Fishing News Books.

Burridge M.E., Goodchild G.A. & Rutland C.L. (1990) A bibliography on fishing with electricity. In: I.G. Cowx & P. Lamarque (eds) *Fishing with Electricity*. Oxford: Fishing News Books, pp 192–243.

Bussell R.B. (1978) *Fish Counting Stations: Notes for guidance in their design and use*. London: Department of Environment.

Cowx, I.G. (ed.) (1990) *Developments in Electric Fishing*. Oxford: Fishing News Books, 358 pp.

Cowx I.G. & Lamarque P. (eds) (1990) *Fishing with Electricity*. Oxford: Fishing News Books, 248 pp.

FAO (1994) Report of the eighteenth session of the European Inland Fisheries Advisory Commission. *FAO Fisheries Report* **509**, 78 pp.

Fewings G.A. (1994) *Automatic salmon counting technologies – a contemporary review*. Pitlochry: Atlantic Salmon Trust.

Gee A.S. & Milner N.J. (1980) Analysis of 70-year catch statistics for Atlantic salmon (*Salmo salar*) in the River Wye and its implications for management of stocks. *Journal of Applied Ecology* **17**, 41–57.

Hamley J.M. (1980) Sampling with gill nets. In: T. Backiel & R.L. Welcomme (eds) Guidelines for sampling fish in inland waters. *European Inland Fisheries Advisory Commission Technical Paper* **33**, 37–53.

Hickley P. & Starkie A. (1985) Cost effective sampling of fish populations in large water bodies. *Journal of Fish Biology* **27 (Supplement A)**, 151–161.

Jensen J.W. (1986) Gill net selectivity and the efficiency of alternative combinations of mesh sizes for some freshwater fish. *Journal of Fish Biology* **28**, 637–646.

LeCren E.D., Kipling C. & McCormack J.C. (1977) A study of the numbers, biomass and year-class strengths of perch (*Perca fluviatilis* L.) in Windermere 1941–66. *Journal of Animal Ecology* **46**, 281–307.

Malvestuto S.P., Davies W.D. & Shelton W.L. (1978) An evaluation of the roving creel survey with non-uniform probability sampling. *Transactions of the American Fisheries Society* **107**, 255–262.

Minotti P.G. & Malvarez A.I. (1991) Newspaper information retrieval to assess the movement of valuable sport-fishing stocks in the Parana River Lower Delta, Argentina. In: I.G. Cowx (ed.) *Catch effort sampling strategies*. Oxford: Fishing News Books, pp 177–183.

National Rivers Authority (1994) *NRA Corporate Plan 94–95*. Bristol: National Rivers Authority.

Nielsen L.A. & Johnson D.L. (1983). *Fisheries Techniques*. Bethesda, Maryland: American Fisheries Society.

North E. & Hickley P. (1989) An appraisal of anglers' catches in the River Severn, England. *Journal of Fish Biology* **34**, 299–306.

Pygott J.R., O'Hara K., Cragg-Hine D. & Newton C. (1990) A comparison of the sampling efficiency of electric fishing and seine netting in two contrasting canal systems. In: I.G. Cowx (ed.) *Developments in Electric Fishing*. Oxford: Fishing News Books, pp 130–139.

Schneider J.C. & Leach J.H. (1977). Walleye (*Stizostedion vitreum vitreum*) fluctuations in the Great Lakes and possible causes, 1800–1975. *Journal of the Fisheries Research Board of Canada* **34**, 1878–1889.

Sharber N.G. & Carothers S.W. (1990) Influence of electrofishing pulse shape on spinal injuries in adult rainbow trout. In: I.G. Cowx (ed.) *Developments in Electric Fishing*. Oxford: Fishing News Books, pp 19–26.

Worthington E.B. (1950) An experiment with populations of fish in Windermere 1939–48. *Proceedings of the Zoological Society of London* **120**, 113–149.

# Chapter 2
# Electric fishing for the assessment of fish stocks in large rivers

J. HARVEY and I.G. COWX *University of Hull, International Fisheries Institute, Hull HU6 7RX, UK*

**Abstract**   Electric fishing is a tool used extensively to catch fish in stock assessment exercises. The method suffers from many limitations primarily related to the depth and width of the water body being surveyed, as well as such factors as conductivity, current velocity and water clarity. This chapter describes a recent development in boom electric fishing using the Wisconsin ring-type array but sequentially firing the anodes to provide an extended electric field of influence with reduced power demands.

The new system caught significantly more fish than conventional gears, although in larger rivers it was still not possible to carry out absolute population estimates because the gear efficiency was unable to overcome the assumptions on which the population models were based. Sampling strategies to overcome this limitation, under a range of scenarios, are described.

KEYWORDS: Electric fishing, stock assessment, sampling strategies

## 2.1     Introduction

Within the sphere of inland fisheries there is an urgent need to obtain information on the status of fish stocks to allow management decisions to be formulated and implemented. The success of any study on fish stocks, particularly where the river is not exploited commercially, depends on effective sampling of a water body. However, conventional sampling techniques are often inefficient, particularly in large river systems, and provide only minimal information about the ecology of fish stocks.

In England and Wales approximately 40% of the fisheries are in waters that cannot be sampled by traditional methods. Thus to obtain information on fish stocks a method of sampling large water bodies, particularly linear systems, on a scientific basis is essential. The method must be cost-effective and not manpower intensive.

Acoustic methods provide information on total biomass but yield little data on species composition, presence of benthic species, or population characteristics (Harvey & Cowx 1995a). Thus some method of actually catching fish is required. Netting techniques are generally not suitable for flowing waters, where there are obstructions in the river, or where extensive weed growth occurs. On slow-flowing or

11

static systems netting techniques can be effective with high catches of fish obtained; however, fish mortalities often occur and the mechanism is manpower intensive (Cowx *et al.* 1990).

Monitoring of anglers' catches, although not manpower intensive, suffers from a bias towards the anglers' target species and is seriously influenced by the fishes' behaviour (Cowx 1990). Furthermore, little information is available on small fish which are a major element in predicting future fish stocks. In the past, electric fishing has been considered selective to fish > 100 mm and damaging to all fish. However, recent gear developments and electronics advances (Novotny 1990) have identified the immense potential of electric fishing in large river systems.

This chapter describes recent developments in the design of electric fishing gear for sampling large lowland rivers. Additionally, information is provided on the development of strategies for stock assessment exercises.


## 2.2    Developments in electric fishing gear

Electric fishing gears have been improved and upgraded immensely in the last decade (Novotny 1990) and raised the importance of safety in designing new gears (Goodchild 1990).

In small streams/rivers conventional hand-held equipment may be operated successfully but in large rivers a disproportionately large amount of effort is required to make small catches of fish (Hickley & Starkie 1985). To overcome the latter problem a linear, multi-electrode array (Cowx *et al.* 1988) mounted on a boat was developed. The design suffered from problems in high conductivity waters, which were subsequently remedied by the sequential energizing of the electrodes (Cowx *et al.* 1990) to produce the same effective field with a smaller anode surface area.

Novotny (1990) advocated that spheres are the ideal electrode shape for attracting and immobilizing fish and it is always advantageous to use the largest electrodes possible within the limitations imposed by physical constraints and the electrical limits imposed by the generator and electrical control system.

However, large spherical electrodes have many disadvantages associated with ease of construction and ability to negotiate obstructions in the water. The development of the 'Wisconsin Ring' array (Novotny & Priegal 1974) in North America resulted in the use of large diameter aluminium rings supporting pendant electrodes which effectively reproduced the field of a single spherical electrode and overcame some of the disadvantages of using spherical electrodes.

To improve on the efficiency of existing gears, a system based on the 'Wisconsin Ring' array was designed and constructed with the aim of creating a large spherical electric field. This design would be further supplemented by electric current characteristics that would maximize capture efficiency, causing minimal damage to fish. Finally the design would have a number of criteria to enable the gear to be versatile and usable in a variety of environmental conditions.

## 2.3    Equipment design

Specifications of the electric fishing gear are detailed elsewhere (Harvey & Cowx 1995a, b). The basic design (Fig. 2.1) consists of two electrode arrays, each consisting of three concentric rings of decreasing radii: 0.8 m; 0.53 m; 0.26 m. Each ring is electrically isolated and supports a number of pendant electrodes constructed from stainless steel wire rope. Each ring array is supported on a boom structure fixed to the front of a suitably stable, preferably cathedral-hulled boat. Electrical power is supplied from a 7.5 kVA Allam generator via a Millstream electric fishing box. In preliminary trials the control box produced an output of 50 or 100 Hz, precise 1/4 sine wave, with variable power control, with each ring energized at the same instant.

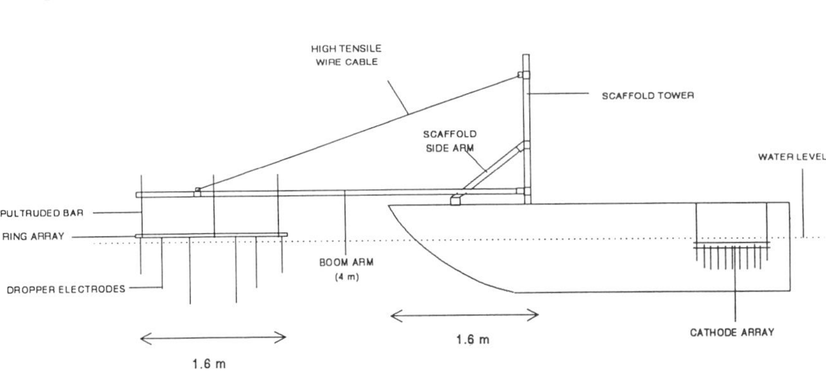


**Fig. 2.1**   General layout of electric fishing boat.

As well as attempting to improve on capture efficiency and reducing damage to fish, certain design criteria were incorporated to improve the overall efficiency of a sampling operation.

The gear conformed to the electrical Health and Safety Regulations and Guidelines demanded by the NRA (NRA 1991). It is relatively simple to construct/maintain, easily transportable, and simple to assemble and dismantle on site. The design is also adaptable for any new electric fishing developments, robust for regular field use and manoeuvrable so as not to pose any hazard or obstruction to other river users.

Another inbuilt advancement in the gear is that all equipment used in the survey can be easily stored in the towing vehicle and boat, thus reducing the number of secondary assistance vehicles. Furthermore all electrode arrays and supporting gears can be removed from the boat quickly and easily thus allowing its use in other survey operations such as setting nets, and so reducing the investment costs in the boat.

## 2.4    Preliminary trials

Preliminary trials with the prototype electric fishing gear were performed on a wide range of river types that were considered representative of those that require sampling for stock assessment and management decisions. The river systems selected (Table 2.1) present a diversity of problems with respect to sampling both by conventional methods and electric fishing. Particular difficulties arise with respect to electric fishing due to the variation in river width and depth, water velocity and water conductivity. To assess the best method of deployment and efficiency of the boom-mounted electric fishing gear a number of surveys were performed.

**Table 2.1** General characteristics of rivers used for field trials.

| Water body | Width (m) | Depth (m) | Flow Regime | Conductivity ($\mu S\ cm^{-1}$) | Trial | Subjective assessment of gear efficiency |
|---|---|---|---|---|---|---|
| River Hull | >20 | 3 | Tidal | 800 | Test ability of gear in high conductivity waters | Depletion sampling possible but accuracy of estimates questionable |
| River Ure | >50 | >2 | Moderate | 500 | Assess ability to determine fish stocks | Improved catches over existing gears but absolute population estimate not possible |
| Beverley Beck | 20 | 3 | Slow | 568 | Determine fish stocks | Depletion sampling possible |
| River Foss | 20 | 3 | Slow | 758 | As above | Depletion sampling possible |
| River Trent | 35 | 2 | Moderate | 1200 | Comparative assessment with linear electrode array | Fishing margins of river only |
| River Eden | >40 | 3 | Fast | 250 | Determine fish stocks where other techniques have failed | Relative assessment of stocks |
| Oxford Canal | 15 | 2 | None | 520 | Comparison against other electric fishing gears and seine netting | Depletion estimates possible; accuracy of estimates tested when compared with net catches |
| River Weaver | 50 | 6 | Slow | 1380 | Comparative assessment with seine netting, angler catches and hydroacoustic techniques | |

The overall aim of these trials was to evaluate the limitations of the electric fishing gear in large rivers, and identify where further improvements in the design and operation of the gear could be made to optimize the use of electric fishing in large rivers.

Fish were caught in all exercises, although the catches in extremely wide rivers, e.g. River Ure (90 m wide), were low and the gear was considered an improvement on existing gears. The gear was capable of providing population estimates using depletion methodology, from isolated sections of slow-flowing rivers, e.g. Beverley Beck. However, violation of the assumptions that govern depletion models often invalidated results, for example on the River Foss, because constant fishing effort was not maintained in two of the depletion runs which resulted in lower catches on the first and third runs than on the second and fourth runs. This invalidation of the assumptions underlying depletion models highlighted the necessity for constant effort to avoid poor population estimates with unacceptable confidence intervals.

On very fast flowing rivers, e.g. River Eden, conventional sampling gears proved ineffective at capturing fish, whereas the electric fishing gear was able to provide information on fish populations present. However, the logistics of utilizing boom-mounted electric fishing gears on very fast flowing rivers needs careful consideration prior to implementation, highlighting the need for sampling strategies. For example, the most effective strategy considered for sampling the River Eden involved electric fishing the deep pool areas between riffle zones. Due to poor access, a major problem with fishery surveys, the most appropriate method of moving the electric fishing vessel from pool to pool involved manhandling the electric fishing boat over the riffle areas.

The preliminary trials highlighted a problem with the pendant electrodes undergoing extensive corrosion and so they were subsequently replaced with 9 mm stainless steel, wire rope which proved efficient at transferring the electric current to the water. Fish mortality was low, but little is known about the potential injuries that occur some time after electric fishing. Thus in order to reduce instant mortalities and possible internal injuries which may affect fishes' subsequent growth and behaviour, a further development in the control box system was deemed appropriate.

## 2.5    Sequential firing control box system

To increase the efficiency of the boom-mounted electric fishing gear a control box capable of sequentially energizing each ring of the electrode array was designed and constructed; a detailed specification is provided elsewhere (Harvey & Cowx 1995a).

The output is **precise** 1/4 sine wave at 50 or 100 Hz at all power settings. This rapid rise and slow decay of current is crucial because any deviation from this as power increases will compromise the catching efficiency. The number of operating anodes (up to 10) can be selected, as can the speed of firing of each anode ring in the electric fishing array. The electrode rings can be energized for 0.2 ms, 0.1 ms, 0.04 ms or 0.02 ms. This development reduces the power demands enormously, with minimal effect on the field characteristics and allowing its use in high conductivity waters.

The basic theory in using the sequential firing control box system is that fish in the immediate vicinity of the anode will be exposed for a considerably reduced time to a compressed intense electric field, which can cause injury or even mortality. The voltage gradient profile of a standard firing control box, i.e. all rings energized, and a sequentially firing control box, i.e. rings energized sequentially, are compared in Fig. 2.2. Close to the anode ring, the standard control box produces a higher voltage gradient than the sequential firing control box which indicates a larger potential danger zone to fish. Furthermore the effective field produced by the sequential firing control box extends to a greater distance (1.9 m) than the field produced by the standard firing control box (1.2 m).

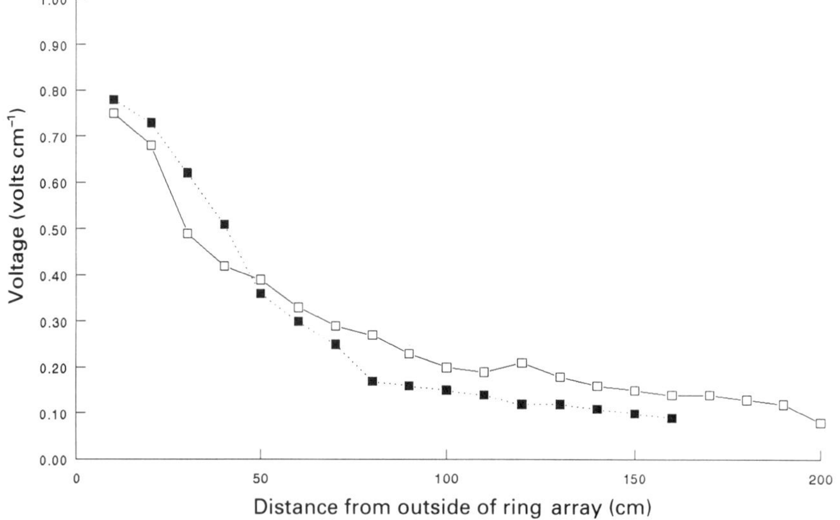

**Fig. 2.2**   Voltage gradient profile of standard control box ■ and sequential-firing control box □

In preliminary tests with the sequential firing system, in the tidal River Hull where conductivities fluctuated (100–800 $\mu$S cm$^{-1}$), mortalities were zero and the output was capable of attracting and immobilizing fish as small as 20 mm for netting, so reducing the selectivity of electric fishing methods. The sequential firing system was able to operate in high and fluctuating conductivity conditions thus alleviating one of the common problems with electric fishing, i.e. high power demands on control boxes.

### 2.6    Comparative trials

Following tests with the sequential firing control box to ensure the new design improved catches and reduced fish mortalities, comparative exercises with conventional sampling gears were performed to identify the advantages and limitations of the electric fishing gear. Comparisons between the electric fishing gear and netting were carried out on two static or very slow flowing channelized river systems/canals, the River Weaver and the Oxford Canal. A comparison with the conventional linear electrode array was performed on a flowing river system, the River Trent.

The exercise on the River Weaver involved a three-catch depletion electric fishing sample between two barrier nets set 200 m apart. Catches were then compared to those from a single haul by a seine net. Few fish were caught from the deep central area of the river, with the majority of fish being caught in marginal areas. As anticipated in a river with little flow, the seine netting exercise removed many more fish in the isolated area than electric fishing. Comparison of roach (*Rutilus rutilus* (L.)) size distribution in the 200 m area (Fig. 2.3) showed the seine netting removed a few larger roach size classes, but was highly selective to fish > 70 mm.

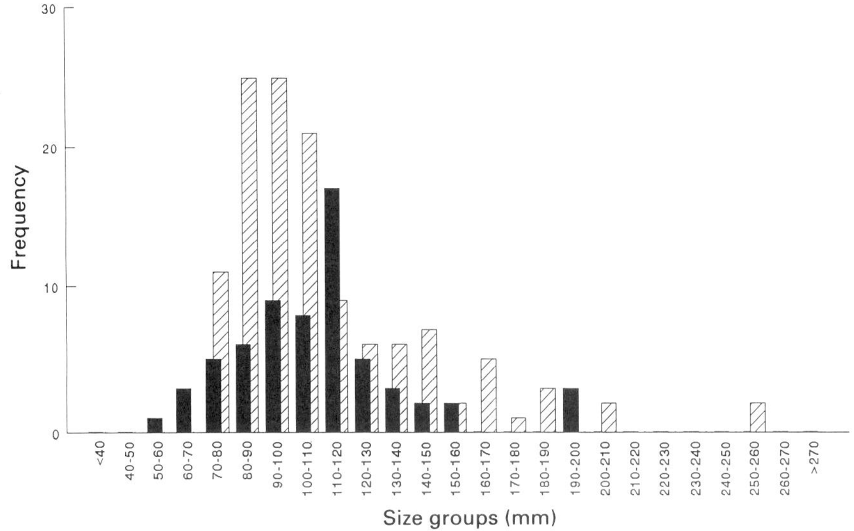


**Fig. 2.3**  Size distribution of roach caught in a 200 m section of the River Weaver by electric fishing ■ and seine netting □.

A further exercise was performed whereby an electric fishing survey of a 1 km length of river margin was compared to the catches from the seine netting exercise (Fig. 2.4). The 1 km survey produced a greater size-class distribution of roach and also revealed an extra species, chub (*Leuciscus cephalus* (L.)), not recorded in the 200 m isolated area.

The netting trial on the Oxford Canal consisted of a similar exercise where a 200 m isolated area was electric fished and subsequently seine netted to give a population estimate and information on size distributions. The seine netting provided a much higher population estimate than the electric fishing gear, with lower confidence limits (Table 2.2). However, catches and hence efficiency of the electric fishing gear were greatly reduced due to the high turbidity encountered in the canal on the day of surveying. The seine net caught more fish in terms of numbers but appeared somewhat selective to fish > 70 mm (Fig. 2.5).

A comparison between a conventional multiple-electrode linear array and the multiple-electrode ring array was performed on the River Trent, Rugeley, to ascertain

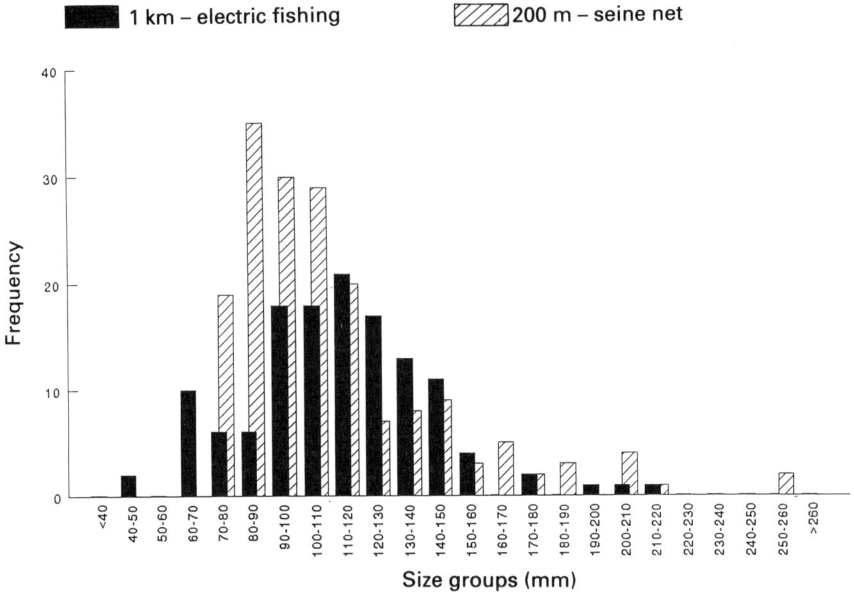


**Fig. 2.4**  Size distribution of roach caught in a 200 m section of the River Weaver by seine netting ☐ and electric fishing 1 km ■.

**Table 2.2**  Total catch, probability of capture (*P*) and population estimates from a 200 m section of the Oxford Canal, using two different sampling methods.

| Gear type | C1 | C2 | Total catch | *P* | Population estimate | 95% CL. |
|---|---|---|---|---|---|---|
| Ring array | 112 | 44 | 156 | 0.60 | 184 | 26 |
| Seine netting | 283 | 114 | 397 | 0.59 | 473 | 44 |

whether the ring electrode configuration was an improvement in terms of catching fish. A two-catch depletion with no barrier nets was performed with the linear array on a 300 m section, and then subsequently electric fished with the ring array 14 days later.

Increased catches of roach, dace (*Leuciscus leuciscus* (L.)) and chub, with similar catches of perch (*Perca fluviatilis* L.), eel (*Anguilla anguilla* (L.)), gudgeon (*Gobio gobio* (L.)), pike (*Esox lucius* L.) and ruffe (*Gymnocephalus cernua* (L.)) were found with the ring array (Table 2.3). Also, two species – bream (*Abramis brama* (L.)) and tench (*Tinca tinca* (L)) – were caught by the ring array which were not captured in the linear array trial. Moreover, the apparent importance of roach to the fishery was not as clearly revealed by the linear array (fewer catches) as by the ring array (Table 2.3). Furthermore, a more representative catch of roach, in terms of size-class distribution, was achieved by the ring array, with the linear array being selective to fish > 100 mm (Fig. 2.6).

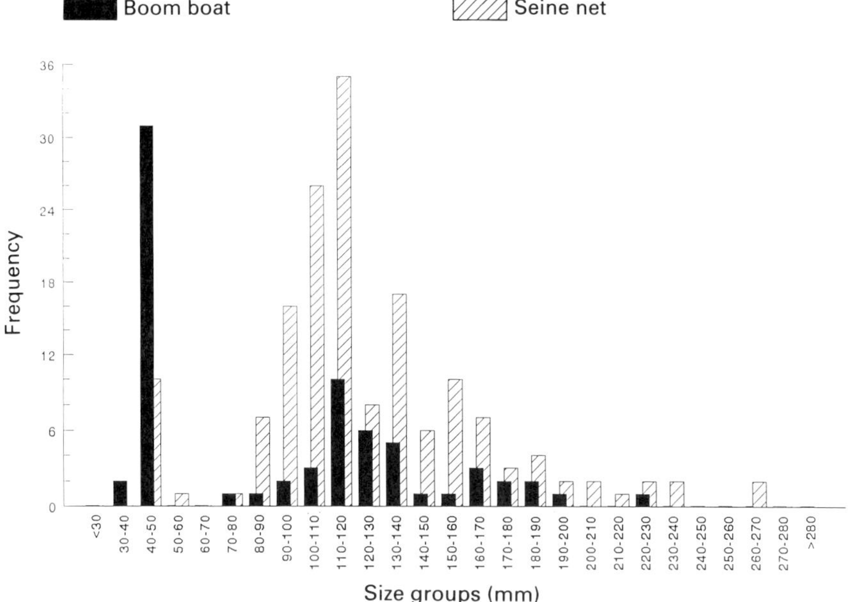


**Fig. 2.5** Size distribution of roach caught in a section of the Oxford Canal by eletric fishing ■ and seine netting □.

**Table 2.3** Catch details from two boat-mounted electric fishing gears on a section of the River Trent, Rugeley.

| Species | Linear array | | Ring array | |
|---|---|---|---|---|
|  | Catch 1 | Catch 2 | Catch 1 | Catch 2 |
| Roach | 30 | 19 | 92 | 37 |
| Dace | 6 | 0 | 12 | 0 |
| Chub | 10 | 1 | 16 | 3 |
| Perch | 10 | 4 | 8 | 3 |
| Eel | 4 | 0 | 5 | 1 |
| Gudgeon | 4 | 2 | 3 | 1 |
| Pike | 0 | 3 | 4 | 0 |
| Ruffe | 0 | 1 | 1 | 1 |
| Bream | 0 | 0 | 1 | 0 |
| Tench | 0 | 0 | 1 | 0 |

A number of conclusions can be drawn from the comparative trials. On flowing rivers, netting techniques are difficult to implement and the development of the electric fishing boom array is more successful than the existing electric fishing gears at removing fish from the water. Limitations with the electric fishing gear are highlighted on very deep rivers in that on slow-flowing rivers/static systems netting techniques are likely to produce better population estimates. However, netting

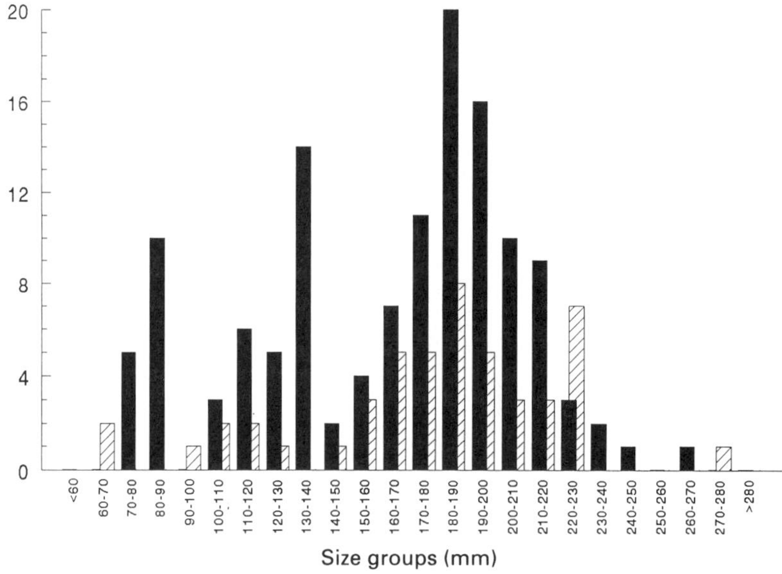


**Fig. 2.6** Size distribution of roach caught in a section of the River Trent by electric fishing with a linear electrode array □ and the sequential-firing ring array ■.

techniques are known to be selective to fish > 80 mm and mortalities much higher than with electric fishing.

A major advantage of using electric fishing gears as an alternative to netting techniques is the reduction in manpower and survey times. For example, 2–3 km of river can be surveyed by three personnel electric fishing while only one or two 200 m areas of river can be surveyed by six personnel using netting techniques over the same time scale (Cowx *et al.* 1990).

## 2.7    Sampling strategies

### 2.7.1    *Introduction*

The multiple anode, sequential firing ring array has been developed to improve sampling efficiency and increase catches. This does not necessarily mean that the sampling efficiency has been improved sufficiently to carry out definitive estimates of fish populations size in large rivers; it merely allows a more representative catch of fish to be made. To avoid the inherent errors arising from any stock assessment exercise using the boom-mounted system a strategic approach to stock assessment which meets the demands of both the manager and the scientist is required. This will ensure that the data collected are both useful and meet the needs of management. These issues have been discussed in detail by Cowx (1995a)

In developing the strategy for any stock assessment exercise a number of questions must be asked before it is carried out. These primarily relate to the objectives of the exercise and the desired precision of the output, and whether the fishing gear and resources are available to meet these needs (Cowx 1995a; Chapter 39).

The objectives of fisheries management fall into a number of categories:

- evaluation of the status of the fish stocks for conservation and enhancement purposes;
- monitoring long-term population changes as a result of natural or anthropogenic activities (i.e. routine monitoring);
- evaluating response of management activities directly targeted at the river system or its fisheries, e.g. restocking or introductions, habitat improvements, water quality improvements, flow regulation;
- assessment of environmental damage, e.g. post-impact appraisal of a pollution incident, fish kill or natural catastrophe such as a drought or flood;
- predicting the impact of development activities on the fisheries, as a function of an environmental impact assessment exercise.

The aim of some of these exercises may be to assess total population size which is an absolute parameter and may require considerable resources to be estimated. Often however, the aim is to assess temporal or spatial changes and trends, such as in environmental monitoring, which may facilitate the estimation of relative parameters (presence/absence) at a reduced cost in time and manpower. So it is important to consider the desired information with respect to individual fish or populations and the accuracy and precision that must be achieved in a stock assessment exercise.

Accuracy is associated with the type of error or bias in the data and poor accuracy tends to lead to assessments that considerably over- or underestimate. Precision is associated with the 'noise' generated by a sampling procedure and is usually reduced by larger sample sizes or repetitive surveys (Southwood 1978).

Bohlin *et al.* (1990) suggested a rough guide for establishing precision levels for fisheries surveys and although the choice of precision level is ultimately the decision of the manager or scientist there must be certain standards that need to be achieved to meet the objective of the survey. Consideration of precision levels is required as data collection exercises tend to be expensive and high precision has to be paid for in time, manpower and financial resources. Also it is often difficult to obtain precise information on the status of the fish populations because the sampling technology is not available.

### 2.7.2  *Sampling methods and strategies*

In UK rivers, the type of electric fishing gear used and the strategy adopted depends almost exclusively on the size of the river being sampled and the current velocity.

Small streams can be sampled by three or four personnel wading with one or two hand-held electrodes, with current output usually pulsed d.c. at 50 or 100 Hz from a small generator, although backpack gear is employed occasionally. In this type of habitat depletion sampling (reviewed by Cowx 1983) a section of river (varying between 20 and 300 m) tends to be reasonably successful, although the population size is probably underestimated due to variable susceptibility of individual fish to electric fishing (Bohlin & Cowx 1990).

In small rivers the same equipment and procedure can be used but fishing is usually based from a boat, except in shallow riffle areas where operators wade, and pull the boat housing the generator. Two electrodes tend to be standard but more manpower is usually required. The population estimates suffer from the same biases as in small streams but accuracy can be further affected by reduced gear efficiency in wider rivers.

In large rivers and canals, boat-mounted, multi-anode boom array electric fishing (Cowx *et al.* 1988, 1990; Harvey & Cowx 1995b) should be used. Isolating sections of most rivers with stop nets is generally difficult because of the current; thus depletion methods using electric fishing as the capture technique are rarely appropriate, although such estimates are often derived. The accuracy of the results is subject to large errors and should be treated with extreme caution. In slow-flowing rivers and canals it may be possible to isolate a section of water and carry out a depletion sampling. However, results from the most efficient electric fishing still tend to be variable and probably do not provide accurate estimates of population size because of poor sampling efficiency. It is also questionable whether the fish community in the isolated section is representative of the reach of river under investigation. When resources are available, it is possible to carry out capture-recapture exercises (reviewed by Cormack 1968, 1972; and Seber 1973), but these have met with varying degrees of success on large rivers (Hunt & Jones 1974; Bowles *et al.* 1990). This approach is generally applicable for estimating population size of fish in large rivers with electric fishing, if the resources are available, sufficient is known about the behaviour and migratory patterns of the target fish species, and marked and unmarked fish are equally susceptible to capture.

Large, still water bodies, such as lakes and reservoirs, can be sampled by electric fishing in the margins using hand-held or boom-mounted electrodes. Population estimates are impractical because of the problems of sampling efficiency, although capture-recapture methods have been attempted.

### 2.7.3    *Gear calibration*

Gear calibration methods have received comparatively little attention yet are possibly the most cost-efficient mechanism for assessing fish abundance. There are two approaches to gear calibration (Bayley 1985): the whole system approach (Serns 1982, 1983); and the point estimate approach (Bayley 1983; Lazauski 1984).

The whole system approach estimates efficiency of the gear in an isolated population of known size (i.e. stocked fish, dewatering or capture-recapture) and the probability of capture is then transposed to the system to be evaluated. However, probability of capture in the test system may not be representative of the probability in the target water.

The point estimate approach aims to calculate gear efficiency in a small area of the habitat to be evaluated by assessing the vulnerable population size in this area with a high efficiency gear. The probability of capture is determined as the proportion of the first catch against the estimated population size.

In both approaches long lengths of river are fished with the calibrated gear and the population abundance ($N$) determined by $N = C/P$, where $C$ is the catch and $P$ the probability of capture.

### 2.7.4    *Relative assessment methods*

An alternative strategy to be considered on large rivers is a measure of relative abundance. This strategy is particularly useful if the manager only wants to assess whether the fishery is changing in terms of species composition or population structure. Large reaches of river are fished and the community structure and size distribution of all species are assessed. If the same gear and effort are used on subsequent sampling occasions, these data are directly comparable and can be used to evaluate any changes in stock structure. The species composition and structure can be examined either by catching all fish in the area to be examined and measuring their lengths, or by a census method which generally involves observers calling out the number of fish of each species that they see according to a scale of sizes. For example, small roach $< 12$ cm, medium roach 12–18 cm, and large roach $> 18$ cm. This information can be interpreted to give an approximation of the size distribution.

The advantages of this strategy are similar to those of gear calibration but even longer reaches of river can be fished because fish are not necessarily caught. It also provides an indirect measure of the status through size distribution. For example, an absence of small fish probably reflects poor recruitment. The observation method can be made somewhat more quantitative by occasionally catching and measuring all fish in a small section of river for calibration of the size structure.

### 2.7.5    *Abiotic and biotic factors*

When deciding which is the best strategy to use consideration should also be given to the abiotic and biotic factors affecting electric fishing.

**Site selection**    The choice of many sites in the river situation is greatly influenced by the ease of access, with many areas on large rivers not being sampled simply because some electric fishing gears cannot gain access to the river. The number of sites and type of sites chosen for a particular river need to reflect the needs of the manager and the scientist.

A single site on the whole river system may reveal an absolute population estimate for that particular area of the river, while a number of sites in the target area will each provide an estimate giving a clearer indication of the stock level in the river.

The size of the site should also be considered. Small sites can be surveyed quickly; however, due to the uneven nature of fish distribution, a representative catch of fish size classes and species present in the area may not always be obtained. A larger length of river may therefore need to be surveyed in order to obtain this representative catch, at extra manpower and time costs.

**Flow**   Large rivers generally have considerable volumes of water flowing through the system, thus affecting the ability to electric fish efficiently. The method of data collection will be affected by the ability to use barrier nets effectively. The ability to sample fish can be affected in rivers with high flows due to immobilized fish being washed downstream of the electric fishing vessel before capture. Flow will also affect the time spent sampling a length of river, with high flows increasing the sampling time.

**Depth**   Depth can be a limiting factor for success of fishing. At depths $> 4$ m, many fish are not in the effective attraction zone of the electric fishing boat. This is especially apparent in channelized rivers for navigation purposes where depths in the centre of the watercourse are too great for effective electric fishing or netting.

**Conductivity**   The effect of conductivity is primarily a gear-related problem. As conductivity increases there is a corresponding increase in electrode current and, consequently, greater power demand from the generator. High conductivity can be overcome by utilizing the multiple anode system in which each electrode is sequentially energized in a rapid series, thus creating a large effective field but with only one or two electrodes being energized at any one time.

**Turbidity**   High turbidity can affect the success of a catch because immobilized fish are difficult to observe. This effect can be compensated for on slow-flowing rivers by the use of a secondary support boat for catching fish slow to rise to the surface behind the boat. However, on fast-flowing rivers, immobilized fish may not come to the surface and cannot be easily netted.

**Width**   Large rivers range in width from 15 m to extremes of several 100 m. Rivers $< 20$ m can be sampled with the electric fishing gear by simply fishing down the centre of the watercourse. However, on wider rivers improved catches of fish will possibly result by sampling a short distance from each margin and then fishing down the centre of the watercourse; the value of this being dependent on the central depths of the river.

**Movements**   One further factor which should be considered when planning a survey is the diurnal and seasonal movements of fish. Many species migrate long distances around spawning time and spawning aggregations can influence the population estimates. Also, fish are known to move towards the surface in the dusk period and it may be beneficial to conduct all sampling in the evening as sampling efficiency may be improved at this time.

## 2.8　Conclusions

Although estimates of fish population abundance are routinely made, critical evaluation of the validity of the results is rare, and greater consideration must be given to the assessment procedures. Selection of the appropriate methodology depends on a number of criteria including efficiency of capture, habitat structure and availability of resources. The whole procedure of stock assessment with electric fishing gears can be improved by strategic planning of the assessment exercise.

## References


Bayley P.B. (1983) Central Amazon fish populations: biomass, production and some dynamic characteristics. PhD thesis, Dalhousie University, Halifax, Nova Scotia.

Bayley P.B. (1985) Sampling problems in freshwater fisheries. *Proceedings of 4th British Freshwater Fisheries Conference*. Univ. of Liverpool, pp 3-11.

Bohlin T. & Cowx I.G. (1990) Implications of unequal probability of capture by electric fishing on the estimation of population size. In: I.G. Cowx (ed.) Developments in electric fishing. Oxford: Fishing News Books, pp 145–155.

Bohlin T., Heggberget T.G. & Strange C. (1990) Electric fishing for sampling and stock assessment. In: I.G. Cowx and P. Lamarque (eds) *Fishing with electricity*. Oxford: Fishing News Books, pp 112–139.

Bowles F.J., Frake A.A. & Mann R.H.K. (1990) The use of boom-mounted multi-anode electric fishing equipment for a survey of the fish stocks of the Hampshire Avon. In: I.G Cowx (ed.) *Developments in electric fishing*. Oxford: Fishing News Books, pp 229–235.

Cormack R.M. (1968) The statistics of capture-recapture methods. *Oceanograhy and Marine Biology Annual Review* **6**, 455–506.

Cormack R.M. (1972) The logic of capture-recapture estimates. *Biometrics* **28**: 337–343.

Cowx I.G. (1983) Review of the methods for estimating fish population size from survey removal data. *Fisheries Management* **14**, 67–82.

Cowx I.G. (1990) Application of creel census data for the management of fish stocks in large rivers in the United Kingdom. In: W.L.T. van Densen, B. Steinmetz and R.H. Hughes (eds) *Management of freshwater fisheries*. Wageningen: Pudoc. pp 526–534.

Cowx I.G. (1995a) Fish stock assessment – biological basis for sound ecological management. In: D. Harper and A. Fergusson (eds) *Biological basis for river management*. London: Wiley & Son, pp 375–388.

Cowx I.G., Wheatley G.A. and Hickley P. (1988) Developments of boom electric fishing equipment for use in large rivers and canals in the United Kingdom. *Aquaculture and Fisheries Management* **19**, 205–212.

Cowx I.G., Wheatley G.A., Hickley P. and Starkie A.S. (1990) Evaluation of electric fishing equipment for stock assessment in large rivers and canals in the United Kingdom. In: I.G Cowx (ed.) *Developments in electric fishing*. Oxford: Fishing News Books, pp 34–40.

Goodchild G.A. (1990) Electric fishing and safety. In: I.G. Cowx and P. Lamarque (eds) *Fishing with electricity*. Oxford: Fishing News Books, pp 157–175.

Hickley P. & Starkie A.S. (1985) Cost effective sampling of fish populations in large water bodies. *Journal of Fish Biology* **27 (Supplement A)**, 151–161.

Harvey J. & Cowx, I.G. (1995a) Electric fishing in large deep waters: User manual. NRA R&D Report 0034/7/ST. Bristol. 46 pp.

Harvey J. & Cowx I.G. (1995b) Can you electric fish large deep rivers? Proceedings of the Institute of Fisheries Management Annual Study Course. Cardiff, September 1993.

Hunt P.C. & Jones J.W. (1974) A population study of *Barbus barbus* (L.) in the River Severn, England. I. Densities. *Journal of Fish Biology* **6**, 255–267.

Lazauski H.G. (1984) An evaluation of pulsed D.C. electrofishing as a method of estimating fish population structure. PhD thesis, University of Auburn, Alabama, 115 pp.

National Rivers Authority (1991) Code of Practice for safety in electric fishing operations. NRA, April 1991, 24 pp.

Novotny D.W. (1990) Electric fishing apparatus and electric fields. In: I.G. Cowx and P. Lamarque (eds) *Fishing with Electricity*. Oxford: Fishing News Books, pp 34–88.

Novotny D.W. and Priegel G.R. (1974) *Electro fishing boats – improved designs and operations guidelines to increase the effectiveness of boom shockers.* Wisconsin Dept Nat Resources Tech Bull, No 73.

Seber G.A.F. (1973) *The estimation of animal abundance.* London: Griffin, 506 pp.

Serns S.L. (1982) Relationship of walleye fingerling density and electrofishing catch per unit effort in northern Wisconsin lakes. *North American Journal of Fisheries Management* **2**, 38–44.

Serns S.L. (1983) Relationship between electrofishing catch per unit effort and density in walleye fingerlings. *North American Journal of Fisheries Management* **3**, 38–44.

Southwood T.R.E. (1978) Ecological methods. London: Chapman & Hall, 524 pp.

# Chapter 3
# The use of resistivity fish counters in fish stock assessment

M.W. APRAHAMIAN, S.N. NICHOLSON and D. McCUBBING *NRA North West, Richard Fairclough House, Knutsford Road, Warrington WA4 1HG, UK*
I. DAVIDSON *NRA Welsh, Shire Hall, Mold, Clwyd, UK*

**Abstract** Validation of the performance of the Logie 2100A fish counter was carried out at Forge Weir on the River Lune, using a video recording system. The efficiency of counting upstream migrating salmonids was 86.9%, and downstream migrants, 79.9%. The efficiency of the counter was size-dependent, increasing for upstream migrants from 40% for fish between 16 and 25 cm in length to approximately 90% for those between 36 and 75 cm. There was a decline in the proportion of fish greater than 75 cm counted; reasons for this are discussed. The pattern for those moving downstream was broadly similar.

The amount of variability in peak signal size that could be accounted for was 79% for upstream and 61% for downstream migrants. The majority of the variability was explained by fish length. However, depending on the direction of movement, water level, conductivity and temperature were also significant ($P < 0.05$).

The temporal pattern of counts was compared with catch per unit effort (CPUE) estimates of abundance from the net and rod fisheries. Significant correlations ($P < 0.05$) existed between the weekly estimates of abundance of salmon determined from the counter and the CPUE estimates from the drift net and rod fisheries, and that of migratory trout and the haaf net CPUE estimates. Where the relationship was not significant, reasons are discussed.

KEYWORDS: resistivity fish counter, stock assessment, *Salmo salar, Salmo trutta*

## 3.1 Introduction

The primary objective of the management of migratory salmonid fisheries in the UK is to ensure the well-being and sustainable exploitation of both salmon (*Salmo salar* L.) and migratory trout (*Salmo trutta* L.). Thus there is a requirement to be able to predict how changes in fishing effort (the principal method of regulating fisheries within the UK) and in water and land usage, affect the population and associated fisheries (Beach & Potter 1987). To achieve these aims it is essential that the size and composition of the population re-entering fresh water are determined. Estimates of the size of the population can be obtained by full river trapping, mark-recapture studies and by the use of acoustic and resistivity fish counters.

The aim of this chapter is to evaluate the potential of a resistivity fish counter in the assessment of the number of salmon and migratory trout migrating into fresh water. There have been a number of studies assessing the performance of various types of resistivity counters at a variety of locations (Hellawell 1973; Beach 1978; Dunkley & Shearer 1982; Fewings 1987; Welton *et al.* 1987; Jones & Strange 1989; Schofield 1989; Dunkley 1991; Reddin *et al.* 1992). These studies, together with those of NRA (North West) and its predecessors, have led to an improvement not only in resistivity fish counter technology, but also in the design of the civil structures on which the counter is deployed. These factors and the environmental conditions, in particular the flow, under which the counter is expected to operate, appear to be the main factors affecting counting performance.


## 3.2    Site Description

### 3.2.1    *Counter Site*

The study was carried out at Forge Weir on the River Lune, situated approximately 4 km upstream of the tidal limit (Fig. 3.1). At this point the river is spanned by an oblique-facing weir on which are situated four counting channels; two in the fish pass (channels 1 and 2) and two on the weir itself (channels 3 and 4). Part of the weir, not covered by the detection electrodes, was fitted with oversails to prevent fish ascending at these positions. A length of weir remains where it is possible for fish to ascend without being counted, but this is considered unlikely because of the position of the weir in relation to the flow.

Channel 1 was used for validation purposes (Fig. 3.2). The counter consists of three electrodes (12 x 12 mm stainless steel) installed in Nitomortar blocks, mixed with aggregate, on the downstream face of the weir (slope 1:7.5) and stretching the full width of the channel (2.0 m). The distance between the crest and the midpoint of the upstream electrode was 12 cm, between the upstream and the centre electrode 38 cm and between the centre and downstream electrode 46 cm. To ensure that the concrete wall on the right hand side of the pass was electrically insulated, a sheet of white polypropylex plastic (9 mm gauge) was fitted and the edges sealed with silicone. A Perspex viewing window was positioned on the left hand side of the pass, to aid with species identification and determination of the presence or absence of an adipose fin.

Trapping facilities exist in the fish pass just upstream of channel 1. The trap was designed as a typical box trap with a set of inscales at the downstream end. The bars were covered with wire mesh (approximately 2.5 x 3.8 cm) to catch the smallest migratory trout.

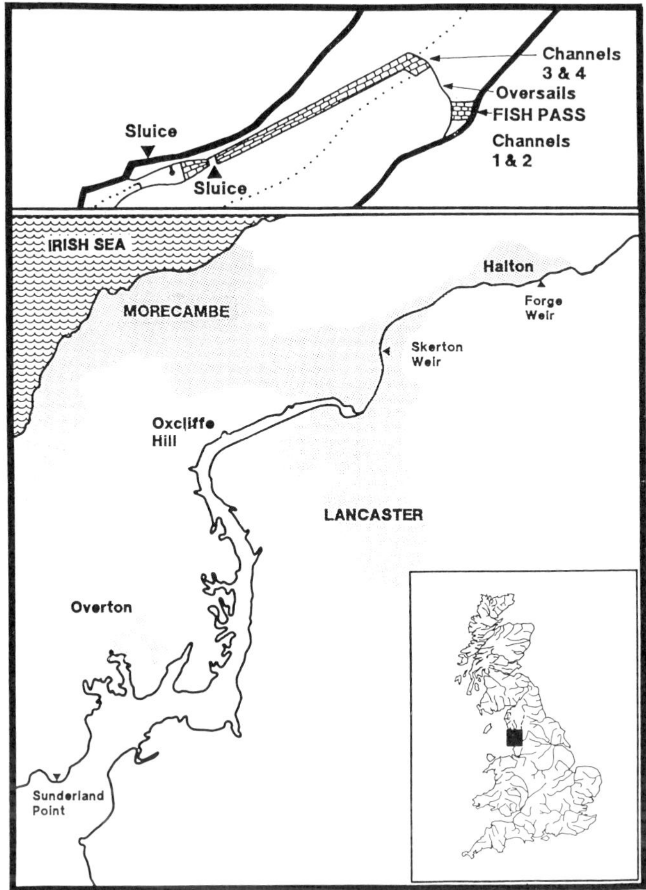


**Fig. 3.1** The study area.

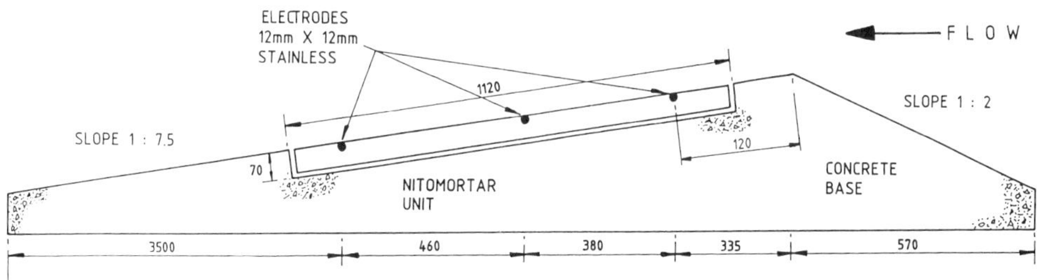


**Fig. 3.2** Side view of the counter site at Forge Weir, channel 1.

### 3.2.2    *Fisheries*

The main net fishery is situated in tidal waters downstream of Lancaster and operates annually between 1 April and the 31 August. The fishery consists of 37 licensed fishermen: 26 haaf/heave nets, 10 drift nets and one seine net. Only the catch data from the haaf and the drift net fisheries were used in this study. The drift net fishery operates seaward of Sunderland point and the haaf net fishery mainly from Oxcliffe Hill to Overton (Fig. 3.1). The size of the mesh is limited to a minimum of 50 mm and 63 mm (knot to knot) in the haaf and drift net fisheries, respectively.

The River Lune supports major salmon and sea trout rod fisheries throughout its catchment, three of which, Upper and Lower Halton and Skerton are owned by the NRA. The Upper and Lower Halton beats are situated immediately upstream and downstream respectively of the counter site at Forge Weir (Fig. 3.1). The Skerton fishery is located in Lancaster close to the tidal limit. The maximum number of anglers allowed to fish for salmon on the Upper and Lower Halton fisheries are 12 and 8 per day, respectively, and for migratory trout 10 and 12 anglers per day, respectively. On the Skerton fishery six permits are available per day for both salmon and migratory trout.

The angling season for salmon runs from 1 February to 31 October and for migratory trout from 1 May to 15 October. A minimum size limit of 30 cm exists for migratory trout on all fisheries.


### 3.3    **Materials and Methods**

The Logie 2100A fish counter was used to interpret changes in electrical resistance between the electrodes. The date, time, conductivity, channel, direction of movement and peak signal size, together with coded values of electrical resistance sampled at 0.01 s intervals (trace data) were recorded by the counter when a change in electrical resistance, above the threshold setting, was detected and the Logie's fish algorithm interpreted a true fish event. On detection of a non-fish event the above information was recorded with the exception of the direction of movement which was represented by the character 'E'. These data were stored on the hard drive of an IBM PC (PS/2 model 55SX).

The threshold was set at 12.7, on a scale of 1–127, for both upstream and downstream events. For events which exceeded the threshold setting the detection limit between two events was 0.5 s.

The data from the Logie fish counter were analysed in relation to that obtained using a video recording system installed over channel 1. The system consisted of two cameras, one mounted overhead on a scaffold frame (Baxall CD6252 shuttered CCD) and the other (Ikegami ICD-42E CCD) situated in the dug-out viewing chamber on the left hand side of channel 1, giving a side view. Two infra-red lamps (240 V, 500 W) were used to illuminate the weir. The outputs from the two cameras were assimilated using a split screen generator. The resultant video signal was combined with the counter output via a 'cashscan' under the control of the PC. The product was then recorded on a Panasonic AG6720A video recorder, using both time lapse (8 frames s$^{-1}$) and real time modes (25 frames s$^{-1}$). A schematic representation of the data acquisition system is presented in Fig. 3.3.

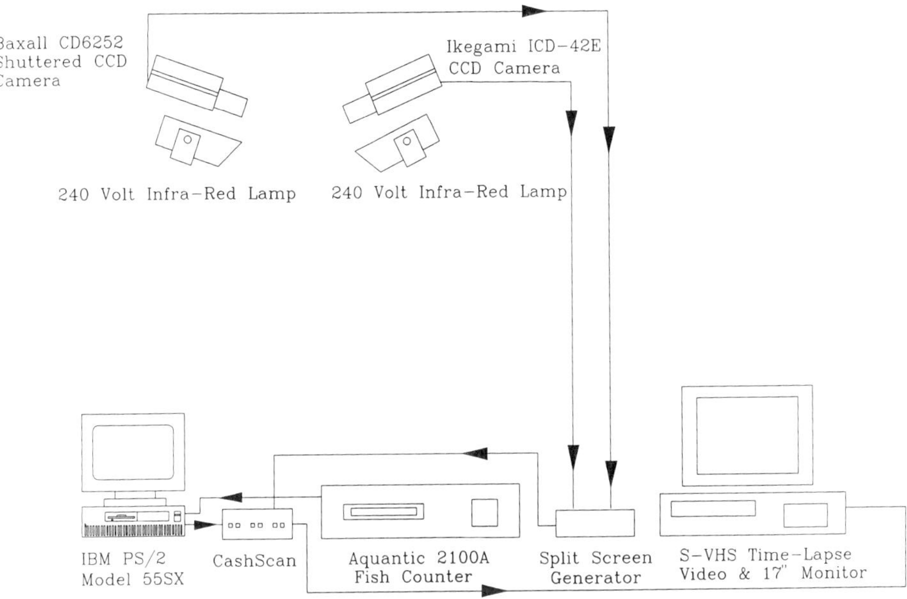


**Fig. 3.3** Schematic representation of the data acquisition system.

Water level, measured at the weir crest on channel 1, and water temperature were recorded at hourly intervals on a Golden River data logger, at Forge Weir.

Trapping was carried out intermittently during the period 4 June–28 October 1993, from approximately 16.00 h to 09.00 h the following day. It was not possible to trap when the depth of water exceeded 0.7 m at the weir crest.

Daily catch and effort data were available for both the net and rod fisheries. These consisted of the number of fish caught per 0.454 g weight category, and effort, both as the number of tides and the number of hours fished that day, for the net fisheries. For the rod fishery the individual weights and the number of hours fished per species were available.

Analysis was carried using the statistical package MINITAB (Ryan *et al.* 1985). Where appropriate, 95% confidence limits were calculated and are presented within brackets throughout the text.

### 3.4    Results

### 3.4.1    *Conditions under which validation was carried out*

Validation was carried out over a range of different water levels from 28 to 66 cm; the level at which each individual fish was recorded is shown in Fig. 3.4. Few observations were possible at high water levels partly because of their relatively infrequent occurrence and partly because of the high water turbidity at levels in excess of 0.54 m, measured at the weir crest. The number of fish caught at different conductivities is displayed in Fig. 3.5.

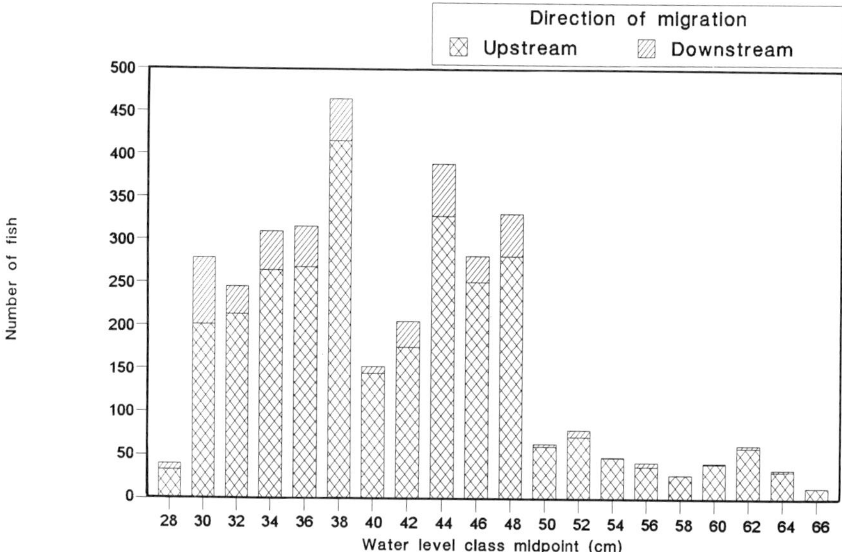


**Fig. 3.4**   Water level at which each fish event was recorded.

### 3.4.2    *Evaluation of counter performance*

*Upstream migrants*

A total of 2769 salmonids were recorded on video migrating over the electrodes (Table 3.1). On 32 occasions two fish passed simultaneously upstream over the counter. As the counter was not designed to count more than one fish at a time, this reduced the expected count to 2737 fish.

Of the counter events, 2378 were from fish traversing the electrodes, giving an accuracy of 86.88% (95% CL 86.85–86.89%). However, 125 fish did not produce a change in electrical resistance above the threshold setting, thus no trace information was recorded. If those fish events below threshold size are excluded, the efficiency of the counter, which generated a signal above the threshold setting, was 91.04% (89.87–92.06%).

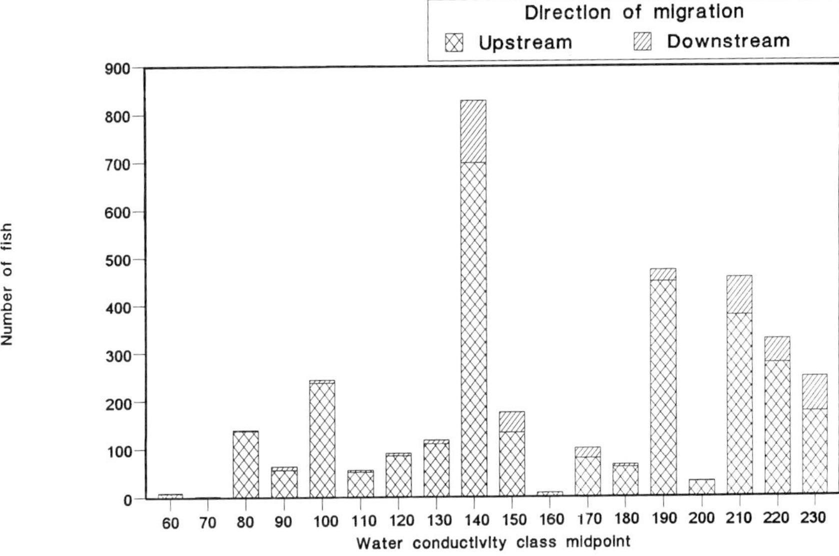



**Fig. 3.5**   Water conductivity at which each fish event was recorded.

**Table 3.1**   Summary of events recorded by the Logie fish counter.

| Event | Upstream | Downstream |
|---|---|---|
| Total number seen on video | 2946 | 471 |
| Total traversing electrode | 2769 | 365 |
| Total not traversing electrode | 177 | 106 |
| Total number of non-fish objects | 0 | 0 |
| Two simultaneous fish* | 32 | 31 |
| Total number of counts | 2445 | 308 |
| Fish traversing | 2378 | 267 |
| Fish not traversing | 35 | 29 |
| Double counts | 3 | 2 |
| Non-fish objects | 0 | 0 |
| Counts for no identifiable reason | 29 | 10 |
| Total not counted | 501 | 144 |
| Fish traversing – with trace | 234 | 50 |
| Fish traversing – without trace | 125 | 17 |
| Fish not traversing – with trace | 141 | 75 |
| Fish not traversing – without trace | 1 | 2 |

* Two fish moving over the electrode array at the same time.

Of the total count, 97.26% (96.57–97.83%) resulted from fish passing upstream over the detection electrodes. The counter recorded relatively few (2.74% (2.15–3.45%)) false counts. It was not possible to identify the reason the counter recorded an event on 29 out of the 67 false counts. From the trace information it was suspected that fish had generated these events, although no fish was seen on video. This situation may arise when a small fish passes close to the side of the counting channel when visibility is poor. A reason was identified for 38 events; 35 of these were fish dropping back downstream having not fully migrated upstream over and away from the counting area, and three from fish which lingered over the electrodes on their upstream passage, producing a double count. Irrespective of this, the false counts were a small percentage of the overall count, and their inclusion only increased slightly the proportion of the upstream migrants accounted for, from 86.88% to 88.3% (88.27–88.30%). If the events below threshold were excluded, the proportion accounted for increased to 93.61% (93.57–93.61%).

*Downstream migrants*

There were 365 salmonids recorded on video migrating downstream (Table 3.1), including 31 occasions where two fish passed over the electrodes simultaneously. Of the potential 334 fish events, 79.94% (75.56–83.66%) were counted. Included in the 365 events, 17 did not exceed the threshold setting. Thus for fish exceeding the threshold value, 84.23% (79.62–87.86%) were counted.

Fish traversing the electrodes generated 86.69% (81.95–89.95%) of the total count. The 41 false counts were mainly generated by fish not fully crossing the electrodes. There were 10 counts for no identifiable reason.

Counter records were made for those 317 fish which produced a trace and traversed the electrodes, representing 84.23% (79.62–87.86%) of the total. The 50 occasions when no count was produced were attributed to migrants not producing a trace recognizable as a true fish event by the counter's fish algorithm. This can arise through aberrant swimming behaviour as opposed to the more typical straight passage exhibited by the majority of the migrants.

**3.4.3**    *Influence of fish size on counter performance*

*Upstream migrants*

Counter performance increased gradually from 40% (34–47%) for the 16–25 cm size group to close to 90% for 95% of the time, for fish ranging in size from 36 to 75 cm (lower 95% confidence limits ranged from 87.5% (size group 66–75 cm) to 96.0% (size group 46–55 cm)) (Fig. 3.6). Fish greater than 75 cm showed a steady decline in the proportion of fish counted, although the sample size was small for this size group.

This decline in performance was attributed to a build-up of water in front of the larger fish (bow wave effect) as they migrated upstream over the counter, and was

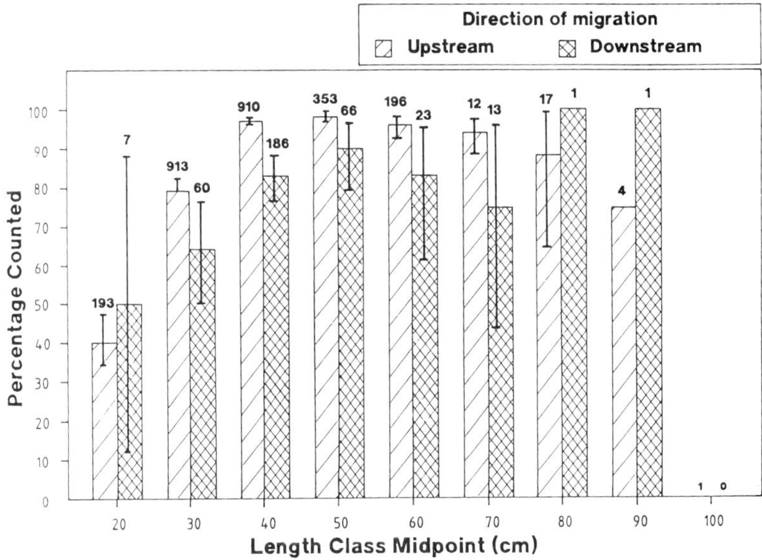



**Fig. 3.6** Proportion of fish counted by the counter in relation to size. Number represents sample size, 95% confidence limits shown where $n > 5$.

particularly prevalent under low flow conditions. The effect was to produce a signal characteristic of a fish moving downstream initially, then without exiting the downstream limit of the counting zone migrating upstream and regaining station (Fig. 3.7). The problem is rejected by the counter.

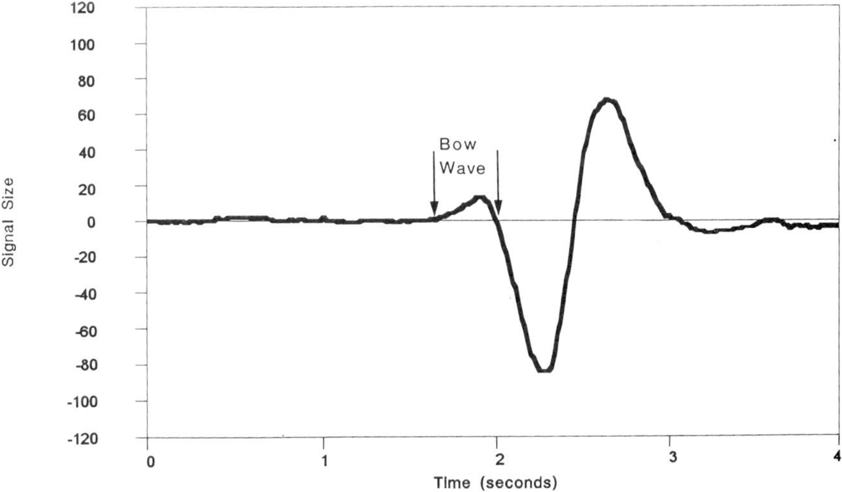



**Fig. 3.7** Typical trace of a fish exhibiting the bow wave effect.

*Downstream migrants*

The pattern for downstream migrants was broadly similar to that evident for the salmonids migrating upstream: an increase in the percentage counted with increasing size, followed by a subsequent decline (Fig. 3.6). However, because of the small sample and associated large confidence limits, for a number of the size classes, the pattern cannot be regarded as definitive.

**3.4.4**    *Relationship between peak signal size and fish length*

A high proportion of the variability in signal size could be accounted for by fish size, with log fish length explaining 69.5% and 57.6% of the variability in log peak signal size for upstream and downstream migrants, respectively. For those fish migrating upstream the inclusion of water level, water temperature and conductivity accounted for an additional 7.4%, 2.6% and 0.08% of the total variability, respectively. For the downstream migrants the inclusion of conductivity ($C$, $\mu$s cm$^{-1}$) accounted for 3.3% of the variability while the addition of water level ($Wl$, m), temperature ($T$, °C) and conductivity did not significantly ($P > 0.05$) improve the relationship.

The relationship between peak signal size ($PSS$) and fish length ($L$ in cm) and certain environmental variables was:

Upstream migrants ($n = 1953$)

$$\log_{10}PSS = -1.39 + 1.74\log_{10}L - 0.557Wl + 0.0248T - 0.00019C \qquad \textbf{(1)}$$
$$(r^2 = 0.79,\ P < 0.001)$$

Downstream migrants ($n = 302$)

$$\log_{10}PSS = -1.31 + 1.62\log_{10}L + 0.00095C \qquad\qquad (r^2 = 0.61,\ P < 0.001)\,\textbf{(2)}$$

The small proportion of the variability accounted for by conductivity indicates that the Logie fish counter adequately compensated for changes in bulk resistance (resistance between the electrodes) resulting from changes in conductivity. As conductivity is inversely related to river flow and thus water depth (Dunkley & Shearer 1982) changes in bulk resistance resulting from changes in water depth will, in part, be compensated for by conductivity. However, as a lag period inevitably existed between a change in river flow and conductivity, it was not possible for the Logie fish counter to compensate fully for changes in river flow through this mechanism. This explains why peak signal size decreased with increasing water depth for fish of a given size. Consequently under high flow conditions there is a higher probability that a small fish was missed by the counter because the resultant signal was less likely to exceed the threshold value.

**3.4.5**    *Comparison of fish counts with CPUE measures of abundance*

*Partitioning counts into salmon and migratory trout*

The upstream and downstream signals recorded by the counter were converted to an estimate of fish size by rearranging equations (1) and (2), above. Numbers within each size group were corrected for counting efficiency using the values presented in Fig. 3.6.

The numbers of salmon and migratory trout within each size group were determined using the values presented in Table 3.2. These proportions were derived from the net and rod fisheries and the trap catch data. The weight (g) measurements from these fisheries were converted to length (cm) using:

$$\log_{10} Weight = -1.44 + 2.71 \log_{10} Length \quad (n = 1132, r^2 = 0.96, P < 0.001) \quad \textbf{(3)}$$

For the months June to August the catch data from the drift and haaf net fisheries were used, while the trap and rod catch per unit effort data were utilized for September and October. For May, because of the small sample size, the June sample was used to convert the counts.

**Table 3.2**   The percentage of salmon in various size classes between June and October (number in brackets refers to sample size). Catch data from the net fisheries were used for the months June–August, and from the trap and rod fishery for September and October.

| Fish length (cm) | June | | July | | August | | September | | October | |
|---|---|---|---|---|---|---|---|---|---|---|
| < 16 | 0 | (0) | 0 | (0) | 0 | (0) | 0 | (0) | 0 | (0) |
| 16–25 | 0 | (0) | 0 | (0) | 0 | (0) | 0 | (0) | 0 | (0) |
| 26–35 | 0 | (6) | 0 | (3) | 0 | (0) | 0 | (23) | 0 | (18) |
| 36–45 | 0 | (81) | 1.5 | (55) | 9.1 | (4) | 0 | (93) | 0 | (18) |
| 46–55 | 1.4 | (222) | 21.8 | (192) | 59.9 | (38) | 3.7 | (15) | 45.3 | (3) |
| 56–65 | 9.7 | (144) | 86.6 | (539) | 99.1 | (798) | 76.3 | (13) | 100 | (18) |
| 66–75 | 31.3 | (62) | 91.3 | (295) | 99.9 | (962) | 92.1 | (47) | 100 | (49) |
| 76–85 | 71.6 | (31) | 93.2 | (91) | 100 | (143) | 100 | (12) | 100 | (12) |
| 86–95 | 100 | (6) | 100 | (57) | 100 | (34) | 100 | (3) | 100 | (5) |
| 96–105 | 100 | (1) | 100 | (15) | 100 | (13) | 100 | (0) | 100 | (2) |
| > 105 | 100 | (0) | 100 | (0) | 100 | (0) | 100 | (0) | 100 | (0) |

*Salmon*

Flow is considered one of the most important factors affecting the movement of salmon into fresh water, and the performance of the rod fishery on the River Lune (Cragg-Hine 1985; Stewart 1969). This was confirmed by a significant correlation between the weekly count estimated as salmon for the period 10 May–31 October 1993 (weeks 19–43; Fig. 3.8) migrating over Forge Weir and the CPUE estimates of

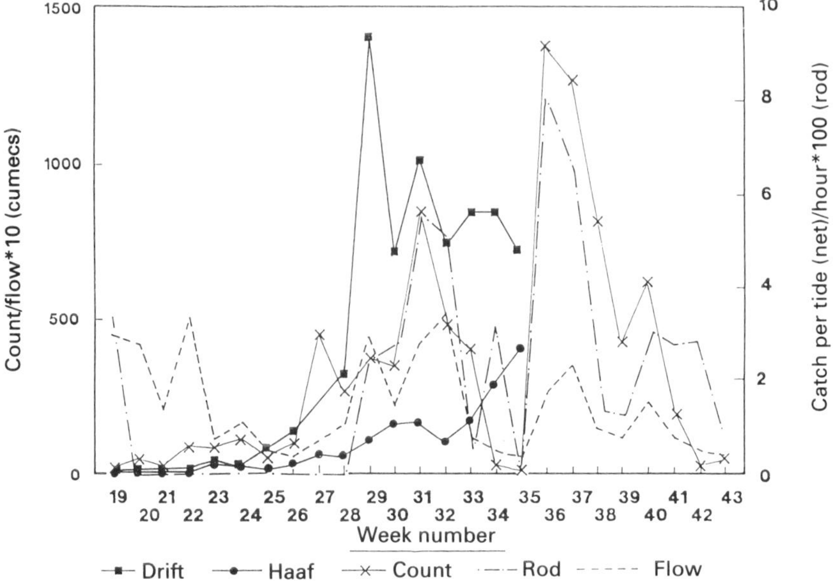


**Fig. 3.8** Comparison of four estimates of abundance of salmon.

abundance from the drift net ($P$ < 0.05; $r = 0.58$) and rod ($P$ < 0.001; $r = 0.75$) fisheries although not from the haaf net fishery ($P$ > 0.05; $r = 0.13$).

For the drift net fishery the relationship between CPUE and count, in particular during weeks 34 and 35, deviated from the overall trend. During these weeks the number of salmon moving over the counter declined sharply although the CPUE of the net fishery remained relatively high and stable. This may be flow-related because flow during these weeks declined, with a consequential reduction in stimulus for fish to migrate up-river. The conditions also appear to have been favourable for the haaf net (weeks 34 & 35) and rod (week 34) fisheries.

It appears that salmon were temporarily resident in the upper estuary and/or lower river because the large number of fish recorded migrating over the counter in week 36 coincided with an elevation in flow.

*Migratory trout*

The CPUE for migratory trout from the net fisheries, between mid-May (week 19) and the end of July (week 30), increased with increasing count (Fig. 3.9). This was not the case for the rod fishery where the CPUE remained low until weeks 31–32. Between weeks 30 and 35 the net CPUE declined to close to zero, although the number of migratory trout estimated by the counter remained relatively high. This decline can, in part, be explained by the increase in the proportion of smaller fish in the population; the proportion of fish estimated to be less than 35 cm increased steadily from 52% at the end of July (week 30) to 98% at the end of August (week 35).

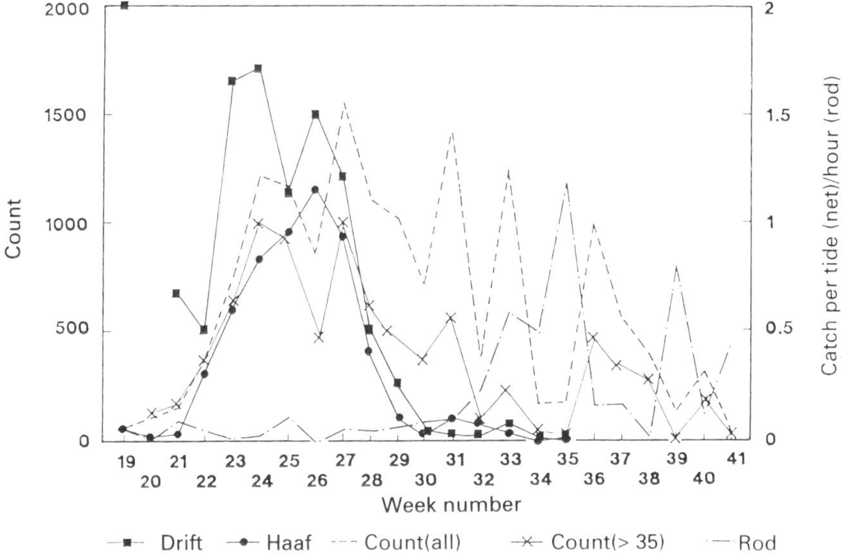



**Fig. 3.9** Comparison of four estimates of the abundance of migratory trout.

A significant correlation existed between the weekly estimates of the total population of migratory trout and that estimated to be $>35$ cm and the weekly CPUE estimate from the haaf net fishery ($P <0.05$, $r = 0.54$; $P <0.05$, $r = 0.78$). For the drift net fishery the correlation with the total population was not significant ($P >0.10$), though for the number estimated as $>35$ cm the relationship was significant at $P <0.10$ ($r = 0.47$). If week 19 was excluded, as only one tide was fished that week and was probably an unrepresentative measure of abundance, the correlation with fish $>35$ cm was significant ($P <0.01$, $r = 0.73$). For the total population the relationship with drift net CPUE, although improved, remained non-significant ($P >0.10$).

The rod fishery was negatively correlated with the number of migratory trout estimated as $>35$ cm ($P <0.05$, $r = -0.51$). The relationship with the estimates of the total population was not significant ($P >0.10$). This may be related to angling conditions. The most successful weeks for angling coincided with low or declining flows (weeks 33–35 and 39), although the exception was during weeks 24–26, despite fish being estimated as abundant.

### 3.5 Discussion

Previously the main factors affecting the performance of open channel resistivity fish counters were: the civil structures on which they were deployed (Welton *et al.* 1987); an inability to compensate for changes in water depth and conductivity (Dunkley & Shearer 1982; Fewings 1987; Jones & Strange 1989); and weed causing a large number of false counts (Hellawell 1973; Welton *et al.* 1987). The latter was not a problem on the River Lune.

The findings of this study support the views of Dunkley (1991) and Reddin *et al.* (1992) that the Logie 2100A fish counter offers, within an acceptable level of accuracy, the possibility of obtaining an assessment of the migratory salmonid stock. Except for fish > 75 cm, for which counting efficiency declined, the performance of the counter was regarded as similar to that for the Logie counter on the North Esk (Dunkley 1991). It would appear to operate successfully under a wide range of flows and was extremely reliable, meeting the specifications for a fish counter (Beach & Potter 1987).

On the whole the Logie fish counter proved to be extremely reliable, lightning strikes being the main reason why a counter may be out of action. In such an event all information stored in the buffer file on the Logie counter may be lost. As a back-up, hourly summary information, on the number of upstream and downstream counts recorded by the counter for the previous hour and from which channel, were downloaded to a printer. In addition, weekly checks on the functioning of the counter were carried out. Modems were installed at all counter sites within North West Region; thus it will be possible to carry out these tasks remotely, ideally on a daily basis and certainly during the main migration period. This ensures that the minimum data are lost in the event of a breakdown.

At Forge Weir there were few counts generated by non-migratory salmonids; the counter was located at the upstream limit of the fish pass. The main problem, other than the effect of water depth on the counting efficiency of the larger fish, occurred when two fish traversed the electrodes at the same time. This is likely to remain a problem as it is unlikely that the algorithm used to analyse the pattern of electrical disturbance (trace data) could be modified to detect two fish events from noise or aberrant fish behaviour. At Forge Weir this particular event occurred relatively infrequently, their exclusion resulting in an under-count of 1.1–6.6%. This may not be the case at other locations, although a similar low occurrence of two fish events was also reported by Dunkley and Shearer (1982).

Although false counts were not a problem on channel 1 at Forge Weir, there were other counting sites operated within North West Region where this was not the case. On those counting channels which regularly dried out false counts were produced when the counting strips first became submerged. False counts can also occur through wind-generated wave action, in particular by a downstream wind when the water depth was low, and from eels (*Anguilla anguilla* L.) migrating in large numbers downstream. Thus under certain conditions these data will require editing. To assist, the trace data were recorded at counter sites and this aided in removing some of the subjectivity in deciding whether the event should be excluded.

On those river systems where it is important to partition the count into salmon and migratory trout it is essential that the counter has some sizing capability. It is also important if salmon are to be separated into their various sea age-classes.

Part of the unexplained variance in peak signal size (21–39%) may be accounted for by variation in swimming height above the electrodes. This directly affects the size of the signal (Fewings 1994), and the measurement of the fish taken from the video (Beach 1978; Dunkley & Shearer 1982; Fewings 1987). Another variable, not measured, which may account for a significant portion of the variance was the weight of the fish. Although weight was closely correlated with length ($r^2 = 0.96$) it will be mass (condition factor) which more directly influences the size of the signal.

The similarity in the pattern between the estimates of abundance of salmon determined from the counter and the drift and rod fisheries, and that of migratory trout and the haaf net fishery estimates, lends support to the contention that the counter was recording true fish events.

For salmon the lack of a significant relationship between counts and the CPUE estimates from the haaf net fishery suggests that factors other than those affecting abundance are controlling the fishery. This may be related to flow, with catchability higher under low-flow conditions (weeks 34 and 35) than at higher flows, by affecting either gear efficiency and/or fish behaviour. Similar conclusions were reported by Clarke *et al.* (1991) for a seine net fishery in the estuary of the River Tywi, attributing higher catchability under low flow conditions to an increase in the time taken for salmon to migrate through the estuary.

For migratory trout the absence of a significant relationship between counts of the total population and the CPUE estimates from the drift net fishery was related to the selectivity of the gear. The fishery exploits fish mainly in excess of 48 cm/1.4 kg, which compares with 42 cm/0.9 kg by the haaf net fishery. For the rod fishery the CPUE remained low until week 32, although the estimates from the haaf net fishery and the counter suggest migratory trout were abundant in the lower river. This low level of exploitation of the larger and thus older migratory trout by the rod fishery can arise either as a result of an inherent low catchability of this proportion of the population and/or because conditions were not suitable for successful angling during May–July. The CPUE then increased steadily until week 35. The increase in CPUE coincided with the entry into fresh water of the 0-sea winter age group of the population, and also with a period of comparatively low flows.

On many river systems there is an overlap in size between salmon and migratory trout. Apportioning the count was based on differences in the relative frequency of fish in various size classes, on a monthly basis. Problems with obtaining an unbiased sample of the population, due to gear selectivity in the net fishery and to size limits for the rod fishery, were evident. There was no corroborative evidence that large numbers of migratory trout, less than 35 cm, were moving upriver during August. On those river systems where no net fishery exists and catch details from the rod fisheries are not available, an additional sampling programme needs to be considered or best judgement must be adopted. The decision is, in part, dependent on the accuracy of the estimate required for the management of the fishery and the cost required to obtain the additional information.

In some situations a fixed size is used to separate salmon from migratory trout – for example, 50 cm by Dunkley and Shearer (1982). Problems can arise in certain situations when using this method. The adoption of a similar size would have underestimated the migratory trout component in all months, though the sample sizes in September and October were small. The adoption of a larger size, of 60 cm, would conversely have underestimated the salmon component by approximately 20% in the period June–August. The preferred scenario would be to operate a full river counting facility with a method of obtaining biological data on the stocks, at a level which would enable the degree of precision required for the management of the fisheries to be obtained.

## Acknowledgements

The authors would like to thank David Cragg-Hine for all his assistance and encouragement throughout the project as well as Ron Shaw, Phil Best, Rebecca Gallagher and Martin Browne for all their help. The views expressed represent those of the authors and not necessarily those of the NRA. Part of the work was carried out under NRA R&D contract (DO3(90)1).

## References

Beach M.H. (1978) Determination of the accuracy of two types of fish counter employing the principles of (i) conductivity change and (ii) acoustic reflection. In: R.B. Bussell (ed.) *Fish Counting Stations: Notes for guidance in their design and use.* London: Department of the Environment Appendix B, pp 1–27.

Beach M.H. & Potter I.C. (1987) MAFF attitude to fish counters. *Counter Workshop.* Atlantic Salmon Trust, Montrose, Scotland, 15–16 September 1987, 18 pp.

Clarke D., Purvis W.K. & Mee D. (1991) Use of telemetric tracking to examine environmental influences on catch effort indices. A case study of Atlantic salmon (*Salmo salar* L.) in the River Tywi, South Wales. In: I.G. Cowx (ed.) *Catch Effort Sampling Strategies: their Application in Freshwater Fisheries Management.* Oxford: Fishing News Books, pp 33–48.

Cragg-Hine D. (1985) The assessment of flow requirements for upstream migration of salmonids in some rivers of north west England. In: J. Alabaster (ed.) *Habitat Modification and Freshwater Fisheries.* London: Butterworth, pp 209–215.

Dunkley D.A. (1991) The use of fish counters in the management of salmonid stocks: the example of the North Esk. *Proceedings of the Institute of Fisheries Management 22nd Annual Study Course.* Aberdeen, 10-12 September 1991, pp 153–158.

Dunkley D.A. & Shearer W.M. (1982) An assessment of the performance of a resistivity fish counter. *Journal of Fish Biology* **20**, 717–737.

Fewings G.A. (1987) The validation of two resistivity fish counters using infra-red telesurveillance at two sites in the North West of England. Unpublished MSc. Thesis, University College of North Wales, Bangor, Wales, 289 pp.

Fewings G.A. (1994) Automatic salmon counting technologies – a contemporary review. Pitlochry, Scotland. Atlantic Salmon Trust, 66 pp.

Hellawell J.M. (1973) Automatic methods of monitoring salmon populations. Proceedings of the International Symposium on the Atlantic Salmon: Management, Biology and Survival of the Species, St Andrews, New Brunswick, Canada, 20-22 September 1972. *Special Publication Atlantic Salmon Foundation* **4**, 317–337.

Jones G.O. & Strange C.D. (1989) Evaluation of the North of Scotland Hydro-Electric Board Mark X fish counter sited at Trostrey weir, River Usk, Wales. NRA (Welsh) Report No *PL/EAE/89/1*, 24 pp.

Reddin D.G., O'Connell M.F. & Dunkley D.A. (1992) Assessment of an automated fish counter in a Canadian river. *Aquaculture and Fisheries Management* **23**, 113–121.

Ryan B.F., Joiner B.L. & Ryan T.B. jr. (1985) MINITAB Handbook. second edition. PWS-KENT, Boston, 386 pp.

Schofield A.J. (1989) An assessment of the performance of the Aquatic 2100A Logie fish counter. Unpublished MAFF Internal Report, 16 pp.

Stewart L. (1969) Criteria for safeguarding fisheries. Fish migration and angling in rivers. *Proceedings of the Institute of Water Engineers* **23**, 39–62.

Welton J.S., Beaumont W.R.C. & Johnson I.C. (1987) Experience of counters in southern chalk-streams. *Counter Workshop*. Atlantic Salmon Trust, Montrose, Scotland, 15–16 September, 1987, 19 pp.

# Chapter 4
# Evaluation of a low disturbance method for estimating population densities of juvenile Atlantic salmon (*Salmo salar* L.) and trout (*Salmo trutta* L.) juveniles in chalk streams

D.J. BIRD, *NRA (South Western Region), Manley House, Kestrel Way, Exeter.*
G.W. LIGHTFOOT and A.P.STREVENS *NRA (South Western Region), Rivers House, Sunrise Business Park, Blandford Forum, Dorset, UK*

**Abstract**   An instream observation technique (snorkel and SCUBA diving) was used to estimate population densities of juvenile Atlantic salmon and trout within quadrat areas. A stratified random sampling approach was used over a wide range of chalk stream habitat types in six contiguous 100 m sections of the River Piddle and one of its tributaries (the Bere Stream) in Dorset. Over the six study sections population densities averaged 0.32 m$^{-2}$ for salmon and 0.11 m$^{-2}$ for trout. The precision of these estimates (expressed as a coefficient of variation) for the six study sites was 0.06 for salmon and 0.13 for trout. The precision for salmon was suitable for most stock assessment purposes, but needs to be improved for trout by increasing the proportion of the site area sampled. Factors affecting accuracy and precision are discussed. The technique is suitable for very young salmonid life-stages and more resistant to the bias caused by channel dimensions than most electric fishing gears. It also appears to cause no mortality with only temporary, very local, displacement of the target species, thereby preserving existing territories and dominance hierarchies. This makes it particularly suitable where repeat assessments of density are sought. In terms of manpower resources the technique compares favourably with other stock assessment practices, although an increased SCUBA requirement in deeper sites will have greater cost implications.

Keywords: Instream observation, population estimate, quadrat, salmonids

## 4.1   Introduction

Information on the distribution and abundance of Atlantic salmon (*Salmo salar* L.) and trout (*Salmo trutta* L.) juveniles is essential if stocks are to be effectively protected and enhanced. The value of such data is determined by *inter alia* their accuracy and precision (Bohlin *et al.* 1989) which are to some extent dependent on the methodologies employed. In the UK, electric fishing is most commonly used to collect this information. However, this method has disadvantages related to cost, disturbance to fish populations, damage to individual fish and the risk of significant bias, particularly when channel dimensions preclude effective sampling (Zalewski & Cowx 1990).

The use of instream observation techniques in behavioural and stock assessment studies was described by several authors (e.g. Keenleyside 1962; Northcote & Wilkie 1963; Heggenes *et al.* 1990).

The majority simply reported the number of the target species encountered in a study length of stream. These techniques required up to five observers (Northcote & Wilkie 1963), presumably to ensure adequate coverage in wide channels. Gardiner (1984) described a more systematic approach whereby salmonids within relatively small marked areas were counted and densities estimated. However, none of these approaches evaluated the precision of the estimated densities.

This chapter describes a development of an instream observation technique that is particularly suited to the estimation of juvenile salmonid abundance. The procedure, which utilizes quadrat theory (Elliott 1977), allows precision to be estimated and, as it creates relatively little disturbance (Heggenes *et al.* 1990), has wider application than most other methods.

Precision estimates were obtained over a wide range of habitat conditions within a chalk stream, to provide a good indication of the overall suitability of the method.

## 4.2     Study area

The River Piddle in Dorset flows 40 km roughly south-east from its spring source to form a common estuary with the River Frome, before discharging into the English Channel via Poole Harbour (Fig. 4.1). As a chalk stream, it is characterized by attributes such as a stable groundwater-fed flow regime, high water clarity, and high productivity (illustrated by significant growth of macrophytes, mainly *Ranunculus* spp. during summer). The substrate is composed largely of fine and coarse gravel, sand and silt. Cobbles and boulders are rare and bedrock is virtually absent from the stream bed.

Anadromous trout and Atlantic salmon are the dominant fish species, whilst eel, *Anguilla anguilla* L., minnow, *Phoxinus phoxinus* (L.), bullhead, *Cottus gobio* (L.), and stoneloach *Barbatula barbatulus* (L.) are common.

Six contiguous 100 m sections of the River Piddle and its principal tributary, the Bere Stream, were selected for the study (Fig. 4.1). It was considered that the full range of habitat types available in this part of the catchment were represented within these 600 m.

## 4.3     Method

Instream observations were undertaken during a period of stable discharge from 29 June to 2 July 1993. Two sites were studied in each day's fieldwork.

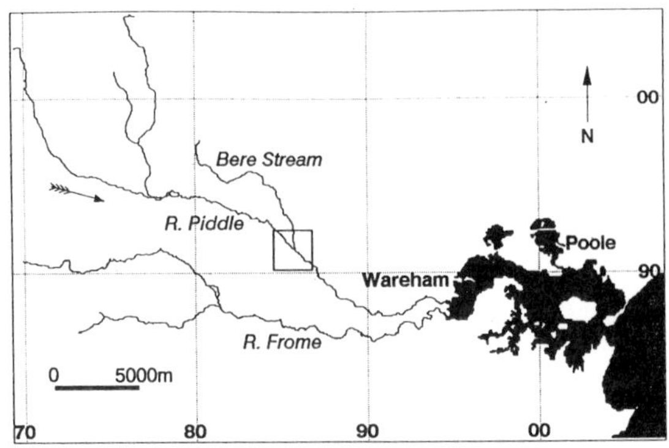


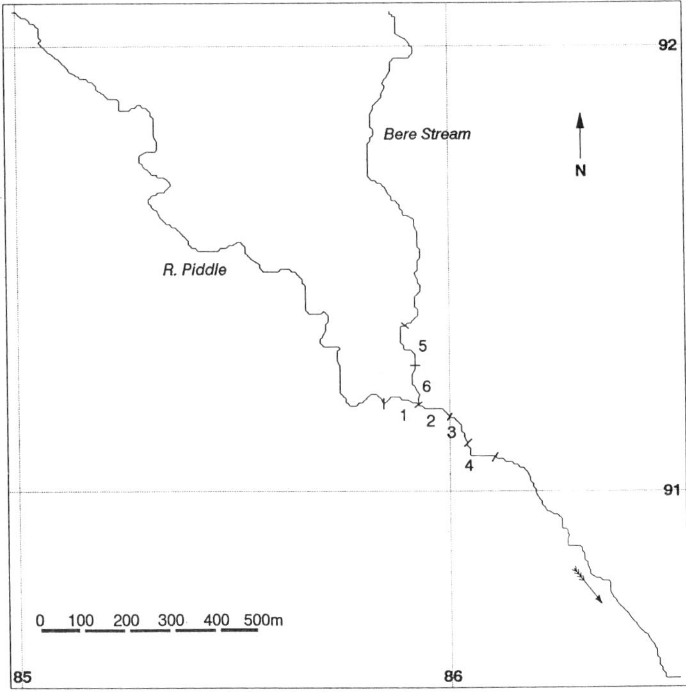


**Fig. 4.1**  Location of study area and sampling site.

The observation strategy at each site was a modification of the stratified random sampling method (Elliott 1977). This unbiased sampling technique, involving examination of marked randomly chosen sampling locations, was originally developed for benthic invertebrates but is also suitable for sedentary, benthic fry (Bovee 1986).

The number of sampling locations examined in each site was proportional to estimated site area and ranged from 79 to 150. The use of larger numbers of sampling locations would produce improvements in the precision of population estimates, if the cost could be justified. To improve sampling efficiency a simplified stratification system was used (Elliott 1977) to classify stream reaches within sites into shallow glide, deep glide, and pool biotopes (Johnson & Elliott 1993).

Prior to the period when observations were made at each site, wet width measurements and estimates of dominant biotope across the channel were made at 1 m intervals. From these data, grids were produced composed of numbered 0.5 m squares within each biotope. Sampling locations from each biotope were then obtained in proportion to their respective total areas by randomly selecting squares.

On the day of observation, sampling locations were marked with an angler's lead weight attached by nylon cord to a float. Both weight and float were numbered to enable identification by the diver and bankside recorder. Markers were placed at the centre of selected squares and a 0.5 m radius search area was defined around the weight position.

To improve precision in the positioning of markers without creating unnecessary disturbance, markers were distributed from a cup attached to a graduated, collapsible pole of varying length (1–7 m).

All observations were undertaken by an experienced diver in daylight hours (between 11.00 h and 16.00 h) under conditions of good underwater visibility ($>2.0$ m), and water temperatures in excess of the minimum 14–15°C considered most favourable for accurate instream counting (Gardiner 1984; Hillman *et al.* 1992). In depths $<0.8$ m, observations were made by snorkelling, but in deeper stretches SCUBA equipment was necessary to search the bottom adequately. In both cases, the diver proceeded upstream through the site, examining each sampling location in turn.

When a fish was encountered within the search area, it was observed for a short period to determine the species and ensure that it was holding station. The relative size of the pectoral fin was the feature most commonly used in identification, whilst body length and colour of adipose fin were also useful. To determine the accuracy of this identification, a sample of fry (21 salmon and 15 trout), which had been classified by the diver, were individually captured by a slurp-gun (DeGraaf & Bain 1986), and examined in detail on the bank.

The precise position of a fish was marked by a small, numbered angling weight attached to a brightly coloured ribbon to facilitate location after the dive. The sampling location, species, and fish marker number were reported to a bankside recorder. This procedure was repeated for all fish observed within the search area.

After the observation period, distances between the sampling location marker and fish location markers were measured so that fish located outside the quadrat zone could be excluded.


**4.3.1**    *Statistical analysis*


*Mean number of individuals per quadrat and variance at the 'site' level*

Population estimates and variances at sites were calculated from the formulae of Elliott (1977):

$$\bar{x}_{SITE} = (n_{SG}\bar{x}_{SG} + n_{DG}\bar{x}_{DG} + n_P\bar{x}_P)/n_{SITE} \tag{1}$$

where $\bar{x}_{SITE}$ is the mean number of individuals per quadrat in a site; $n_{SG}, n_{DG}, n_P$ are the number of randomly chosen sampling locations in shallow glide, deep glide, and pool respectively; $\bar{x}_{SG}, \bar{x}_{DG}, \bar{x}_P$ are the mean number of individuals per quadrat in shallow glide, deep glide, and pool strata respectively; and $n_{SITE}$ is the total number of randomly chosen sampling locations in a site.

$$SE(\bar{x}_{SITE}) = (1/n_{SITE})\sqrt{\left(n_{SG}S_{SG}^2 + n_{DG}S_{DG}^2 + n_P S_P^2\right)\left(1 - n_{SITE}/N_{SITE}\right)} \tag{2}$$

where $SE(\bar{x}_{SITE})$ is the standard error of $\bar{x}_{SITE}$; $S_{SG}^2$, $S_{DG}^2$, $S_P^2$ are the variance of the number of individuals per quadrat in shallow glide, deep glide, and pool strata respectively; $N_{SITE}$ is the total number of potential sampling locations in a site; and $(1 - n_{SITE}/N_{SITE})$ is the finite population correction included in the formula because the sampling fraction exceeded 10% in all sites.

$$V_{SITE} = SE(\bar{x}_{SITE})^2 \tag{3}$$

where, $V_{SITE}$ is the variance of $\bar{x}_{SITE}$; 95% confidence limits were assigned to $\bar{x}_{SITE}$ using student's t distribution.

### 'Stock' level estimates of precision

The six sites sampled represented the target area (Bohlin 1990). 'Stock' level estimates of precision were made using a simple random sampling (SRS) version of the 'Bohlin equations' (Bohlin 1990) (R.Wyatt pers. comm.):

$$\bar{x}_{STOCK} = (\sum_{}^{n} \bar{x}_{SITE\,i})/n_{STOCK} \tag{4}$$

where $\bar{x}_{STOCK}$ is the mean number of individuals per quadrat at the 'stock' level; $\bar{x}_{SITE\,i}$ is the mean number of individuals per quadrat at the $i$th site; and $n_{STOCK}$ is the number of sites.

$$V_t = (\sum_{}^{n}(\bar{x}_{SITE\,i} - \bar{x}_{STOCK})^2/(n_{STOCK} - 1) \tag{5}$$

where, $V_t$ is the total variance of the $\bar{x}_{SITE}$ estimates.

$$V_w = \sum_{}^{n} V_{SITE\,i}/n_{STOCK} \tag{6}$$

where, $V_w$ is the within-site variance, i.e. component of total variance $(V_t)$, which is due to the imprecision $(V_{SITE})$ of each $\bar{x}_{SITE}$ estimate; $V_{SITE\,i}$ is the variance of $\bar{x}_{SITE\,i}$.

$$V_b = V_t - V_w \tag{7}$$

where, $V_t$ is the total variance; $V_b$ is the between-site variance; $V_w$ is the within-site variance.

$$(\bar{x}_{STOCK}) = V_w/n_{STOCK} \tag{8}$$

where $(\bar{x}_{STOCK})$ is the variance of $\bar{x}_{STOCK}$

$$C = \sqrt{V(\bar{x}_{STOCK})}/\bar{x}_{STOCK} \qquad\qquad (9)$$

where $C$ is the coefficient of variation or precision level.

## 4.4     Results

Analysis of the sample of fry captured by the slurp-gun, showed that species identification by the diver was achieved with 100% accuracy. It was found that trout (mean fork length 65 mm) were significantly larger than salmon (mean fork length 52 mm) (U-test, $P < 0.05$).

Population estimates and 95% confidence limits at the 'site' level show overall Atlantic salmon were significantly more abundant than trout (U-test, $P < 0.05$) (Table 4.1). The abundance of both species was closely related to the macrohabitat characteristics of the sites. Thus, highest densities were found on the shallow glide-dominated Bere Stream (sites 5 and 6) characterized by a high proportion of suitable gravel substrate, *Ranunculus* spp. cover, and a plentiful invertebrate drift supply. By contrast, the lowest densities were found in site 4, which was deep glide-dominated due to impoundment and was consequently slower flowing with silted gravel.

'Stock' level parameters show that the coefficient of variation for salmon was approximately half that estimated for trout (Table 4.2).

**Table 4.1**   Population estimates and 95% confidence limits for salmon and trout at the site level.

| Site | Salmon | | Trout | |
|---|---|---|---|---|
| | per quadrat | per m$^2$ | per quadrat | per m$^2$ |
| 1 | $0.18 \pm 0.08$ | $0.23 \pm 0.10$ | $0.06 \pm 0.05$ | $0.08 \pm 0.07$ |
| 2 | $0.18 \pm 0.06$ | $0.23 \pm 0.08$ | $0.04 \pm 0.03$ | $0.05 \pm 0.04$ |
| 3 | $0.23 \pm 0.07$ | $0.29 \pm 0.09$ | $0.05 \pm 0.03$ | $0.06 \pm 0.04$ |
| 4 | $0.09 \pm 0.05$ | $0.12 \pm 0.06$ | $0.02 \pm 0.02$ | $0.02 \pm 0.03$ |
| 5 | $0.45 \pm 0.13$ | $0.61 \pm 0.17$ | $0.18 \pm 0.08$ | $0.22 \pm 0.10$ |
| 6 | $0.36 \pm 0.02$ | $0.46 \pm 0.02$ | $0.17 \pm 0.07$ | $0.22 \pm 0.09$ |
| Mean | 0.25 | 0.32 | 0.09 | 0.11 |

## 4.5     Discussion

The selection of shallow, fast-flowing riffle sections of streams by juvenile salmonids has been widely reported (see Morantz *et al.* 1987 for review). It is widely believed that, in this habitat, electric fishing gives accurate and precise estimates of juvenile salmonid abundance (Bohlin *et al.* 1989). The use of electric fishing for the whole study area, however, would have introduced depth bias (Bohlin *et al.* 1989), as several individuals, particularly salmon, were observed in pools $> 2$ m in depth.

Approaching the quadrat marker correctly was important and dependent on water

**Table 4.2** 'Stock' level parameters for salmon and trout in the study reach.

| Parameter | Salmon | Trout |
|---|---|---|
| $V_t$ | 0.02003 | 0.00496 |
| $V_w$ | 0.00148 | 0.00068 |
| $C_w$ | 0.15263 | 0.30833 |
| $V_b$ | 0.01854 | 0.00428 |
| $C_b$ | 0.53996 | 0.77336 |
| $\bar{x}_{STOCK}$ | 0.25220 | 0.08460 |
| $V(\bar{x}_{STOCK})$ | 0.00025 | 0.00011 |
| $C$ | 0.06230 | 0.12591 |

clarity. In very shallow water ($<0.2$ m) it is notoriously difficult to sample by instream observation (Keenleyside 1962; Gardiner 1984), accounting for the need to observe the quadrat from a distance downstream in good visibility conditions ($\geqslant 2$ m). Nevertheless, in general, if the quadrat marker was approached carefully, the location of all salmonid fry could usually be marked with relative ease and reliable accuracy.

Potential bias created by fry that were utilizing cover and therefore not readily visible, was reduced by systematic searching through any cover within the quadrat area. Otherwise the problem of undetected fry was not considered to be important due to the paucity of large substrate as refuges in chalk streams. Isolated fry would often flee to a new position (rarely more than 1 m distant), but the original station could be marked and the individual then pursued for identification. The intrusion of disturbed fry into the quadrat zone was considered to be negligible and readily apparent to the diver.

Throughout this study, very little disturbance of salmonid fry was observed (Keenleyside 1962; Heggenes *et al.* 1990). Occasionally, the marker weight could be positioned without the fish moving position and it was rare for a fish that had moved not to return to its station soon after the diver had retreated. These observations suggest that territories and dominance hierarchies were maintained and that there was no mortality associated with this method. This is unique as other stock assessment methodologies often cause direct mortality and probably have marked effects on behaviour. This technique is therefore considered useful in studies that require repeated estimates of density. It may be particularly valuable in production studies and in describing important early mortality patterns in salmonids (Elliott 1990), particularly as electric fishing can be less efficient for capture of fish $<100$ mm (Zalewski & Cowx 1990). It can also be considered as an alternative methodology if a fishery owner does not consent to electric fishing because of fears of damage to fish.

In practice, the precision associated with a stock assessment programme is largely determined by the proportion of the target area that is sampled (Bohlin 1990). In the case of a whole target area being sampled, as in this study, the overall precision achieved is a function of the within-sites variation. The results indicate that, for

Atlantic salmon fry, the method estimates abundance with a degree of precision acceptable for most management purposes. By contrast, the precision associated with trout fry was probably unacceptable for making stock-level statements. This difference between species was probably due to the relatively low density of trout in the study area coupled with their more contagious distribution. It should be possible to improve precision by enlarging the quadrat, increasing the proportion of quadrats sampled, or a combination of the two. Both approaches, however, have cost and practical implications. Increasing the proportion of sampled quadrats is probably the better means of improving precision, as enlarging quadrat area may introduce sampling error at high fish densities. During this study, the maximum number of salmonid fry recorded within a single quadrat area was seven. It is possible that even the small disturbance associated with positioning the first few marker weights, may affect the accuracy with which further markers can be placed within a single quadrat area.

The cost of this method compares favourably with other techniques, although the exact expenditure is more dependent on stream dimensions. Where SCUBA is not required, 100 m of stream can be sampled for the equivalent of 1.3 man-days (including a provision for initial site measurements, stratification and randomization). When SCUBA was used, the cost increased, due to the Health and Safety requirements of a standby diver and a dive supervisor (Health and Safety Executive 1981), who cannot become involved in any other aspect of the work whilst the dive is in progress. On these occasions the cost rose to approximately 1.9 man-days 100 m$^{-1}$. By comparison, quantitative electric fishing would require approximately 1.2 man-days 100 m$^{-1}$ to obtain a similar estimate. The capital cost of the gear also compares well with that of other methods. The equipment used in this study would cost approximately £3000 (including SCUBA gear for two personnel), with negligible recurring expenses.

The method described in this chapter is only suitable for juvenile salmonids. The benthic preference of fry eliminated the need to search further in the water column. Older salmonid life-stages, having more variable habitat preferences, may exist throughout the water column and are less diver tolerant. Similarly non-salmonids, such as cyprinids, may be present over a range of depths and often occur in large shoals which would be difficult to enumerate.

The method described has both advantages and disadvantages compared with more traditional techniques such as electric fishing. It is presented as an alternative quantitative approach that may be considered when others are unsuitable. It is suggested that its use may be extended to other habitat types, particularly those where high water clarity and a relatively smooth substrate prevail.


**Acknowledgements**

We wish to thank our colleagues of the National Rivers Authority (South Western Region) who contributed during the fieldwork phase of this study. The views expressed here are those of the authors and not necessarily those of the National Rivers Authority.

## References


Bohlin T. (1990) Estimation of population parameters using electric fishing: aspects of the sampling design with emphasis on salmonids in streams. In: I.G. Cowx (ed.) *Developments in Electric Fishing*. Oxford: Fishing News Books, pp 156–73.

Bohlin T., Hamrin S., Heggberget T.G., Rasmussen G. & Saltveit S.J. (1989) Electrofishing – theory and practice with special emphasis on salmonids. *Hydrobiologia* **173**, 9–43.

Bovee K.D. (1986) Development and evaluation of habitat suitability criteria for the Instream Flow Incremental Methodology. *Instream Flow Information Paper 21*. United States Fisheries and Wildlife Service, Biological Report **86**(7), 235 pp.

DeGraaf D.A. & Bain L.H. (1986) Habitat use by and preferences of juvenile Atlantic salmon in two Newfoundland rivers. *Transactions of the American Fisheries Society* **115**, 671–681.

Elliott J.M. (1977) Some methods for statistical analysis of samples of benthic invertebrates. *Freshwater Biological Association Scientific Publication* No 25, 160 pp.

Elliott J.M. (1990) Mechanisms responsible for population regulation in young migratory trout, *Salmo trutta* L. III The role of territorial behaviour. *Journal of Animal Ecology* **59**, 803–818.

Gardiner W.R. (1984) Estimating population densities of salmonids in deep water in streams. *Journal of Fish Biology* **24**, 41–49.

Health and Safety Executive (1981). *Diving Operations at Work Regulations*. H.S.E. SI 1981/399, HMSO.

Heggenes J., Braband A. & Saltveit S.J. (1990) Comparison of three methods for studies of stream habitat use by young Brown trout and Atlantic salmon. *Transactions of the American Fisheries Society* **119**, 101–111.

Hillman T.W., Mullan J.W. & Griffith J.S. (1992) Accuracy of underwater counts of juvenile Chinook salmon, Coho salmon and Steelhead. *North American Journal of Fisheries Management* **12**, 598–603.

Johnson I.W. & Elliott C.R.N. (1993) *River Allen Habitat Mapping Survey*. Report to the National Rivers Authority Wessex Region. Wallingford: Institute of Hydrology, 12 pp.

Keenleyside M.H.A. (1962) Skin-diving observations of Atlantic salmon and brook trout in the Miramichi River, New Brunswick. *Journal of the Fisheries Research Board of Canada* **19**, 625–634.

Morantz D.L., Sweeney R.K., Shirvell C.S. & Longard D.A. (1987) Selection of microhabitat in summer by juvenile Atlantic salmon (*Salmo salar*). *Canadian Journal of Fisheries and Aquatic Science* **44**, 120–129.

Northcote T.G. & Wilkie D.W. (1963) Underwater census of stream fish populations. *Transactions of the American Fisheries Society* **92**, 146–151.

Zalewski, M. & Cowx, I.G. (1990) Factors affecting the efficiency of electric fishing. In: I.G. Cowx & P. Lamarque (eds) *Fishing with electricity*. Oxford: Fishing News Books, pp 89–111.

# Chapter 5
# Assessment of the deep water trawl fisheries of the South East Arm of Lake Malawi using exploratory surveys and commercial catch data

M. BANDA, T. TOMASSON *Fisheries Research Unit, P.O. Box 27, Monkey Bay, Malawi*

D. TWEDDLE *JLR Smith Institute for Ichthology, Private Bag 1015, Grahamstown, RSA. Correspondence to D. Tweddle.*

**Abstract**  At least 150 species are commonly caught in the demersal trawl fishery in the South East Arm of Lake Malawi, with none comprising more than 10% of the overall catch in the long term. Biomass surveys gave useful indicators of the size of the stocks, and treating the stocks as single species and applying surplus production models to commercial catches to assess potential yields allowed the development of a sustainable, profitable fishery. However, catch rates in the deepwater (40 to 100 m depth) trawl fishery of the South East Arm of Lake Malawi have declined in recent years. Surplus production model results indicate overfishing. Research surveys indicate a decline in biomass, even in untrawled areas. A trawl survey with a new, more powerful research vessel has shown that there are differences in catch rates over the area, with the southern parts nearest the commercial trawlers' bases having lower catches than areas further north. Species composition and size data show that there is little difference between areas of equivalent depth, although for some species there is a trend towards larger mean size in the more lightly fished northern areas. In conjunction with the evidence from catch rates, this suggests the main impact of trawling is limited to a smaller area than that licensed for fishing.

In other, shallow water trawling areas of the South East Arm, catch and effort data, biomass surveys and species composition of the catches all show evidence of heavy exploitation. In future, management will be based not only on catch and effort data but on monitoring the health of the stock through regular surveys of size and species composition of the stocks in the trawling areas.

KEYWORDS: Malawi, trawling, overfishing, surplus production models

### 5.1    Introduction

Lake Malawi is a deep rift lake approximately 600 km long and up to 80 km wide, with a maximum depth of about 700 m. At the southern end of the lake, the gently shelving smooth lake bed provides extensive demersal trawling areas (Fig. 5.1). Experimental trawling began in the mid-1960s as a way of exploiting small cichlid species which could not be easily or economically fished by existing methods. Following these experiments, commercial trawl fishing began in 1968 and rapidly expanded. An initial report on the fishery (Tarbit 1972) was followed by detailed investigations into the changes taking place as the fishery developed (FAO/UNDP 1976; Turner 1977a,b; Tweddle & Turner 1977). Recommendations were implemented which restricted the number of licences to be issued for demarcated areas.

There were large changes in species composition of catches as the trawl fishery intensified, with larger cichlid species of localized occurrence disappearing from the catches, while small species actually increased in abundance (FAO/UNDP 1976; Turner 1977a). The large catfish species were also greatly reduced in abundance.

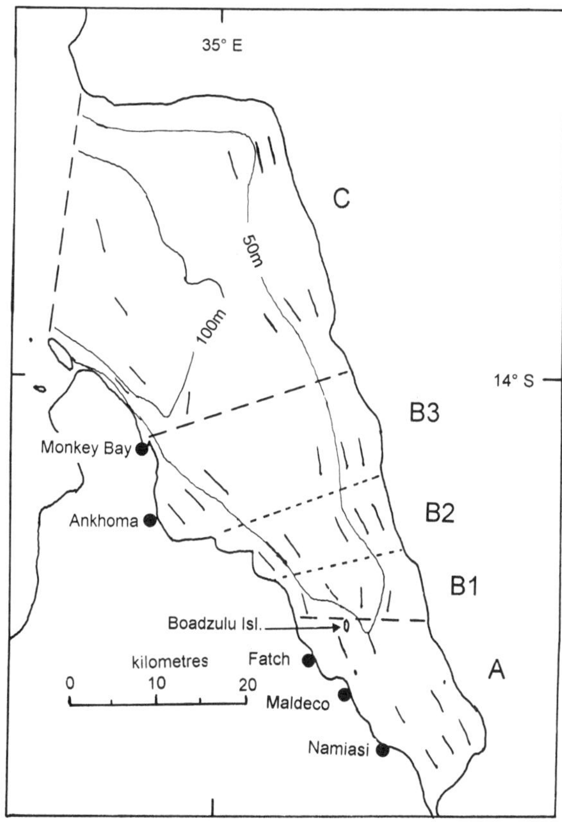


**Fig. 5.1**  The South East Arm of Lake Malawi, showing the trawling areas A, B and C, the main landing sites and places mentioned in the text, depth contours, and the stations used for the *Ethelwynn* biomass surveys.

The South East Arm of the lake was divided into three areas for stock assessment and licensing of the trawl fisheries (Fig. 5.1). A is the shallow area (maximum depth 50 m) south of Boadzulu Island, B is the area from Boadzulu to Monkey Bay and C is the area from Monkey Bay north to Makanjira. Areas B and C have since been divided into deep ( > 50 m) and shallow ( < 50 m) areas for stock assessment based on species composition studies of catches in the shallow water pair trawl fishery and the deeper water stern trawl fishery (FAO 1993; Turner *et al.* 1995). At present, three stern trawlers are licensed for Areas B and C, and six pair trawlers for Areas A, B and C. The demersal trawl catch averages 2200 t yr$^{-1}$ (mean 1988–1992) out of a total catch for all fisheries in the South East Arm of approximately 13 000 t yr$^{-1}$ over the same period.

From the mid-1970s to the mid-1980s, monitoring of commercial catches continued (Tweddle & Magasa 1989; Lewis & Tweddle 1990), although species changes were not investigated. Because of concern over the changes which appeared to be occurring in fishing practices, and the need to reassess the species and size composition of the catches, a new assessment project was initiated and became operational in 1989. In the shallow grounds south of Boadzulu Island, where most of the pair trawl fishery is concentrated, larger species have declined further since the mid-1970s, but there was little evidence of any change in the deeper waters to the north of Boadzulu Island (Turner *et al.* 1995).

Based on commercial catch records, FAO (1993) contended that there has been a collapse of the fishery in the deeper waters to the north of Boadzulu Island, and recommended that fishing effort be strictly controlled and no expansion allowed.

Given that the area fished by the deeper water commercial trawlers was limited to the area immediately north of Boadzulu Island and that little or no fishing activity was observed further north in the South East Arm, a survey of the area was undertaken using the new research vessel, *MV Ndunduma*, to assess the validity of the FAO recommendations. In addition to assessing biomass, data on species composition and size distribution of the catches were collected to assess the exploitation status of the stocks in the areas affected by various fishing regimes. Area B was further subdivided into three areas for this purpose (Fig. 5.1). The results of this investigation were combined with those of the previous study started in 1989 and with an assessment of the commercial catch data to provide a comprehensive picture of the current status of the South East Arm trawl fishery.

## 5.2     Methods

### 5.2.1     *Biomass estimates*

Two different vessels were used to assess the biomass of the exploited stocks using the swept area method.

**MV Ethelwynn Trewavas**     Surveys were conducted using the research vessel *Ethel-*

*wynn*, a 14 m trawler powered by a 90 HP engine. A trawl with 25 mm mesh codend, the same as that employed in previous surveys (FAO/UNDP 1976), was used and the stations fished were the same as in the earlier surveys (Fig. 5.1). Biomass for each trawling area was estimated from the mean catch of several 30 min hauls in the area multiplied by the ratio of total fishing area to area covered by the trawl (Alverson & Pereyra (1969) as described by FAO/UNDP (1976)).

In the South East Arm, trawl areas A, B and C (Fig. 5.1) were divided into three depth ranges for area and biomass estimation, 0–50 m, 51–100 m and 101 m +. From May 1989 to January 1994, 13 surveys were conducted. Surveys were intended to be quarterly but equipment and financial problems led to some surveys being omitted.

**MV Ndunduma**   The new 17.5 m long, 380 HP research vessel *Ndunduma* towed a larger net than the *Ethelwynn* at almost twice the speed, 6.5 km h$^{-1}$ (= 3.5 knots) v. 3.7 km hr$^{-1}$ (= 2 knots). Biomass estimates were made from two surveys, in May/August and in December, 1993. The surveys were carried out from 10 to 110 m depth during daylight using an otter trawl of 23 m headrope length with a 38 mm mesh codend. One hundred and forty hauls of 30 min and 1 h duration were made during the two surveys, 90 hauls during the first survey and 50 hauls during the second survey. Each trawl was made parallel to the shore line at the chosen depths, using a GPS system to mark the exact position of the haul. Biomass estimates were made for the same areas as those covered by the *Ethelwynn*. Earlier trawling surveys revealed that most cichlid species have restricted vertical distributions (Eccles & Lewis 1979; Eccles & Trewavas 1989). Since there is a gradient of increasing depth from south to north in the South East Arm, the areas were further subdivided into 20 m depth bands, i.e. 10–30 m, 31–50 m, 51–70 m, 71–90 m and 91 + m. In addition, Area B was subdivided into B1, B2 and B3 to investigate differences in stock density in relation to distance from the trawlers' bases. The subdivided area data are presented in terms of mean catch per unit effort (CPUE).

### 5.2.2   *Species composition*

For the purposes of this study, only data on species composition from the surveys conducted on the *Ndunduma* are presented. The results of the two surveys on the *Ndunduma* were kept separate bearing in mind the seasonal nature of trawl catches in the lake (Tweddle & Magasa 1989; Lewis & Tweddle 1992). The catch from each trawl was sorted into four categories (Turner 1977a), i.e. small fish species (predominantly cichlids), clariid catfish, kampango (*Bagrus meridionalis* Günther), and others which are primarily composed of larger species, e.g. cyprinids, mormyrids and cichlids such as *Oreochromis* and *Buccochromis* spp. Each category was then weighed. Further catch details were recorded for approximately half of the May–August survey and all the December survey. In these hauls, the last three categories were sorted into individual species, measured (total length nearest cm) and weighed (kg). For small fish, a sample of about 15 kg was collected from different parts of the catch, sorted into species, measured and weighed. In the May/August survey, sorting was carried out on these samples in the laboratory after preservation in formalin. The sampled hauls were used to extrapolate species composition for all hauls in each area. In the December survey, all sampling was carried out on fresh fish on board. Each catch was checked for rarities in case they were missed in the subsampling. Catches for all hauls were adjusted to catch per hour for comparison of hauls.

### 5.2.3     *Catch and effort data*

All trawl units in Lake Malawi have submitted complete daily catch (by species group and by weight) and effort records monthly since the inception of the trawl fisheries in 1968. FAO/UNDP (1976) standardized effort in terms of boat days, using the first unit to fish commercially as one standard boat. CPUE was directly related to engine horse power. Ratings of the pair trawlers ranged from 0.38 to 0.49 standard boats, and a pair trawl day was standardized at 0.5 boat days. Although the horse power of most of the pair trawlers has since been increased, from 22 HP to 26 or 30 HP per boat, the efficiency of these vessels varies considerably, with frequent breakdowns and erratic fishing effort. Fluctuations in efficiency in the various units more than outweigh the increased power: hence the pair trawl day continues to be classified as 0.5 standard boat days.

The deep water trawlers of the Maldeco Company have also increased in HP over the years, although there is again a question about the efficiency of these now elderly vessels. The *MV Dennis Sanudi* is rated at one standard boat. This vessel had an 85 HP engine from 1972 to 1984 and 125 HP thereafter, with occasional spells with the old engine during overhauls. Maldeco have occasionally used two other trawlers in the area, while another fishing company, F. Fatch, has one vessel. Two fisheries department-owned trawlers fished in Areas B and C intensively during the 1980s. In this study, effort for the stern trawl fishery of Areas B and C has been standardized annually against that of the *Dennis Sanudi* using the formula:

*Total effort for the year = Total catch/Dennis Sanudi catch × Dennis Sanudi effort*

Sustainable yields were estimated from the data for the fishing areas using the Schaefer (1954) surplus production model, using effort based on the mean of the year of fishing and the previous year following the method of Gulland (1961).

### 5.3     **Results**

### 5.3.1     *Biomass estimates*

**MV Ethelwynn Trewavas**  To compare with the estimates from the early 1970s, all biomass estimates from 1989 to 1993 (Figs. 5.2 & 5.3; Table 5.1) were combined for

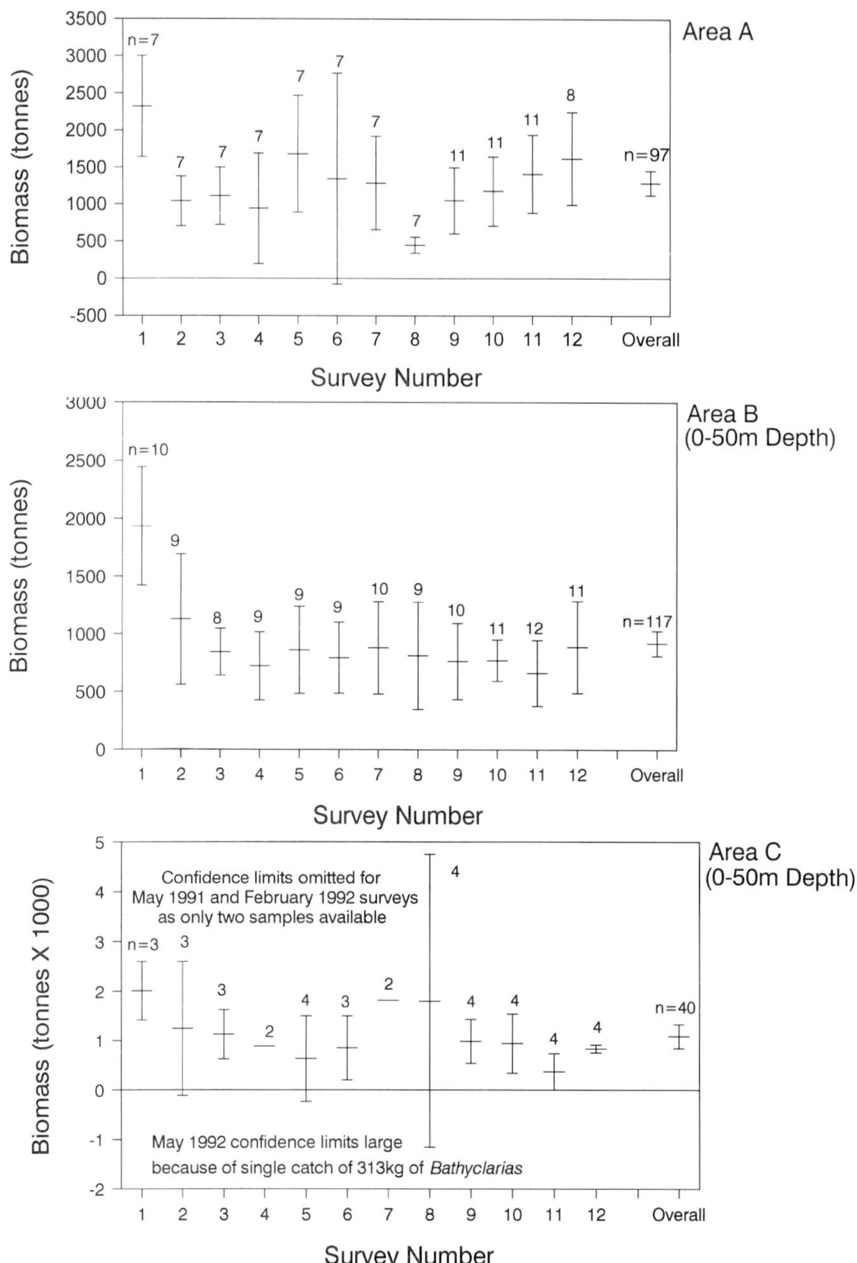



**Fig. 5.2** Biomass estimates made from *Ethelwynn* trawl surveys for the shallow water areas in the South East Arm. The mean estimate and the 95% confidence limits are shown for each survey. A mean estimate for the entire period from 1989 to 1994 is also shown for each area. The surveys were conducted on the following dates: 1 = Sep 1989, 2 = Jun 1990, 3 = Oct 1990, 4 = May 1991, 5 = Aug 1991, 6 = Nov 1991, 7 = Feb 1992, 8 = May 1992, 9 = Feb 1993, 10 = Jun 1993, 11 = Sep 1993, 12 = Jan 1994.

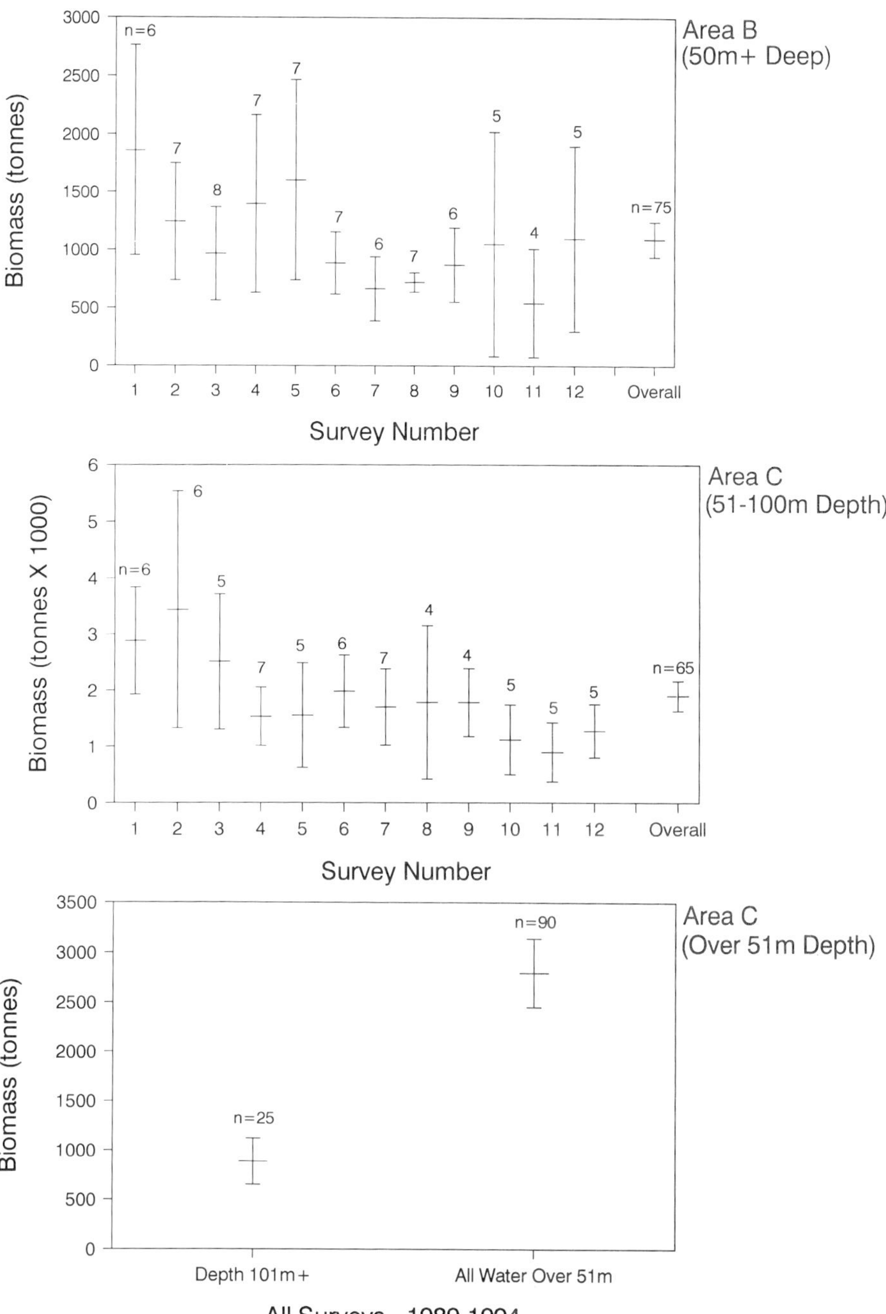


**Fig. 5.3** Biomass estimates made from *Ethelwynn* trawl surveys for the deep water areas in the South East Arm. The mean estimate and the 95% confidence limits are shown for each survey. A mean estimate for the entire period from 1989 to 1994 is also shown for each area. As few trawls were made in water over 100 m deep, an overall estimate only is shown. In the same chart, an overall estimate is included for all waters in Area C over 50 m depth, i.e. the data for 51–100 m and for 101 + m combined. The surveys were conducted on the dates shown in Fig. 5.2.

**Table 5.1**   Biomass estimates (in metric tonnes) for the trawling areas of the South East Arm. The *Ethelwynn* estimate for 1971/73 is taken from FAO/UNDP (1976) but corrected for errors in measurements, the *Ethelwynn* estimate for 1989/94 is the mean for all samples between September 1989 and January 1994, while the *Ndunduma* estimate is based on the May/August and December 1993 surveys.

| Fishing zone | Depth zone m | Area (km²) | *Ethelwynn* estimate 1971/73 | *Ethelwynn* estimate 1989/94 | *Ndunduma* estimate 1993 |
|---|---|---|---|---|---|
| Area A | Total | 221 | 2900 | 1290 | 1640 |
| Area B | 0–50 | 231 | 2310 | 920 | 1490 |
|  | 50–100 | 233 | 2330 | 1090 | 1110 |
| Area C | 0–50 | 256 | 1940 | 1080 | 2380 |
|  | 51–100 | 538 | 3100 | 1900 | 5200 |
|  | 101 + | 263 |  | 890 |  |

each area to achieve an average for the whole period and thus eliminate short-term fluctuations. The 1970s data were corrected for errors in area measurement and area covered by the trawl found in the FAO/UNDP (1976) report (Fig. 5.4). These data show a decline in biomass in all areas since the 1970s.

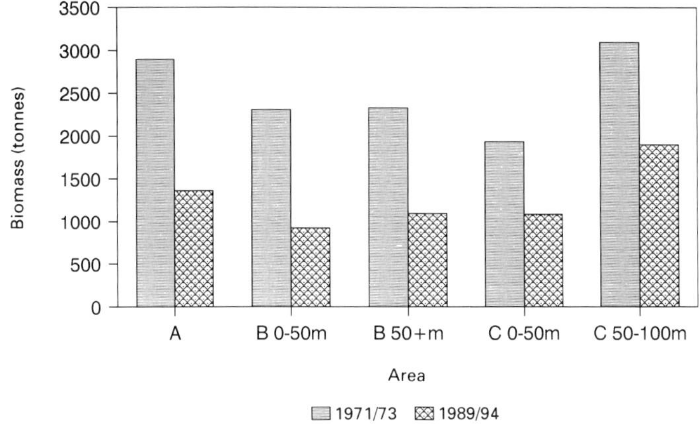


**Fig. 5.4**   A comparison of the biomass estimates for the different areas and depth ranges of the South East Arm for the 1970s and the 1990s surveys. The same vessel, gear and stations were used in both sets of surveys.

In addition to the changes since the 1970s, the apparent downward trend in biomass for each area from 1989 to 1994 (Figs. 5.2 & 5.3) was significant except for Area A (Table 5.2). As most of the decline was apparently between the first two surveys, regressions were also calculated omitting the first survey results. These resulted in a loss of significance for all areas except for the deeper parts of Area C, where the declines remained significant at the 1% level.

**MV Ndunduma**   Biomass estimates made from *Ndunduma* catches compared with

**Table 5.2**  Results of linear regressions of biomass estimates for each haul against date of survey for *Ethelwynn* surveys from 1989 to 1994, showing significant declines in biomass in all areas except Area A. The percentage decline from the first survey to the last is based on calculated biomasses from the regression results.

| Area and depth range (m) | No. of hauls | Correlation coefficient | Significance ($P$) | Percentage decline in biomass from 1989 to 1994 |
| --- | --- | --- | --- | --- |
| A | 108 | 0.105 | n.s. | 20 |
| B 0–50 | 117 | 0.378 | 1% | 55 |
| B 51+ | 75 | 0.332 | 1% | 52 |
| C 0–50 | 40 | 0.328 | 5% | 53 |
| C 51–100 | 65 | 0.559 | 1% | 69 |
| C 101+ | 25 | 0.623 | 1% | 77 |
| C 51–100+ | 90 | 0.578 | 1% | 71 |

n.s. not significant

those from the *Ethelwynn* were in reasonable agreement for Areas A and B (Table 5.1). Values from *Ndunduma* ranged from 2% to 60% higher, which reflected the increased catches of larger fish such as catfish, cyprinids and large cichlids as a result of the faster trawling speed. The difference in codend mesh size, 38 mm on *Ndunduma* and 25 mm on *Ethelwynn*, did not have a major impact on these results. The *Ndunduma's* codend was enclosed in a heavy-duty 100 mm mesh bag which appeared to restrict the mesh opening. The two gears have identical selectivity curves (Fig. 5.5) for

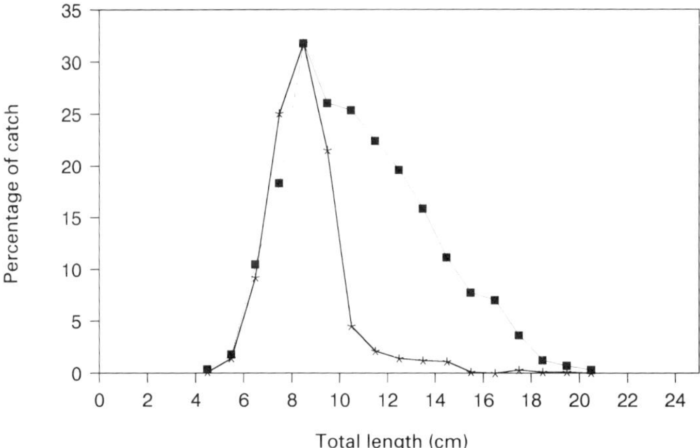


**Fig. 5.5**  Size selection by the 25 mm mesh used on the *Ethelwynn* (shown by the heavy line and asterisks) and 38 mm mesh on the *Ndunduma* (thinner line and squares) in 1993. The data are combined length frequencies for samples of 'small' cichlids taken after the larger species were sorted from the catch (see Methods). The data for the *Ethelwynn* are the actual percentage of the catch by number in each size interval. The *Ndunduma's* catch has been scaled up to match the *Ethelwynn's* maximum percentage to facilitate comparison. This shows identical selectivity curves for the small fish but increased catches of larger cichlids in the *Ndunduma's* trawl.

the size at first capture, but the higher trawling speed of the *Ndunduma* improved catches of larger fish and gave higher biomass estimates.

In Area C, biomass estimates from the *Ndunduma's* catches were 2.2 to 2.7 times greater than estimates based on *Ethelwynn* data. This is partially due to large catches of *Copadichromis* cf. *mloto/virginalis*, which are schooling, zooplanktivorous cichlids, by the *Ndunduma*.

There was considerable variation of catch between the two surveys with no marked trend with depth (Fig. 5.6). Similarly, comparison of overall catch rates between areas A, B and C showed no statistical difference (t-test, $P > 0.1$), although there were marked differences within areas. The catch rates in the shallowest depth range (10–30

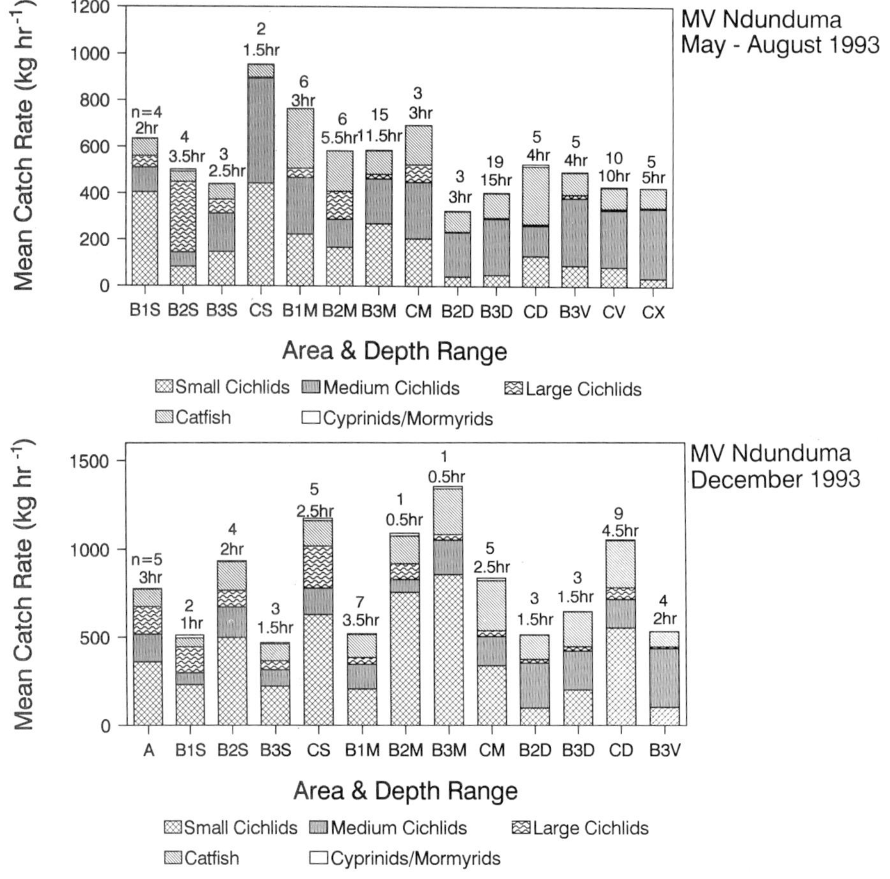

Fig. 5.6 Catch rates of the various species groups in the two *Ndunduma* surveys, showing the differences between the areas and depth ranges fished. The codes for each sample are as follows: A = Area A; B1 to B3 are the three areas into which Area B was divided to investigate differences in catch in relation to distance from the trawlers' bases; C = Area C. Depth ranges are: S = 10–30 m, M = 31–50 m, D = 51–70 m, V = 71–90 m, and X = over 91 m, thus e.g. B3M = Area B3 between 31 and 50 m deep. The numbers above the bars are the number of trawls (= n) and the number of hours trawled.

m) of Area B were particularly low in both the May/August and December surveys. Similarly, the catch rate in December in the depth range 31–50 m was lowest in Area B1. The deeper zone (51–70 m) also showed low catch rates in Area B2 with increasing CPUE to the north through B3 to C in both surveys.


### 5.3.2   *Species composition*

One hundred and twelve species categories were identified from 87 hauls during the two surveys of the South East Arm of Lake Malawi conducted by the *Ndunduma*. Some of these, e.g. *Rhamphochromis* spp., *Diplotaxodon* spp. and *Bathyclarias* spp., contained several species as the taxonomy of the groups is still confused. Forty three genera in six families were represented in the catches. The six families were Cichlidae, Clariidae, Cyprinidae, Bagridae, Mochokidae and Mormyridae. The family Cichlidae was represented by 102 species and species groups and comprised about 75% of the total catch by weight. The utaka species *C.*, cf. *mloto/virginalis*, alone made up 19% of the overall catch. The catfish *B. meridionalis*, *Synodontis njassae* Keilhack and *Bathyclarias* spp. were also important at about 20% of the catch. Other species made up the remaining 5%.

The cichlids were divided into three size ranges based on known maximum size for the species to assess whether heavy fishing affected species balance, as Turner (1977a) and Turner *et al.* (1995) showed marked declines in large cichlid abundance, particularly in Area A. There was no apparent trend in abundance of small, medium and large cichlids by area for each depth range (Fig. 5.6). The relatively high abundance of large cichlids in Area A and B shallow waters, however, is explained by large catches of *Oreochromis* spp., which are not truly demersal cichlids.

There were marked differences in depth range for most species, e.g. *Copadichromis* cf. *eucinostomus* and *Lethrinops longimanus* Trewavas were frequently caught in shallow waters ( < 40 m depth) and *Lethrinops gossei* Burgess & Axelrod and *Alticorpus mentale* Stauffer & McKaye in deeper waters ( > 50 m). However, other species had wider depth ranges, e.g. *Copadichromis* cf. *mloto/virginalis*, *Bathyclarias* spp., *B. meridionalis* and *S. njassae* were frequently caught at most depths.

In general, there was little change in species size with area. However, there were exceptions where there was evidence of a trend to increasing mean size from south to north (Table 5.3). No species showed the reverse trend with increased size towards the south.

The relationship between size and depth was also investigated as larger areas of deeper water towards the north could bias results if species size increased with depth, e.g. if juveniles occupied the inshore habitats and moved offshore as they grew. This is known to occur in Lake Malawi *Oreochromis* spp. (Lowe 1952; FAO 1993). The results showed that for *Lethrinops longipinnis* Eccles & Lewis, *Sciaenochromis spilostichus* (Trewavas) and *Otopharynx speciosus* (Trewavas), the south–north trend to increased size was real (Table 5.3). With large sample sizes for *S. spilostichus* in the 51–70 m depth range, for *L. longipinnis* in waters over 30 m and for *O. speciosus* in

**Table 5.3**  The relationship between mean length (in cm) of some fish species from the *Ndunduma* samples with depth and with trawling area. The number in brackets is the number of fish in the sample.

| Species | Depth (m) | A | B1 | B2 | B3 | C |
|---|---|---|---|---|---|---|
| *S. spilostichus* | 10–30 | 11.8 (16) | | 17.6 (1) | 17.0 (1) | 19.6 (1) |
| | 31–50 | | 13.0 (56) | 14.2 (8) | 11.8 (2) | 12.9 (8) |
| | 51–70 | | 9.2 (11) | 13.6 (52) | 13.8 (188) | 15.4 (57) |
| | 71+ | | | | | 13.6 (20) |
| *O. speciosus* | 10–30 | 18.4 (41) | 22.4 (61) | 25.0 (19) | 24.4 (8) | 25.9 (11) |
| | 31–50 | 19.5 (22) | 19.1 (491) | 22.2 (136) | 20.1 (98) | 21.0 (254) |
| | 51–70 | | 16.7 (55) | 18.9 (214) | 19.7 (294) | 20.3 (370) |
| | 71+ | | | | | 20.2 (76) |
| *L. longipinnis* | 10–30 | | 11.5 (1) | 12.6 (16) | 14.1 (44) | 15.5 (13) |
| | 31–50 | 13.4 (7) | 12.4 (58) | 13.2 (9) | 14.9 (64) | 15.7 (189) |
| | 51–70 | | | 8.2 (1) | 14.7 (64) | 16.0 (118) |
| | 71+ | | | | | 14.2 (36) |
| *L. longimanus* | 10–30 | 8.5 (75) | | 10.1 (73) | 11.1 (22) | |
| | 31–50 | 12.4 (4) | 11.0 (253) | 17.1 (4) | 12.1 (2) | 15.0 (35) |
| | 51–70 | | 13.0 (20) | 9.2 (274) | 10.5 (561) | 14.1 (329) |
| *C. cf. mloto/virginalis* | 10–30 | 9.0 (163) | 11.2 (331) | 8.1 (66) | 9.3 (16) | 11.7 (45) |
| | 31–50 | 6.9 (50) | 11.2 (214) | 10.3 (327) | 10.8 (449) | 11.5 (370) |
| | 51–70 | | 9.9 (85) | 11.5 (120) | 12.1 (199) | 12.2 (582) |
| | 71+ | | | | | 12.4 (201) |
| *B. meridionalis* | 10–30 | 44.2 | 46.7 | 52.4 | 51.2 | 42.2 |
| | 31–50 | 42.8 | 35.1 | 44.0 | 40.7 | 41.0 |
| | 51–70 | | 38.7 | 32.6 | 38.3 | 35.5 |
| | 71+ | | | | | 39.6 |

waters deeper than 50 m in particular, the trend was also clear. Sample sizes were erratic and the trends less clear for *L. longimanus*. C. cf. *mloto/virginalis* only showed a definite south–north trend in water over 50 m depth. *O. speciosus* showed a trend to decreasing size with depth in all parts of Area B and C, the opposite of the trend shown in *Oreochromis* spp.

*B. meridionalis* and *Bathyclarias* are large species and a small number of large specimens could affect the results. For instance *Bathyclarias* of 10 to 20 kg were fairly common and one specimen weighed 50 kg. Nevertheless, a definite trend was noted in the December survey with catfish catches increasing from south to north in each depth range (Fig. 5.6). In each area, the mean length of *B. meridionalis* declined with depth, due to higher numbers of juveniles in deeper water (Table 5.3).

The number of species in each area was investigated by recording the mean number of species per pull (Fig. 5.7). The number of species in the catch (recognizing that some 'species' are in fact complexes of closely related species) decreased slightly with depth. There were similar numbers in the 10–30 m depth range in all areas, with a suggestion of a slight decline from south to north. In 31–50 m there was no clear

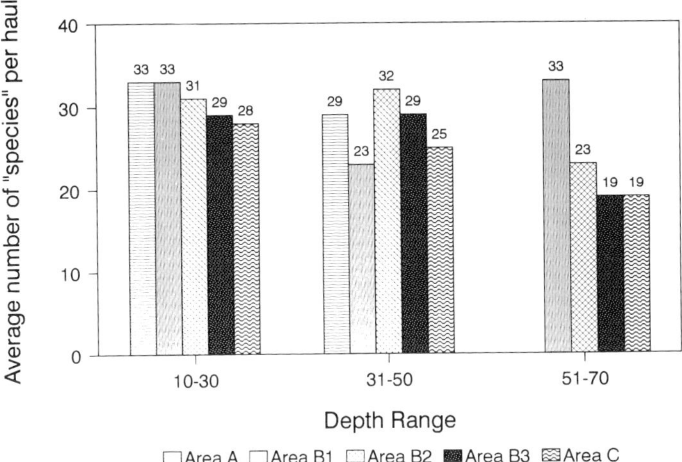


**Fig. 5.7** The average number of 'species' per trawl haul in the different areas and depth ranges. Some 'species' are actually species groups, e.g. *Diplotaxodon* spp.

trend, while in 51–70 m there was only one sample for Area B1 and two for B3, thus no conclusions could be drawn.

### 5.3.3  *Commercial fishery catch and effort data*

**Area A** The whole of Area A is shallow, fished by pair trawlers, and treated as a unit (FAO/UNDP 1976; Tweddle & Magasa 1989; FAO 1993; Turner *et al.* 1995). FAO (1993), in the latest analysis of this fishery, suggested that the fishery had recently gone through three distinct phases and was now in a state where increased effort could cause a collapse, while reduced effort would lead to a very high CPUE. They stated that the fishery was in a dangerously unstable state. However, the CPUE and effort data on which these conclusions were based were constrained to a narrow range (Figs 5.8A, B). This is mainly because effort was fixed between 450 and 600 standard boat days (= 900 to 1200 pair trawl days) from 1978 onwards by the licence regulations. Given the variations in ability of the various fishing units, their frequent breakdowns, their tendency to move from area to area depending on catch rates and their perceptions of how other units are fishing, and the fluctuations in stocks which have been apparent throughout the history of the fishery (particularly natural fluctuations in *C.* cf. *mloto/virginalis* abundance), the interpretation of three phases is of doubtful validity.

However, the combined data from biomass surveys (Fig. 5.9) and species composition studies (Turner *et al.* 1995), together with the illegal fishing methods used to maintain catch rates, suggested that the fishery was in a vulnerable state, and in 1991 CPUE was at its lowest since mesh size was increased from 25 mm to 38 mm in 1977.

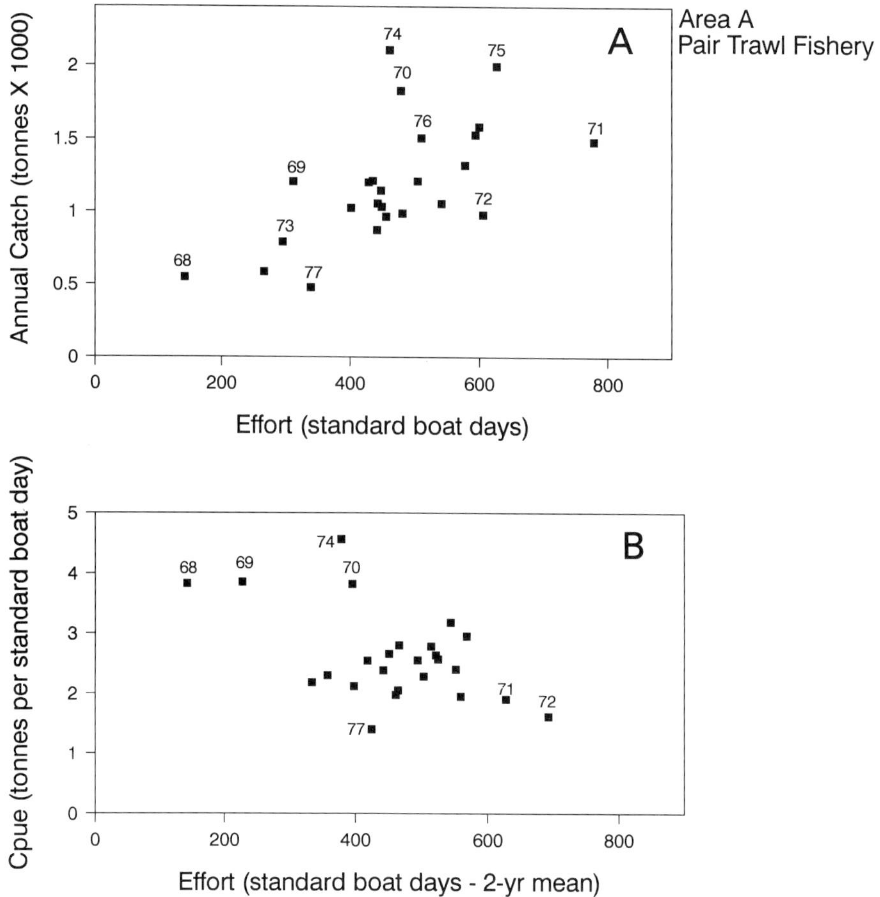



**Fig. 5.8**   The relationships between catch (A) and CPUE (B) with effort for the Area A pair trawl fishery. The data points for the earliest years are marked to show the large initial changes in the fishery. The data show high initial catch rates from 1968 to 1970 and a subsequent sharp decline. A very strong utaka year class gave high catches in 1974. The 1977 CPUE was very low because of the change in legal codend mesh size from 25 to 38 mm. The points since 1978 are grouped in a very small area of the graphs.

Action was therefore taken and a one-year closure from November 1992 led to a four-fold increase of biomass over the year.

**Areas B and C shallow**   Units licensed for Area C fish entirely in Area B on the western shore. No fishing has taken place in Area C, apart from one unit for a very brief period in the early 1980s. The data therefore are for the southern and particularly the western part of Area B. Effort and catches were erratic (Figs 5.10A & B) and the data do not provide any indication of sustainable yields. Ankhoma Fishing Company started fishing in the west of Area B3 in 1987. CPUE has declined by two thirds since 1989 and is now much lower than the CPUE in the Area A fishery (Fig. 5.10C).

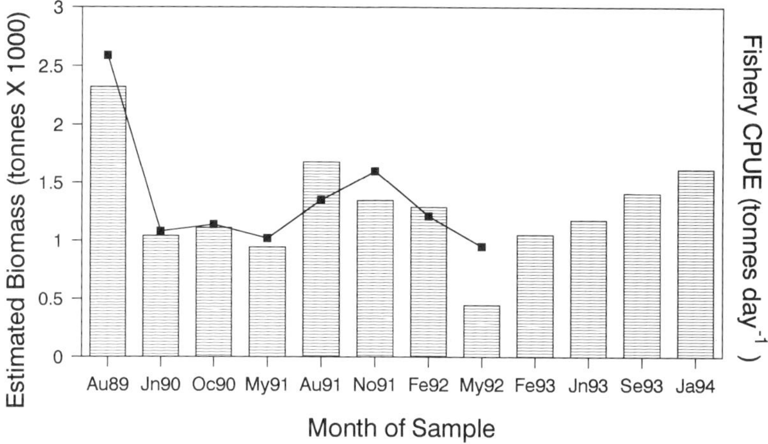


**Fig. 5.9**  The relationship between the CPUE of the pair trawl fishery (shown by the line and squares) and the biomass estimates made using *Ethelwynn* survey data for the same months (bars). In this figure, the CPUE is that for the pair trawlers and is not converted to standard boat days. In Figs 5.8, 5.10 and 5.11, effort is given in standard boat days.

**Areas B and C deep**  There is no natural division between these areas and bottom trawl licences cover both areas. Most effort was in Area B in waters from 40–70 m depth (FAO 1993), although Fisheries Department vessels fished in both areas in the early 1980s.

From the fishery data it was possible to determine the relationships of catch and CPUE with effort since 1973 (Fig. 5.11). Until 1987 there was a clear relationship between CPUE and effort ($r = 0.84$). The estimate of maximum sustainable yield (MSY) based on this relationship was 1860 t at an annual effort of 610 trawl days (Fig. 5.11A). Effort exceeded this from 1983 to 1989, resulting in lower catches. The MSY estimate of 1860 t is approximately 30% of the virgin biomass estimates of Areas B and C over 50 m depth calculated using research data from the early 1970s.

Since 1988, the fishery appears to have entered a different phase. Reduction in effort through the withdrawal of the Fisheries Department vessels and Maldeco's other two vessels has not resulted in the expected improvement in CPUE.

## 5.4     Discussion

The rule of thumb that '$MSY = 0.5\ M \times$ virgin stock' (Gulland 1971) was considered to be over-optimistic by Sparre *et al.* (1989) who suggested replacing 0.5 with 0.2. Tweddle and Turner (1977) estimated $M$ to average 1.3 for some of the trawl-exploited cichlids of the South East Arm, with $M = 0.9$ for the most abundant species, *C.* cf. *mloto/virginalis*. Initial estimates for the Area A trawl fishery using the Schaefer (1954) surplus yield model suggested that Gulland's rule of thumb held reasonably true for the Lake Malawi trawl fisheries (FAO/UNDP 1976;

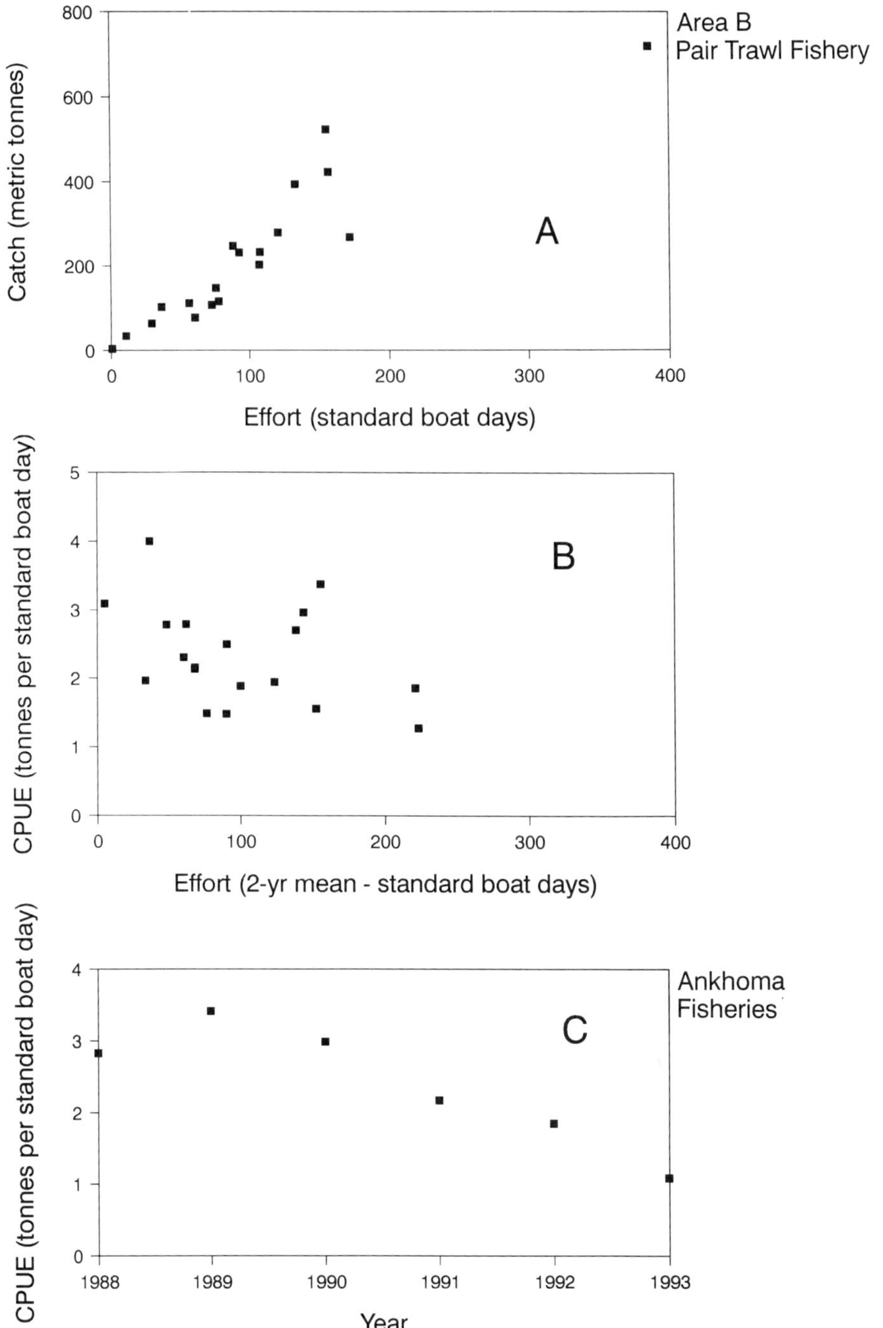



**Fig. 5.10**   (A) The relationship between catch and effort, and (B) between CPUE and effort for the Area B pair trawl fishery. (C) The decline in catch rates for Ankoma Fishing Co., which fishes in Area B3.

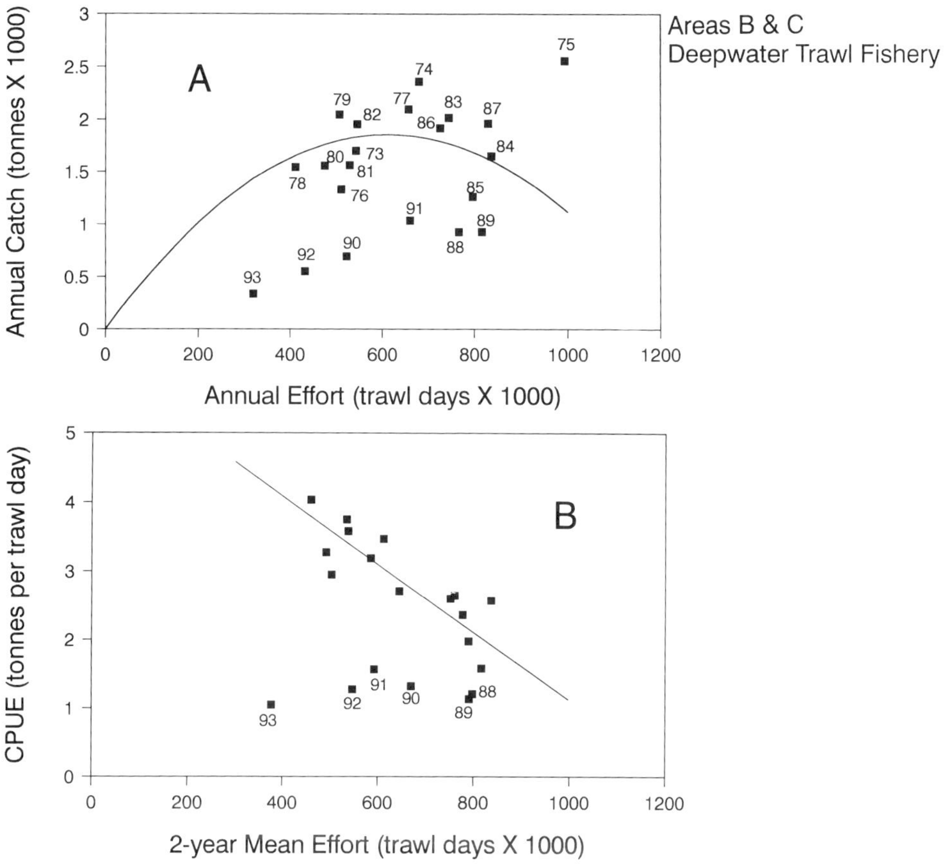



**Fig. 5.11**   (A) The relationship between total annual catch and effort for the deepwater trawl fishery of Areas B and C. (B) The relationship between CPUE and effort (mean of the year of fishing and the previous year) for the fishery. The regression line was calculated using only the data to 1987 as the fishery since then has clearly followed a different pattern. The Schaefer sustainable yield curve in (A) was calculated from the regression in (B).

Tweddle & Magasa 1989), with MSY estimated to be about 45% of the stock (which may now be corrected to about 40% taking into account errors in FAO's sizing of Area A and calculation of swept area of trawl). CPUE, and thus biomass, by the early 1980s, when the fishery was fully exploited, was approximately half that of the initial fishery.

Using this measure of sustainability, Area A fishery was overfished by the late-1980s. Biomass was down to about 40% of the early 1970s stock levels, which were already affected by heavy fishing pressure and therefore not representative of a virgin stock (the mean annual catch in Area A for 1970/72 was 1364 t) and fishermen in the 1980s were resorting to illegal methods (small meshes, doubled codends, and fishing close inshore) to sustain economic catch rates. The boats all had more powerful engines, and effort increased from 1987 in excess of the 500 standard boat day optimum. The trend in biomass from 1989 to 1992 was downwards, while CPUE declined by 30% from 1989 to 1991.

The low catches, loss of some species (Turner *et al.* 1995) and illegal fishing practices led to a decision to close Area A to trawling for a year from November 1992, after which the re-opened fishery would be more closely controlled to ensure sustainability in future. The closure of the fishery had a marked impact on the stocks. Biomass rapidly increased as the fish grew, with estimates after one year of closure more than four times that prior to closure. The species composition after the year of closure was similar to that of Areas B and C in the same depth range.

The *Ethelwynn* biomass estimates for Area B in less than 50 m depth were less than half that of the early 1970s. Trawling only started in Area B in 1971, thus the 1970s estimate of 2655 t for the area was probably close to the virgin stock size. The *Ethelwynn* data suggested that shallow waters in Area B are heavily exploited. The *Ndunduma* data confirmed the low biomass estimates in Area B shallow. Both *Ndunduma* surveys showed low catch rates (Fig. 5.6) for the area, particularly in B3 where pair trawlers have been based in recent years.

FAO (1993) stated that the deep water bottom trawl fishery of Areas B and C collapsed after being seriously over-exploited since 1983. The area in question is over 1000 km$^2$. Fig. 5.11 shows the current situation in this fishery, with only two trawlers now operational in the area. With reduced effort since 1989, CPUE has not recovered, as would be expected from the previous pattern in the fishery.

The data on CPUE and, to a lesser extent, size distribution of certain species from the *Ndunduma* surveys suggest that the area near to Boadzulu Island (Area B1) was most heavily exploited with less pressure on more northerly areas. Catch rates in water deeper than 30 m improved from south to north. This is most noticeable in the December data for the 31–50 m depth range from Areas B1 to B3 and for the 51–70 m depth range from B2 to C in both the May–August and the December surveys (Fig. 5.6). This is opposite to expectation from the well-documented higher productivity in the south (Eccles 1974; Bootsma 1993; Mwanyama 1993). The biomass figures from the *Ethelwynn* agree with these findings, with Area B deep water data indicating fairly heavy exploitation since the 1970s while the Area C data showed a less severe decline.

Turner *et al.* (1995) showed considerable changes in species composition in Area A which can be attributed to fishing, but only minor changes in Area B. *Lethrinops microdon* Eccles & Lewis had declined in abundance but not in the proportion of catches in which it occurred, while *C.* cf. *mloto/virginalis* also declined. *C.* cf. *mloto/virginalis* has always, however, been subject to large natural fluctuations in abundance, and the present study showed that this species was again abundant in 1993, comprising as much as 90% of individual trawl catches and averaging 24% in December, the highest proportion ever recorded in trawl surveys. The low catches of utaka in 1991/92 were not due to over-exploitation.

Thus exploitation levels over the areas as a whole may not be as severe as were implied by FAO (1993). Omitting *C.* cf. *mloto/virginalis* from estimates, as the species is erratic in abundance and cannot be relied on for sustained high catches, the biomass figures still show stocks of nearly 5000 t for the areas of B and C in excess of 50 m depth. Yields higher than the 1500 t MSY claimed by FAO (1993) may thus be sustainable if effort is well distributed. The limited range of the current vessels appears to have resulted in localized overfishing which was assumed by FAO (1993) to apply to a much larger area. More detailed study of not only the species composition but also the size composition of catches over the whole area shows that fishing effort overall may not be excessive. If the heavily fished areas close to Boadzulu Island are rested, stocks should recover and allow an overall improvement in CPUE and profitability.

The dramatic declines in *Ethelwynn* biomass estimates between 1989 and 1994 in all areas except Area A (where a one-year closure reversed a decline) (Table 5.2) give cause for concern. There has been no change in gear or crew on the *Ethelwynn* since 1990, and the good correlation between the biomass estimates and actual CPUE of the pair trawl fishery in Area A (Fig. 5.9) gives some confidence in the estimates, at least for that area. Reasonable agreement was also observed between *Ethelwynn* and *Ndunduma* biomass estimates for Areas A and B, allowing for the greater power of the latter vessel and the small sample size. However, the *Ndunduma* had much higher catch rates in Area C. Its catches were boosted by high catches of utaka, *C.* cf. *mloto/ virginalis*, shoaling fish which were missed in the *Ethelwynn's* surveys. Utaka accounted for 23% of the *Ndunduma's* Area C biomass estimate overall, and 34% in the December survey. Utaka also occurred in the other areas, however (19% of the total Ndunduma catch), so even without utaka the difference between the *Ndunduma* and *Ethelwynn* biomass estimates for Area C was greater than for Areas A and B.

Evidence for major changes in species composition of the South East Arm, especially Area A, was presented by FAO/UNDP (1976) and Turner (1977a). The changes consisted of a decline in some of the previously abundant medium and large cichlid species such as *Lethrinops stridei* Eccles & Lewis and *L. macracanthus* Trewavas and an increase of small cichlids such as *Otopharynx argyrosoma* (Regan), *Pseudotropheus livingstonii* (Boulenger) and *Lethrinops auritus* (Regan). A decline in the large catfish *Bagrus meridionalis* and *Bathyclarias* spp. was also noted. All these changes were attributed entirely to the effect of fishing. Turner *et al.* (1995) reported further declines in large species such as *Taeniolethrinops furcicauda* (Trewavas), *T. praeorbitalis* (Regan) and *Ctenopharynx* spp., and possible localized extinction of *Lethrinops mylodon* Eccles & Lewis and *L. macracanthus*.

In the present study, no further major changes in species composition were detected. It was found that there was little difference in the species composition of Areas B and C in the same depth ranges. Similarly, Area A catches had a similar species composition to the equivalent shallow areas of Area B and C. The slight trend to decreased number of fish species from Area A to Area C in the shallower waters is probably a result of the larger area in that depth range in the south.

The size range of species being landed in a given area is a sensitive indicator of events happening to the resource (Caddy & Bazigos 1985). Factors that affect mean size of fish are growth, recruitment and fishing. There were differences in the mean size of some species between areas (Table 5.3) with an increase in mean size from south to north, while other species showed a fairly consistent size range throughout. It is possible that those species which do show a south/north trend in size distribution are relatively sedentary species and more vulnerable to localized fishing pressure, while the species that do not show a clear trend are wider ranging. Further studies with larger sample sizes over a longer period of time would help to clarify this picture. The data, although preliminary, corroborate the data on CPUE in the areas fished and help to confirm that the southern areas are more heavily fished than areas further north.

Tweddle *et al.* (1994) documented a statistically significant decline in the catfish catches from the South East Arm fisheries as a whole between 1976 and 1989. The December *Ndunduma* survey showed a trend to increased catfish catches from south to north at all depth ranges, which may be another indicator of heavy pressure by trawling in the south. However, the May/August data do not show such a trend so further investigation is needed into catfish abundance over a wider area to assess how great the impact of trawling is on the stocks.

In addition to fishing pressure, various factors can influence catch rates, e.g. natural fluctuations in stocks and variations in efficiency of fishing units. In the past, both types of fluctuation have been documented in Lake Malawi fisheries. For example, Tweddle & Magasa (1989) showed natural fluctuations in both chambo (*Oreochromis* spp.) and trawl fish stocks and variations in the efficiency of trawl units.

The research vessel *Ethelwynn* has been completely renovated and there has been continuity of gear and crew. The decline in *Ethelwynn* catch rates in the relatively unfished Area C is thus unlikely to be a result of inefficiency. The small sample sizes and wide confidence limits in each survey indicate that caution must be used when interpreting the results. Nevertheless, the declines were significant (Table 5.2) in both deep and shallow areas. Natural stock fluctuations may be involved such as the well-documented changes in stocks of utaka (FAO/UNDP 1976; Turner *et al.* 1995). Utaka abundance may have considerable influence on stock assessment, as was recognized by Turner (1977b).

Bootsma (1993) documented considerable differences in primary productivity of Lake Malawi under the influence of hydrological and meteorological conditions. In 1990/91, for example, mean photosynthetic rate at a station off Nkhata Bay halfway up the lake was approximately half that measured at the same station in 1980. There was considerable annual variation in algal growth on rocks at the water's edge as the level recedes in the dry season (pers. obs.). In some years the rocks were carpeted in long green filamentous algae; in others they were bare. This must be attributable to limnological fluctuation from year to year. There was very little algal growth on the rocks in the past five years. Productivity variations will influence fish stocks, particularly primary consumers.

Turner *et al.* (1995) pointed out that the artisanal fishery now also exploits the small cichlid stocks and cautioned that this should be taken into consideration. Certainly, in Area A and the inshore shallow areas elsewhere, small meshed beach seines take quantities of small cichlids. Catches of small cichlids from seines ranged from about 100 t to 400 t $yr^{-1}$ from 1976 to 1988 (Tweddle *et al.* 1994) but have since increased markedly. From 1989 the catch of small cichlid species from seines averaged over 1000 t, largely as a result of increasing effort by beach seines. With the increase in length of some beach seine nets and the illegal inshore activity of the pair trawlers, there is now more overlap than in the 1970s (Linn & Campbell 1992). However, this overlap is restricted to the inshore areas, and the deeper waters of Areas B and C with their different species composition are not exploited by the artisanal fishery. The deepwater trawl fishery can be treated as an independent fishery, at least for the cichlids, while the shallow water trawl stocks still appear to respond to changes in trawling effort independently of the artisanal fishing effort. The situation, however, will require close monitoring in future.

FAO (1993) stated that the fish biomass in waters over 70 m depth was low and implied that these waters will not be exploited. The *Ndunduma's* catch figures showed that these waters yield catches almost as high as those in the medium depth range currently fished in Area B (Fig. 5.6). FAO (1993) are therefore unduly pessimistic in recommending no expansion and possibly further reductions in effort and even temporary closure of the fishery.

This study illustrates a pragmatic approach to management of the trawl fishery. The high number of species, some of which are stenotopic and some wide-ranging, means that virtually all standard methods of stock assessment have limitations. However, by combining long-term research surveys assessing catch rates, species composition and size range of catches, and assessment of catch and effort data from the fisheries, it is possible to implement sound management strategies. The essential element is continued long-term monitoring. More information is needed on natural fluctuations in productivity and on the fishes' biology and distribution. The taxonomy of the fishes needs attention to remove the problems experienced in assessing species changes from surveys conducted several years apart, where many species are undescribed and are given inconsistent temporary names by different workers. It is also not possible to predict the final species composition and whether or not changes in species will affect the sustainable yields from the fisheries. It appears that after initial changes in Area B species composition (Turner 1977a) there has been little further change despite the localized heavy fishing pressure. The intention is to attempt to maintain the species balance while continuing to monitor catches and identify potential for further expansion should improved catch rates be achieved and sustained through improved management.

## Acknowledgements

This chapter presents the results of the efforts over many years of Fisheries Department staff, too numerous to name, in the research section, the statistics section, the fishing and workshop sections at Monkey Bay. Ramsey Makwinja and James Magasa are particularly acknowledged for their work on the Demersal Fisheries Reassessment Project, funded by ODA. ODA and ICEIDA funded the rehabilitation of the vessel *Ethelwynn Trewavas*. ICEIDA provided the department with the research vessel *Ndunduma* and back-up services including vessel operating expenses.


## References


Alverson D.L. & Pereyra W.T. (1969) Demersal fish exploitation in the northwestern Pacific Ocean – an evaluation of exploratory fishing methods and analytical approaches to stock size and yield forecast. *Journal of the Fisheries Research Board of Canada* **26**, 1985–2001.

Bootsma H.A. (1993) Algal dynamics in an African Great Lake, and their relation to hydrographic and meteorological conditions. PhD thesis, University of Manitoba, 311 pp.

Caddy J.F. & Bazigos G.P. (1985) Practical guidelines for statistical monitoring of fisheries in manpower limited situations. *FAO Fisheries Technical Paper* **257**, 86 pp.

Eccles D.H. (1974) An outline of the physical limnology of Lake Malawi (Lake Nyasa). *Limnology and Oceanography* **19**, 730–42.

Eccles D.H. & Lewis D.S.C. (1979) A taxonomic study of the genus *Lethrinops* Regan (Pisces: Cichlidae) from Lake Malawi. Part 3. *Ichthyological Bulletin of Rhodes University* **38**,1–25.

Eccles D.H. & Trewavas E. (1989) Malawian cichlid fishes: the classification of some haplochromine genera. Lake Fish Movies, Herten, West Germany, 335 pp.

FAO/UNDP (1976) An analysis of the various fisheries of Lake Malawi. Based on the work of J. Turner. Rome, FAO, *FI:DP/MLW/71/516 Technical Report 1*, 73 pp.

FAO (1993) Fisheries management in south-east Lake Malawi, the Upper Shire River and Lake Malombe, with particular reference to the fisheries on chambo (*Oreochromis spp.*). *CIFA Technical Paper 21*, 113 pp.

Gulland J.A. (1961) Fishing and the fish stocks of Iceland. *Fishery Invest., Lond., Ser.* 2 (23) 1–52.

Gulland J.A. (ed.) (1971) *The fish resources of the ocean*. Oxford: Fishing News Books, Ltd, 255 pp.

Lewis D.S.C. & Tweddle D. (1990) Breeding seasonality in some commercially important trawl-caught fish species in southern Lake Malawi. *Collected Reports on Fisheries Research in Malawi. Occasional Papers* **1**, 23–36.

Linn I.J. & Campbell K.L.I. (1992) Interactions between white-breasted cormorants *Phalacrocorax carbo* (Aves: Phalacrocoracidae) and the fisheries of Lake Malawi. *Journal of Applied Ecology* **29**, 619–634.

Lowe R.H. (1952) Report on the Tilapia and other fish and fisheries of Lake Nyasa 1945–47. Part 2. London, *Colonial Office Fishery Publications*, **1**(2), 126 pp.

Mwanyama N.C. (1993) Relations entre les ressources alimentaires, l'alimentation et la reproduction des Chambos (poissons tilapias du genre *Oreochromis*) dans les lacs Malawi et Malombe (Afrique centrale). PhD thesis, Université de Provence – Aix – Marseille, 158 pp.

Schaefer M.B. (1954) Some aspects of the dynamics of populations important to the management of commercial marine fisheries. *Bulletin of the Inter-American Tropical Tuna Commission* **1**, 27–56.

Sparre P, Ursin E. & Venema S.C. (1989) Introduction to tropical fish stock assessment. Part 1 – Manual. *FAO Fisheries Technical Paper* **306/1**, 337 pp.

Tarbit J. (1972) Lake Malawi trawling survey – interim report 1969–1971. *Malawi Fisheries Bulletin* **2**, 16 pp.

Turner G.F., Tweddle D. & Majwinja R.D. (1995) Changes in demersal cichlid communities as a result of trawling in southern Lake Malawi. In: T.J. Pitcher & P.J.B. Hart (eds) *The Impact of Species Changes in African Lakes*, London, Chapman and Hall.

Turner J.L. (1977a) Changes in the size structure of cichlid populations of Lake Malawi resulting from bottom trawling. *Journal of the Fisheries Research Board of Canada* **34**, 232–8.

Turner J.L. (1977b) Some effects of demersal trawling in Lake Malawi (Lake Nyasa) from 1968 to 1974. *Journal of Fish Biology* **10**, 261–71.

Tweddle D., Alimoso S.B. & Sodzapanja G. (1994) Analysis of catch and effort data for the fisheries of the South East Arm of Lake Malawi, 1976–1989, with a discussion of earlier data and the inter-relationships with the commercial fisheries. Malawi Fisheries Department, Traditional Fisheries Assessment Project Working Paper, TFAP/2 (1991). *Malawi Fisheries Bulletin* **13**, 34 pp.

Tweddle D. & Magasa J.H. (1989) Assessment of yield in multi-species cichlid fisheries in the Southeast Arm of Lake Malawi, Africa. *Journal du Conseil International pour l'Exploration de la Mer* **45**, 209–222.

Tweddle D., & Turner J.L. (1977) Age, growth and natural mortality rates of some cichlid fishes of Lake Malawi. *Journal of Fish Biology* **10**, 385–98.

# Chapter 6
# Selectivity of Breder traps for sampling fish fry

J. KUBECKA *Hydrobiological Institute, CAS, Na sadkach 7, 370 05 Ceske Budejovice, Czech Republic*

**Abstract**   The species and size composition of Breder trap catches were compared with the catches from quantitative shore seines in Elbe Backwater and Klicava reservoir (Czech Republic). Catches by Breder traps were highly selective, the proportion of 1 + bitterling, perch, ruffe and gudgeon fry was overestimated while 0 + bitterling, roach, rudd, chub, bream and *Leucaspius* fry were underestimated. Breder traps caught smaller 1 + bitterling and bigger 0 + perch than seines. Diurnal swimming activity of fry monitored by Breder traps was similar to the results obtained by lift-nets. Catch per unit of effort of Breder traps fell faster than the fish density in littoral zones during summer and autumn.

KEYWORDS: Breder traps, seine nets, selectivity, fish fry

## 6.1   Introduction

Traps are considered to be a selective sampling gear (Craig 1980), but under certain circumstances, trap catches have been found to be accurate indicators of species and size composition (Penaz *et al.* 1978; Jacobsen & Kushlan 1987; Hayes 1989). Plexiglas Breder traps (Breder 1960) have been used in many studies, but little is known about their selectivity for coarse fish fry.

## 6.2   Material and Methods

Two localities were used to determine the selectivity of Breder traps:

- Poltruba Backwater in the Elbe inundation area 25 km north-east of Prague;
- Klicava Reservoir 50 km south-west of Prague.

Breder traps, modified according to Penaz (1971), were placed in the littoral at both localities at depths between 0.2 and 1 m. The traps were set for between 24 and 28 h in the Poltruba Backwater on 27 July 1982, 26 September 1982 and 9 August 1983 and in the Klicava Reservoir on 2 August 1982 and 7, 17 and 24 September 1982. Catches of traps were inspected every 2–4 hours during the 24-hour cycle to assess swimming activity of inshore fry. All catches were identified and standard length (nearest mm) was measured. Differences between mean lengths were compared using the student

76

t-test. An adjacent area, 10–20 m from the trapping sites, was fished using a fry seine (13.5 m long, 1.8 m high, mesh 1 mm with additional weights on the ends of net). Total catch was 3078 fish by the traps and 3368 by the seine in the Poltruba Backwater and 1060 fish by traps and 1514 fish by the seine in the Klicava Reservoir. A 1 m$^2$ grey-green lift net (1 mm mesh) was used as an alternative measure of swimming activity. The nets were lifted 2 min after setting, and ten lifts were made every 2 h on 9 August 1983. Total catch by the lift net was 424 fry.

## 6.3     Results

### 6.3.1     *Species composition*

The inshore fry community of Poltruba Backwater in 1982 and 1983 was dominated, in both trap and seine catches, by bitterling, *Rhodeus sericeus ammarus* (Bloch) (Fig. 6.1). Sunbleak, *Leucaspius delineatus* (Haeckel), which represented about 25% of fry seine catch was not caught in the traps. Perch, *Perca fluviatilis* L., was a regular by-catch of Breder traps in Poltruba, but the affinity of bitterling to traps was so high that the proportion of perch was usually lower in traps than in the seine.

In the Klicava Reservoir, perch showed the greatest affinity to traps, especially in the August sample, roach, *Rutilus rutilus* (L.) was indifferent while rudd, *Scardinius erythrophthalmus* (L.) and chub, *Leuciscus cephalus* (L.) were rarely found (Fig. 6.2). By contrast rudd was an extremely important part of the seine net catches (Fig. 6.2).

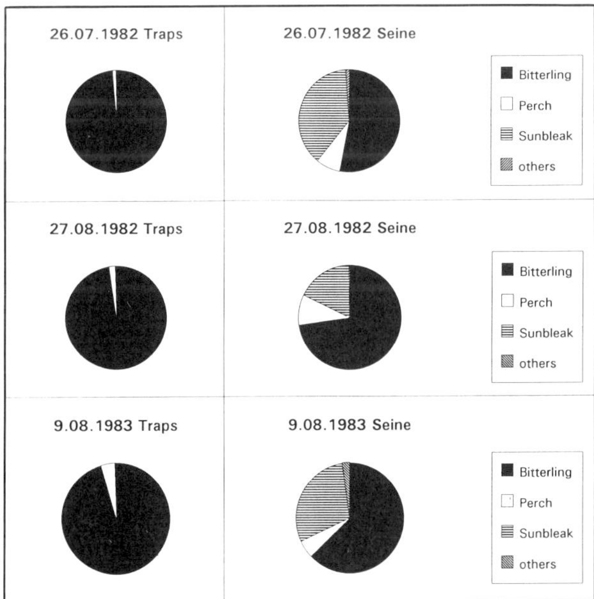


**Fig. 6.1**  Species composition of different fry catches in Poltruba Backwater by traps and seine.

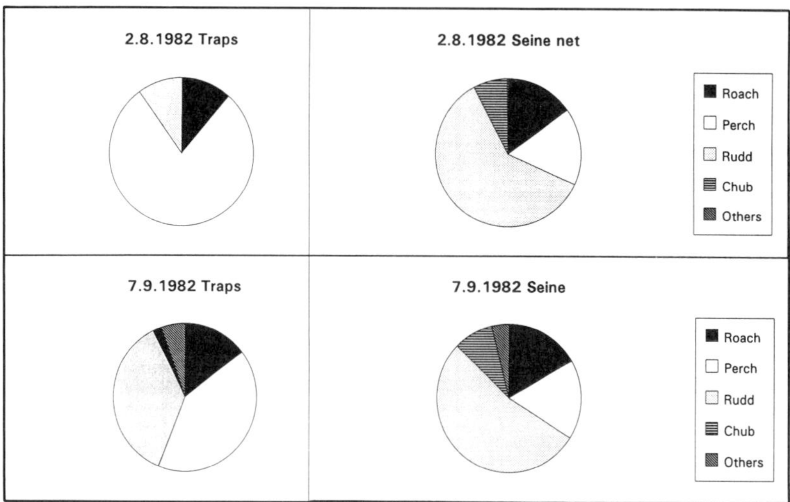

**Fig. 6.2**  Species composition of different fry catches in Klicava Reservoir by traps and seine.

### 6.3.2    *Length composition*

Although 0+ bitterling were seen around the traps and were caught by the seine, they were not recorded in traps in the Poltruba Backwater (Fig. 6.3). Length composition of >0+ bitterling trap and seine catches showed bimodality associated with spring and summer hatched 1+ fish. Average length of seine catches was significantly higher ($P$ <0.005).

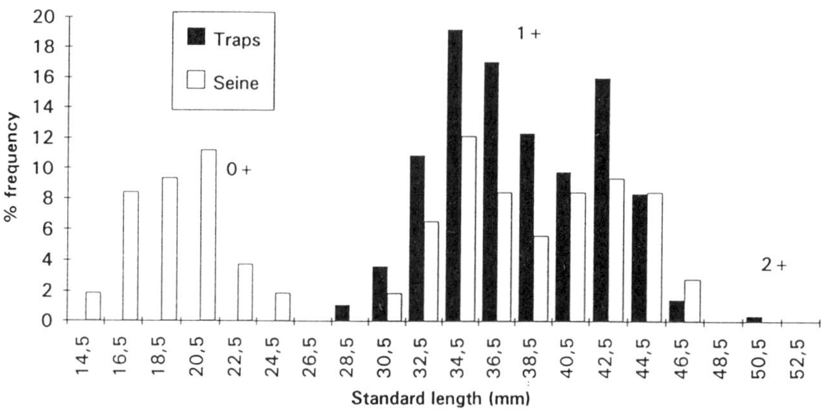

**Fig. 6.3**  Length frequency distribution of traps and seine catches of bitterling in Poltruba Backwater 26/27.7, 1982.

The average length of perch caught by traps in Klicava was significantly bigger ($P$ <0.005) than fish caught by the seine net (Fig. 6.4) in four out of five cases. Average lengths of cyprinids were not significantly different ($P$ >0.05).

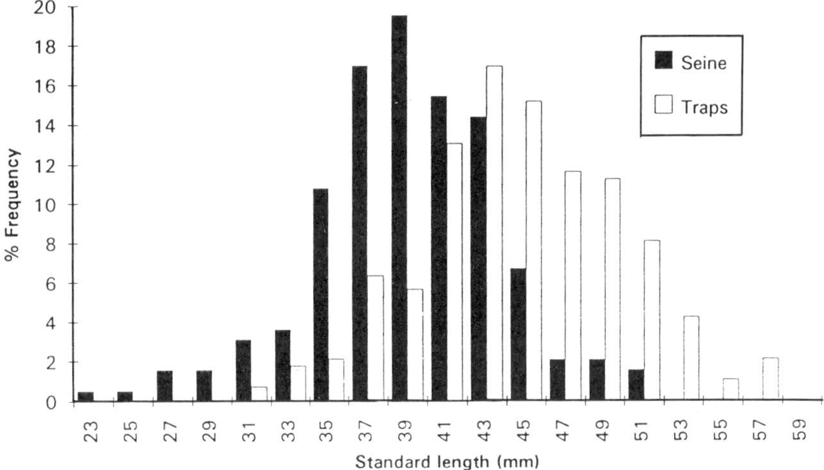

**Fig. 6.4**  Length frequency distribution of traps and seine catches of Perch in Klicava Reservoir 2/3.8, 1982.

### 6.3.3    *Catch per unit effort*

The catch per unit of effort (CPUE) of bitterling by Breder traps and lift nets in the Poltruba Backwater showed similar diurnal activity, with peaks in the early morning and evening (Fig. 6.5). All coarse fish fry capture by Breder traps at both localities also showed diurnal activity.

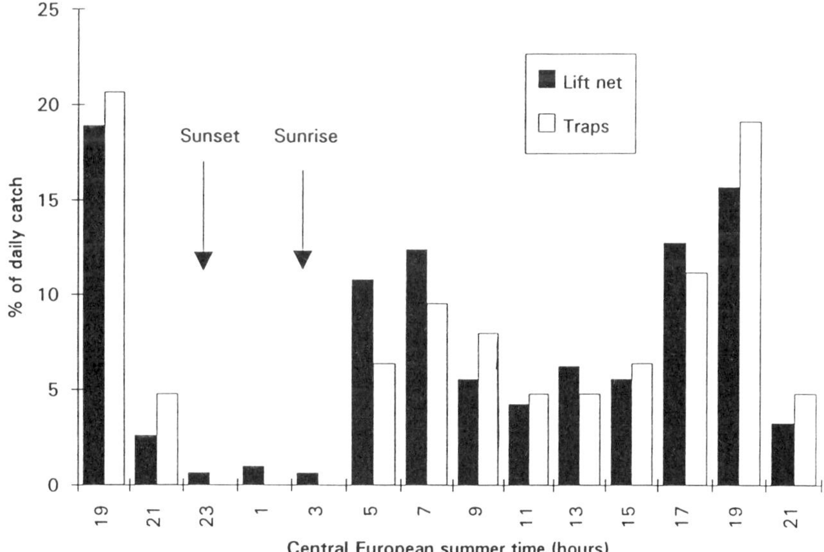

**Fig. 6.5**  Swimming activity of bitterling in Poltruba Backwater during 24 h monitored by traps and lift-nets 9–10.8, 1983. Positions of sunset and sunrise indicated by arrows.

The total catch for 10 traps over a 24 h cycle was compared with catches from the seine net each month for the July–October period in Klicava Reservoir (Fig. 6.6). The CPUE for traps fell off faster in time compared with CPUE of seine net in the same habitat. Sampling by traps during September and October was inefficient as bigger fry appeared to avoid the traps more successfully. Increase of the trap mouth aperture from 6 mm to 10 mm had no effect.

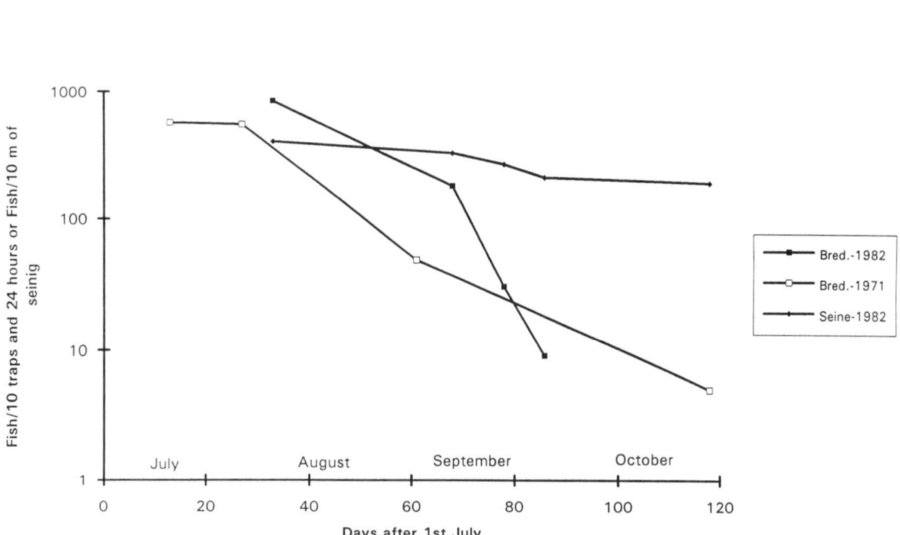



**Fig. 6.6**   Changes of CPUE of Breder traps and seine in the littoral of Klicava Reservoir during summer and autumn. Bred. 1971 represents the time course of CPUE of traps in earlier experiment (Cerny 1973) showing similar pattern.

### 6.4    Discussion

The small seine net used as the reference sampling tool has been shown not to be 100% efficient (Penczak & O'Hara 1983; Lyons 1986; Parsley *et al.* 1989; Pierce *et al.* 1990). Most of the potential escape was accounted for by avoidance by benthic fish below the bottom line during dragging of the net above the coarse substratum. This escape route was virtually excluded during this study because optimal sections (smooth substrate) for shore seining were selected. The possibility of fry avoidance around the net before the margins were pulled ashore remained but this was not observed as fry tended to concentrate in the middle of the seine during the haul.

The main source of selectivity of traps is fish behaviour before entering the trap

(Kanehiro *et al.* 1985). It was possible to observe shoals of cyprinid fry, other than bitterling, actively changing direction immediately after encountering the trap. Marked fry placed in traps also showed that it was possible to escape.

Accepting seine catches as representative (Leslie & Timmins 1992), it was possible to distinguish fish species with high and low affinity to Breder traps. The first group contained 1+ bitterling, perch, ruffe, and gudgeon, *Gobio gobio* (L.), (Penaz *et al.* 1978, also recorded very high affinity of the last species to Breder traps); the second group comprised sunbleak, 0+ bitterling, rudd, chub and bream, *Abramis brama* (L.). Affinity of fish to traps seems to be strongly age-dependent during the first summer and autumn of life. All the problems mentioned complicate the interpretation of Breder trap catches. They can only be used reliably to monitor swimming activity and relative abundance of species with high affinity during the early stages of their life cycle.

## Acknowledgements

The author is grateful to Dr M. Svatora and Prof. O. Oliva for their help during field works, and to an anonymous referee for valuable suggestions.

## References

Breder C.M.Jr. (1960) Design for a fry trap. *Zoologica N.Y.* **45**, 155–159.

Cerny, K. (1973) Locomotory activity of fish fry in Klicava Water reservoir, *Acta. Mus. Reginae Hredecensis, Ser. A, Scientium Nat*, 105–120.

Craig J.F. (1980) Sampling with traps. In: T. Backiel & R.L. Welcomne (eds) Guidelines for sampling fish in inland waters. *EIFAC Technical Paper* **33**, 55–70.

Hayes J.W. (1989) Comparison between a fine mesh trap net and five other fishing gears for sampling shallow lake fish communities in New Zealand. *New Zealand Journal of Marine and Freshwater Research* **23**, 321–324.

Jacobsen T. & Kushlan J.A. (1987) Sources of sampling bias in enclosure fish trapping: effects on estimates of density and diversity. *Fisheries Research* **5**, 401–412.

Kanehiro H., Suzuki M. & Matuda K. (1985) An analysis of the fish behaviour to model trap net. *Bulletin of Japanese Society for Scientific Fisheries* **51**,1983–1988.

Leslie J.K. & Timmins C.A. (1992) Beach seine collections of freshwater larval fish at the shore of lakes. *Fisheries Research* **15**, 243–251.

Lyons J. (1986) Capture efficiency of a beach seine for seven freshwater fishes in a North-Temperate Lake. *North American Journal of Fisheries Management* **6**, 288–289.

Parsley M.J., Palmer D.E. & Burkhardt R.W. (1989) Variation in capture efficiency of a beach seine for small fishes. *North American Journal of Fisheries Management* **9**, 239–244.

Penczak T. & O'Hara K. (1983) Catch-effort efficiency using three small seine nets. *Fisheries Management* **14**, 83–92.

Penaz M. (1971) Locomotor activity of gudgeon fry. *Journal of Moravian Museum, Trebic* **8**, 59–65.

Penaz M., Prokes M. & Wohlgemuth E. (1978) Fish fry community of the Jihlava River near Mohelno. *Acta Scienciarum Naturalium Academiae Scientiarum Bohemoslovacae, Brno* **XII**, 5, 1–36.

Pierce C.L., Rasmussen J.B. & Leggett W.C. (1990) Sampling littoral fish with a seine: corrections for variable capture efficiency. *Canadian Journal of Fisheries and Aquatic Sciences* **47**, 1004–1010.

# Chapter 7
# Estimation of migrant yellow eel stock in large rivers through the survey of fish passes: a preliminary investigation in the River Meuse (Belgium)

E. BARAS, J.Cl. PHILIPPART and B. SALMON *Laboratory of Fish Demography and Aquaculture, Department of Ethology, University of Liège, 10 Chemin de la Justice, B-4500 Tihange, Belgium*

**Abstract**   The migration dynamics of yellow eels in the River Meuse were investigated at the Ampsin navigation weir, using fish passes as tools for discriminating resident and migrating eels. In 1993, 9548 eels (11–65 cm, mean: 29.5 cm, pseudonormal distribution) were captured (daily maxima = 818–823 eels) in the two Denil fish passes on the weir. The periodicity of migration was consistent with that in previous years (1989–1993: 44 ± 7 days, $P_{50}$ on 1 June). Eels migrated in waves, almost independently of environmental parameters; 3724 eels were marked (Panjet inoculator) and released downstream of the weir. Recaptures (2.1%) suggest an annual rate of migration of 45 km yr$^{-1}$ and a migration flux of ± 445 000 fish (16.5 t), with most eels probably migrating through the sluices. Utilization of fish passes to clarify eel stock distribution within the R. Meuse Basin is discussed.

KEYWORDS: eel, *Anguilla anguilla*, fish pass, migration periodicity, River Meuse, stock assessment

## 7.1    Introduction

Should the European eel *Anguilla anguilla* (L.) be considered as a threatened species? The question has been raised by Brusle (1989) and Moriarty (1990) based on literature reviews and catch statistics. Although not universal (e.g. River Thames: Naismith & Knights 1990) the decline of eel catches is more or less widespread throughout Europe. The causes are pollution, introduced parasites, overfishing and hydraulic management (Moriarty 1987). Stock assessment of a migratory species that is exploited at almost all stages of its life cycle (leptocephali, elvers, yellow and silver eels) implies a characterization of stock renewal at all these stages. Paradoxically, yellow eels, that represent most of the freshwater life of *Anguilla anguilla*, have been less extensively studied than elvers or silver eels (Tesch 1977; Deelder 1984; Moriarty 1987 for synthesis). Some authors (e.g. Aprahamian 1988) have documented the variations in

82

sex, size and age structure along rivers but, with few exceptions (e.g. Moriarty 1986a,b; Vøllestad & Jonsson 1988), the dynamics and intensity of migration are not well known, especially in large river ecosystems. This is partly due to the difficulties in discriminating between migratory and resident fractions of the yellow eel population when using conventional sampling techniques (e.g. fyke nets; Poole 1990).

The aim of this study was a preliminary assessment of yellow eel migrations in the River Meuse. At the beginning of the twentieth century there were around 70 active eel fisheries in the Dutch River Meuse (Deelder & Van Drimmelen 1960), but a major reduction has been recorded in most of its Belgian tributaries (Huet & Timmermans 1966; Philippart & Vranken 1983). The anadromous migration of yellow eels has been studied since 1989 using fish passes on obstacles impeding free migration in the river (Moriarty 1987; Legault 1987). The specific problems dealt with are seasonal periodicity of migration, population structure and intensity of the ascent of yellow eels in the Belgian R. Meuse in 1993.

## 7.2     Material and Methods

The River Meuse has a catchment of 36 011 km$^2$. It flows from France (506 km) through Belgium (182 km) and The Netherlands (183 km) into the North Sea. The Belgian stretch of the R. Meuse had an original slope of 0.23‰ before its transformation into a canalized river during the last 150 years (Philippart *et al.* 1988). Sixteen locks and navigation weirs were built, making the Meuse accessible to boats with a capacity of 1000–4000 t (but up to 9000 t below Liège). The R. Meuse is connected with the harbour of Antwerpen (River Schelde estuary) through the Albert Canal (Fig. 7.1). The navigation weir at Ampsin is located 44 km upstream of the Belgium/ Netherlands border and 0.9 km downstream of the heated effluent from the Tihange nuclear power plant (2600 MW), causing a 3–4°C water temperature increase in all seasons. Flow of the R. Meuse at this site ranges from 16 to 2800 m$^3$ s$^{-1}$ (Belgian Navigation Office, 1988–1993). The weir was initially (1958) equipped with sluices (right bank) and with two fish passes on the sides of the spillway. Each pass is composed of three successive Denil (1909) ladders (7.5 × 1.4 m; slope: 24%) equipped with 22 multiple-planes baffles spaced 35 cm apart and set at an angle of 45° to the axis of the channel. The ladders are separated by 7 m$^3$ intermediate pools. In 1964, the weir was equipped with a 10 000 kW hydroelectric plant (catching 250 m$^3$ s$^{-1}$), on the left bank of the river.

The fish passes were controlled three times a week outside of the migratory period (January to late April and late June to mid-July) and daily as soon as the first migration peak was detected (late April to mid-June). Controls always took place between 8:00 and 10:00 (GMT +1 then GMT +2). The upstream outlets of the fish passes were equipped with a steel grid and a coated wire mesh net preventing fish from leaving the pass. The superior pool (30 m$^3$) was emptied by a valve system and the intermediate pools with an electric pump (30 l s$^{-1}$). Eels were captured with dipnets, anaesthetized with 2-phenoxyethanol (0.4 ml l$^{-1}$), counted and measured. A repre-

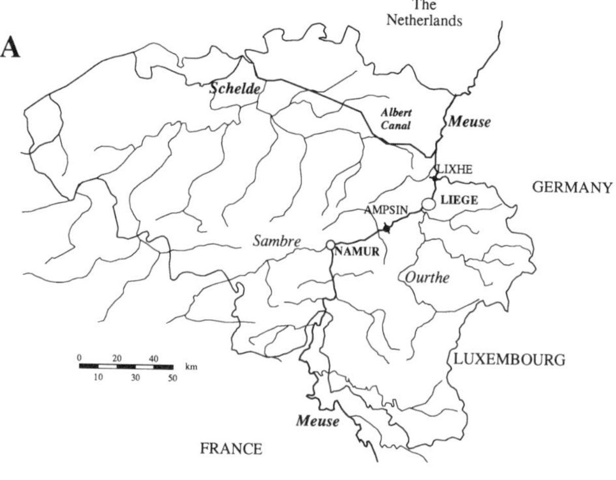


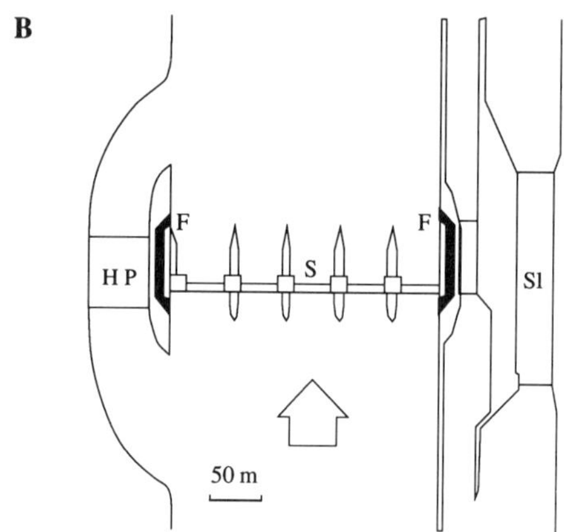


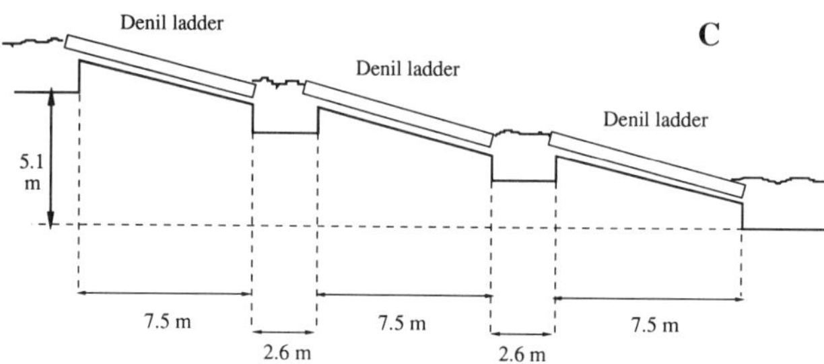


**Fig. 7.1**  Location of the navigation weirs at Ampsin and Lixhe. B Diagrammatic view of the weir at Ampsin: F = fish pass; S = spillway; HP = hydroelectric plant; Sl = sluices. C Longitudinal section of Denil fish passes.

sentative sample was weighed for length–weight relationship and biomass estimates. An eel trap (modified from Legault 1987) that was non-selective towards eel size (Baras *et al.* 1994) was plunged at the downstream entrance of the pass (from 3 June 1993 onwards) to ensure that the fish passes captured a representative size distribution of eels. Water temperature was measured daily. Stepwise regression analyses were used to account for the variations of catch per unit effort (CPUE: total amount of eels caught daily in a fish pass) depending on environmental variables (values, daily and weekly variations). Null hypotheses are rejected at $P < 0.05$.

For mark-recapture experiments and assessment of migration speeds, eels were marked with dyes (alcian blue, acrylic red and Indian ink) using a Panjet inoculator (Hart & Pitcher 1969). Batches of 200 eels captured on the same day were marked and released, with each batch identified by a unique combination of dyes and site of injection. Eels were released in the afternoon following their capture, at various distances (0.20 to 0.75 km, except for the first week: 0.75 to 3.0 km) downstream of Ampsin weir. The distances were based on the minimum observed migration speeds of yellow eels (10–15 km yr$^{-1}$; Aprahamian 1988). The Petersen estimate was used for mark–recapture data, based on plots of mark loss and mortality with time from feasibility studies (Salmon 1993).

## 7.3     Results

The two fish passes of the Ampsin navigation weir were each controlled 75 times from 5 January to 16 July 1993; 9548 yellow eels (114–645 mm) were captured, of which 8344 (87.4%) were in the right fish pass. The length-frequency distribution was pseudonormal (mean: 295 mm, SD: 43 mm, N = 7767), 36.7 % of the eels ranging between 280 mm and 320 mm and 97.9 % between 200 and 400 mm (Fig. 7.2). The

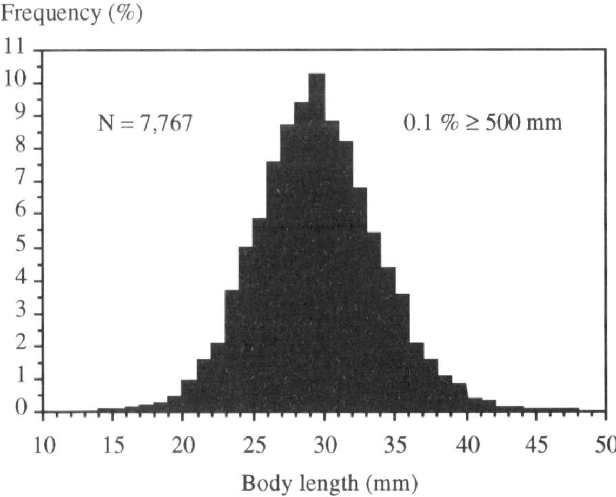


**Fig. 7.2**  Length frequency distribution (10 mm class interval) of the eels captured in the fish passes at Ampsin in 1993.

length–weight relationship was described by: log $W$ (g) $= -6.59 + 3.30$ log $L$ (mm) ($r$ $= 0.97$, 705 df).

The length distributions of eels captured in the eel traps (pooled; N $= 895$; mean length: 306 mm) and in the fish pass (pooled; N $= 4335$; mean length: 292 mm) were significantly different (Student's t-test t $= 8.6$, $P < 0.05$). However, the difference between the mean length in the two samples was low and the size range of eels in the fish pass (114–645 mm) encompassed the size range of eels in eel traps (176–615 mm). These results suggest that the selectivity of the Denil fish passes towards eel size was very low and that the length frequency distribution (Fig. 7.2) was representative of that of migrant yellow eels in this part of the River Meuse.

In 1993, the CPUE ranged between 0 and 818–823 eels per day (8 and 12 June); 90% of the eels were captured between 27 April and 13 June (47 d), with the $P_{50}$ of the catch distribution on 3 June. These values were close to those observed in 1989–1993 (Table 7.1): $44 \pm 5$ d (SD) and 1 June $\pm 7$ d (SD), respectively for seasonal 90% range and $P_{50}$. The seasonal 99% range did not extend to more than 60 days. From the distributions of CPUE and eel lengths, four migration periods were detected in 1993 (Fig. 7.3). During periods I (26 April–7 May) and IV (6–16 June), the mean length of eels decreased significantly, from $324 \pm 46$ mm to 214 mm $\pm 82$ mm ($r = -0.92$, 5 df) and $312 \pm 39$ mm to $278 \pm 33$ mm ($r = -0.92$, 10 df). A similar, but non-significant, size decrease was observed in period III (27 May–5 June; $r = -0.552$, 7 df; $P > 0.05$).

At the seasonal level, neither flow nor temperature accounted for the daily variations of CPUE (stepwise regression, $P > 0.05$). The relative independence of eel migration towards temperature was substantiated by the distribution of eel catches along the thermal axis in 1989–1993 (90% range from 16.9°C to 20.9°C in 1991 and from 22.8°C to 29.0°C in 1993; Table 7.1). However, during period I (April 26–7 May), the CPUE was dependent on daily variations of water temperature ($r = 0.864$; 8 df).

**Table 7.1** Periodicity of yellow eels migrations in the fish passes at the Ampsin navigation weir in 1989–1993 (1989–1991: left fish pass; 1992–1993: left and (right) fish passes). Controls from early January to mid- or late July.

| Year | Total Number | Seasonal $P_{50}$ | Seasonal 90% range (d) | Thermal $P_{50}$°C | Thermal 90% range °C |
|---|---|---|---|---|---|
| 1989 | 53 | 8 June | May 15–July 1 (47 d) | 22.6 | 19.8–25.0 |
| 1990 | 735 | 23 May | May 6–June 19 (44 d) | 23.0 | 21.9–24.2 |
| 1991 | 454 | 27 May | May 5–June 10 (36 d) | 18.8 | 16.9–20.9 |
| 1992 | 616 (4208) | 28 May | May 15–June 29 (45 d) | 24.2 | 21.5–26.8 |
| 1993 | 1208 (8344) | 3 June | April 27–June 13 (47 d) | 26.0 | 22.8–29.0 |

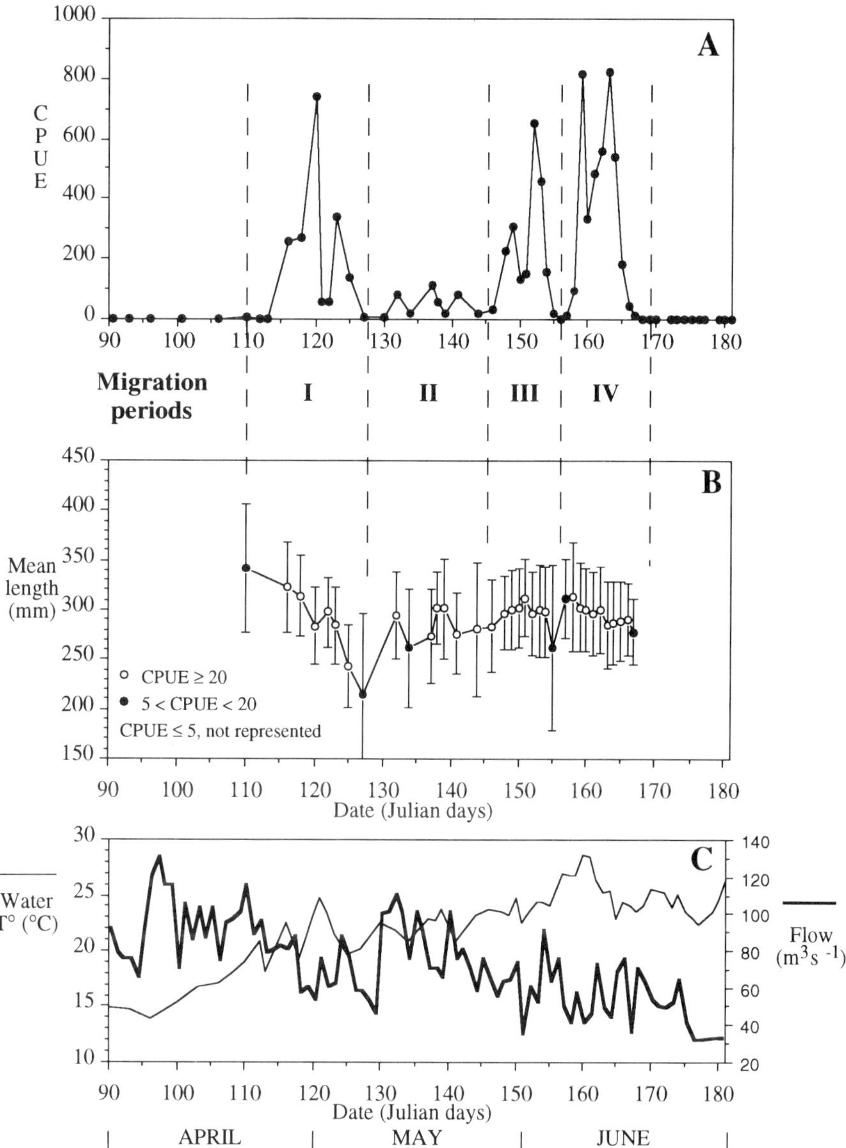



**Fig. 7.3**  A Variations of catch per unit effort (CPUE); B daily mean length of eels at Ampsin in 1993 (fish pass on the right side of the weir). Error bars on B = SD of length; C Variations of river flow and temperature during the study; I–IV: migration periods suggested from distributions of CPUE and lengths.

In the period 26 April to 11 June 1993 3724 eels were marked. The first groups (26 and 30 April) were released at distances ≥750 m downstream of the weir and, with two exceptions, no eel was recaptured within the two first weeks. From this observation, it was decided to release marked eels at a closer range (125, 400 and 750 m). In total, 72 eels were recaptured of which 60 were released at close range. No eel was recaptured after 16 June 1993. The delays between marking and recapture did not differ significantly between the three distances $\leqslant 750$ m (Kruskal-Wallis, $H = 1.7$, $P > 0.05$), most of the eels (18 out of 60) being recaptured the following morning. These observations suggest that eels can migrate 750 m in a single night. From the periodicity and apparent duration of migrations (maximum seasonal 99% range: 60 days), the annual migration rate was estimated to be about 45 km.

The apparent recapture rate (72 out of 3724 marked eels) was corrected for mortality and tag loss parameters, depending on dye used and time between marking and recapture and estimated at $2.1\% \pm 0.5\%$ (95% confidence interval). Using the Petersen estimate, the number of yellow eels migrating through the River Meuse at Ampsin in 1993 was 445 000. Considering the apparent absence of selectivity of the fish passes, that is equivalent to 16.5 t, based on the length frequency distribution (Fig. 7.2).

### 7.4    Discussion

In the River Meuse at Ampsin, the periodicity of eel migration was relatively constant throughout 1989–1993 (around 2 June; see Table 7.1). Moriarty (1986b) observed a higher variation in dates of migration of smaller eels (mode: 10–15 cm) in the low River Shannon, where migration started between 17 May and 24 June in 1973–1983. He found that the onset of migration was correlated with water temperature of 13–14°C. This was consistent with Sörensen (1950, in Deelder 1984) who observed that eels did not migrate until water temperature reached 15°C. Mann (1963, in Deelder 1984) found a similar increase of eel migration with temperature (up to 22°C) and decrease as temperature dropped. At Ampsin, the role of temperature was restricted in 1993 to the beginning of the migration (late April–early May). Previous records at the same site (1989–1992) indicated that the effect of temperature on eel migration is highly variable (Table 7.1). This was probably related to the unnaturally high thermal regime of the R. Meuse at Ampsin, resulting from the warm effluent of the Tihange nuclear power plant. In this context, where the minimum temperature for eel migration was reached early in the season (early to mid-March), the role of temperature would be secondary and the influence of time of the year (suggested by Moriarty 1986b) would be more obvious.

In 1989–1993, yellow eels stopped their upstream migration at Ampsin in late June–early July, much sooner than observed in other rivers (Moriarty 1986b; Vøllestad 1986). Considering the possible density-dependent migration pressure (Moriarty 1986b) and associated mortality risks (Vøllestad & Jonsson 1988), this precocious end of migration could be related to high seasonal or annual migration rates. The estimated migration rate (45 km yr$^{-1}$) in the R. Meuse in 1993 was much higher than observed by Moriarty (1986b) and Aprahamian (1988): 10–15, 15 and 20–30 km yr$^{-1}$, respectively in the rivers Dee (Wales), Shannon (Ireland) and Severn (England). As hypothesized by Aprahamian (1988), higher migration rates could be expected in low gradient rivers like the R. Meuse (80 m/230 km), although Hussein (1981) observed that eels could migrate on the average 46 km yr$^{-1}$ in a much steeper gradient (167 m/ 92 km). The estimated migration rate in the River Meuse was consistent with the length frequency distributions of eels captured in 1993 at Ampsin and Lixhe (44 km downstream; Fig. 7.4): the modes of the two length distributions were 4 cm apart, corresponding to the average growth of yellow eels reported in literature (2–6 cm yr$^{-1}$, Sinha & Jones 1975). The age of the eels captured at Ampsin was not determined but the smallest eels (110–120 mm) probably spent less than 2 years in fresh water, suggesting that the 45 km yr$^{-1}$ migration rate may be underestimated for young individuals. If these small eels had taken the 'short cut' through the Albert Canal (150 km v. 230 km; Fig. 7.1a), their migration rate would have been 75 km yr$^{-1}$. The observation by Sörensen (1950 in Deelder 1984), that smaller eels are less inhibited by light and are more inclined to continue their upstream movement in daylight, could possibly account for this very high migration rate.

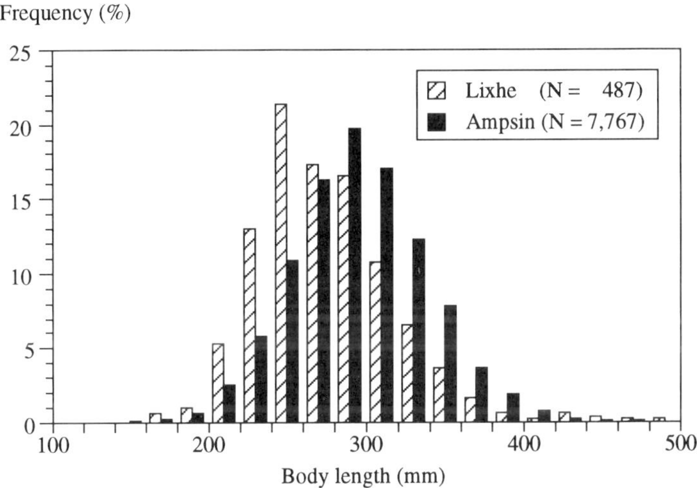



**Fig. 7.4**  Length-frequency distributions (20 mm class interval) of eels captured in the fish passes of the weirs at Lixhe and Ampsin (R. Meuse, January–July 1993). The distance between the two sample sites (44 km) corresponds to the estimated annual migration rate ($\pm45$ km yr$^{-1}$) and the length difference between the two distributions to the average annual growth of European eels (2–6 cm yr$^{-1}$, Sinha & Jones 1975).

The length frequency distribution of migrant yellow eels in Ampsin in 1993 was pseudonormal; the eels ranged from 11–65 cm and averaged 29–30 cm. Similar size or age distributions were observed elsewhere (Aprahamian 1988; Vøllestad & Jonsson 1988), as well as in the River Meuse Basin (Philippart & Vranken 1983; Vriese 1992) although these distributions referred to the whole population (migratory and resident fractions). Dahl (1983) and Moriarty (1986b) sampled migratory yellow eels in the vicinity of dams and observed a comparable size range (10–50 cm, up to 10 years old). The test on the representativity of the size frequency distribution of eels captured in the fish passes showed that the pseudonormal distribution observed at Ampsin did not originate from the selectivity of the Denil fish passes towards eel size (suggested for eel trawls by Deelder 1984). It may either reflect growth variability (2–6 cm yr$^{-1}$; Sinha & Jones 1975) or variable migratory tendencies of eels of different ages, but the absence of age determination does not permit these hypotheses to be substantiated. A marked variation of yellow eel size was also observed throughout the season. Moriarty (1986b) showed that the size of eels in the River Shannon decreased throughout the season, due to a later and shorter migration period of small eels ( < 10 cm). In this study, the size variations were structured differently and corresponded to variations of catch per unit effort. From these observations, it is suggested that yellow eels migrate in waves, these waves possibly accounting for the apparent stochastic migration – beyond seasonal limits – observed at Ampsin. Such waves were documented in elvers of *Anguilla anguilla* (Deelder 1958) and of *Anguilla rostrata* L. (Sörensen & Bianchini 1986) that accumulate before starting their movement into inland waters. At present, it is not known if the waves observed in the R. Meuse correspond to distinct eel groups of different origins (distances from the navigation weir) and respond simultaneously to the same environmental stimulus or to a clumping of eels of different origins that would migrate as a consequence of population pressure. These questions could be partly answered by the recapture in the coming years of the eels tagged at Ampsin in 1993.

Only 72 out of the 3724 eels marked in 1993 were recaptured. The recaptures were considered complete since the last recapture took place on 16 June 1993. Even when corrected for mortality and mark loss parameters, the recapture rate was only 2.1%, much lower than in most studies on *Anguilla* spp. in lakes (Hurley 1972; Beumer 1979) but similar to that obtained by Moriarty (1.2%; 1986a) on a similar sample ($N$ = 3602) of *A. anguilla* in Meelick Bay (Ireland). These results suggest that most eels ($\sim$98%) migrate through the sluices that represent the only alternative to fish passes and are operational at a time of the day (from 6:00 GMT + 2) when yellow eels were still active (Legault 1987; Baras *et al.* 1994). This hypothesis tends to be supported by the ratio between the section of fish passes and the section of sluices, being around 2%, precisely the estimated recapture rate. Based on this recapture rate, the stock of yellow eels migrating past Ampsin in 1993 was estimated as $\pm$445 000 eels for a biomass of 16.5 t. Considering the paucity of comparative data in large river ecosystems and in the R. Meuse Basin itself (excepted R. Ourthe Basin: 17 kg ha$^{-1}$; Philippart 1980), these values are mainly indicative. From the variations of catches recorded from 1989 to 1993 in Ampsin, it was suggested that yellow eels migrating through the Ampsin navigation weir in 1993 were probably much more abundant than in the previous years and that the estimates presented are probably higher than the average renewal rate.

In conclusion, the study conducted in 1993 in the R. Meuse confirmed the effectiveness of surveys of fish passes for estimating yellow eel migration dynamics in a large regulated river. It also showed a constant periodicity of eel migration over 5 consecutive years and migration wave dynamics apparently independent of environmental variables beyond seasonal limits. Further understanding of eel migration dynamics and stock in such an environment would imply a survey in coming years of more numerous fish passes on dams and weirs in France, Belgium and The Netherlands.

## Acknowledgements

The 'Eel' research programme was financed by the Commission Provinciale de Liège du Fonds Piscicole and supported by the Région Wallonne (Ministry of Nature Conservation; migratory species, Salmon 2000 Project). We are indebted to the Belgian Navigation Office who allowed the access to navigation weirs and fish passes, and to the Belgian Royal Institute of Meteorology for environmental data. Special thanks to J.M. Lambert and G. Rimbaud for field support.

## References

Aprahamian M.W. (1988) Age structure of eel, *Anguilla anguilla* (L.) populations in the River Severn, England and in the River Dee, Wales. *Aquaculture and Fisheries Management* **19**, 365–376.

Baras E., Salmon B. & Philippart, J.C. (1994) Evaluation of a eel trap sampling method for the assessment of migrant yellow eels *Anguilla anguilla* (L.) in the River Meuse. *Bulletin Français de la Pêche et de la Pisciculture*, **335**, 7–16.

Beumer J.P. (1979) Feeding and movement of *Anguilla australis* and *A. reihnardtii* in Macleods Morass, Victoria, Australia. *Journal of Fish Biology* **14**, 573–592.

Brusle J. (1989) L'anguille européenne (*Anguilla anguilla*), une espéce jugée commune jusqu'à la dernière décennie, mérite-t-elle d'acquérir aujourd'hui le statut d'espèce menacée ? *Bulletin de la Société Zoologique de France* **114**, 61–70.

Dahl J. (1983) Some observations on the ascent of young eels at the Tange Power Dam, River Gudena. EIFAC Working Party on Eel, Drottningholm. *mimeo*, 11 pp.

Deelder C.L.(1958) On the behaviour of elvers (*Anguilla vulgaris* Turt.) migrating from the sea into freshwater. *Journal du Conseil. Conseil International pour l'Exploration de la Mer* **24**, 136–146.

Deelder C.L.(1984) Synopsis of biological data on the eel *Anguilla anguilla* (Linnaeus, 1758). *FAO Fisheries Synopsis* **80 (1st revision)**, 73 pp.

Deelder C.L., and Van Drimmelen, E.D. (1960) The decline of fish stocks in the Netherlands' section of the rivers Rhine and Meuse. In: U.I.C.N., 7th Technical Meeting, Vol. IV. Athenes (Greece), September 1958, pp 170–185.

Denil G. (1909) Les échelles à poissons et leur application aux barrages de Meuse et l'Ourthe. *Bulletin de l'Académie des Sciences de Belgique* **1909**, 1221–1224.

Hart P.J.B. & Pitcher T.J. (1969) Field trials of fish marking using a jet inoculator. *Journal of Fish Biology* **1**, 383–385.

Huet M. & Timmermans J.A. (1966) La population piscicole de l'Ourthe (grosse rivière belge de la zone à ombre et du type supérieur de la zone à barbeau). *Verhandlungen der Internationalen Vereinigung für Theoretische und Angewandte Limnologie* **16**, 1192–1203.

Hurley D.A. (1972) The American eel (*Anguilla rostrata*) in eastern Lake Ontario. *Journal of the Fisheries Research Board of Canada* **29**, 535–543.

Hussein S.A. (1981) The population density, growth and food of eels *Anguilla anguilla* L. in some tributaries of the River Tweed. In: Proceedings of the Second British Freshwater Fisheries Conference, University of Liverpool, pp. 120–128.

Legault A. (1987) L'anguille dans le bassin de la Sèvre Niortaise. *Publications du Département Halieuttique* **6**. *Ecole Nationale Supérieure Agronomique de Rennes*, 295 pp.

Mann H. (1963) Beobachtungen über den aalaufstieg in der aalleiter an der staustufe geesthacht im jahre 1961. *Fischwirt* **13**, 182–186.

Moriarty C. (1986a) Observations on the eels of Mellick Bay, Lough Derg, 1981–1984. *Vie et Milieu* **36**, 279–283.

Moriarty C. (1986b) Riverine migration of young eels *Anguilla anguilla* (L.). *Fisheries Research* **4**, 43–58.

Moriarty C. (1987) The European eel – Discoveries and developments. In: W.W. Crozier and P.W. Johnston (eds) Proceedings of the 17th Annual Study Course of the Institute of Fisheries Management (University of Ulster, Coleraine, 9–11 Sept. 1986), pp 100–110.

Moriarty C. (1990) European catches of elver of 1928–1988. *International Review Gesamt Hydrobiology* **75**, 701–706.

Naismith L.A. & Knights, B. (1990) Modelling of unexploited and exploited populations of eels, *Anguilla anguilla* (L.), in the Thames estuary. *Journal of Fish Biology* **37**, 975–986.

Philippart J.C. (1980) Essai d'évaluation des ressources ichtyologiques actuelles et potentielles dans le bassin de l'Ourthe (bassin de la Meuse) en Belgique. In: J.H. Grover (ed.), *Allocation of Fishery Resources*, Proceedings of the Technical Consultation on Allocation of Fishery Resources, Vichy (France), April 1980, pp 298–307. FAO, Rome.

Philippart J.C., Gillet A. & Micha, J.C. (1988) Fish and their environment in large European river ecosystems: the River Meuse. *Sciences de l'eau* **7**, 115–154.

Philippart J.C., & Vranken, M. (1983) L'Anguille. In: Atlas des Poissons de Wallonie. Distribution, écologie, éthologie, pêche, conservation. *Cahiers d'Ethologie appliquée*, **3(supplement 1-2)**, pp 55–61.

Poole W.R. (1990) Summer fyke nets as a method of eel capture in a salmonid fishery. *Aquaculture and Fisheries Management* **21**, 259–262.

Salmon B. (1993) Structure et intensité du flux migratoire de l'anguille jaune *Anguilla anguilla* (L.) en Meuse en 1993. BSc Thesis Zool. Sci., University of Liège, December 1993, 41 pp + ann.

Sinha V.R.P. & Jones, J.W. (1975) *The European Freshwater Eel*. Liverpool: Liverpool University Press, 146 pp.

Sörensen J. (1950) An investigation of some factors affecting the upstream migration of the eel. *Report of the Institute of Freshwater Research of Drottningholm* **32**, 126–172.

Sörensen P.W. & Bianchini M.L. (1986) Environmental correlates of the freshwater migration of elvers of the American eel in Rhode Island Brook. *Transactions of the American Fisheries Society* **115**, 258–268.

Tesch F.-W. (1977) *The Eel. Biology and Management of Anguillid Eels*. London: Chapman and Hall, 434 pp.

Vøllestad L.A. (1986) Growth and production of female yellow eels (*Anguilla anguilla* L.) from brackish water in Norway. *Vie et Milieu* **36**, 267–271.

Vøllestad L.A., & Jonsson, B. (1988) A 13 year study of the population dynamic and growth of the European eel *Anguilla anguilla* in a Norwegian river: evidence for density-dependent mortality, and development of a model for predicting yield. *Journal of Animal Ecology* **57**, 983–997.

Vriese T. (1992). De visstand in de Grensmaas. Report of the project 'ecological rehabilitation of the River Meuse'. OVB Research Report, **6** (May 1992), 95 pp + ann.

# II CATCH EFFORT METHODS

# Chapter 8
# Catch census of coarse fish anglers in England and Wales

R. STEEL, K. O'HARA and P.A. SMITH *SGS Environment, Yorkshire House, Chapel Street, Liverpool L3 9AG, UK*

**Abstract**   The majority of surveys of the catches of coarse fish anglers in the UK have concentrated on the collection of data from fishing matches. Many of these studies have provided useful long-term data which have described population trends. Few systematic studies of the anglers fishing for pleasure rather than competitively have been undertaken although these represent the majority of angler visits. Preliminary results are reported on rivers where the results of angler interviews and examination of catches have been checked by the use of alternative sampling techniques. The design of the surveys is described.

KEYWORDS: Creel census, management, recreational fisheries

## 8.1   Introduction

At the present time no generally agreed suitable variations of standard fish sampling techniques have been developed for effective stock assessment of large lowland rivers in the UK. Previous studies have attempted to utilize angler match catch records as a sampling method and some of these have identified long-term changes in both abundance and species composition of coarse fish stocks (Axford 1979,1991; Hickley & North 1981; Cowx 1990, 1991).

In the UK, little attention has focused on the use of anglers fishing for sport (pleasure anglers), rather than competitively, as a stock assessment tool, despite these anglers representing the majority of angler visits (NOP 1980). Roving creel censuses of pleasure anglers are widely undertaken in the USA and are combined with population data as a stock assessment technique (e.g. Bayley & Austin 1989). The USA techniques are well developed but differ from any technique which could be applied to the coarse fisheries of the UK in three respects.

- The US recreational fishery venues are physically large compared to the UK and often can only be partially surveyed within a surveying day. Consequently, careful consideration needs to be paid to the planning and implementation of the roving creel census to provide sufficient precision to assessments of effort and catch. UK fisheries are small by comparison and can usually be completely surveyed within one day.

- US anglers rarely adopt a catch-and-release policy with fish being captured for the table. Within the UK, coarse anglers have a catch-and-release policy at all times.
- UK pleasure coarse anglers regularly catch small fish and species as part of their sport (Cooper & Wheatley 1981) which can be compared to the pursuit of larger sports fish in the USA (Santucci & Wahl 1991).

This chapter describes the application of a roving creel census technique to two UK coarse fisheries and presents the preliminary results. The study formed part of the National Rivers Authority (NRA) R & D contract entitled 'The Use of Catch Statistics to Determine Stock Size' which is ongoing.

## 8.2    Methodology

### 8.2.1    *Selection of survey sites*

The rivers Weaver and Dane, in north-west England, were selected for the surveys of recreational anglers.

The River Weaver is a slow-flowing channelized river of approximately 35 m width and a relatively uniform 4.5 m depth in the central navigation channel. The river has off-channel backwaters and navigation locks. The following areas were selected for surveying: Hartford Bridge (NGR SJ 647 714); Hartford Locks (NGR SJ 642 705); and Bradford Mill (NGR SJ 652 688).

The River Dane is a small river of approximately 10 m width and of varying depths resulting from the sand- and gravel-bottomed pool and riffle habitat. Dense stands of *Ranunculus* sp. are present along with fish habitat features associated with a natural river, e.g. overhanging trees and natural instream structures (Swales & O'Hara 1980). The following areas were selected for surveying: Cotton Hall (NGR SJ 742 677); Byley Bridge (NGR SJ 715 675); Croxton Hall (NGR SJ 698 673).

Both rivers are popular recreational and competition fisheries which are frequented by anglers throughout the UK coarse fish season (16 June–15 March inclusive). The areas selected on both rivers were of a size which allowed all sites on a river to be surveyed in a single day. If low numbers of anglers were encountered on a particular day then both rivers would be surveyed.

### 8.2.2    *Timing of surveys*

Surveys were undertaken between July and September 1993 on a near daily basis and a further survey was undertaken in November 1993. The work was undertaken by a single field surveyor usually working between the hours of 09.00 h and 18.00 h, although surveying times varied slightly during the census. Surveys were carried out during the summer period as peak angler effort was anticipated at this time. Additionally, all species, including warm-water eurytherms such as tench *Tinca tinca* (L.) and carp *Cyprinus carpio* (L.), would be actively feeding. However, the feeding of such species would be checked by temperature in the November survey.

### 8.2.3   *Angler census form*

A survey questionnaire was devised for interviewing anglers (Fig. 8.1). The information to be collected was based on a pilot survey carried out in 1992 on a range of stillwaters and rivers in England and Wales. Data were collected on angling methods and bait, location of the angler on the fishery, fishing effort and the anglers' perception of their catches in terms of species, size and numbers.

### 8.2.4   *Tray photographic method*

In association with the census form, examination of angler catches was made where the fish had been retained in a keepnet. The fish were placed in a large, shallow, plastic tray (60 × 40 × 10 cm) and covered with a thin perspex sheet. Photographs were taken directly above the tray using an automatic compact camera. Large catches of fish were divided and the whole catch recorded over several photographs. A number and date were placed in the tray to allow the resulting photographs to be identified and matched to a particular angler. The four corners of the tray were marked with a dark spot for analysis purposes. Preliminary tests showed a yellow tray provided photographs with the best definition.

Photographs were analysed using a Microstation GIS System. The marked corners of the tray in the photograph were selected with a cross-hair mouse and tablet and defined as the corners of an 'imaginary tray' of scaled dimensions on the computer screen. The fork length of the fish could then be determined by selecting the head and tail fork of the fish on the photograph with the mouse. Fork lengths estimates were provided by the computer, based on scaling of the linear measurement of the fish in relation to the 'imaginary tray'.

### 8.3   **Results**

Survey statistics obtained for the River Weaver and Dane are presented in Table 8.1.

### 8.3.1   *Angler behaviour*

The anglers' start and expected finishing times for each fishing session were noted. The peak start time of angling sessions on the rivers Weaver and Dane was 09.00 h, with 90% of anglers having started by 12.00 noon (Figs 8.2a & 8.2b). The peak finishing times for angling sessions were 16.30 h on the River Weaver and 15.30 h on the River Dane (Figs 8.2c & 8.2d).

Coarse Fish Questionnaire                    ref: 2360/c/93

VENUE ______________   GRID REF: ______________   DATE: ____/____/ 93

TYPE OF FISHERY:     Free     Day ticket     Private

DISTANCE TO NEAREST ACCESS POINT: ______________ (metres)

TIME STARTED FISHING: __________   PRESENT TIME: __________   NUMBER OF HOURS     EXPECTED LEAVING
                                                              FISHED: __________   TIME __________

NUMBER OF RODS USED:  1  2  3

ARE YOU FISHING FOR A PARTICULAR SPECIES? _____

FISHING METHOD:     ROD 1 __________     ROD 2 __________     ROD 3 __________

BAIT USED:     ROD 1__________     ROD 2 __________     ROD 3 __________

IS ANGLER USING A KEEPNET?

YES          NO

FORK LENGTH MEASUREMENT OF FISH (mm.)
(enter number caught in each size category)

| SPECIES | 0–49.9 | 50–99.9 | 100–149.9 | 150–199.9 | 200–249.9 | 250–299.9 | 300–349.9 | >350 |
|---|---|---|---|---|---|---|---|---|
| ROACH | | | | | | | | |
| BREAM | | | | | | | | |
| DACE | | | | | | | | |
| CHUB | | | | | | | | |
| CARP | | | | | | | | |
| CRUCIAN | | | | | | | | |
| TENCH | | | | | | | | |
| RUDD | | | | | | | | |
| BARBEL | | | | | | | | |
| GUDGEON | | | | | | | | |
| BLEAK | | | | | | | | |
| RUFFE | | | | | | | | |
| PERCH | | | | | | | | |
| PIKE | | | | | | | | |
| EELS | | | | | | | | |

WAS THE CATCH EXAMINED?

YES                          NO

ARE YOU CURRENTLY INVOLVED WITH TNE NRA LOG BOOK SCHEME

YES     NO

WOULD YOU LIKE TO BE INVOLVED IN NRA LOG BOOK SCHEME?

YES     NO

COULD WE CONTACT YOU WITH FOLLOW-UP QUESTIONNAIRE LATE IN SEASON?

YES     NO

NAME:

ADDRESS:

DO YOU CONSIDER YOURSELF TO BE:

1 PLEASURE FISHING                     [ ]
2 SPCIMEN FISHING                      [ ]
3 PRACTISING FOR FISHING MATCHES       [ ]

IS THIS YOUR USUAL FISHING METHOD?

    YES      NO (GIVE NUMBER)      [ ]

EFFORT INFORMATION

HOW MANY DAYS A WEEK DO YOU FISH ON AVERAGE IN:

SPRING ___   SUMMER ___   AUTUMN ___   WINTER ___

(MAR-JUN)   (JUN-SEPT)   (SEPT-DEC)   (DEC-MAR)

WHAT DAYS DO YOU USUALLY FISH?:

MON   TUES   WED   THURS   FRI   SAT   SUN

EXPERIENCE

|  | Inexperienced |   |   |   | Experienced |   |
|---|---|---|---|---|---|---|
| Interviewer | 1 | 2 | 3 | 4 | 5 | 6 |

DO YOU CONSIDER YOUR CATCH TO BE TYPICAL?

POOR          FAIR          GOOD

HOW DO YOU RATE THIS FISHERY?

**Fig. 8.1**  Angler census form.

**Table 8.1**  Basic survey statistics for creel censuses on the rivers Weaver and Dane.

|                                   | Weaver | Dane  |
| --------------------------------- | ------ | ----- |
| Number of survey occasions        | 53     | 62    |
| Number of anglers encountered     | 500    | 587   |
| Number of catches photographed    | 55     | 46    |
| Number of man-days of angling     | 229.5  | 337.5 |

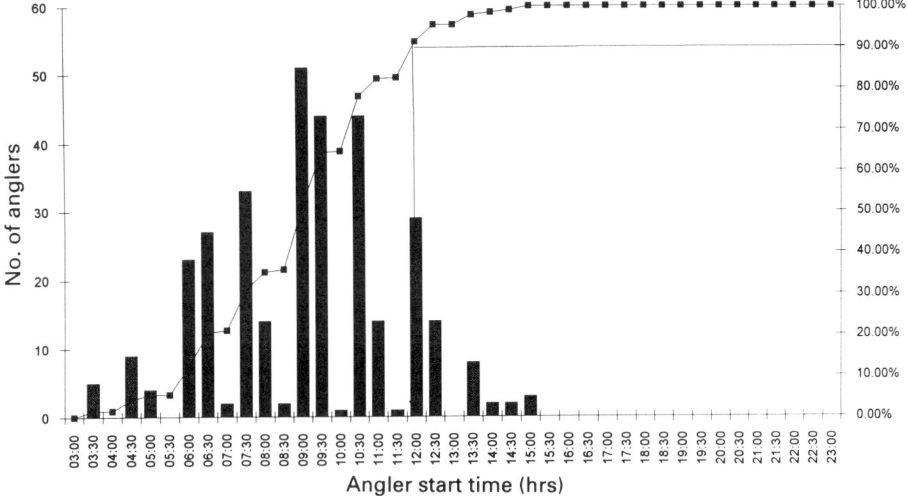

**Fig. 8.2**  (a) Angler start time on the River Weaver.

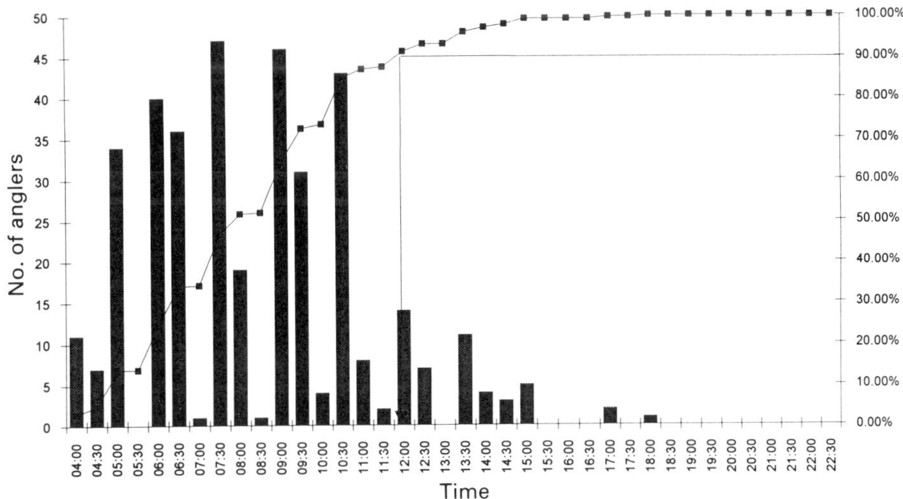

**Fig. 8.2**  (b) Angler start time on the River Dane.

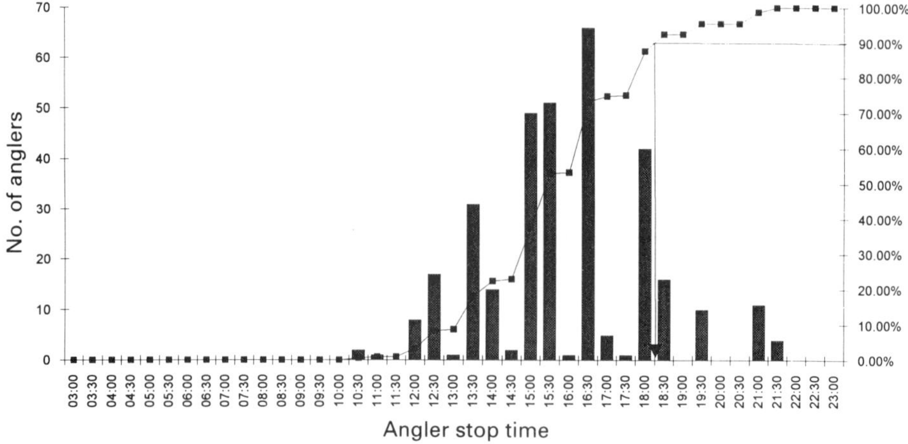

Fig. 8.2   (c) Angler finishing time on the River Weaver.

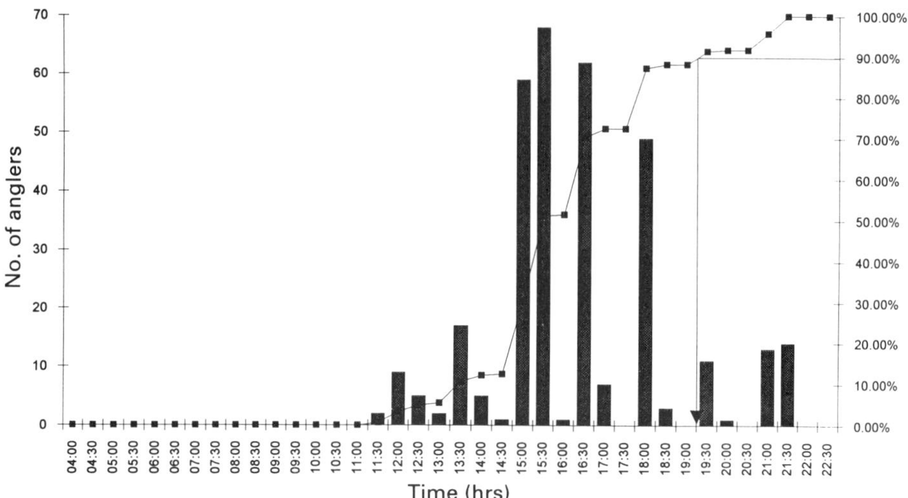

Fig. 8.2   (d) Angler finishing time on the River Dane.

The mean time periods for angling sessions on the rivers Weaver and Dane were 6 h 50 min and 7 h 20 min, respectively.

Anglers were also asked to recollect the number of days spent fishing through the different seasons of the year (Figs 8.3a, 8.3b, 8.3c & 8.3d). This indicated that the majority of angling visits took place through the summer period.

Analysis of the location of anglers with respect to the access points to the fishery demonstrated a contagion of angler distribution and thus corresponding effort on both fisheries; 90% of anglers were interviewed within 200 m of an access point on the River Weaver and 550 m on the River Dane (Figs 8.4a & 8.4b).

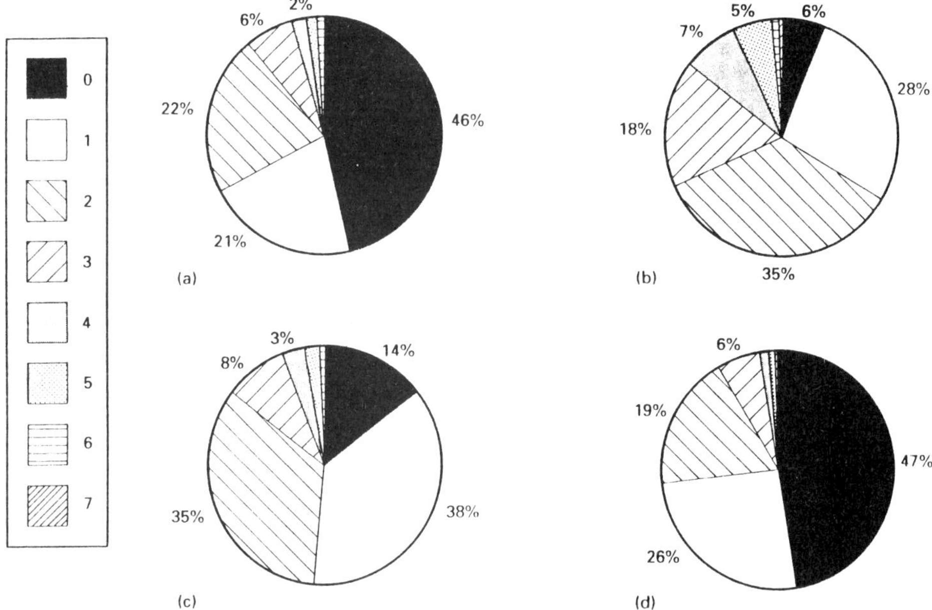

**Fig. 8.3**  Anglers' recollections of number of days fished per week in (a) spring (b) summer (c) autumn and (d) winter.

### 8.3.2   *Angler catch results*

Overall results of the anglers' recollection of their catches, for both the rivers Dane and Weaver (Figs 8.5a & 8.5b), were compared to photographed and examined catches. This demonstrated that recollection was generally poor. Anglers tended to overestimate both the size and numbers of fish caught. Misidentification of juvenile

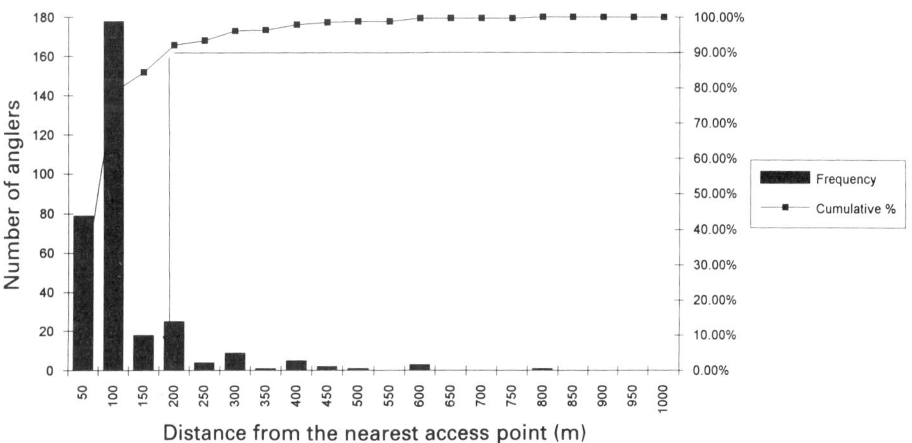


**Fig. 8.4**  (a) Distance of angler from the nearest access point for the River Weaver.

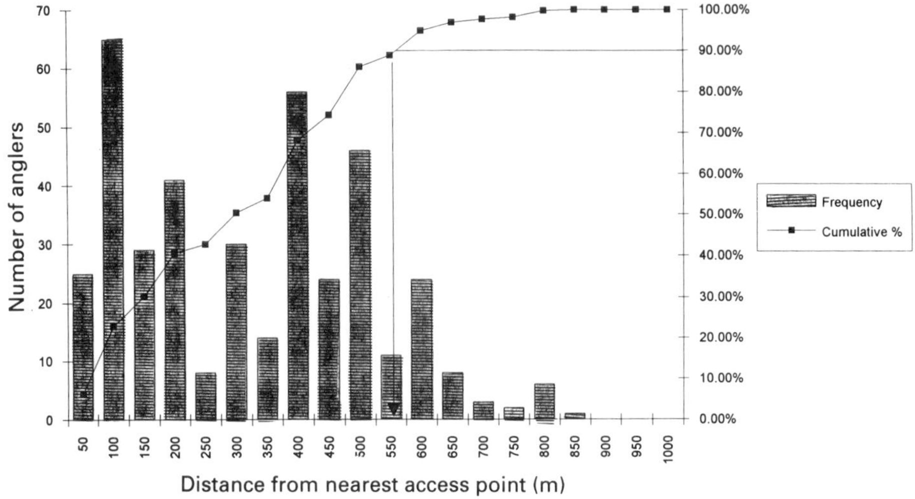


**Fig. 8.4**   (b) Distance of angler from the nearest access point for the River Dane.

cyprinids was also a common error with the anglers' recollection of their catch (Figs 8.5c, 8.5d & 8.5e).

Analysis of the photographed catches by the GIS system showed that both species composition and fork length can be obtained. Regression analysis was carried out between actual measured length and estimated length from the GIS (Fig. 8.6). These data were obtained from attended angling matches where each catch was photographed and every fifth angler catch measured.

Length frequency histograms of data obtained from photographed catches are presented in Figs 8.7a, 8.7b, 8.7c and 8.7d.

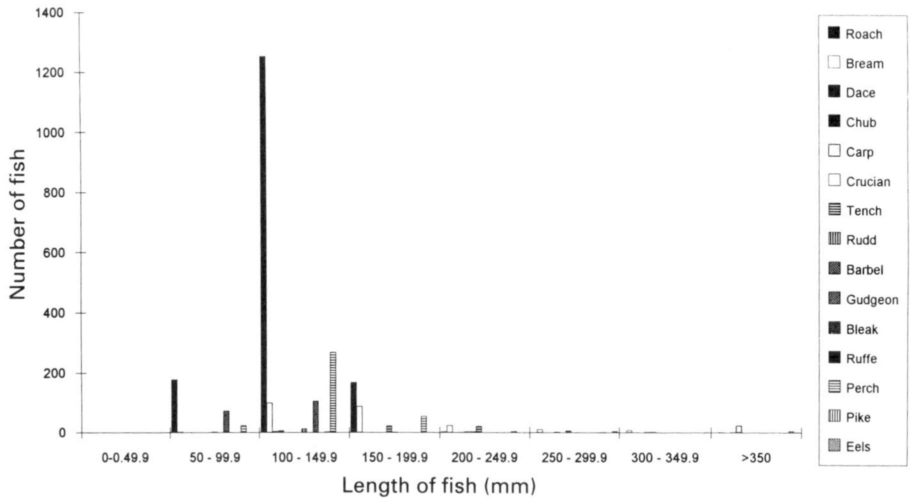


**Fig. 8.5**   (a) Total angler recollected catch by size and species on the River Weaver.

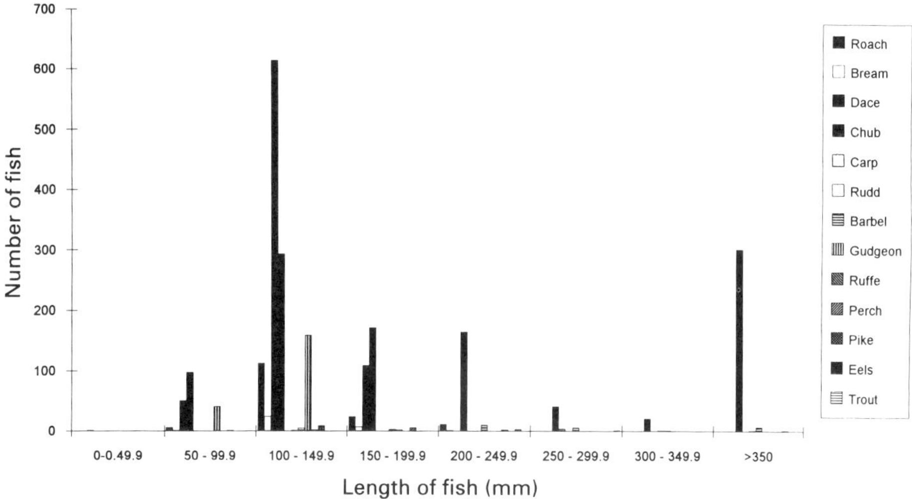


**Fig. 8.5**   (b) Total angler recollected catch by size and species on the River Dane.

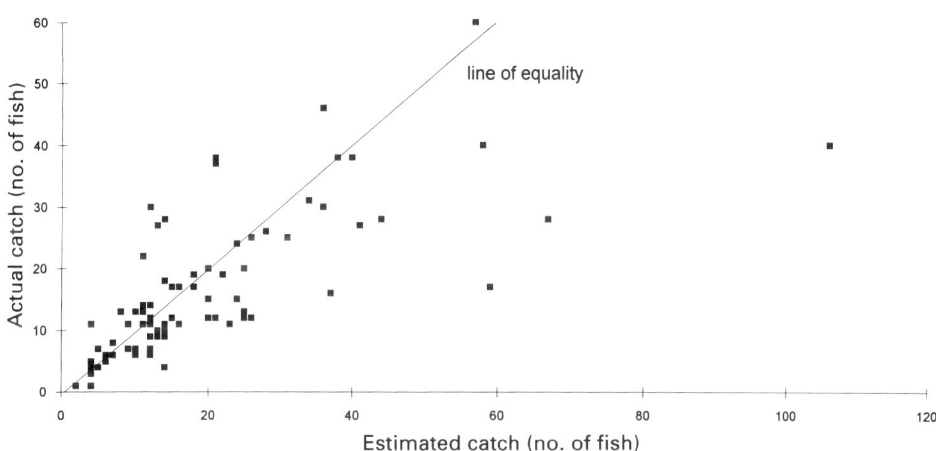


**Fig. 8.5**   (c) Relationship between recollected number of fish caught for anglers fishing the Rivers Dane and Weaver compared to their actual catch held in their keepnet.

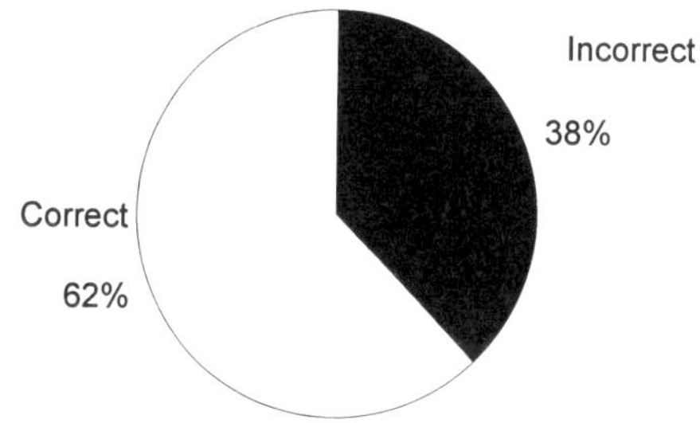


**Fig. 8.5**   (d) Proportion of anglers fishing the River Weaver who correctly recollected the species composition of their catch.

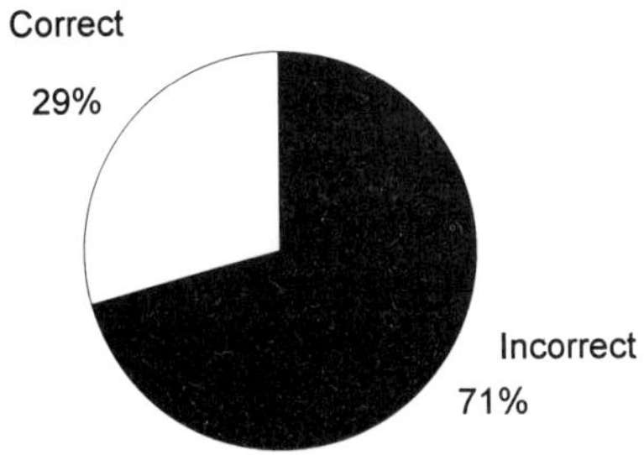


**Fig. 8.5**   (e) Proportion of anglers fishing the River Dane who correctly recollected the species composition of their catch.

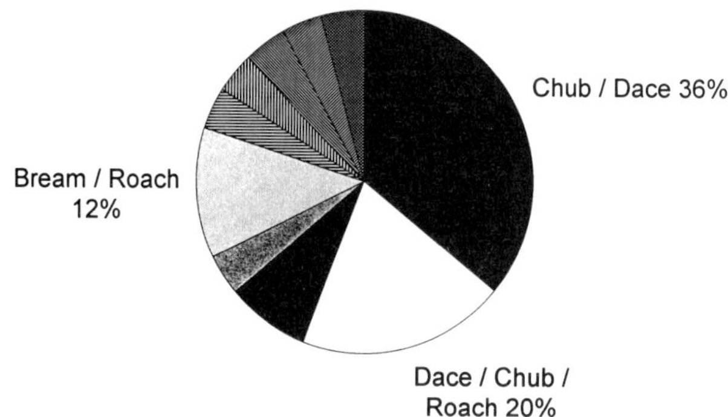


**Fig. 8.5**   (f) Types of error in identification of fish species by anglers fishing the Rivers Dane and Weaver.

### 8.4    Discussion

The objective behind this study was to provide a reliable cost-effective method using anglers to determine stock size of coarse fish populations. Methods such as seine netting and electric fishing can readily provide both qualitative and quantitative data on small rivers, pools and lakes. It is the larger systems that have proved impossible to sample quantitatively or qualitatively by any technique other than angling. New methods such as sonar counting have been developed (Kubecka, Chapter 13), but these are only likely to be of use in rivers with a good depth and relatively uniform channel. The costs of electric fishing and other survey methods are likely to be examined in the future by bodies such as the NRA, together with logistics of carrying them out.

The use of angling catch data has proved of great benefit in following population trends (Cowx 1990, Axford 1991) and these data have been used to obtain quantitative estimates (Cooper & Wheatley 1981). It is, however, unlikely that quantitative estimates of large fisheries will be a realistic objective, and qualitative sampling to determine major population trends and monitor year class strength fluctuations will be adequate for most management objectives (Mills & Mann 1985; Cowx 1990).

The use of competition anglers to collect data on fish stocks has certain limitations.

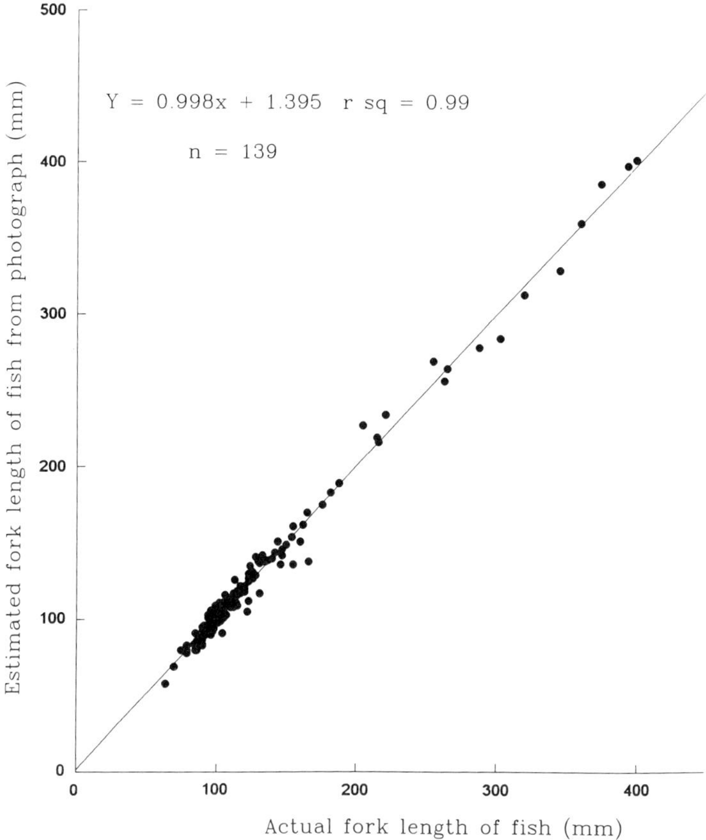


**Fig. 8.6**   Regression of estimated fish length from photographs against actual length.

There are time and cost constraints in that a survey team must be available, usually on a Saturday or a Sunday, and a large number of captured fish have to be processed quickly. Additionally, anglers, although effective in their catch rate (O'Hara & Williams 1991), are often highly selective, and as a group only comprise about 10% of the angling population (NOP 1980).

Pleasure anglers are fishing not necessarily to maximize the weight of fish caught, as in competitions. An analysis of their objectives will be appraised in future surveys but as a generalization they are more likely to be fishing for any species and size of fish using a range of methods. The exceptions to this are specialist anglers who will be concentrating on a particular species.

The data collected in the present survey showed that anglers catch a wide size range of fish and species, particularly small individuals. Of particular interest are the presence of small fish in catches, i.e. below 10 cm, which would be difficult to capture by electric fishing (Pygott *et al.* 1990). Seine netting which has been used successfully on large canalized lowland rivers can give a full range of sizes and species but requires a large team to carry out the method and cannot easily deal with contiguously

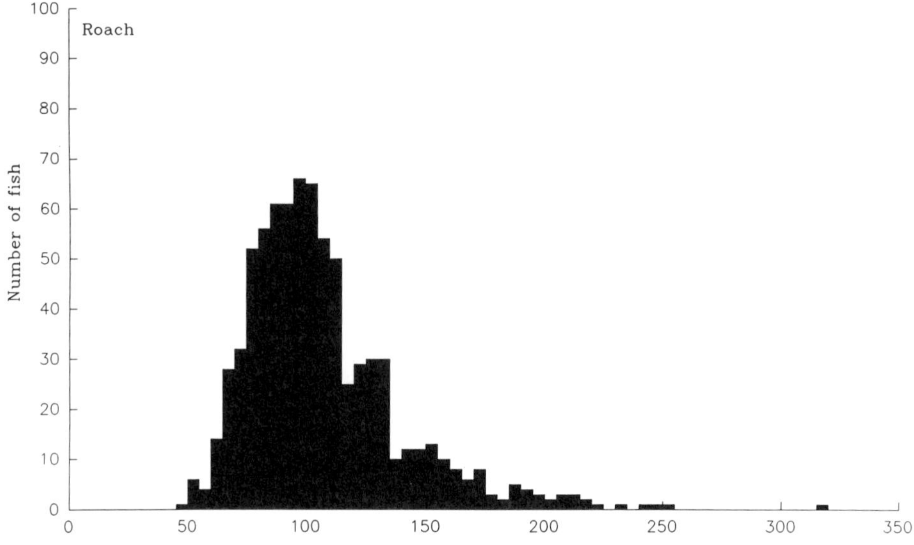



**Fig. 8.7**   (a) Length frequency histogram of angler catches from photographs, River Weaver.

distributed fish (Kell 1991). The benefits of the technique used are its detection of all species and sizes and low manpower requirements. Only one person was used to collect all of the field data presented.

During surveying there is a requirement to record catches accurately with the minimum of effort and disturbance to the angler. The photographic tray method was developed for this purpose and allowed both the size and species composition of catches to be established. The data thus obtained were important for determination of population structures for different species. Analysis of the photographs was also undertaken rapidly by this technique.

Clear differences in seasonal effort were detected and it is suggested that it is best to concentrate the sampling effort in the summer months. The efficient timing of surveys may be determined from the census method such that the maximum number of anglers, fishing effort and stock data can be encountered within the surveying day. On linear systems, such as those surveyed, it was found that there was a contagion of anglers around access points to the fishery. Therefore, on fisheries with restricted distribution, it may be necessary to survey large areas beyond the immediate access area only on a limited basis due to the low encounter rate with anglers. This may be considered a disadvantage when compared to attendance at large angling competitions where anglers are evenly distributed along long sections of river. Furthermore, it may be questioned as to whether the distribution of fish on non-competition fished linear systems is influenced by the contagious distribution of anglers.

The survey will be extended into 1994 and will be compared with angling competition data and alternative quantitative and qualitative studies on the rivers Weaver

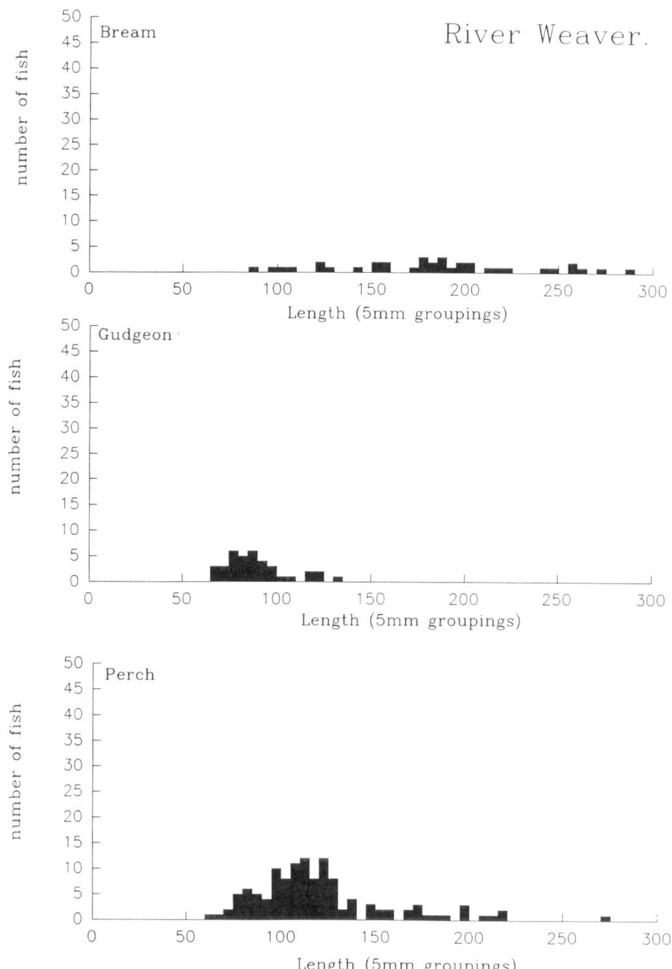



**Fig. 8.7**   (b) Length frequency histogram of angler catches from photographs, River Weaver.

and Dane. The preliminary conclusion from the study was that it provided an easily applicable tool for obtaining data quickly. This, combined with a rapid measurement method, will allow determination of general fish species abundance and year class strengths. Some backup data on scales must be collected and validation of lengths for each census should be carried out.

A growing development which could affect the use of the method is the trend for pleasure anglers not to use keep nets to retain their catch on conservation grounds. The present study showed that 79% of the anglers were using keep nets but examination of the catch was only allowed by 23% of anglers. It was demonstrated by this study and other survey groups that anglers have a poor recollection of their catch. If this technique is developed as a nationally applied census tool, it may be necessary to

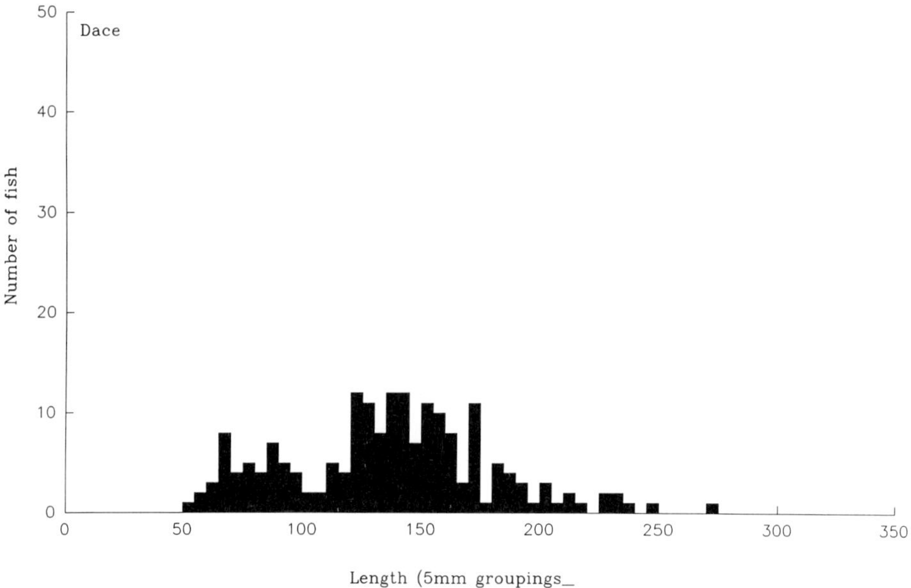


**Fig. 8.7**   (c) Length frequency histogram of angler catches from photographs, River Dane.

have monitoring river stretches and lakes where anglers are requested to use a keep net.

It is believed that with further refinement and application of statistical rigour to the sampling programme, the approach has the potential to provide a sampling technique for water bodies unsuitable for sampling by standard methods.

## References

Axford S.N. (1979) Angling returns in fisheries biology. *Proceedings of the 1st British Freshwater Fisheries Conference*, University of Liverpool, pp 252–271.

Axford S.N. (1991) Some factors affecting angling catches in Yorkshire rivers: In: I.G. Cowx (ed.) *Catch effort sampling strategies*. Oxford: Fishing News Books, pp 143–153.

Bayley P.B. & Austin D.J. (1989) The fisheries analysis system (FAS). *American Fisheries Symposium* **6**, 199–205.

Cooper M.J. & Wheatley G.A. (1981) An examination of the fish population in the River Trent, Nottingham, using angler catches. *Journal of Fish Biology* **19**, 539–556.

Cowx I.G (1990) Application of creel census data for management of fish stocks in large rivers in the United Kingdom: In: W.L.T van Densen, B Steinmetz & R Hughes (eds) *Management of freshwater fisheries*. Wageningen: Pudoc, pp 526–534.

Cowx I.G. (1991) The use of angler catch data to examine the potential fishery management problems in the lower reaches of the River Trent, England. In: I.G. Cowx (ed.) *Catch effort sampling strategies*. Oxford: Fishing News Books, pp 154–165.

Hickley P. and North E. (1981) An appraisal of anglers' catch composition in the barbel reach of the River Severn. Proceedings of the second British Freshwater Fisheries Conference, University of Liverpool, pp 94–100.

Kell L. (1991) A comparison of methods for coarse fish population estimation: In: I.G. Cowx (ed.) *Catch effort sampling strategies*. Oxford: Fishing News Books, pp 184–201.

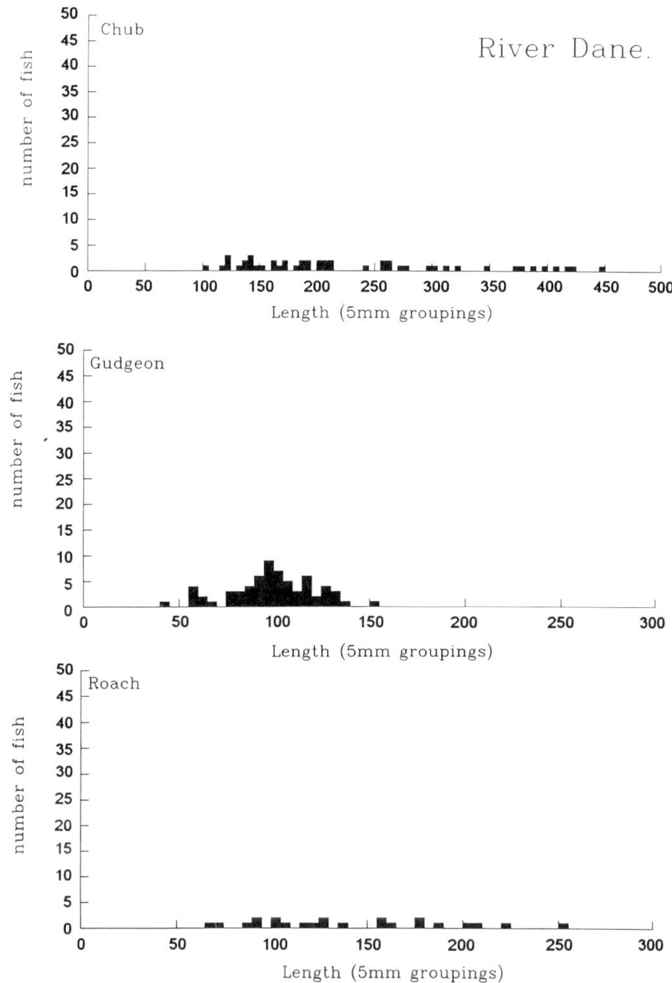


**Fig. 8.7**   (d) Length frequency histogram of angler catches from photographs, River Dane.

Mills C.A. & Mann R.H.K. (1985) Environmentally-induced fluctuations in year class strength and their implications for management. *Journal of Fish Biology* **(Suppl A) 27**, 209–227.

NOP (National Opinion Poll) (1980) National Angling Survey. Summary report. Water Research Centre, Medmenham. 19 pp.

O'Hara K. and Williams T.R. (1991) Analysis of catches from the British National Angling Championships. In: I.G. Cowx (ed.) *Catch effort sampling strategies.* Oxford: Fishing News Books, pp 214–223.

Pygott J.R., O'Hara K., Cragg-Hine D. & Newton C. (1990) A comparison of the sampling efficiency of electric fishing and seine netting on two contrasting canal systems: In: I.G. Cowx (ed.) *Developments in electric fishing.* Oxford: Fishing News Books, pp 130–140.

Santucci Jnr V.J. & Wahl D.H. (1991) Use of a creel census and electrofishing to assess Centrachid populations. *American Fisheries Symposium* **12** pp 481–491.

Swales S. & O'Hara K (1980) Instream habitat improvement devices and their use in freshwater fisheries management. *Journal Environmental Management* **10**, 167–79.

# Chapter 9
# Log books as a mechanism for assessing long-term trends in salmonid fisheries, with particular reference to the sea trout stock of the River Tywi

D.M. EVANS *University of Hull, International Fisheries Institute, Hull HU6 7RX, UK*

**Abstract**   Log books were distributed annually to salmonid anglers fishing the Tywi catchment between 1990 and 1993. Annual return rates ranged between 37% and 51%, and anglers recorded a mean of between 18.6 and 21.8 trips per season. Catch per unit effort (CPUE) demonstrated that the rod fishery performed best in 1992 (0.1397 fish per angler hour) followed by 1993 (0.1149), 1991 (0.1029) and 1990 (0.0853), respectively. Monthly analysis displayed a general increase in CPUE between March and July but a decrease thereafter. The pattern of CPUE, and effort, was better related to month of fishing than river flow; possible reasons for this are discussed.

## 9.1   Introduction

It has long been recognized that total annual catch returns may not reliably indicate trends in fishery performance due to variations in factors such as fishing effort and reporting rates (Churchward & Hickley 1991). Log books are increasingly used to gather detailed information to improve the quality of catch data, both for annual shifts and also for temporal trends within season.

The River Tywi, with an annual declared rod catch of 5194 fish (1987–1991), supports the most prolific recreational sea trout *Salmo trutta* L. fishery in Wales, and England. During this period the declared rod catch from the River Tywi was 26.2% of the total declared Welsh catch, and 15.8% of the total declared catch for England and Wales combined.

## 9.2   Materials and methods

At the beginning of each season (1990–1993) anglers were sent log books and asked to record details of fishing trips in the Tywi catchment. Details requested for each trip are shown in Fig. 9.1; night time was defined as the period during which anglers considered that car headlamps would be required for driving. At the end of each season the anglers were sent pre-paid envelopes in which to return their log book. Relevant details from each completed book were extracted and the log books were returned to the anglers.

110

REMEMBER: ONLY ONE TRIP TO BE RECORDED ON EACH PAGE

| RIVER | | LOCATION | |
|---|---|---|---|

| DATE | PLEASE CIRCLE APPROPRIATE BOXES | | | | | | | | | | |
|---|---|---|---|---|---|---|---|---|---|---|---|
| DATE | 1 | 2 | 3 | 4 | 5 | 6 | 7 | 8 | 9 | 10 | 11 |
| | 12 | 13 | 14 | 15 | 16 | 17 | 18 | 19 | 20 | 30 | 31 |
| | 21 | 22 | 23 | 24 | 25 | 26 | 27 | 28 | 29 | | |
| MONTH | JAN | FEB | MAR | APR | MAY | JUN | JUL | AUG | SEP | OCT | |

| FISHING EFFORT | HOURS |
|---|---|
| HOURS FISHING FOR SALMON ONLY | |
| HOURS FISHING FOR SEA TROUT (SEWIN) ONLY | |
| HOURS FISHING FOR BOTH SPECIES AT THE SAME TIME | |
| HOURS FISHING AT NIGHT | |

| CATCH | | | | | | | | IF NIL TICK BOX ——————> | | | |
|---|---|---|---|---|---|---|---|---|---|---|---|
| FISH No. | SPECIES (TICK BOX) | | WEIGHT | | CAPTURE TIME (TICK) | | METHOD | | | TAG No. | |
| | SALMON | SEWIN | LBS | OZ | DAY | NIGHT | FLY | SPNR | BT | | |
| 1 | | | | | | | | | | | |
| 2 | | | | | | | | | | | |
| 3 | | | | | | | | | | | |
| 4 | | | | | | | | | | | |
| 5 | | | | | | | | | | | |
| 6 | | | | | | | | | | | |
| 7 | | | | | | | | | | | |
| 8 | | | | | | | | | | | |
| 9 | | | | | | | | | | | |
| 10 | | | | | | | | | | | |

COMMENTS . . . . . . . . . . . . . . . . . . . . . . . . . . . . . . . . . . . . . . . . . . . . . . . . . . . . . . . . . . . . . . . . . . . . . . . . . . . .

. . . . . . . . . . . . . . . . . . . . . . . . . . . . . . . . . . . . . . . . . . . . . . . . . . . . . . . . . . . . . . . . . . . . . . . . . . . .

**Fig. 9.1**  An example page from a log book.

The data were classified into relevant sections of the river including the lower river (downstream of Dryslwyn bridge), the middle river (Dryslwyn bridge to Llangadog bridge), the upper river (above Llangadog bridge), the River Gwili and the River Cothi (Fig. 9.2).

Mean daily flows for the four-year period were obtained from the Capel Dewi (Nantgaredig) gauging station.

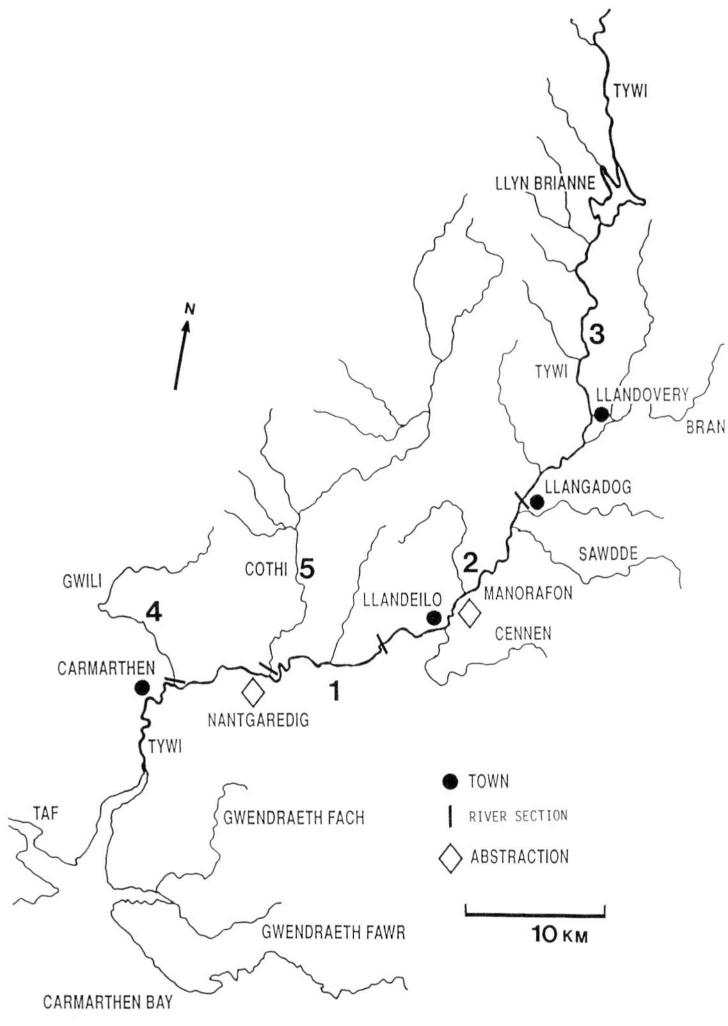


**Fig. 9.2**  The Tywi catchment and the location of the different river sections.

## 9.3    **Results**

### 9.3.1    *Anglers' response*

The return rate of books varied between 37% and 51%, but not all books were properly completed (Table 9.1). Between 18.6 and 21.8 fishing trips were made per season (Table 9.1). The number of trips for sea trout varied between 16.6 and 19.8 per season, with the remaining trips allocated to salmon *Salmo salar* L. fishing. The mean length of trips for sea trout in each year ranged from 3.3 to 3.4 h.

**Table 9.1**    Summary of catch and effort statistics for the Tywi catchment combined, 1990–1993.

| Statistic | 1990 | 1991 | 1992 | 1993 |
|---|---|---|---|---|
| Books out | 297 | 343 | 330 | 325 |
| Books in (usable) | 151 (116) | 155 (105) | 144 (104) | 120 (94) |
| % returned | 51 | 45 | 44 | 37 |
| Total trips | 2156 | 2261 | 2265 | 1769 |
| Total trips per angler | 18.6 | 21.5 | 21.8 | 19.2 |
| *Sea trout* | | | | |
| No. anglers | 115 | 104 | 102 | 90 |
| Total trips | 1907 | 2054 | 1991 | 1592 |
| Total hours | 6399.75 | 6763.5 | 6735.25 | 5188.5 |
| Total hours at night | 1580.5 | 1739.75 | 1559.75 | 1027 |
| Mean length of trip | 3.36 | 3.29 | 3.38 | 3.26 |
| Trips per angler | 16.6 | 19.8 | 19.5 | 17.7 |
| Catch | 546 | 696 | 941 | 596 |
| Catch per angler | 4.75 | 6.69 | 9.23 | 6.62 |
| Catch per trip | 0.286 | 0.339 | 0.473 | 0.374 |
| Catch per hour | 0.0853 | 0.1029 | 0.1397 | 0.1149 |
| Average weight (kg) | 1.23 | 1.04 | 1.09 | 1.39 |

### 9.3.2    *Catch and effort for sea trout*

The overall seasonal catch of sea trout per hour from the Tywi catchment was 0.0853, 0.1029, 0.1397 and 0.1149 in 1990, 1991, 1992 and 1993 respectively (Table 9.1; Fig. 9.3). This implies that the sea trout fishery performed best in 1992, followed by 1993, 1991 and 1990 respectively.

The monthly catch per hour for each year (Fig. 9.4) displayed a steady increase in CPUE from April onwards, with a peak in July. In most years there was then a decline in CPUE in the last 3 months of the season. The monthly pattern of CPUE for sea trout was again reflected by a similar pattern of effort for sea trout (Fig. 9.5).

The daily CPUE analysis for each year (Figs 9.6a-9.9a) demonstrated the early timing of the sea trout run, with fish captured in March and April. A steady increase in CPUE occurred between March and a peak in July; then CPUE declined until the end of the season. This pattern was modified in 1991 and 1992 by secondary peaks in

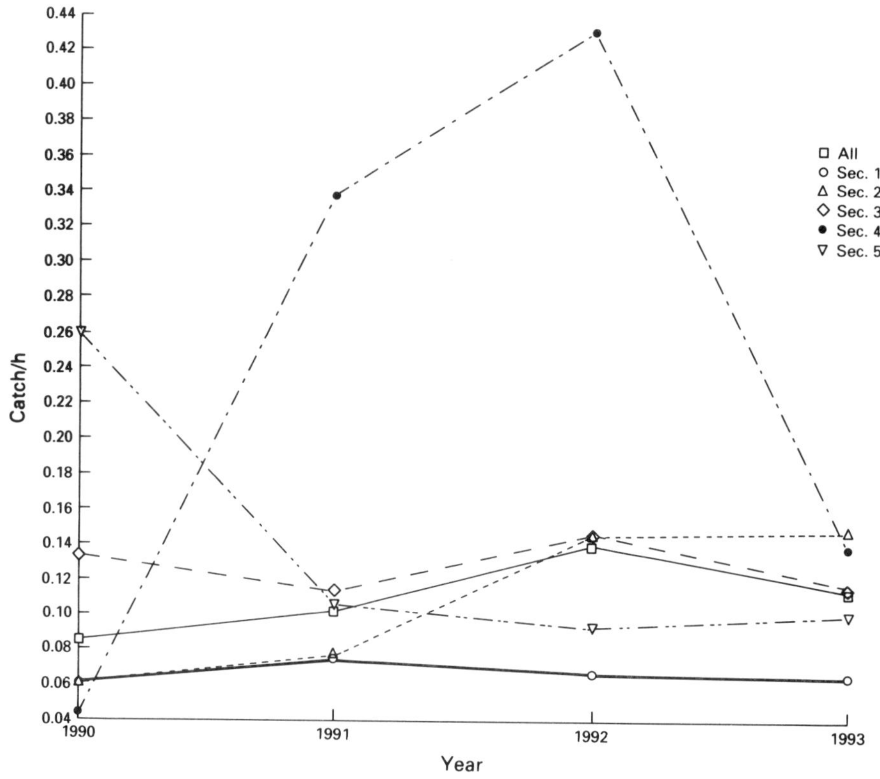



**Fig. 9.3**  Annual sea trout CPUE from the River Tywi, indicating the differences between river sections.

September and October, respectively (Figs 9.7a & 9.8a). Daily CPUE did not coincide with peaks in flow; the highest CPUE occurred during low flows. Daily effort for sea trout demonstrated a similar pattern to CPUE with effort typically increasing from March to July followed by a decrease until the end of the season. However, in both 1990 and 1991 increases in effort occurred in October. Freshets were observed to have some influence upon patterns of effort, but the high levels of effort during low flows in July/August demonstrated that time of the year was the over-riding influence upon effort.

There was a marked difference in the spatial CPUE within the catchment (Fig. 9.3). The main river displayed the highest CPUE in the upper river, followed by the middle and lower river sections, respectively. Whilst the upper and lower river sections showed relative stability over the four years, the middle river displayed a marked increase from 1990 to 1993. The Cothi displayed a decreasing trend over the four years with a CPUE generally less than the middle river section. The Gwili displayed a low CPUE in 1990 and 1993 but large increases in CPUE in 1991 and 1992.

A large proportion of sea trout fishing was undertaken at night and in each year the CPUE at night was higher than during daylight (Table 9.2). Night-time CPUE was typically two to three times higher than day-time CPUE.

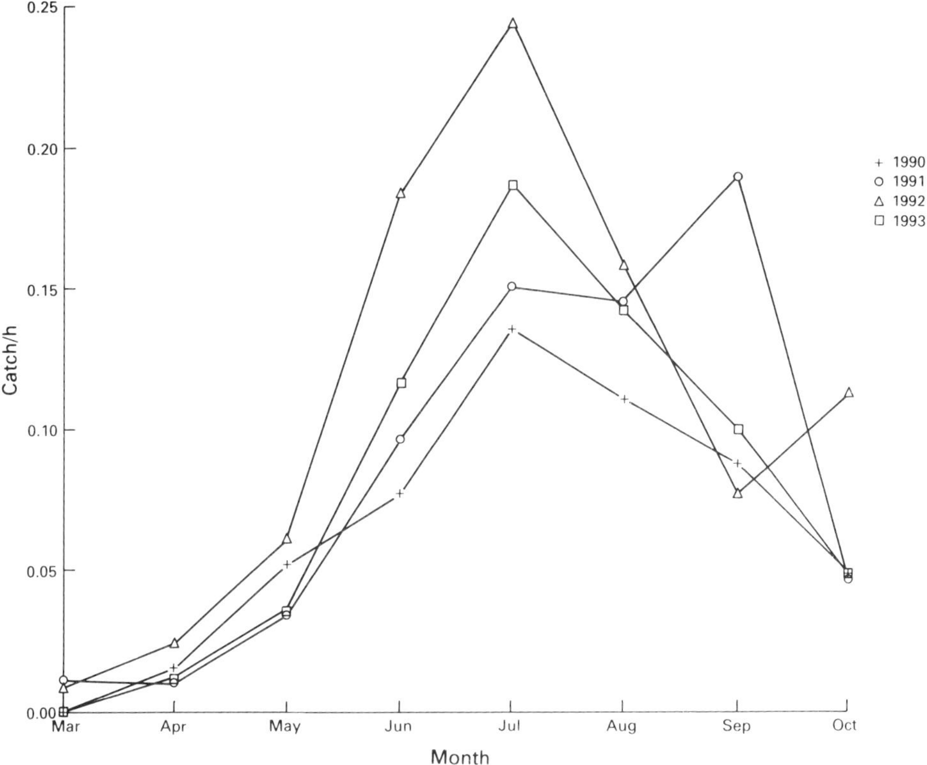



**Fig. 9.4**   Monthly CPUE for the Tywi catchment for 1990–1993.

### 9.3.3    *Sea trout CPUE in terms of weight*

The CPUE given above is in terms of the number of sea trout caught per hour. However, there was a marked difference in the mean weight of sea trout caught between 1990 and 1993 (Table 9.1), with lower mean weights in 1991 and 1992. When annual fishery performance (CPUE) was measured in terms of weight of sea trout caught per hour (Fig. 9.10), the yield increased steadily from 1990 to 1993.

### 9.3.4    *Variation in catch between anglers*

In all years most anglers were unsuccessful in catching sea trout. The overall mean CPUE in each season (from the distribution of the CPUE of individual anglers) varied from 0.1063 to 0.1286 (Table 9.3).

### 9.3.5    *Weight distribution of sea trout*

Over the four-year period weights were obtained from 2959 sea trout. The smallest weight recorded was 6 oz (170 g) and the largest fish weighed 15 lb (6.81 kg). The

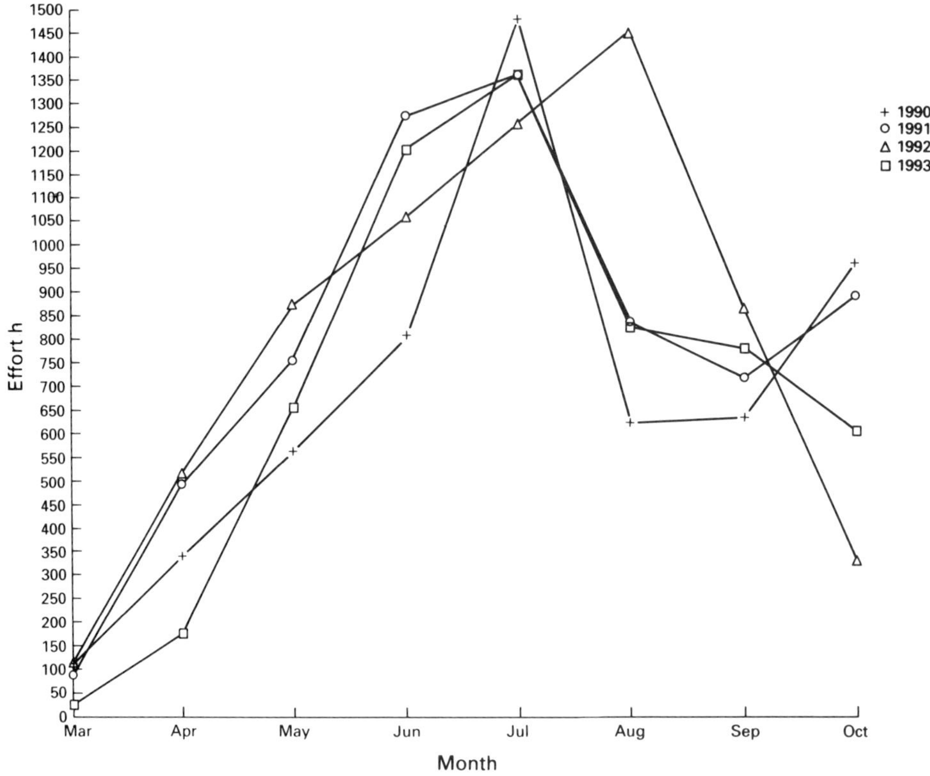


**Fig. 9.5**   Monthly effort (h) per 100 anglers for the Tywi catchment 1990–1993.

overall weight distribution (Fig. 9.11) demonstrated the dominance (43.8% of catch) of small fish ($\leqslant$ 1.5 lb, 0.7 kg) (Table 9.4). These smaller fish correspond to the whitling or . + sea age group of sea trout (Mee pers. comm.; Evans unpublished). The relative abundance of these fish varied between years with lower relative abundance in 1993, reflected in the high average weight for the year (Table 9.1).

Appreciable numbers of fish in the 1.5–4 lb, (0.7–2 kg) class were also recorded, representing 35.5% of the overall sample. This weight class corresponds primarily to the .1 + or one winter post-smolting age class (Evans unpublished). From the overall sample 20.7% of sea trout were observed to be > 4 lb (2 kg) in weight.

## 9.4    Discussion

### 9.4.1    *Catch and effort*

The return rate of log books was similar to or higher than that observed in other studies (e.g. Wightman 1987; Aprahamian 1993). Participation in the study was voluntary and it is assumed that the chance of returning a book was not influenced by the

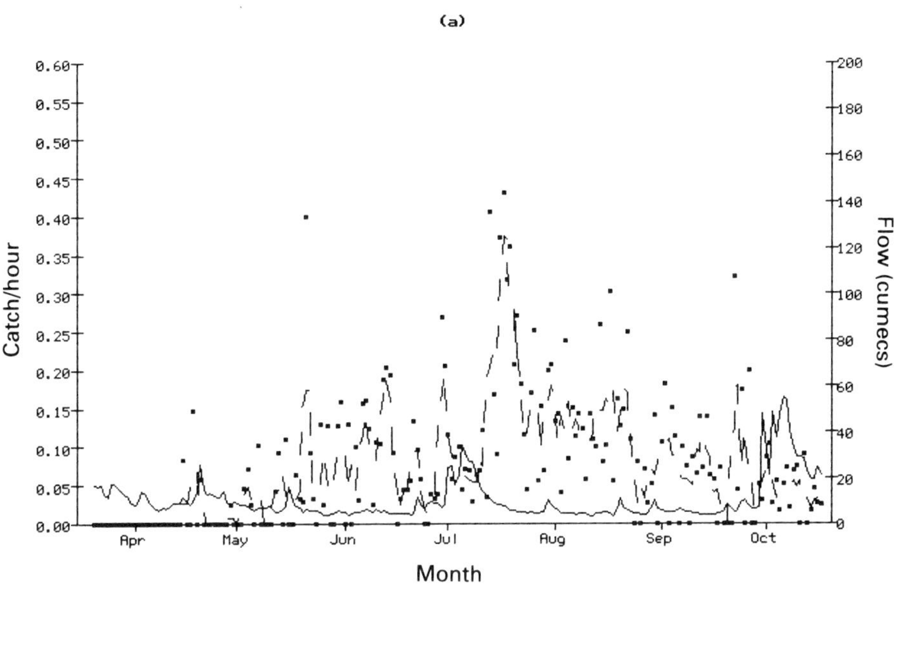

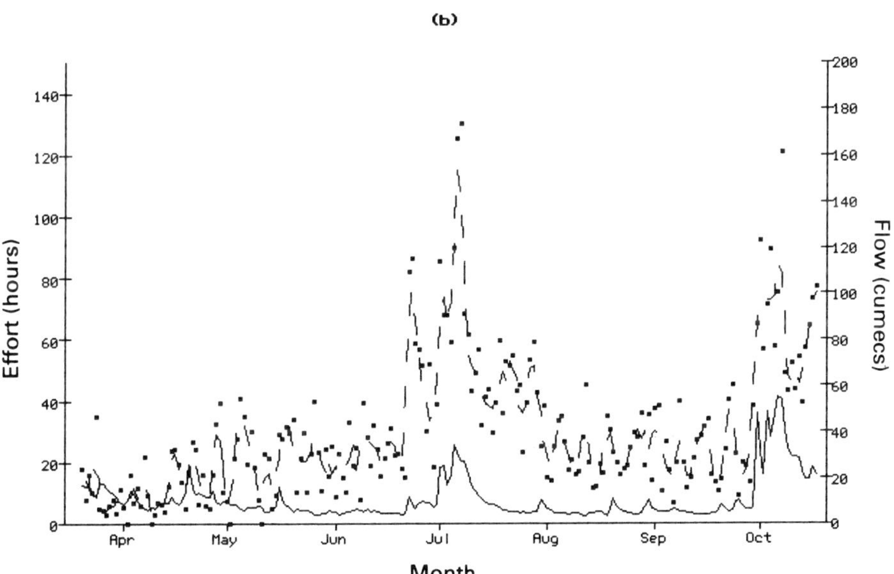

**Fig. 9.6** (a) Daily CPUE for the Tywi catchment in 1990 (dot) with a three-point moving average (dash) and the corresponding mean daily flow (solid, Y2 axis); (b) Daily effort (h) for the Tywi catchment in 1990 (dot) with a three-point moving average (dash) and the corresponding mean daily flow (solid, Y2 axis).

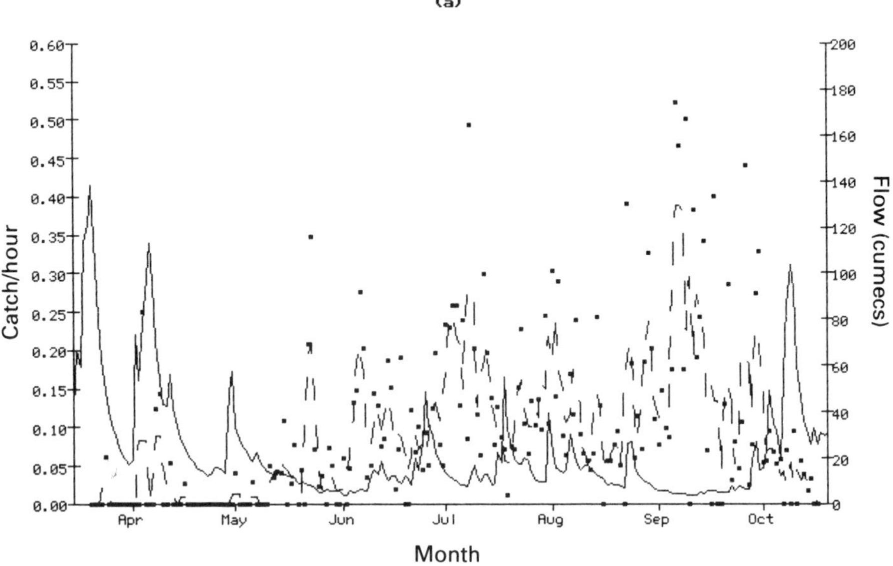

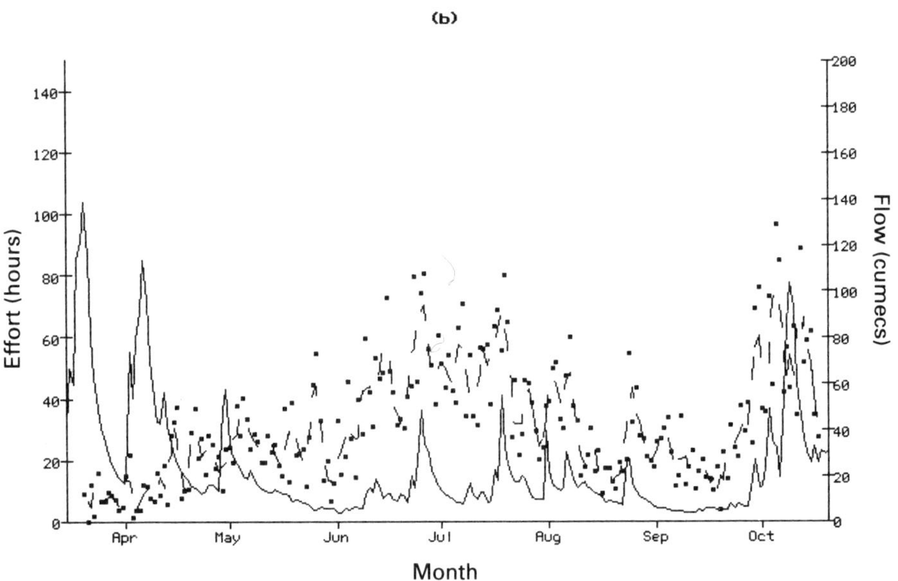

**Fig. 9.7** (a) Daily CPUE for the Tywi catchment in 1991 (dot) with a three-point moving average (dash) and the corresponding mean daily flow (solid, Y2 axis); (b) Daily effort (h) for the Tywi catchment in 1991 (dot) with a three-point moving average (dash) and the corresponding mean daily flow (solid, Y2 axis).

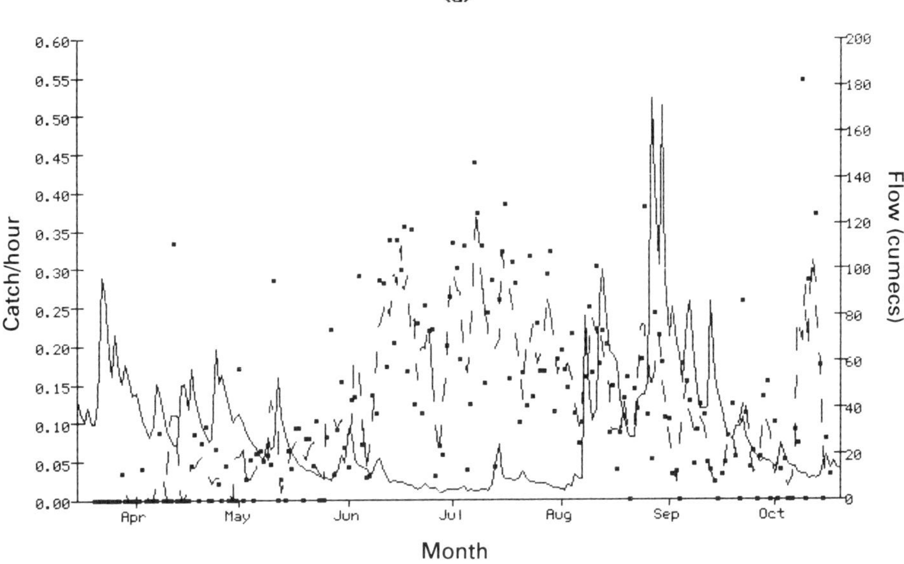


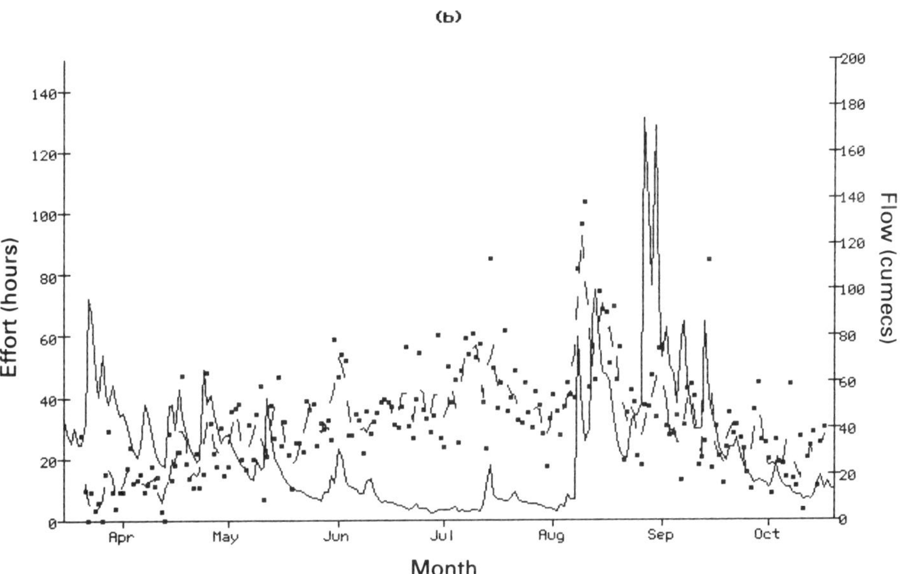


**Fig. 9.8** (a) Daily CPUE for the Tywi catchment in 1992 (dot) with a three-point moving average (dash) and the corresponding mean daily flow (solid, Y2 axis); (b) Daily effort (h) for the Tywi catchment in 1992 (dot) with a three-point moving average (dash) and the corresponding mean daily flow (solid, Y2 axis).

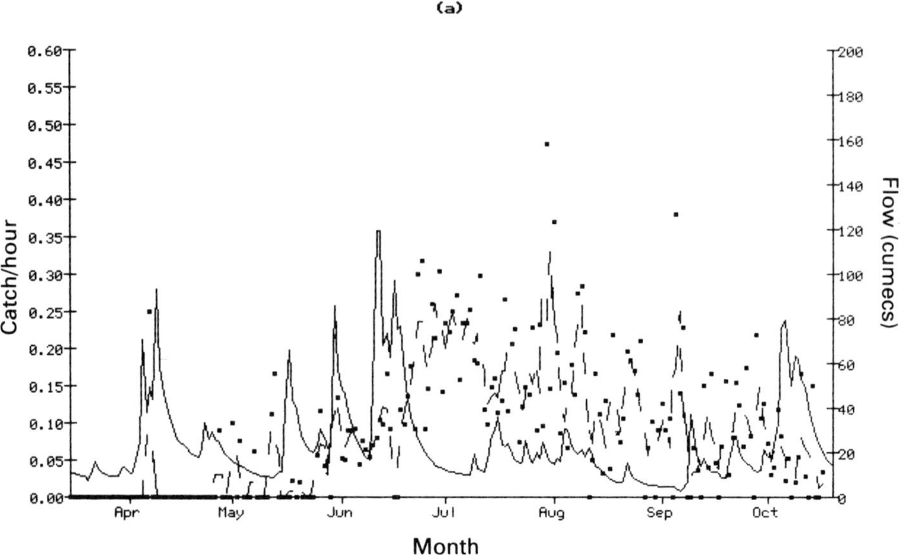

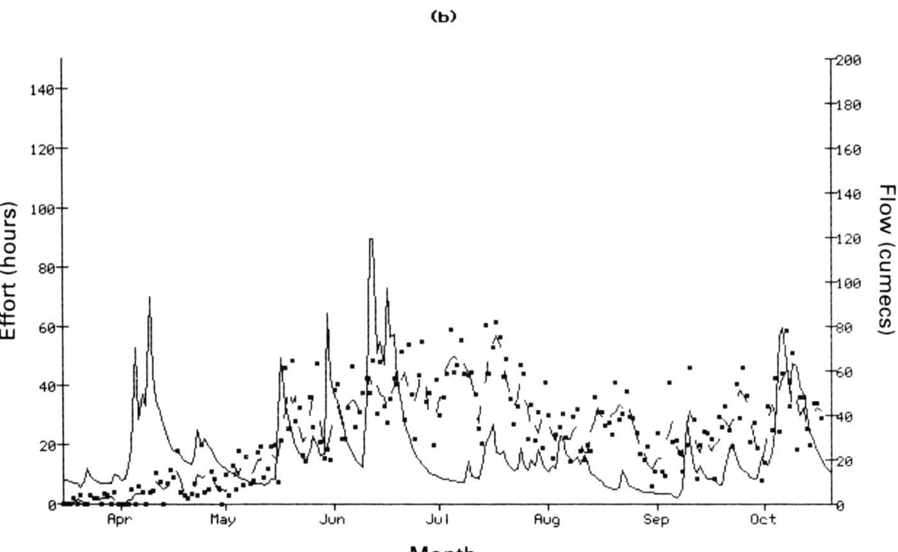

**Fig. 9.9**   (a) Daily CPUE for the Tywi catchment in 1993 (dot) with a three-point moving average (dash) and the corresponding mean daily flow (solid, Y2 axis); (b) Daily effort (h) for the Tywi catchment in 1993 (dot) with a three-point moving average (dash) and the corresponding mean daily flow (solid, Y2 axis).

**Table 9.2**  The overall season catch per hour (CPUE) for the Tywi catchment split between day and night fishing.

| Year | Overall CPUE | Day CPUE | Night CPUE |
| --- | --- | --- | --- |
| 1990 | 0.0853 | 0.0542 | 0.1803 |
| 1991 | 0.1029 | 0.0802 | 0.1684 |
| 1992 | 0.1397 | 0.1018 | 0.2654 |
| 1993 | 0.1149 | 0.0906 | 0.2132 |

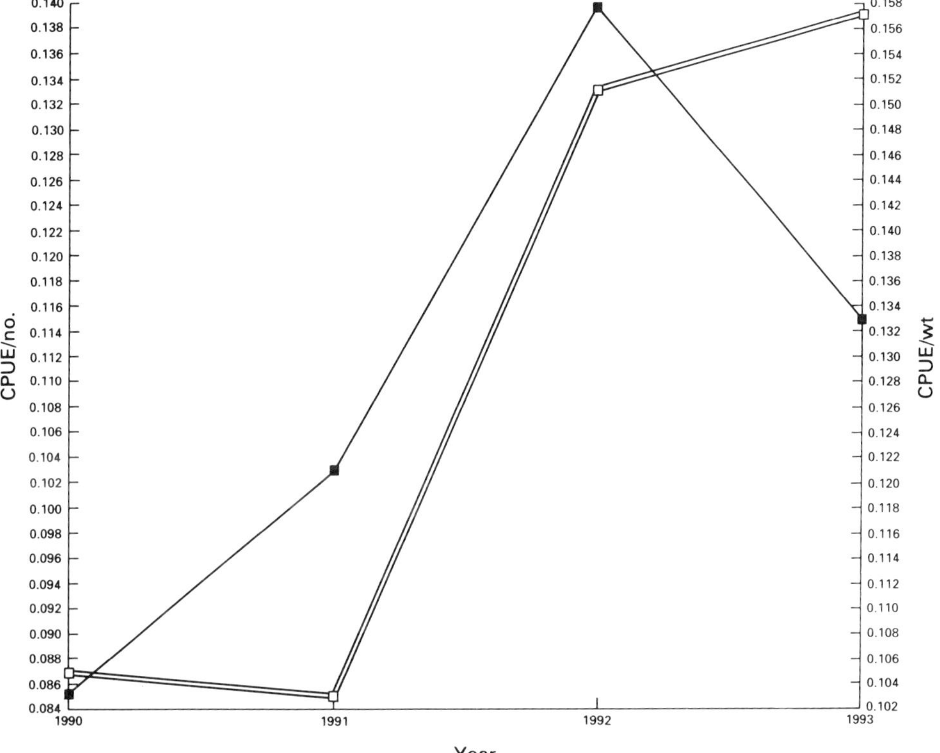

**Fig. 9.10**  Tywi catchment catch effort data 1990–1993, CPUE in terms of number of fish per hour (single) and weight of fish per hour (double, Y2 axis).

**Table 9.3**  The mean CPUE from individual anglers for sea trout in 1990–1993. Poisson 95% CI as mean ±1.96 (SQR (mean/no. observations)).

| Year | Mean | 95% CI |
| --- | --- | --- |
| 1990 | 0.1063 | 0.0596 |
| 1991 | 0.1111 | 0.0641 |
| 1992 | 0.1286 | 0.0696 |
| 1993 | 0.1093 | 0.0683 |

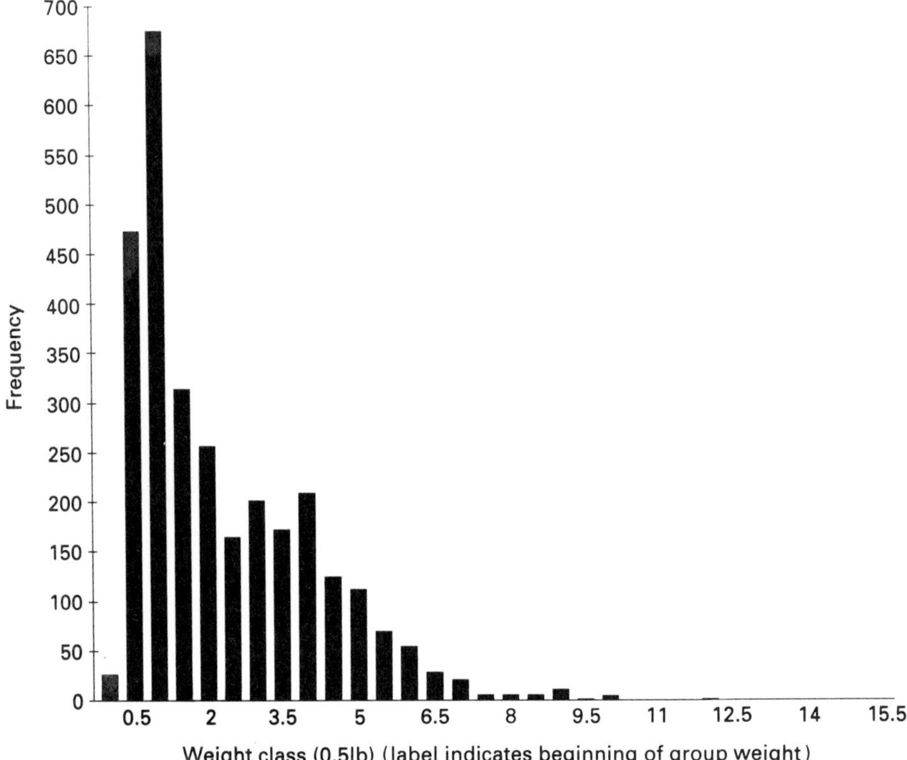



**Fig. 9.11**  Weight distribution of sea trout from anglers' catches 1990–1993 (*n* = 2959).

**Table 9.4**  Weight distribution (%) of sea trout in the classes ≤ 1.5 lb, 1.5 lb–4 lb and > 4 lb. (Overall *n* = 2959).

| Class | 1990 | 1991 | 1992 | 1993 | Overall |
| --- | --- | --- | --- | --- | --- |
| ≤ 1.5 lb | 43.6 | 51.5 | 45.4 | 33.6 | 43.8 |
| 1.5 lb–4 lb | 30.1 | 34.5 | 37.5 | 37.9 | 35.5 |
| > 4 lb | 26.3 | 14.0 | 17.1 | 28.5 | 20.7 |

anglers' success rate. The high frequency of anglers returning books with no recorded catch suggests that this may the case, and the highly negatively skewed distribution of individual anglers' CPUE is similar to that described elsewhere (Small 1991).

Sea trout CPUE will indicate trends in fishery performance, but this will only be representative of the part of the stock entering within the fishing season. On the Tywi the majority of sea trout enter during the angling season (D. Mee pers. comm.) and this implies that angling CPUE may be a reliable indication of the stock.

Although low flows have been demonstrated to reduce the success of entry into rivers by larger sea trout (D. Clarke pers. comm.), many of the smaller fish (whitling) entered in July–August when flows were lowest. This confused any relationship between flow and availability. Sea trout also remain available to capture throughout the season, independent of flow (D. Mee pers. comm.), further confusing relationships between flow and availability.

The month of the year was found to have a greater influence upon fishing effort and CPUE than river flow; both effort and CPUE generally increased from March to July and then decreased to October. The CPUE data suggest that flow may have a negative relationship with CPUE since CPUE was typically highest during the lower flows of July. Here angling method further confused the issue of sea trout capture and flow. The principle method used during low flows is fly fishing at night, and night time CPUE was demonstrated to be two or three times higher than day time CPUE. This implies that sea trout are either more catchable at night or that the more successful anglers fish at night.

The relationship between catch, effort and abundance is given by:

$$Catch = Effort \times Catchability \times Abundance \qquad \textbf{(1)} \text{ (Paloheimo \& Dickie 1964)}$$

$$CPUE = Catchability \times Abundance \qquad \textbf{(2)}$$

The relationship between CPUE and abundance is therefore influenced by catchability. If catchability is constant between years, or follows the same temporal pattern between years, then CPUE is linearly related to abundance. However, other studies indicate that this may not be the case and that catchability increases with decreasing stock levels (Mills *et al.* 1986). Small (1991) suggested a proportional relationship between CPUE and abundance that lays between the cube and square roots of the number of fish entering a river. This implies that to double the CPUE would require a four to eight-fold increase in the sea trout run. A better understanding of the relationship between angling CPUE and fish abundance will be available through studies where abundance can be assessed through trap or counter systems (Aprahamian, Nicholson, McCubbing & Davidson, Chapter 3; I. Davidson pers. comm.)

Anglers in this study used all available legal fishing methods: fly, spinning and bait. As well as the variation in anglers' skill there may be a difference in the efficiency of the different gears *per se*, and this will affect catchability and therefore the catch for each unit of effort. Whilst the effort recorded in this study was not separated between gears, changes to future recording methods could enable this. If differences do occur between methods then this could be used to assess any changes in future patterns of fishing methods or any regulatory restriction such as that currently used for spring salmon (Mawle 1992).

Log books allow a measure of fishery performance accounting for the variation in effort not afforded by total catch figures. However, this principle assumes that each unit of effort is independent and it does not consider the possible 'congestion externalities' that may occur. Some fisheries are heavily fished and it is therefore possible that gear saturation occurs and anglers 'interfere' with each other, reducing the probability of capture for the next angler through disturbance of the water. This would reduce CPUE within a given area with increasing angling effort; thus with the same abundance of fish in a given area, reducing the angling effort (e.g. private/ syndicate waters) may increase CPUE. This has implications for annual CPUE data if major changes occur in the pattern of usage over time. A combined index of CPUE and the total number of anglers fishing a river may therefore indicate long-term trends better than CPUE alone, though assessing the total number of anglers may be difficult.

A further factor that also needs to be considered is the constancy of availability of fish within the rod fishery throughout the season. This is particularly important when assessing the proportion of fish caught or catchability from abundance estimates using traps or counters. Often it is assumed that once fish have entered the river they will remain within the fishery, but in latter months, especially for sea trout, fish may migrate into larger tributaries where rod fisheries do not occur (Evans pers. obs.). This effectively decreases the abundance of sea trout within the rod fishery.

The number of different sea ages at return is a peculiar facet of sea trout biology. It has been stated that rods may not represent the proportion of large fish present (LeCren 1985), and therefore a knowledge of whether different catchabilities exist between different sea-age fish is needed. This will aid the interpretation of the changes in annual CPUE with the relative differences of these sea ages within the stock.

The difference in spatial CPUE within the catchment has important implications for constancy in annual surveys. For future surveys the same proportion of anglers from each section should be used since bias towards more anglers from one or other of the sections may bias the annual catchment CPUE. Spatial catch and effort information can provide patterns of usage throughout the catchment although factors such as flow may influence spatial availability.

CPUE is measured to assess changes in stock abundance and therefore infer changes in spawning stock and deposition rates. Due to the different traits in age at return/maturation of sea trout, an index of abundance in terms of numbers may not reflect changes in the spawning stock. Since egg deposition rates rely upon the size of fish, CPUE in terms of weight assesses changes in spawning stock more effectively. This is particularly important when sex ratios are considered since whitling from the Tywi have a female:male ratio close to unity and non-whitling have a sex ratio of 2.4:1 in favour of females (Evans unpublished).


**Acknowledgements**

Many thanks are extended to all the anglers who participated in the survey and to Dr D.R.K. Clarke for commenting on the manuscript. This work was carried out as part of a project co-sponsored by NERC, Atlantic Salmon Trust, Fishmongers Company and the NRA (Welsh Region).

## References


Aprahamian M.W. (1993) An analysis of migratory salmonid catch effort data, derived from anglers' log books, 1991. Warrington: NRA Report No. NRA/NW/FTR/93/3, 86 pp.

Churchward A.S. & Hickley P (1991) The Atlantic salmon fishery of the River Severn (UK). In: I.G. Cowx (ed.) *Catch effort sampling strategies.* Oxford: Fishing News Books, pp 1–15.

LeCren E.D. (1985) The biology of the sea trout. Summary of a symposium held at Plas Menai. Atlantic Salmon Trust, Pitlochry, 42 pp.

Mawle G.W. (1992) Usk salmon: Recommendations for action. Cardiff: NRA Technical Fisheries Report No.1, pp 69.

Mills C.P.R., Mahon G.A.T. & Piggins D.J. (1986) Influence of stock levels, fishing effort and environmental factors on anglers' catches of Atlantic salmon, *Salmo salar* L., and sea trout, *Salmo trutta* L. *Aquaculture and Fisheries Management* **17**, 289–297.

Paloheimo J.E. & Dickie L.M. (1964) Abundance and fishing success. *Rapp. R-V. Reun. Cons. Int. Explor. Mer* **155**, 152–163.

Small I. (1991) Exploring data provided by angling for salmonids in the British Isles. In: I.G. Cowx (ed.) *Catch effort sampling strategies.* Oxford: Fishing News Books, pp 81–91.

Wightman R.P.W. (1987) The results of an angler log-book survey on the River Tawe, 1986 season. Welsh Water Authority, Llanelli. Report No. SW/87/14.

# Chapter 10
# Estimating historical trends of migratory fish runs into rivers without counters

IAN SMALL *Dept of Environmental and Evolutionary Biology, University of Liverpool*

**Abstract** The patterns of catchability, rod and in-river net catches, obtained from annual records from migratory fisheries in rivers where traps or counters have been installed, were modelled as simple linear trends for the period 1970–1990. Rod effort, where not recorded, can be modelled as simple linear or curvilinear patterns, and the trend of catch per unit effort obtained. The estimated trends of annual total runs into estuaries of rivers without counters are calculated from: *(Rod CPUE/catchability model) + in-river net catch.*

KEYWORDS: Atlantic salmon, catch, effort, *Salmo salar*

## 10.1   Introduction

Throughout the United Kingdom and Ireland there are more than 300 major river basins up which salmonids migrate to reach their spawning grounds. However, traps or counters have been installed to monitor the numbers of fish migrating on only 30 of these rivers. Prior to 1950 it was possible to assess the changes in runs surviving the net fisheries, by recording the catches by rod fisheries (Grimble 1913), as the rate of change in rod effort was very slow. However, after 1950 the rod effort increased rapidly on most large waters accessible to the public (Gee & Milner 1980; Small 1991a) and it became necessary to take this into account and consider catch per unit effort (*CPUE*) as an indicator of change in stock levels (Small 1991b). This is confirmed by the data considered in this paper where, on nearly 40% of the waters, the annual rod catch records do not correlate with the runs of fish.

As so few rivers are being monitored by fishery-independent methods it would be convenient for fishery managers if a method could be devised to determine the trends of runs of fish in the remaining waters, from available records. This chapter indicates some methods to achieve this objective based on UK salmon, *salmo salar* L., and sea trout, *Salmo trutta* L., rivers.

## 10.2   Annual data methods

A search of published papers and unpublished reports showed that the probable origin of a suitable model to relate catch (*C*) with effort (*f*), catchability (*q*) and abundance of fish ($N$), lies in the hypothesis of Paloheimo and Dickie (1964) – 'That for most fish stocks, $q$ varies inversely with stock abundance and geographical area occupied by the stock'. This results in the *CPUE*, $C/f$, being curvilinearly related to stock abundance, which is difficult to check without fishery-independent data on abundance.

The model is usually formulated (critically reviewed by Crecco & Overholtz 1990, also Bannerot & Austin 1983) as:

$$C/f = qN$$

where $q = a N^b$, giving

$$C/f = aN^{(b+1)} \tag{1}$$

### 10.2.1   *Counter data*

Suitable equipment for counting salmon + grilse has been installed at various hydro-electric schemes in Ireland and Scotland since the early 1950s. These were followed by counters or traps on selected non-hydro rivers elsewhere since about 1970. Records from traps can usually be taken as close to reality but those from electronic devices are better treated as indices of changes in abundance, due to the difficulties with calibration (Holden 1987) and differential accuracy in respect of fish length (Nicholson & Aprahamian 1993).

This chapter is based on the published records in Annual Reports and papers, supplemented in many cases by information made available by official bodies and the authors, from the following fisheries where reasonably complete data exist for effort (rod and net), catch and abundance: The Burrishoole (Salmon Research Agency of Ireland 1970–91); Foyle (Foyle Fisheries Commission 1952–92; Elson & Tuomi 1975); Bush (Dept. of Agriculture, Northern Ireland 1970–91); Conon (The Scottish Office Agriculture and Fisheries Dept. & North of Scotland Hydro-Electric Board 1952–84); Tummel (Pitlochry Angling Club 1975–92); Welsh Dee & Usk (NRA Welsh Region 1976–91); Severn (NRA Severn & Trent Region 1977–91; Churchward & Hickley 1991); Frome (Beaumont *et al.* 1991); Ellidaar (Iceland) (Mundy *et al.* 1978).

In addition there are datasets, particularly from Ireland, Scotland and in NW England, where count and catch data are available but trends of effort must be assessed by other means (Table 10.1).

### 10.2.2   *The relation of in-season* CPUE *to annual runs including the close season*

It has been demonstrated that Equation (1) is a satisfactory model for use with data obtained from many waters in Britain and Ireland, (Small 1991a; Beaumont *et al.* 1991). Investigations to date indicate that for migratory fish, the exponent (b + 1) for the relation of in-season CPUE to *runs*, including those when the rods and nets are not in operation, has the range of values shown in Table 10.2.

For the waters considered in this chapter the proportion of the total annual run

**Table 10.1**  Migratory fish catchability: Linear rates of change with time, 1970–90. Annual data: $q = c +/- d$ (Year: 1980) Benchmark year: 1980.

| Water | Effort unit | Period | Rod | | | Net | | | Effort |
| --- | --- | --- | --- | --- | --- | --- | --- | --- | --- |
| | | | $c$ | $d$ | Ratio $d/c$ | $c$ | $d$ | Ratio $d/c$ | |
| Foyle | Seas.lic. | 1970–92 | 9.660e-5 | 2.626e-6 | 0.0283 | 3.903e-3 | 9.423e-5 | 0.0246 | Drift net lic. |
| Bush | Rod.day | 1973–90 | 1.291e-4 | −3.927e-6 | −0.0304 | – | – | – | – |
| Furnace | Rod.day | 1979–91 | 1.630e-3 | −8.180e-5 | −0.050 | – | – | – | – |
| Feeagh | Rod.day | 1979–91 | 2.790e-3 | −1.780e-4 | −0.064 | – | – | – | – |
| Conon | Rod.day | 1970–90 | 1.231e-4 | 3.897e-6 | 0.0317 | – | – | – | – |
| | | 1961–72 | 1.200e-4 | −9.067e-6 | 0.078 | – | – | – | – |
| Tummel | Rod.day | 1975–92 | 4.262e-5 | 1.547e-6 | 0.036 | – | – | – | – |
| W.Dee | Seas.lic. | 1975–90 | 1.115e-4 | −5.718e-6 | −0.051 | 9.602e-3 | −4.090e-4 | −0.051 | Net lic. |
| Usk | All.lic. | 1975–91 | 2.020e-4 | −7.082e-6 | −0.0351 | – | – | – | – |
| Severn | Seas.lic. | 1979–90 | 2.287e-4 | −4.895e-9 | 0.0001 | 1.643e-3 | 3.652e-4 | 0.222 | Putcher |
| | | 1970–83 | – | – | – | 1.890e-3 | −8.798e-5 | −0.0359 | Putcher |
| Frome | Rod.H | 1973–89 | 2.645e-5 | −9.245e-7 | −0.035 | – | – | – | – |
| Ellidaar | Rod.H | 1950–77 | 5.808e-5 | −1.777e-6 | −0.031 | – | – | – | – |
| *Salmon: Annual effort modelled* | | | | | | | | | |
| Faughan | Seas.pmt | 1967–87 | 2.190e-4 | 6.771e-6 | 0.031 | – | – | – | – |
| Beauly | Rod.day | 1980–91 | 3.541e-5 | −6.299e-7 | −0.0197 | – | – | – | – |
| N.Esk | Rod.day | 1981–91 | 1.963e-5 | 5.018e-7 | 0.0255 | 0.0248 | 4.936e-4 | 0.0199 | Net.lic. |
| Tay | – | – | – | – | – | 3.928e-3 | −5.526e-5 | −0.015 | Crew.mth |
| Coquet | Seas.lic. | 1964–83 | 2.120e-4 | 1.577e-5 | 0.0360 | – | – | – | – |
| Lune | Rod.day | 1975–86 | 7.720e-5 | −3.830e-6 | −0.0375 | 1.558e-3 | −5.160e-4 | −0.331 | Net.lic. (1974–83) |
| | | 1981–92 | – | – | – | 1.596e-3 | 5.080e-4 | 0.318 | Net.lic. |
| Ribble | Rod.day | 1974–83 | 3.620e-4 | 3.878e-6 | 0.011 | 3.331e-2 | −1.188e-3 | −0.036 | Net.lic. |
| Mean & SD | | | | | 0.035 | 0.118 | | 0.133 | 0.134 |
| *Sea trout: Full annual data* | | | | | | | | | |
| Furnace | Rod.day | 1979–91 | 4.240e-3 | −2.140e-4 | 0.050 | – | – | – | – |
| Feeagh | Rod.day | 1979–91 | 2.350e-3 | −1.060e-4 | 0.045 | – | – | – | – |
| *Sea trout: Annual effort modelled* | | | | | | | | | |
| Coquet | Seas.lic. | 1968–83 | 5.448e-4 | −2.101e-7 | −0.039 | – | – | – | – |
| Lune | – | 1981–91 | – | – | – | 1.596e-3 | 5.080e-4 | 0.318 | Net.lic. |
| Ribble | Seas.lic. | 1974–81 | 1.530e-4 | 2.287e-5 | 0.019 | 2.067e-2 | 2.902e-3 | 0.140 | Net.lic. |
| Hodder | Seas.lic. | – | 6.880e-4 | −4.455e-5 | 0.0038 | – | – | – | – |
| Mean & SD | | | | | 0.031 | 0.020 | | | |

**Table 10.2**  Relation of in-season *CPUE* to annual runs including the close season.

| Species | Datasets | Run | Mean ($b + 1$) | 90% CL |
| --- | --- | --- | --- | --- |
| Salmon rod | 23 | Above nets | 0.51 | 0.22–0.80 |
| Salmon nets | 12 | Below nets | 0.68 | 0.40–0.96 |
| Sea trout rod | 4 | Above nets | 0.55 | 0.40–0.70 |

taking place in the rod close season, in non-drought years, lies in the range 0.002 to 0.10. In smaller rivers in drought years the proportion may rise to about 0.25, but inclusion of these years in the datasets still produces significant correlations between in-season *CPUE* and *Total run*$^{(b+1)}$.

River flow is a major determinant in the rank order of magnitude of the migration of salmonids. i.e. the bigger the river the greater the likelihood of large runs of fish. Between-river differences in annual flow seem to correlate with angling effort and therefore catches, (Bunt 1991). However, there is no evidence to suggest that annual mean daily flow or *log$_e$ flow*, when used in conjunction with *CPUE* or *log$_e$ CPUE*, are significant predictors of runs of fish up the Foyle, Bush, Conon and Tay. Therefore flow is not considered further in estimating the trends of annual runs.

**10.2.3**     *Estimating trends of annual runs into estuaries from incomplete data*

The model, (Equation (1)) can be restated as:

$$Catch = (effort\ model) \times (catchability\ model) \times (run\ model) \tag{2}$$

and

$$Run = (catch/effort\ model)/(catchability\ model) \tag{3}$$

Before Equation (3) can be put into effect, it may be necessary to make adjustments to the recorded data, and to obtain a model for catchability, the unknown quantity.

**Rod effort models**     Where catch and count are recorded the trend of effort can be estimated by the best fit method described in Gardiner (1991). If only catch data are available it will be necessary to assess the likely trend by comparison with that from a similar water. For example, on large waters reasonably accessible to all anglers (Wye, Severn, Dee, Foyle, Tyne & Tay), rod effort increased approximately linearly from 1950 to 1980 by a factor in the range 3.4 to 5.3. Over the period 1979–1991 rod hours at the Burrishoole fishery increased by a factor of 3.5. On rivers with restricted access, the factors are: Hampshire rivers 3.0; Usk 2.0; Coquet 3.7; Beauly 2.6; Conon 1.4; Thurso 1.2; Ellidaar 1.8. From 1980 to 1985 rod effort on several large rivers, the Foyle, Conon, Tay and Welsh Dee, tended to become static and then resume its upward trend.

Local beat and association records should give indications of a suitable model. If these are in the form of truncated counts of catches, i.e. no records of blank days, then analyses by the method described in Small and Downham (1985) may indicate major changes in the pattern of effort. Benchmark values of effort in particular years can be derived from known local records of *CPUE* on a typical section of the river.

**Adjusting rod catch records**     Recorded catches seldom include 100% of the catch, but it is possible in areas where rod licences or permits are issued, given a reliable record of the proportion of anglers making a return, to adjust the recorded catch by the method described in Small (1991a). Limited trials have shown that the adjustment method can be applied to net catches and other species of fish.

**Counts**    Where possible the records should be examined month by month to find periods when the trap or counter was out of action. Reasonable mean or interpolated values might be inserted or that year's value omitted.

In this study where the trap or counter is not located at or close to the tidehead, fishery officers were asked to assess the proportion of the total annual rod catch obtained below the counter, also where spawners are active below a counter, to assess an approximate number in a benchmark year. Numbers in other years were estimated by a model derived from Fahy and Fluskey (1987):

$$Spawners = S \times (count\ or\ rod\ catch)^{0.5} \tag{4}$$

The estimated catch and numbers of fish were then added to the count.

When estimating trends from catch data alone, benchmark values for the run can be estimated from exploitation rates obtained from mark and recapture surveys, e.g. Laughton (1991) or suitable mean values obtained from other sources (Solomon and Potter 1992).

### 10.2.4    *Catchability (q)*

For this chapter catchability is defined as *CPUE/run*, i.e. the proportion of the annual run taken by one unit of effort. This generally increases with decreasing runs of fish (Small 1991a) but may rise or fall with changes in fishing method, gear or skill.

The current series of analyses of rod and net fisheries with counters, produce simple models for catchability which show near linear trends in the period 1970–1990 with slopes ranging $+/- (0.10 \times mean\ catchability)$ per year. The mean factor is close to 0.035 *SD* 0.02 (Table 10.1).

A long-term decrease in rod catchability may result from a dilution of anglers' skill with increasing numbers fishing. However, a dominant increasing trend may arise from decreasing numbers of fish (Peterman & Steer 1981; Small 1991a). If information on the likely trend is absent, try three models, one increasing, a mean value at a benchmark year and one decreasing, and note how the overall estimated trend of the run into the river is affected by the different models.

Benchmark values of $q$ can be obtained from *CPUE*/benchmark values of the run: preferably for a year in the middle of the period under consideration. Inserting the models for $q$, and where necessary use of suitable benchmark values, into Equation (3) gives three estimates of the annual runs of fish. Judgement must be exercised as to whether one or more of the estimated trends is reasonable. If this assessment can be made, the trends of relative exploitation rates are easily obtained.

Where the trends estimated from Equation (3) are based on rod *CPUE*, the recorded in-river net catches must be added to give the overall trends at the mouth of the river.

## 10.3    Annual data results

The analyses were based on data from 14 fisheries where trends of annual effort, catch and counts were known; and 13 where effort was not recorded but could be modelled. A check on the efficacy of the procedures is shown in Fig. 10.1, where the trend for the runs of salmon into the Foyle estuary from 1965 to 1992, estimated from catch per drift net licence, is compared with the observed trend. The correlation was highly significant ($P < 0.01$). For the waters analysed the correlations between the observed values of annual runs and those estimated via a catchability model are listed in Table 10.3. Where they are significant, they tend to be at the 1% level or better ($P < 0.01$). When not significant, and a trend curve is fitted to the observed values, the correlations of trends are highly significant ($P < 0.001$). The best correlations were obtained from subsets of data, not longer than about 20 years, either side of a major break or change in the temporal records of catch or *CPUE*.

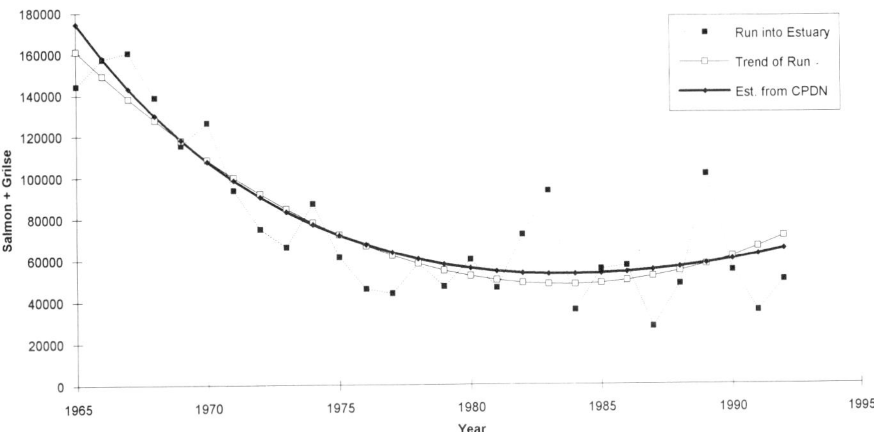


**Fig. 10.1**   Foyle Basin. Runs of salmon into the estuary: comparison of observed and estimated trends.

If such major changes are not readily apparent in the plots of recorded data it may be prudent to examine deviations from the long-term mean by CUSUM techniques, to check if and when hidden changes occur, perhaps due to effort, different methods of collecting or recording the data, etc. ( A.S. Champion, 1993, pers. comm.)

As these results seem to demonstrate that the trend of the annual total runs of fish, i.e. including the run in the close season, can be estimated from *CPUE*, recorded or modelled, the method was applied to five other fisheries where rod and net catches only are available: the Spey, Aberdeen Dee, Tweed, Nith, Tyne and Wye, with plausible results.

An example for the River Tweed is illustrated in Figs 10.2 to 10.6. Here a benchmark value of 41 720 rod days in 1989 was obtained from The Commissioners' survey of angling effort and a rate of increase of about 860 rod days per year assessed for the

**Table 10.3**   Correlations between estimated and observed trends of runs.

| Water | Period | Rod or net | $q$ trend | $r$ | $P$ |
|---|---|---|---|---|---|
| *Salmon: full annual data* | | | | | |
| Foyle | 1965–92 | net | inc. | 0.880 | 0.0001 |
| | | rod | inc. | 0.735 | 0.0001 |
| Bush | 1973–91 | rod | dec. | 0.981 | 0.0001 |
| Furnace | 1981–91 | rod | dec. | 0.933 | 0.001 |
| Feeagh | 1981–91 | rod | dec. | 0.778 | 0.0048 |
| Conon | 1953–91 | rod | mean | 0.542 | 0.0007 |
| | 1970–91 | rod | inc. & dec. | 0.945 | 0.0001 |
| Tummel | 1975–92 | rod | inc. | 0.819 | 0.0001 |
| Welsh Dee | 1975–90 | rod | dec. | 0.810 | 0.0001 |
| | | net | inc. | 0.863 | 0.0001 |
| Frome | 1973–89 | rod | dec. | 0.767 | 0.0003 |
| Ellidaar | 1950–77 | rod | dec. | 0.681 | 0.0001 |
| *Salmon: annual effort modelled* | | | | | |
| Faughan | 1967–87 | rod | inc. | 0.507 | 0.0191 |
| Beauly | 1968–91 | rod | mean | 0.872 | 0.0001 |
| N. Esk | 1981–91 | rod | inc. | 0.936 | 0.0001 |
| | | net | inc. | 0.779 | 0.0047 |
| Tay | 1971–90 | rod | inc. | 0.865 | 0.0001 |
| | | net | mean | 0.677 | 0.0001 |
| Coquet | 1968–83 | rod | inc. | 0.824 | 0.001 |
| Lune | 1974–91 | rod | inc. | 0.906 | 0.0001 |
| | | net | dec. | 0.839 | 0.0001 |
| Ribble | 1974–91 | rod | inc. | 0.966 | 0.0001 |
| | | net | mean | 0.688 | 0.0046 |
| *Sea trout: full annual data* | | | | | |
| Feeagh | 1979–91 | rod | dec. | 0.858 | 0.0031 |
| *Sea trout: annual effort modelled* | | | | | |
| Coquet | 1968–83 | rod | dec. | 0.875 | 0.0001 |
| Ribble | 1979–88 | rod | dec. | 0.715 | 0.0006 |

Note: catchability trend: inc. = increasing; dec. = decreasing; mean = no change.

period 1967–1992. Assuming a mean rod exploitation rate of 0.20 gave a benchmark value in 1980 for the run surviving the nets. For catchability models the rate of increase/decrease was assessed as $\pm 0.03$ of the estimated benchmark value in 1980.

The estimated total run of salmon into the harbour at Berwick included the recorded net and coble catch (Small 1991c). It will be noted that while the recorded rod catch and estimated catch per rod day both indicated an upward trend since 1980, in the same period the estimated runs of fish into the harbour showed reasonably steady or downward patterns. However, these trends may have to be interpreted in the light of changed timing of the different age components of the runs.

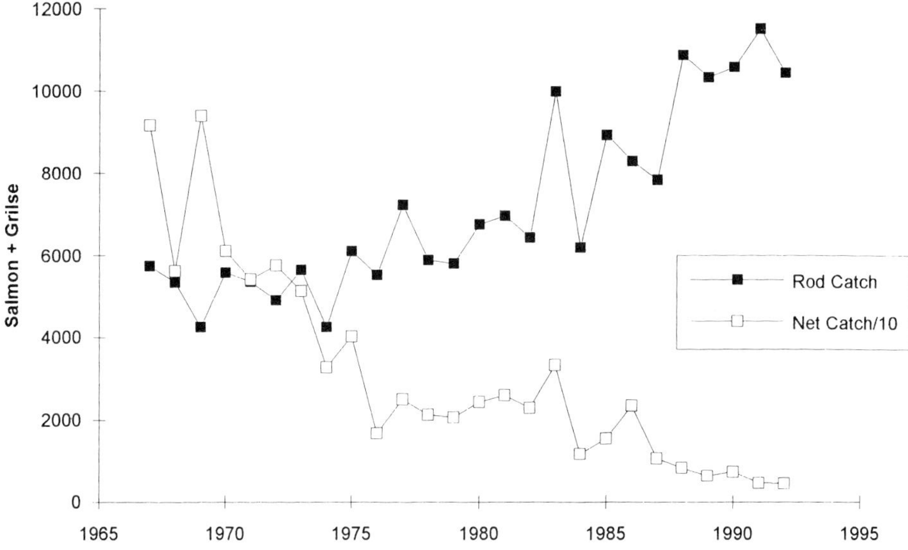

Fig. 10.2   River Tweed: trends of rod and net catches.

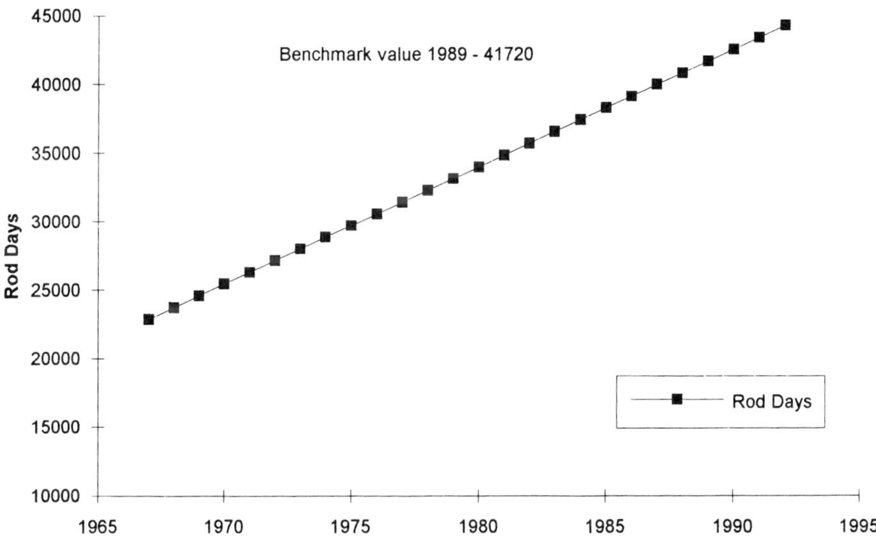

Fig. 10.3   River Tweed: rod effort model.

## 10.4     Monthly data

Initial surveys of within-season rod effort on several major rivers in Scotland, and the records provided by anglers in England and Wales volunteering to keep log books, have opened up a new field for analyses. If the distribution of rod effort can be modelled then the published records of monthly catches from returns, and

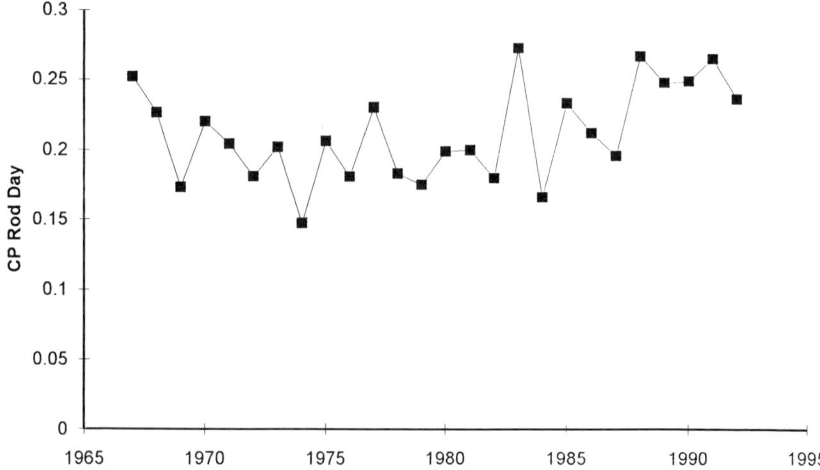

**Fig. 10.4**   River Tweed: estimated catch per rod day.

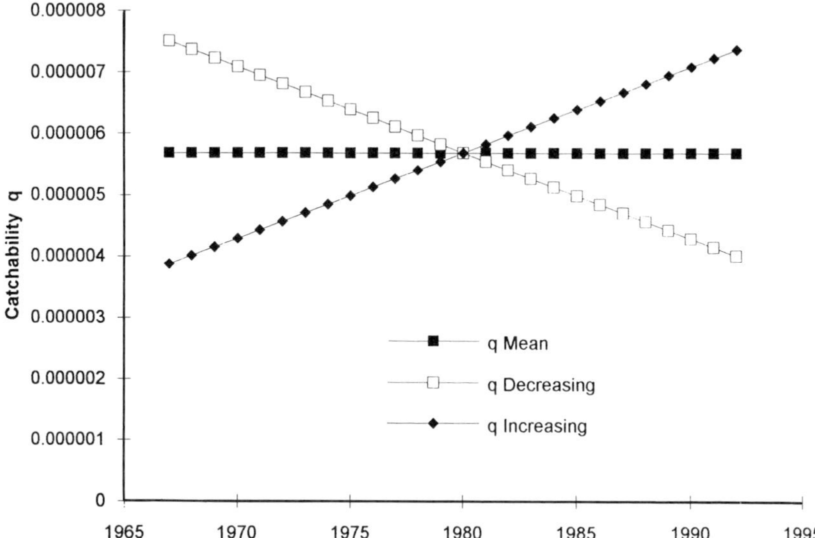


**Fig. 10.5**   River Tweed: rod catchability models.

catches of spring, summer and autumn fish, can be reassessed in terms of *CPUE* and flow.


### 10.4.1    *Monthly proportions of annual rod effort*

Exploratory analyses of the log book records (Aprahamian 1993) and other datasets enabled the proportion of annual total rod effort exerted in each month to be determined. The patterns emerging were curvilinear: cubic or parabolic for salmon

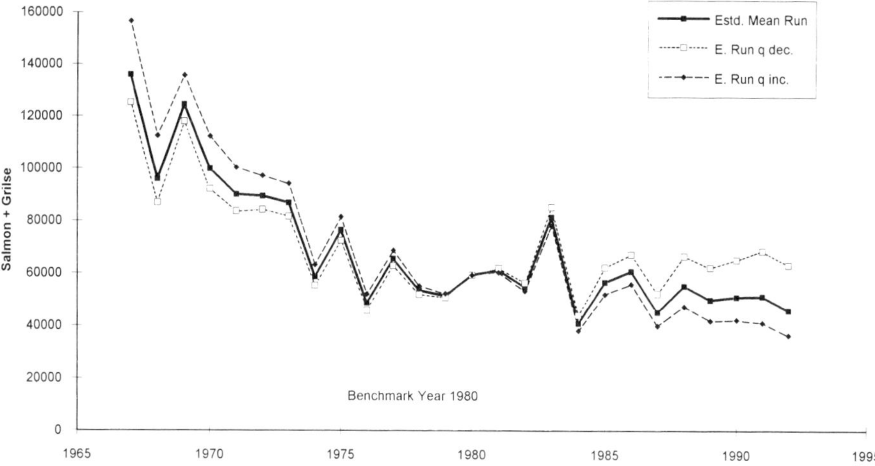



**Fig. 10.6**   River Tweed: estimated trends of runs of salmon and grilse into Berwick Harbour.

and parabolic for sea trout, (Fig. 10.7). The parameters for the curves available are shown in Table 10.4.

Proportions have altered through the years, probably following the changes in the timing of runs. Where good records of effort do not exist for previous decades it may be possible to estimate values from the relationships between effort and rod catch indicated in the second part of Table 10.4.

The evidence indicates that the monthly distribution of rod effort is likely to be specific to rivers, fisheries and species (see also Mee *et al.* 1993). The patterns obtained for five North West English rivers were sufficiently similar to justify the pooling of the data; therefore regional patterns are a possibility for a group of smaller rivers.

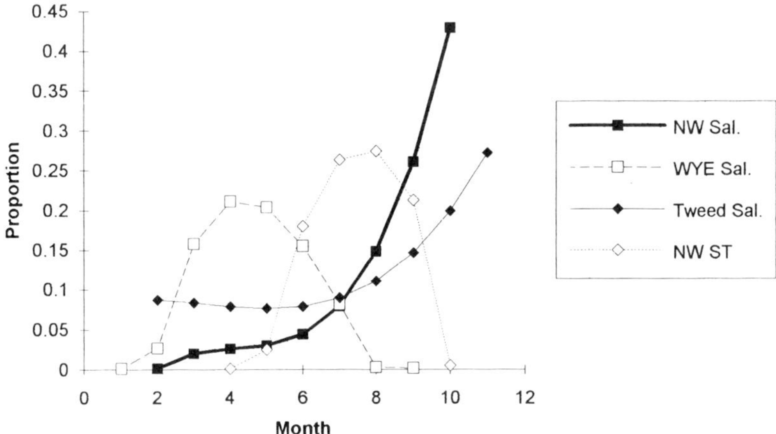



**Fig. 10.7**   Sample proportions of monthly rod effort.

**Table 10.4**  Monthly proportions of total annual rod effort (Proportion $= a + bM + cM^2 + dM^3$ where $M$ has a value 1 to 9, 10 or 11) and relation of effort to catch (*Effort* $= c \times$ *(rod Catch$^e$)*). North west rivers: pooled data from Eden, Derwent, Kent, Lune & Ribble.

| River | Period | Unit of effort | $a$ | $b$ | $c$ | $d$ | $r^2$ | $c$ | $e$ | $r^2$ |
|---|---|---|---|---|---|---|---|---|---|---|
| North west rivers | 1991 | Rod h | −0.121 | 0.101 | −0.024 | 0.0019 | 0.98 | 112. | 0.64 | 0.96 |
| Welsh Dee | 1990–2 | Rod h | −0.030 | 0.070 | −0.014 | 0.0009 | 0.46 | 405. | 0.39 | 0.85 |
| Upper Wye | 1977 | Rod h | −0.541 | 0.404 | −0.066 | 0.0030 | 0.82 | 187. | 0.33 | 0.75 |
| Middle Wye | 1977 | Rod h | −0.893 | 0.685 | −0.132 | 0.0077 | 0.77 | | NS | |
| Lower Wye | 1977 | Rod h | 0.140 | −0.469 | 0.057 | −0.0046 | 0.85 | 321. | 0.27 | 0.45 |
| Tweed | 1989 | Rod day | 0.079 | 0.012 | −0.005 | −0.0005 | 0.84 | 326. | 0.37 | 0.97 |
| Tummel | 1991 | Rod day | −0.081 | 0.087 | −0.0077 | — | 0.82 | 28. | 0.32 | 0.86 |
| Bush | 1972–3 | Day tkt | −0.405 | 0.179 | −0.0138 | — | 0.56 | 48. | 0.27 | 0.33 |
| | 1990–1 | Day tkt | 0.900 | −0.510 | 0.1000 | −0.0060 | 0.63 | 61. | 0.05 | NS |
| *Sea trout* | | | | | | | | | | |
| North west rivers | 1991 | h | −1.830 | 0.557 | −0.0360 | — | 0.80 | 50. | 0.67 | 0.95 |
| Welsh Dee | 1990–2 | h | −0.190 | 0.080 | −0.0043 | — | 0.52 | 55. | 0.36 | NS |

NS: not significant

### 10.4.2  *Relation of monthly CPUE to counts and cumulative counts of fish*

The significance of the relation of monthly rod catch rates of salmon and sea trout to counts and cumulative counts, was tested using the following modifications of Equation (1):

$$CPUE = a \times (count \text{ or } cumulative\ count^{(b+1)}) \tag{5}$$

$$CPUE = a \times (count \text{ or } cum.\ count^{(b+1)}) \times (monthly\ mean\ daily\ flow^f) \tag{6}$$

Trials of these equations seem to indicate that pooled data from several river-years give significant solutions to Equation (5). However, when 34 individual river and year records were examined, in only 9% was *CPUE* significantly related to count, but 68% were related to cumulative count, with or without the flow term, the exponent $(b+1)$ often approaching the value 1.0. The constant of proportionality $a$ varies considerably from river to river and year to year. Linear multiple variable regressions seem to follow the same pattern.

### 10.4.3  *Trends of rod catchability (q)*

These appear to be similar to the general pattern of annual data, where $q$ is high at low counts. However, when individual waters were examined it seems that there was no common temporal pattern as with annual data, the trends varying from constant to parabolic. Therefore to estimate monthly runs reliance must be placed on using suitable values for the parameters in model equations.

### 10.4.4     *Monthly mean daily flow*

Correlation analyses indicate that for monthly data, mean daily flow usually has a significant effect, at the 1% level or better, on anglers' effort and catches, but is an uncertain predictor of *CPUE*, counts and cumulative counts of salmon and sea trout. Results show that the effect on the accuracy of estimated trends of runs may be measurable but minor.

It seems that highly significant relationships between flow, or flow events, and movements of fish, counts, rod effort and catches were more easily obtained if daily data were analysed, a subject with its own rapidly increasing series of publications (e.g. Gee 1980).

### 10.5     Method of estimating cumulative counts on rivers without counters

It seems that there is a possibility of estimating the cumulative counts for salmon and sea trout from:

$$Cumulative\ count\ =\ CPUE/a \tag{7}$$

$$Cumulative\ count\ =\ (CPUE/(a*(MDF^f)))^{1/(b+1)} \tag{8}$$

Trials on the data for individual years from the Derwent, Kent, Lune, Welsh Dee and the Bush give correlations between estimated and observed cumulative counts significant at the 5% level or better (e.g. Fig. 10.8). Many more analyses are required to see if sufficiently reliable values for the parameters in Equations (7) and (8) can be estimated for use on rivers without counters.

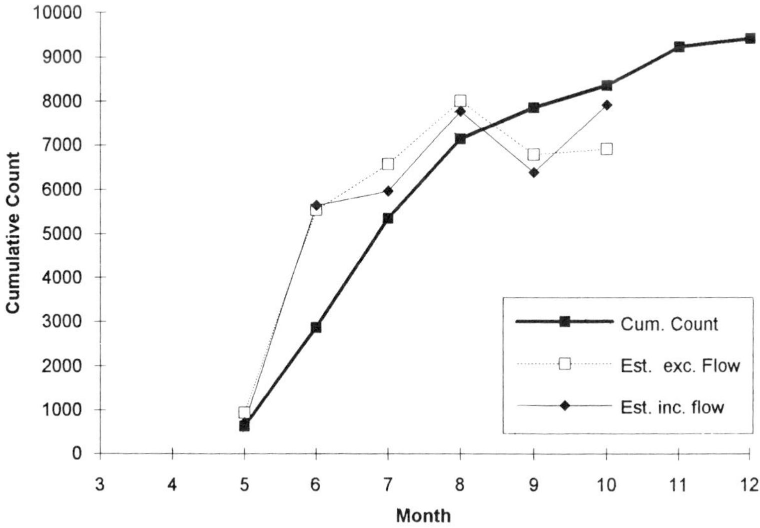


**Fig. 10.8**   River Lune: sea trout monthly data 1991, observed and estimated cumulative counts.

### 10.6    Discussion

'We know that any model assuming fixed parameters will be "wrong" in the strict sense, i.e. it will make numerical predictions that are not precisely correct. Therefore the issue in choosing a model … is not whether it is right or wrong. Rather, the practical issue is deciding which simplifying conveniences are good enough … given whatever limited historical information is available' (Hilborn & Walters 1992).

At present the changes in annual total runs of migratory fish into rivers are judged mainly by the recorded rod and net catches. Although the number of traps and electronic counters is increasing year by year, the total number installed is likely to be confined to a small minority of waters. In a period when angling effort is increasing and netting activity is sharply diminishing it is necessary to take this into account when assessing changing trends of runs into the majority of rivers.

The procedures outlined may assist in this process. The key to obtaining reasonable estimates of trends lies in the choice of models for the trends of catchability, and where required, effort. The values shown in Table 10.1 and local knowledge may help, but trials of high positive and low negative rates of change in effort combined with high positive and negative rates of change in catchability will indicate practical upper and lower bound models.

It must be emphasized that the methods will only produce trends. The division of a very volatile quantity, *CPUE*, by a small number, catchability, can induce big errors in estimates of runs in particular years so the procedures are not suitable for estimating absolute numbers of fish in individual years. Any analysis should be treated as an aid to interpretation of the trends of recorded catches and local biological information.

As for monthly records, now that the performance of anglers is being sampled by log book and other surveys, if a database of the patterns of monthly proportions of annual rod effort is established for individual river basins, this may enable more detailed study of *CPUE*, and therefore estimated runs of spring, summer and autumn runs of salmon, on waters where rod and net effort is changing rapidly. Future studies may show that regional variations in patterns of monthly effort should be taken into account.

The exploratory analyses of monthly data also indicate that it may be possible to estimate the cumulative count of migratory fish runs from monthly mean rod *CPUE*.


### Acknowledgements

The author is greatly indebted to the University of Liverpool for support provided during the past 12 years, and to The Environmental Advisory Unit Ltd for data gathering and material help since 1992. Special thanks go to the many colleagues, fishery officers, scientists and others who provided information and guidance in this study. M. Aprahamian, R. Campbell, R. Gardiner and K. O'Hara kindly found time to read the manuscript and made helpful suggestions.

## References


Aprahamian M. (1993) An analysis of migratory salmonid catch effort data, derived from anglers' log books 1991. Bristol NRA Report NRA/NW/FTR/33/3.

Bannerot S.P. & Austin C.B. (1983) Using frequency distributions of catch per unit effort to measure fish stock abundance. *Transactions of the American Fisheries Society* **112**, 608–617.

Beaumont W.R.C., Welton J.S. & Ladle M. (1991) Comparison of rod catch data with known numbers of Atlantic salmon (*Salmo salar*) recorded by a resistivity fish counter in a southern chalk stream. In: I.G. Cowx (ed.) *Catch effort sampling strategies. Their application in freshwater fisheries management.* Oxford: Fishing News Books, pp 49–60.

Bunt D.A. (1991) Use of rod catch effort data to monitor migratory salmonids in Wales. . In: I.G. Cowx (ed.) *Catch effort sampling strategies. Their application in freshwater fisheries management.* Oxford: Fishing News Books, pp 15–32.

Churchward A.S. & Hickley P. (1991) The Atlantic salmon fishery of the River Severn (UK). In: I.G. Cowx (ed.) *Catch effort sampling strategies. Their application in freshwater fisheries management.* Oxford: Fishing News Books, pp 1–14.

Crecco V. & Overholtz W.J. (1990) Causes of density-dependent catchability for Georges Bank Haddock, *Melanogrammus aeglefinus. Canadian Journal of Aquatic Sciences* **47**, 385–394.

Elson P.F. & Tuomi A.L.F. (1975) The Foyle fisheries: new basis for rational management. The Foyle Fisheries Commission.

Fahy E. & Fluskey R. (1987) The redds on the bed. *Salmon & Trout Magazine* **234**, p 71.

Gardiner W.R. (1991) Modelling rod effort from catch and abundance data. In: I.G. Cowx (ed.) *Catch effort sampling strategies. Their application in freshwater fisheries management.* Oxford: Fishing News Books, pp 92–99.

Gee A.S. (1980) Angling success for Atlantic Salmon (*Salmo salar*) in the River Wye in relation to effort and river flows. *Fisheries Management* **11**, 131–8.

Gee A.S. & Milner N.J. (1980) Analysis of 70-year catch statistics for Atlantic salmon (*Salmo salar*) in the River Wye and implications for management of stocks. *Journal of Applied Ecology* **17**, 41–57.

Grimble A. (1913) *The salmon rivers of Scotland, Ireland, England.* 3 Vols. London: Kegan Paul.

Hilborn R. & Walters C.J. (1992) *Quantitative Fisheries Stock Assessment. Choice, Dynamics and Uncertainty.* London, New York: Chapman Hall, 533 pp.

Holden A.V. (1987) The automatic counter – a tool for the management of salmon fisheries. Pitlochry: The Atlantic Salmon Trust.

Laughton R. (1991) The movements of adult Atlantic salmon (*Salmo salar* L.) in the River Spey as determined by radio telemetry during 1988 and 1989. Scottish Fisheries Report No. 50. Pitlochry: SOAFD.

Mee D., Evans D. & Davidson I. (1993) Salmon & sea trout – 'a matter of effort'. *Glas-y-Dorlan* **17**, p 8 NRA, Welsh Region.

Mundy P.R., Alexandersdottir M. & Eiriksdottir G. (1978) Spawners – recruit relationship in Ellidaar. *Journal of Agricultural Research Iceland* **10**, 47–56.

Nicholson S.A. & Aprahamian M. (1993) Field validation of the Logie counter at Forge Weir on the River Lune, 1992. NRA Report NW/FTR/93/5.

Paloheimo J.E. & Dickie L.M. (1964) Abundance and fishing success. *Rapp. R-V. Revu. Cons. Int. Explot. Mer* **155**, 152–163.

Peterman R.M. & Steer G.J. (1981) Relation between sport-fishing catchability coefficients and salmon abundance. *Transactions of the American Fisheries Society* **110**, 585–593.

Small I. & Downham D.J. (1985) The interpretation of anglers' records (trout, seatrout and salmon.) *Aquaculture and Fisheries Management* **16**, 151

Small I. (1991a) Exploring data provided by angling for salmonids in the British Isles. In: I.G. Cowx (ed.) *Catch effort sampling strategies. Their application in freshwater fisheries management.* Oxford: Fishing News Books, pp 81–91.

Small I. (1991b) A review of some methods of enhancing and interpreting fishery records. Paper presented to the Joint Atlantic Salmon Trust and Royal Irish Academy Workshop on The Measurement and Evaluation of the Exploitation of Atlantic Salmon. Dublin, April 1991.

Small I. (1991c) Report to the River Tweed Commissioners: Exploratory estimates of the trends of salmon runs entering the river at Berwick Harbour.

Solomon D.J. & Potter E.C.E. (1992) The measurement and evaluation of the exploitation of Atlantic Salmon. The Atlantic Salmon Trust.

# Chapter 11
# Interpreting trends in adult Atlantic salmon (*Salmo salar* L.) populations from angling records: an analysis of catch and effort data from the River Spey


G. W. SMITH *SOAFD Marine Laboratory, PO Box 101, Victoria Road, Aberdeen, UK*
R. LAUGHTON *Spey District Fishery Board, 121 High Street, Forres, UK*
S.J. DORA *SOAFD Marine Laboratory, PO Box 101, Victoria Road, Aberdeen, UK*



**Abstract**   Trends in salmon stock dynamics may be inferred from an analysis of rod catches. Interpretation may be confounded by changes in the quality and quantity of fishing effort employed, however. In 1991 a study was set up on the River Spey to standardize the collection of both catch and effort data. This chapter reports seasonal patterns of catch and effort in the 1992 and 1993 angling seasons and examines the extent to which angling effort and the identity of individual anglers may be used to explain variations in rod catch.

There was no indication that angling effort varied seasonally in response to seasonal changes in salmon catches. This resulted in seasonal patterns of catch per unit effort (CPUE) similar to the catch data. Most variation in fishing effort between estates was in the month when the angling season began.

On two estates where the identities of anglers were recorded, 5.5% and 15% of anglers were responsible for 50% of all catches. There was a significant relationship between the total effort employed by individual anglers and their total catch. Variation in individual fishing effort did not entirely account for differences in angler success. There was a wide range of CPUE determined for individual anglers, with angler identity explaining between 26 and 47% of the variation in daily CPUE from May to August.

KEYWORDS: Angling effort, angler identity, Atlantic salmon, rod catch, CPUE


## 11.1   Introduction

With the recent decline in net fisheries for Atlantic salmon (*Salmo salar* L.) in Scotland (Shearer 1992), an increasing proportion of the Scottish catch is taken in recreational rod fisheries. Dunkley *et al.* (1993), for example, note that rod catches increased from 19% of the total reported Scottish catch in 1981 to 45% in 1991. Rod catches are thus an increasingly important source of information on the status of salmon stocks as well as making a significant contribution to the Scottish economy. At its simplest, successful management of recreational angling does not require estimates of the absolute size of the exploited fish population. In principle, reliable measures of the size and age structure of the catch over a number of years may be sufficient to derive underlying trends in the available fish stocks.

Identifying trends in salmon stocks from catch statistics is fraught with difficulty, however (e.g. Lakhani 1986; Shearer 1986; Bielak & Power 1988). For example, although returning adult salmon enter many rivers throughout the year (Shearer 1992), they are only sampled over a defined angling season. There is also increasing evidence that catchability of the fish may vary during the riverine migration (Laughton 1991; Smith & Laughton 1993). The size of the catch will also depend upon environmental factors such as river flow and temperature (Alabaster 1970; Alabaster 1990; Gee 1980).

A number of authors have also emphasized that some measure of angling effort is required if underlying trends in stock dynamics are to be inferred with any degree of confidence from trends in angling catch (e.g. Shearer 1986). Angler success may also vary considerably between individuals (Cryer & Maclean 1991) and trends in the catch may thus result from changes in the quality as well as the quantity of fishing effort employed. In 1991 the Spey District Fishery Board and the Scottish Office Agriculture and Fisheries Department set up a project to standardize the collection of both catch and effort data over a wide area of the River Spey, to determine the information required to interpret catch statistics in relation to salmon stocks and to set up a long-term data set from which future trends in the Spey stock dynamics may be reliably monitored.

This chapter reports seasonal patterns of catch and effort for 10 estates in the 1992 and 1993 angling seasons. In 1992, the latest year for which data were published, these estates accounted for 37.4% of the total rod catch of the river. In addition, the relationship between catch and effort in individual estates was assessed in conjunction with variations in the contribution made by individual anglers.


## 11.2    Materials and methods

The daily catch and angling effort employed on 10 estates were recorded over the whole of the angling season, February to September (Table 11.1). Fishing effort was expressed as the time spent fishing by each individual angler. All estates are situated on the main stem of the River Spey, except estates 7 and 8 which are situated on a major tributary, some 50 km from the mouth of the river.

As with much of the reporting of salmon catches in Scotland (Shearer 1992), the sea ages of the fish were largely estimated on the basis of their weights; no scale samples were taken for subsequent confirmation of age. Daily information on the time, to the nearest 30 minutes, that each angler spent fishing and the number and weight of each

**Table 11.1**  Relative positions on the River Spey of the 10 estates participating in the study and the years in which catch and effort data were collected from each.

| Estate | Distance upriver (km) | Years included |
| --- | --- | --- |
| 1 | 11.0 | 1992 |
| 2 | 18.9 | 1992 & 1993 |
| 3 | 22.3 | 1992 & 1993 |
| 4 | 25.1 | 1993 |
| 5 | 35.0 | 1993 |
| 6 | 44.2 | 1992 & 1993 |
| 7 | 50.5 | 1992 & 1993 |
| 8 | 83.2 | 1992 & 1993 |
| 9 | 68.2 | 1992 & 1993 |
| 10 | 90.1 | 1993 |

putative multi-sea-winter (MSW) salmon (fish remaining at sea for more than one winter before returning to fresh water) and grilse (1 sea-winter fish) was recorded.

In addition, the identity of each angler fishing in estate 8 was recorded and this information was used to assess the variation in angler success between individuals. Historical records from estate 2, which included the daily catch of individually identified anglers for the years 1970 to 1991, were also analysed. This latter record contained no information on anglers failing to catch fish on any given day, nor were the times each angler spent fishing recorded.

## 11.3    Results

### 11.3.1    *Identification of sea age classes of salmon*

The weight distributions of fish classified as MSW salmon and grilse are shown in Figure 11.1. Grilse catches were restricted to the period between May and September, and analysis of variance (ANOVA) indicated a significant increase in the weight of grilse as the season progressed (ANOVA: $N = 3670$, $F = 14.189$, $P < 0.001$).

Rod caught MSW salmon also showed a significant increase in weight during the early months of the angling season (ANOVA, February to June: $N = 1251$, $F = 23.772$, $P < 0.001$). However, the weights of MSW salmon showed a marked decline between June and July and remained significantly lower in the period between July and September than earlier in the season (Mann Whitney test: $N = 3408$, $\chi^2$ approximation $= 24.349$, d.f $= 1$, $P < 0.001$). This may have arisen because there was significant misclassification of grilse as the more prized MSW fish from July onwards, 'grilse error' (Shearer 1985). In the absence of independent confirmation of the sea age of the fish, catches classified as both MSW salmon and grilse were aggregated for the present analysis.

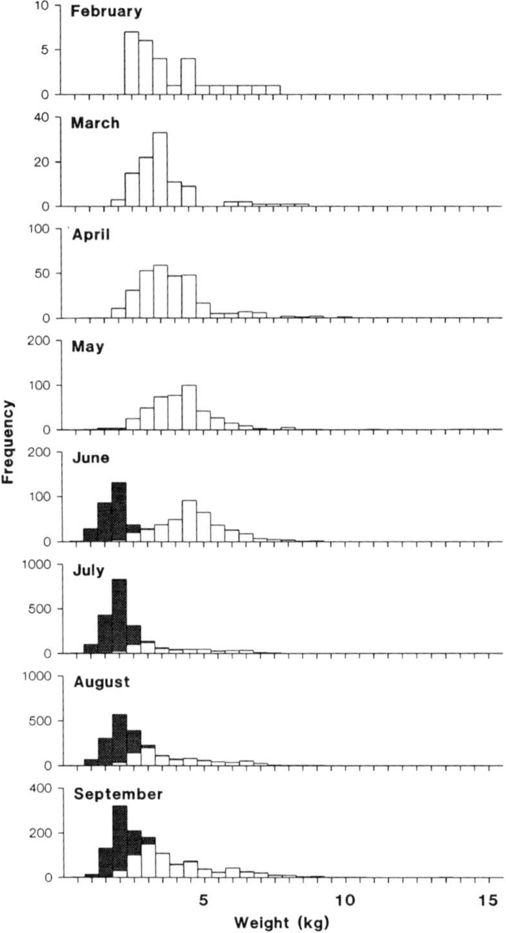

**Fig. 11.1**  Monthly frequency distributions of the weights of putative MSW salmon (white bars) and grilse (black bars). Catches from all estates sampled in 1992 and 1993 have been combined. Scales on the Y axis have been varied from month to month in order that patterns of weight distributions can be seen most clearly.

### 11.3.2  *Seasonal patterns of catch, effort and catch per unit effort*

In the first five months of the angling season (February to June) there was a slight rise in catches averaged over all estates, followed by a sharp increase in July (Fig. 11.2). There was no indication from the data that angling effort responded to these seasonal changes in the catch. Fishing effort did not closely mirror catch rates, and although maximum fishing effort was expended in June, angling effort remained relatively stable over much of the angling season which resulted in a seasonal pattern of catch per unit effort (CPUE) which was similar to the catch data.

Most variation in fishing effort between estates was in the month when the angling season began (Fig. 11.3). Fishing effort was consistent over the angling season in most

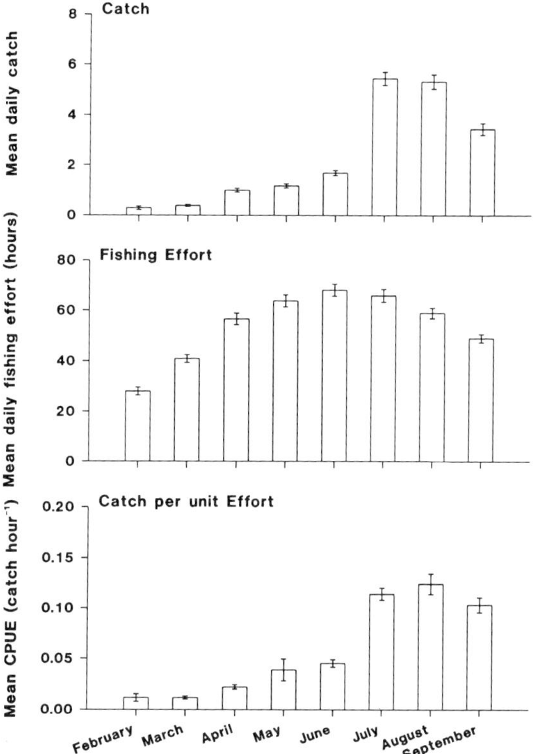



**Fig. 11.2**  Seasonal patterns of catch, angling effort and catch per unit effort. Monthly data has been averaged over all estates and both years; the error bars indicate the standard error of each mean.

of the lower estates, but was more variable in the uppermost estates on the river. Angling effort did not appear to vary in response to changes in CPUE on any estate (Fig. 11.4). Similar to the general pattern (Fig. 11.2), CPUE in the lower sections of the river (up to estate 6 [44.2 km]) showed a general rise in the first five months of the angling season, followed by a sharp increase in July and a general decline in September (Fig. 11.4). By contrast, CPUE rose more generally throughout the season above estate 6, reaching a peak in August or September.

Comparisons between 1992 and 1993 (Fig. 11.5) showed significant increases in both catch and CPUE in June and July and significant reductions in September. Fishing effort was relatively stable between years; a significant increase in fishing effort only occurred in August.

### 11.3.3    *Catches of individual anglers*

A total of 920 successful anglers were identified from estate 2 records between 1970 and 1991. A total of 40 anglers fished estate 8 for between 13 and 235 hours in 1992 and 1993. Analysis of both data sets indicates that much of the catch was attributable

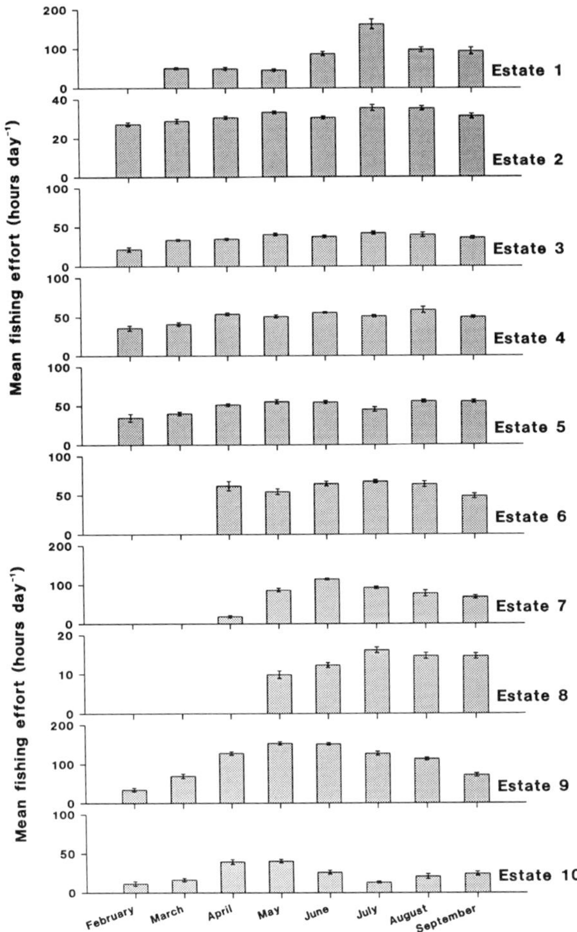



**Fig. 11.3** Seasonal patterns of angling effort for each estate. Mean daily fishing effort for each month has been averaged over 1992 and 1993; the error bars indicate the standard error of each mean. All estates are situated on the main stem of the Spey except for estates 7 and 8.

to relatively few anglers. After ranking anglers in order of their aggregate catch over the study periods (Fig. 11.6), 5.5% of anglers in estate 2 and 15% of anglers in estate 8 were responsible for 50% of all catches. Much, though not all, of the differences in catch between anglers can be attributed to differences in angling effort. Regression analysis of complete effort records from estate 8 (Fig. 11.7) indicated a highly significant linear relationship between catch and effort ($N = 40$, $r^2 = 0.554$, $P < 0.001$).

Variation in fishing effort between anglers on estate 8 was accounted for by calculating CPUE both from the daily records (Fig. 11.8a) and from the aggregate catch and effort data recorded for each individual angler (Fig. 11.8b). Much of the resulting variation in daily CPUE was explained by the identity of the individual angler. Angler

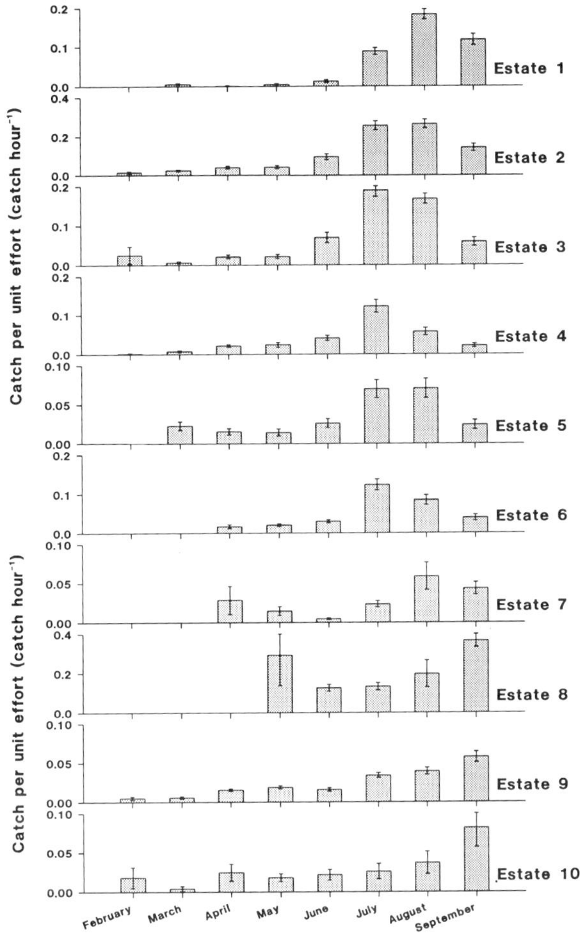



**Fig. 11.4**  Seasonal patterns of CPUE for each estate. Mean daily CPUE for each month has been averaged over 1992 and 1993; the error bars indicate the standard error of each mean. All estates are on the main stem of the Spey except for estates 7 and 8.

identity was a significant factor associated with daily CPUE in each month of the angling season except September (Table 11.2).

Previous studies have suggested that one factor explaining differences in CPUE between anglers was the degree of fishing effort employed, increasing effort (and therefore familiarity with the water) being associated with a significant increase in CPUE (Alabaster 1986; Cryer & Maclean 1991). In the present study, analysis of CPUE in estate 8 demonstrated no significant association between CPUE and the total effort employed by individual anglers (ANOVA: $N = 40$, $F = 1.991$, $P = 0.198$).

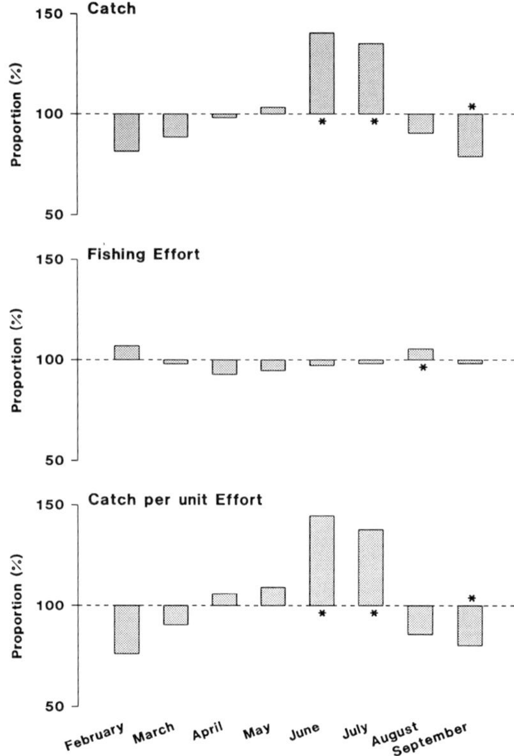


**Fig. 11.5**  Month by month changes in catch, angling effort and CPUE between 1992 and 1993. Mean monthly data in 1993 is expressed as a proportion of the 1992 data aggregated over all estates collecting information in both years. Months showing a significant difference (Mann Whitney test: $P < 0.05$) between years are indicated by an asterisk.

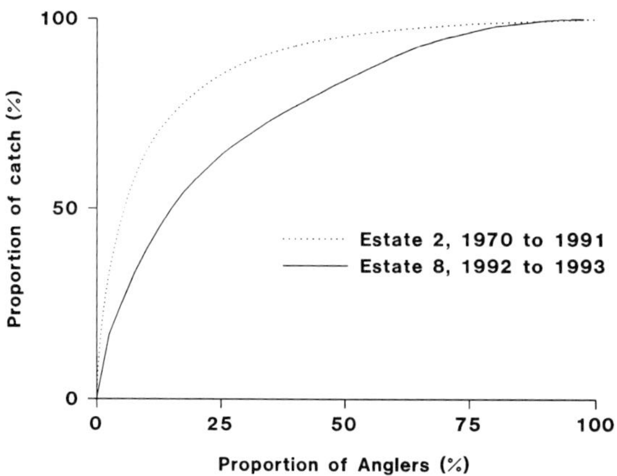


**Fig. 11.6**  Cumulative proportion of the aggregate catch accounted for by individual anglers, ranked according to their total catch over each study period.

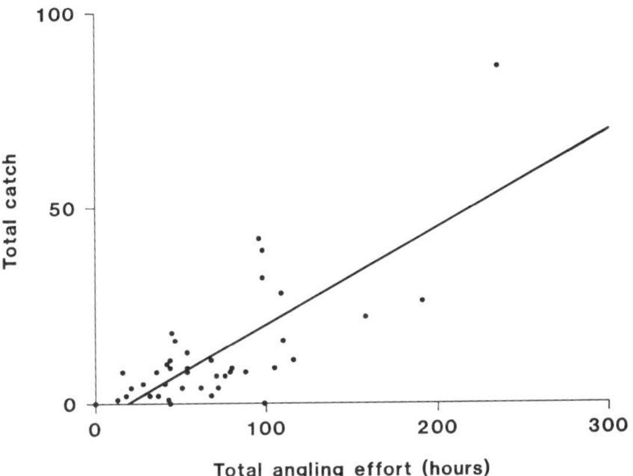

**Fig. 11.7**   The relationship between total catch and effort for each individual angler recorded fishing in estate 8. The line calculated by least squares regression (*catch* = 0.25*effort* – 4.79; *N* = 40, $r^2$ = 0.554, *P* < 0.001) is also indicated.

## 11.4     Discussion

### 11.4.1     *Using rod catches to monitor salmon stocks*

The development of reliable resistivity fish counters has provided a powerful tool for estimating the total numbers of salmon entering a given river system (Bussel 1978; Dunkley & Shearer 1982). Such counters generally need to be mounted on a weir or similar physical structure at a suitable site and are expensive to install, particularly for large river systems. On many occasions, therefore, rod catches may be the only source of information from which trends in available fish stocks may be estimated.

In some instances the analysis of rod catch data may also be a more reliable index of salmon available to a particular fishery than the total counts of fish in a river. Adult salmon are migratory and home to particular spawning locations. As such they become available to a succession of angling sites during their riverine migration; and, although Beaumont *et al.* (1991) found a significant relationship between the numbers of fish ascending a fish counter and the CPUE in a neighbouring rod fishery, current knowledge does not yet permit general predictions to be made on the relationship between data from fish counters and catches at particular fisheries throughout a river system. Thus, while total counts of fish entering a river may be of use when estimating the total spawning stock, such data may be of little value if those fish available to a rod fishery at a particular site on the river system are of interest.

### 11.4.2     *Identifying the sea age class of rod caught fish and interpreting trends in stock components*

Salmon enter the larger east coast Scottish rivers throughout the year, and Shearer (1990) has demonstrated that timing of entry to the River North Esk is related to both

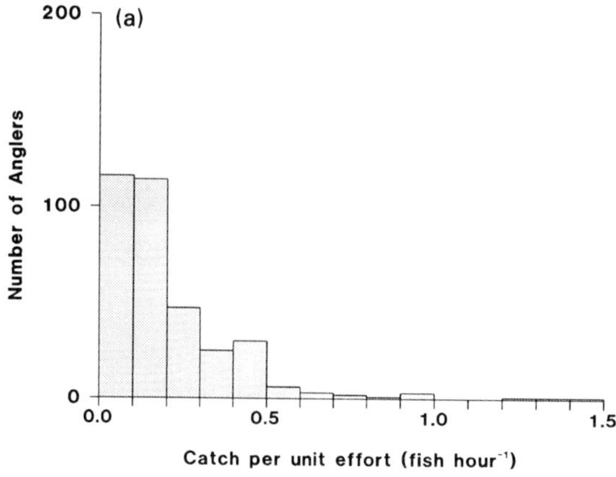


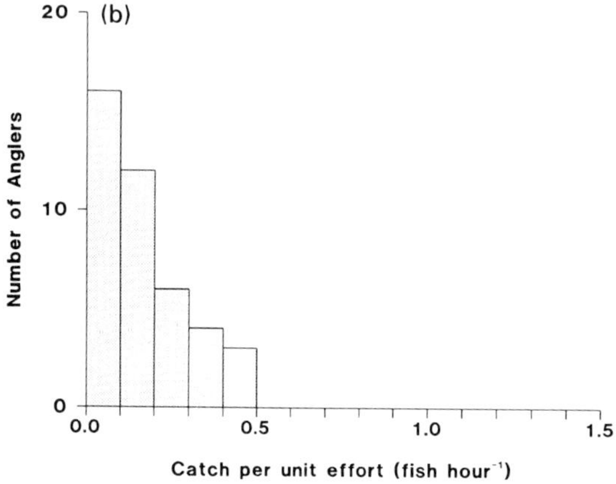


**Fig. 11.8**   Distribution of CPUEs of individually identified anglers fishing in estate 8 over the study period. a) CPUE calculated from daily angling records; b) CPUE calculated from records for each individual angler aggregated over the study period.

sea and river age of the fish. In turn, riverine migration patterns and spawning positions of fish within each sea age class component of the stock reflects its return time (Laughton & Smith 1992; Smith & Laughton 1993) and thus different components of the returning salmon stock may become available to particular rod fisheries along a river system at different times of the angling season. Different stock components may also exhibit different trends in their stock dynamics over time. For example, Shearer (1988) and Laughton & Smith (1993) identified a decrease in catches of spring MSW salmon (MSW salmon caught before 1 May) and an increase in the catches of grilse in the Spey from the 1950s to the present day.

Trends in annual catch statistics may therefore conceal as much as they reveal, and detailed catch records which allow analysis to be stratified according to time of year are required if an accurate indication of developments in each stock component is to be derived. Differentiating accurately between particular sea age classes is also

**Table 11.2**  Results of an Analysis of Variance (ANOVA) to test the degree of significance of the relationship between angler identity and daily CPUE in estate 8 for each month of the angling season.

| *Month* | *N* | *F* | *P* |
|---|---|---|---|
| May | 29 | 3.297 | 0.018 |
| June | 77 | 2.877 | 0.006 |
| July | 79 | 5.036 | < 0.001 |
| August | 83 | 2.057 | 0.031 |
| September | 82 | 1.246 | 0.285 |

important if trends in the catch data are to be interpreted reliably. The present analysis suggests that visual inspection of fish is not a reliable method of estimating sea age. Dunkley *et al.* (1993) reported that 'grilse error' resulted in catches of grilse being underestimated by an average of 28.3% from 1981 to 1991. Thus, scale sampling (Anon. 1984) would appear to be a necessary component of any programme which sets out to monitor angler catch and effort. Increases in workload may be kept to a minimum if preliminary sampling is carried out to identify those times of the year when 'grilse error' is likely to be a problem and also which size classes of fish are most vulnerable to misclassification.

### 11.4.3   *Angling effort*

Seasonal patterns of angling effort were relatively stable within estates and did not vary in response to seasonal changes in either catch or CPUE, an observation which may, in part, reflect the structure of salmon fishing in Scotland (Williamson 1991). As with many of Scotland's major salmon rivers, the total number of anglers fishing particular estates is limited. The anglers in the present study often have to book months or years in advance and are thus constrained on where and when they can fish. Thus their ability to vary fishing effort either seasonally or in response to perceived short-term changes in CPUE may be severely limited. Although angling clubs, which allow members greater access to the river, exercise fishing rights on some sections of the River Spey, none are represented in the present study.

### 11.4.4   *Angler identity*

In common with previous studies (Alabaster 1986; Cryer & Maclean 1991), a small proportion of anglers in estates 2 and 8 accounted for much of the catch in both estates. Analysis of fishing effort in estate 8 suggested that much of this difference in angling success was explained by variation in the effort expended by individual anglers. This supports data presented by Alabaster (1986) who showed that when ranking anglers in a trout fishery in order of aggregate catch, the top 10% of anglers expended approximately 50% of the total recorded effort. However, accounting for variations in individual fishing effort did not entirely eliminate differences in angler success rate. There was a wide range of CPUE determined for individual anglers in estate 8, with angler identity explaining between 26 and 47% of the variation in daily CPUE between May and August.

It is not clear from the present study whether differences in CPUE between individual anglers can be explained by quantifiable features of angler behaviour. Both Cryer & MacLean (1991) and Alabaster (1986) note that variation in angler familiarity with the area, measured by the degree of angling effort expended, was related to CPUE. Analysis of CPUE for each angler fishing on estate 8 in 1992 and 1993 demonstrated no such association between CPUE and the total effort employed by individual anglers. Alabaster (1986) analysed catches from the first four years of a newly established fishery where the anglers were probably unfamiliar with the area. By contrast, the estates on the River Spey constitute well established fisheries where anglers may have become familiar with the area over a number of previous years and thus there may be little reason to expect a relationship between angler familiarity and total effort for any given year.

Angler identity thus appears to be a significant factor in accounting statistically for individual differences in angler CPUE which at present cannot be accounted for by other measurable features of angler behaviour. Thus, particularly in fisheries where angler composition may vary significantly from year to year, recording the identity of individual anglers may be important if trends in the available stock are to be inferred from rod catches.

## Acknowledgements

We are grateful to all proprietors, ghillies and anglers who put in a considerable effort in helping us to both set up the study and also collect the data. The Spey District Fishery Board provided encouragement for the project and Dr Tony Hawkins, Dr Dick Shelton and John Webb provided useful comments on a draft of this chapter. The Marine Laboratory typists prepared the final manuscript.

## References

Alabaster J.S. (1970) River flow and upstream movement and catch of migratory salmonids. *Journal of Fish Biology* **2**, 1–13.

Alabaster J.S. (1986) An analysis of angling returns for trout, *Salmo trutta* L., in a Scottish river. *Aquaculture and Fisheries Management* **17**, 313–326.

Alabaster J.S. (1990) The temperature requirements of adult Atlantic salmon *Salmo salar* L., during their upstream migration in the River Dee. *Journal of Fish Biology* **37**, 659–661.

Anonymous (1984) Atlantic salmon scale reading. Report of the Atlantic salmon scale reading workshop, Aberdeen, Scotland, 23–28 April 1984. International Council for the Exploration of the Sea, 17 pp.

Beaumont W.R.C., Welton J.S. & Ladle M. (1991) Comparison of rod catch data with known numbers of Atlantic salmon (*Salmo salar*) recorded by a resistivity counter in a southern chalk stream. In: I.G. Cowx (ed.) *Catch effort sampling strategies, their application in freshwater fisheries management.* Oxford: Fishing News Books, pp. 49–60.

Bielak A.T. & Power G. (1988) Catch records – facts or myths. In: D. Mills & D. Piggins (eds) *Atlantic Salmon: Planning for the Future.* London: Croom Helm, pp 235–255.

Bussell R.B. (1978) *Notes for guidance in the design and use of fish counting stations*. London: Department of the Environment, 70 pp.

Cryer M. & MacLean G.D. (1991) Catch for effort in a New Zealand recreational trout fishery – a model and implications for survey design. In: I.G. Cowx (ed.) *Catch effort sampling strategies their application in freshwater fisheries management*. Oxford: Fishing News Books, pp 61–71.

Dunkley D.A., MacLean J.C. & Milne J.M.A. (1993) The effect of misreporting grilse as salmon on the reported Scottish national catch. International Council for the Exploration of the Sea, Anadromous and Catadromous Fish Committee, CM 1993/M:40.

Dunkley D.A. & Shearer W.M. (1982) An assessment of the performance of a resistivity fish counter. *Journal of Fish Biology* **20**, 717–737.

Gee A.S. (1980) Angling success for Atlantic salmon (*Salmo salar*) in the River Wye in relation to effort and river flows. *Fisheries Management* **11**, 131–138.

Lakhani K.H. (1986) Salmon population studies based upon Scottish catch statistics: statistical considerations, pp 116–120. In: D. Jenkins & W.M. Shearer (eds) *The status of the Atlantic salmon in Scotland* (ITE Symposium No. 15). Abbots Ripton: Natural Environment Research Council, Institute of Terrestrial Ecology, pp 16–20.

Laughton R. (1991) The movements of adult Atlantic salmon (*Salmo salar* L.) in the River Spey as determined by radio telemetry during 1988 and 1989. *Scottish Fisheries Research Report* **50**, 35 pp.

Laughton R. & Smith G.W. (1992) The relationship between the date of river entry and the estimated spawning position of adult Atlantic salmon (*Salmo salar* L.) in two major Scottish east coast rivers. In: I.G. Priede & S.M. Swift (eds) *Wildlife Telemetry*. London: Ellis Horwood, pp 423–433.

Laughton R. & Smith G.W. (1993) Recent trends in the rod catches of Atlantic salmon (*Salmo salar* L.) from four sites on the River Spey. *Aquaculture and Fisheries Management* **24**, 671–679.

Shearer, W.M. (1985) Salmon catch statistics for the River Dee 1952-83. In: D. Jenkins (ed.) *The Biology and Management of the River Dee* (ITE Symposium No. 14). Abbots Ripton: Natural Environment Research Council, Institute of Terrestrial Ecology, pp 127–141.

Shearer W.M. (1986) An evaluation of the data available to assess Scottish salmon stocks. In: D. Jenkins & W.M. Shearer (eds) *The Status of the Atlantic Salmon in Scotland* (ITE Symposium No. 15). Abbots Ripton: Natural Environment Research Council, Institute of Terrestrial Ecology, pp 91–111.

Shearer W.M. (1988) Fisheries in the Spey Catchment. In: D. Jenkins (ed.) *Land Use in the River Spey* (ACLU Symposium No. 1). Aberdeen: Aberdeen Centre for Land Use, pp 197–212.

Shearer W.M. (1990) The Atlantic salmon (*Salmo salar* L.) of the North Esk with particular reference to the relationship between both river and sea age and time of return to home waters. *Fisheries Research* **10**, 93–123.

Shearer W.M. (1992) *The Atlantic salmon. Natural history, exploitation and future management*. New York: Halsted Press, 244 pp.

Smith G.W. & Laughton R. (1993) The relationship between rod catches and the distribution of adult Atlantic salmon (*Salmo salar* L.) during the initial phase of riverine migration. *Aquaculture and Fisheries Management* **24**, 681–683.

Williamson, R. (1991) *Salmon fisheries in Scotland*. Pitlochry: Atlantic Salmon Trust, 42 pp.

# Chapter 12
# Catch and effort data and the management of the commercial fisheries of Itaipu reservoir in the upper Paraná river, Brazil

E.K. OKADA and Â.A. AGOSTINHO *UEM-NUPELIA Bloco H90, Avenida Colombo, 5790, 87020-900 – Maringá (PR), Brazil*
M. PETRERE Jr. *UNESP – Departamento de Ecologia, CP 199, 13506-900 – Rio Claro (SP), Brazil*

**Abstract**   The results of 7 years of catch and effort data from Itaipu reservoir, the largest reservoir in the Paraná river basin, are discussed. The data suggest slight overfishing. General strategies for fisheries management are proposed.

KEYWORDS: Itaipu reservoir, reservoir fisheries management, tropical fisheries

## 12.1   Introduction

Presently 90% of the energy consumed in Brazil is from hydroelectric sources and 70% of the reservoirs are concentrated in the southeast region, in the Paraná river basin (Fig. 12.1) (Paiva 1982).

Riverine fisheries are an important economic activity in Brazil. The total landings of commercial fisheries in 1990 was 800 000 t, of which 210 000 t (26%) were freshwater species. Professional fisheries are becoming important in Brazil but artisanal fisheries remain significant, not least in providing protein for indigenous peoples in remote places, but also in the newly formed reservoirs and dams scattered across the country.

The Paraná/La Plata basins have a drainage system of 2 800 000 km$^2$, draining all south-central South America, from the borders of the Andes to the Serra do Mar along the Atlantic Coast. Its main river, the Paraná is 4695 km long, being the tenth largest river in the world, and the second largest in South America, after the Amazon. Its major tributary is the Paraguay River (2550 km). River discharge in its upper course varies from 8400 m$^3$ s$^{-1}$ to 13 000 m$^3$ s$^{-1}$ (minimum of 2550 m$^3$ s$^{-1}$ and maximum 33 740 m$^3$ s$^{-1}$). It is formed by the rivers Paranaíba and Grande (Fig. 12.1).

Among the main river basins in South America, the Paraná is the one most intensively dammed. By the end of the decade, it is expected that 69 hydroelectric impoundments with areas greater than 200 ha will be created in the Brazilian portion of the basin alone. The 45 existing reservoirs in the basin transformed the main

154

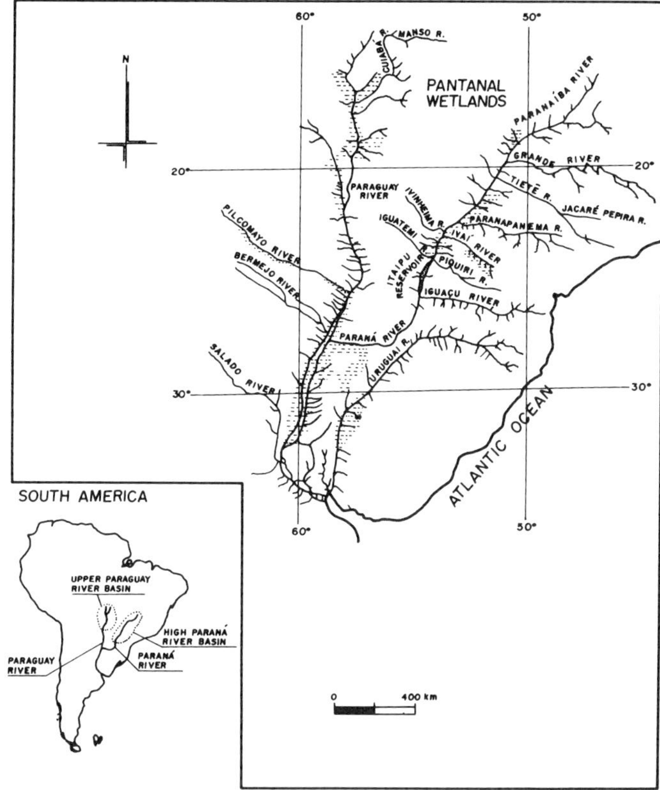


**Fig. 12.1**   The Paraná river basin and its location in South America.

Paraná river and its tributaries (Grande, Paranaíba, Tietê, Paranapanema, Iguaçu) into a succession of lakes. Of the 809 km of the Paraná flowing in the Brazilian territory, 483 km are running water. These will be impounded by more than 50% in 1995 by the building of Porto Primavera reservoir. Later the building of Ilha Grande reservoir will eliminate the last lotic stretch of the river, as the remaining Brazilian 30 km below the Itaipu reservoir will also be dammed by the Argentinean–Paraguayan reservoir of Corpus. The inundated area of the large reservoirs in the Paraná basin, at the beginning of the 1990s was 10 000 km$^2$ but by the end of the century will reach about 17 000 km$^2$, just in the Brazilian territory. The dams, combined with intense human occupation of the Paraná basin which has an amenable climate and very rich soil with easy access, have contributed to the reduction of landings and the disappearance of large migratory fish species, mainly in the higher courses of the Paraná river (Agostinho *et al.* 1994).

The reservoir of Itaipu (Fig. 12.2) (1460 km$^2$, nominal generating capacity of 12 600 MW, mean water residence time = 40 d) was closed in 1982 and since then a very intensive commercial fishery has flourished in its different biotopes. Balon (1974) suggested that the yield of tropical reservoirs is initially high, decreasing afterwards and

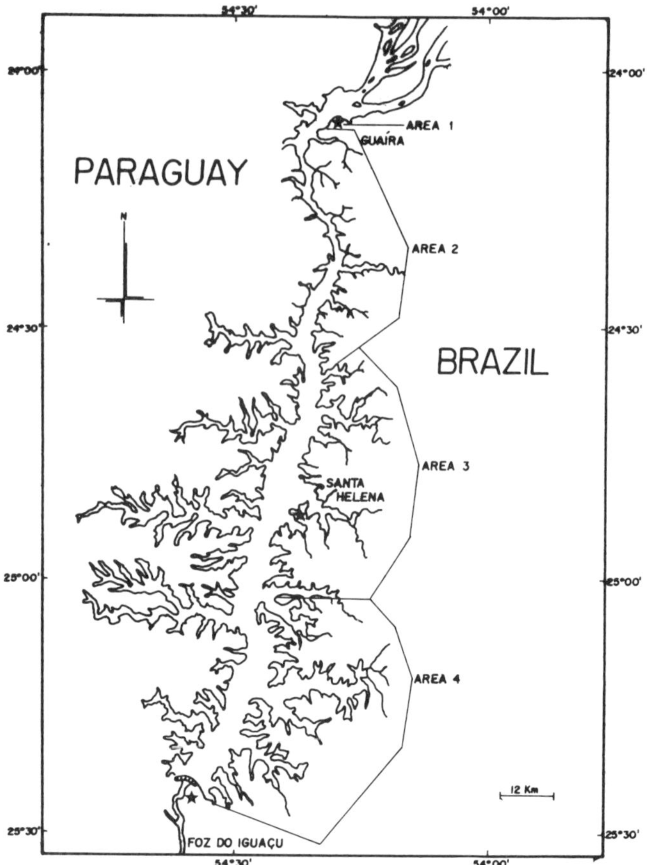


**Fig. 12.2** Itaipu Reservoir and its fishing areas.

stabilizing at a level usually higher than the original river stretch where it was built. Comparisons between the commercial catches in the Itaipu reservoir with other lotic stretches upriver, between 1986 and 1988 showed that the catches in the reservoir were higher, averaging 0.436 kg m$^{-2}$ of gillnet 24 h$^{-1}$ when compared to the main channel (0.277) or in channels of the inundated floodplain (0.307), albeit being surpassed by the catches in the marginal lagoons (0.578) (Agostinho and Júlio in press).

Petrere and Agostinho (1993) and Agostinho *et al.* (1994) give more detailed information about the fish fauna and commercial fisheries of Itaipu reservoir. The aim of this chapter is to describe the fish stock assessment procedures used on the reservoir.

## 12.2    Materials and methods

Catch and effort data were collected from about 220 000 individual fishing trips between 1987 and 1993 The data were collected according to year, season, high and low water, and landing site. The high water season extends from October to March, when temperatures are higher with intense rains. In this chapter the total catches are analysed, without considering particular fish species, in an attempt to detect patterns of local abundance. The effort data were calculated as the number of fishermen multiplied by the days of effective fishing (*nfdf*).

Fish were caught and landed in localities broadly classified in four categories (Fig. 12.2):

- *Area 1:* in the vicinity of the town of Guaíra, a predominantly lotic environment, where the fisheries are mostly concentrated upon the stocks of cascudo preto *Rhinelepis aspera* and cascudo abacaxi *Megalancistrus aculeatus* caught by cast-nets;
- *Area 2:* the predominantly lotic environment where the fisheries are mostly concentrated upon the catch of armado *Pterodoras granulosus* using longline and upon the curimba *Prochilodus lineatus* with gillnets;
- *Area 3:* the transitional zone where the fisheries are mostly concentrated upon the stocks of sardela *Hypophthalmus edentatus*, curvina *Plagioscion squamosissimus* and mandi *Pimelodus maculatus*, using gillnets with small mesh sizes (3.5, 4 and 5 cm, knot to knot mesh sizes);
- *Area 4:* the predominantly lentic environment in the body of the reservoir, where the fisheries are mostly concentrated upon the stocks of sardela, curimba and traíra *Hoplias malabaricus*, using gillnets with a mesh size of 4 cm.

The Schaeffer model was used to evaluate the relationship between catch (*C*) and effort (*f*):

$$C = af - bf^2 \tag{1}$$

After inspecting catch effort scatter plots of the original data, the following tentative statistical linear model was developed (Petrere 1986a,b; Ribeiro & Petrere 1990):

$$\ln(C, \text{kg})_{i,j,k} = \mu + year_i + season_j + area_k + interactions + \\ b \ln(no.\ fishermen \times fishing\ days) + e_{i,j,k} \tag{2}$$

where *i* is the year of capture, *j* the season and *k* the area.

This analysis of covariance model was considered suitable for examining the details in the data structure, the interactions between the covariates and the parameter variables (year, season and area) for detecting parallelism, and inspecting the residuals of the final reduced model (Petrere 1986a,b).

### 12.3     Results

The catch each year (Table 12.1) showed considerable variation between fishing season and area. When related to total commercial catch and effort for the years 1987 to 1993, the catch remained constant each year but the fishing effort increased, nearly

**Table 12.1**  Catch from different zones of Itaipu Reservoir for the period 1987 to 1993.

| Year | Area 1 | Area 2 | Area 3 | Area 4 | Total |
|---|---|---|---|---|---|
| 1987-SS | 45594.7 | 284768.4 | 478857.8 | 114851.3 | 924072.2 |
| 1987-AW | 51463.3 | 177924.3 | 342927.8 | 66482.8 | 638798.2 |
| 1988-SS | 84610.1 | 196850.0 | 497583.4 | 109704.7 | 888748.2 |
| 1988-AW | 58132.9 | 125430.1 | 343561.0 | 83828.2 | 610952.2 |
| 1989-SS | 90579.4 | 241636.1 | 453320.9 | 142787.4 | 928323.8 |
| 1989-AW | 96846.1 | 209731.9 | 377464.0 | 114258.9 | 798300.9 |
| 1990-SS | 54743.5 | 202906.8 | 362163.9 | 156657.7 | 776471.9 |
| 1990-AW | 65197.2 | 151098.7 | 308768.3 | 125354.5 | 650418.7 |
| 1991-SS | 47873.6 | 240469.2 | 441327.1 | 113874.3 | 843544.2 |
| 1991-AW | 50728.8 | 201613.5 | 385736.7 | 107118.0 | 745197.0 |
| 1992-SS | 36834.8 | 245990.4 | 443803.9 | 116789.9 | 843419.0 |
| 1992-AW | 36620.3 | 238172.7 | 357893.4 | 96682.0 | 729368.4 |
| 1993-SS | 26251.0 | 268008.5 | 387671.5 | 154599.1 | 836530.1 |
| 1993-AW | 26335.7 | 240137.7 | 312324.0 | 126732.5 | 705529.9 |
| Total | 771811 | 3024738 | 5493404 | 1629721 | 10919675 |

SS = spring and summer season; AW = autumn and winter season
Total per season SS = 6 041 109 kg; AW = 4 878 565 kg

doubling over the study period (Table 12.2). This relationship was described by the Schaeffer model as:

$$C = 0.034f - 1.75 \times 10^{-7} f^2 \qquad (\text{gl} = 5, \ r^2 = 99.7\%, \ F_{1,5} = 719^{**}) \qquad (3)$$

The maximum sustainable yield (MSY) was 1609 t yr$^{-1}$, corresponding to an $f_{op}$ of 95 895 fishermen × days of fishing. Note: in 1993 this optimum level of effort appeared to have been exceeded.

Following inspection of the scatter plots, and stepwise application of the covariant model (2), the model was reduced to the form

$$\ln(C, \text{kg})_{i,j,k} = \mu + year_i + area_k + (year \times season)_{ij} + b \ln(nfdf) + e_{i,j,k} \qquad (4)$$

where $\ln(nfdf)$ is the ln(number of fishermen × days of fishing)

**Table 12.2**  Total catch (*t*) and total effort (*f*) per year for Itaipu Reservoir, expressed in number of fishermen multiplied by the respective days of fishing for the period 1987 to 1993.

| Year | Catch (t) | Effort (f) |
|---|---|---|
| 1987 | 1563 | 67525 |
| 1988 | 1500 | 74332 |
| 1989 | 1727 | 89562 |
| 1990 | 1427 | 89609 |
| 1991 | 1589 | 99911 |
| 1992 | 1573 | 99072 |
| 1993 | 1542 | 120817 |

All the relationships were found to be log linear and the residual variance showed a strong positive kurtosis ($g_2 = 1.174$, $n = 336$, $P < 0.01$). Eight observations appeared to be outliers, but were included in the model.

The test of parallelism showed that the interaction between $\ln(nfdf) \times$ area was strong ($F_{3,309} = 6.090$, $P < 0.001$), but the interactions $\ln(nfdf) \times$ year and $\ln(nfdf) \times$ season were not significant ($P > 0.05$; Table 12.3).

**Table 12.3** Analysis of covariance of the variables used in the catch effort model.

| Source of variation | SS | df | MS | F | P |
|---|---|---|---|---|---|
| Year | 45.783 | 6 | 7.631 | 35.7 | 0.001 |
| Area | 6.649 | 3 | 2.216 | 10.4 | 0.001 |
| Year × Season | 4.128 | 6 | 0.688 | 3.2 | 0.004 |
| $\ln(nfdf)$ | 7.396 | 1 | 7.396 | 34.6 | 0.001 |
| Error | 68.182 | 319 | 0.214 | | |

$CV = 6.3\%$

## 12.4  Discussion

Petrere and Agostinho (1993) showed that among the reservoirs of the Paraná basin, the ones with higher catches (kg ha$^{-1}$) were those with free fluvial stretches upstream (i.e. the Itaipu and Barra Bonita) or with large tributaries draining directly into the reservoir (e.g. Jupiá). This shows the importance of the floodplain as a determining factor for reservoir biomass enhancement.

Catches in the spring and summer were expected to be higher than in the autumn and winter (Table 12.1) because most fisheries are based in passive gillnets and fish are less active at the lower temperatures. Furthermore fish form shoals in the spring and summer as they prepare to migrate upstream.

The zones between the fluvial and lentic reaches are also deposition zones and higher catches are expected because of the increased primary production (Kimmel *et al.* 1990). This is probably the main reason for 50% of the catches being taken from Area 3 (Table 12.1).

Despite the shortcomings of analytical models applied to short-term catch and effort data (Hilborn & Walters 1992), the model developed probably reflects the situation in Itaipu Reservoir. The reservoir has probably stabilized as total catches are fairly constant, despite fishing effort steadily increasing (Table 12.2). Most of the fisheries are heavily dependent upon the situation upriver, as many fish species, e.g. curimba, armado, jaú *Paulicea luetkni*, pintado *Pseudoplatystoma corruscans*, cascudo preto, migrate upstream for reproduction. Most of these species grow in the reservoir area and the backwaters of its tributaries. Juveniles of armado, jaú, pintado and barbado *Pinirampus pirinampu*, all migratory catfishes, are fairly common in the commercial landings, and there are suggestions of growth overfishing. The cascudo preto stock in Guaira appears overfished, as only small-sized individuals now appear in the castnet catches, and since 1990 catches have halved and CPUE has decreased by 70%. This situation has probably also worsened due to siltation around Guaira, arising from the increased deposition of sediments in the reservoir upper limit, because cascudo preto, the main species caught, is a demersal fish.

The results (Equation (4) and Table 12.3) suggest that the reservoir is a heterogeneous environment, due to the interactions between year and season. The additional interaction between the fishing effort and fishing area suggests that fishing strategy is not uniform across the reservoir, probably because the fishing gears used in each area are different. Mesh regulations introduced in 1991, allowing gillnets of 3.5 cm knot to knot mesh size, may also be responsible for the last interaction, as it provoked an increase in the catch of sardela. Further multiple comparisons between adjusted means need to be treated with caution because the effects are not the same in different levels of treatment and fishing effort.

Future considerations for the management of the reservoir fish stock should include the following activities.

(1)    Divert excess fishing effort towards the stocks of curvina, using gillnets of 4 cm. Curvina is the only introduced fish species in the reservoir, being responsible for 17.5% of total catches, 25% of which is caught with mesh 4 cm.

(2)    More than 50% of the armado sold in the market are immature specimens, caught in the backwaters of the tributaries of areas 2 and 3, with 3.5 cm mesh gillnets originally set for sardela. The armado is the third most important fish species caught in the reservoir (15.7%). If the marketing of immature fish were forbidden, fishermen would probably release them, but they can survive for 4 days entangled in the gillnets. The 3.5 cm mesh gillnet could also be restricted to the open water surface for catching the zooplankton filter-feeding sardela, without affecting the juveniles of other fish species.

(3)    Professional fisheries in the Paraguayan side of Itaipu should be permanently forbidden, because the margin areas in their country comprise lagoons and inundated forest, which act as a reserve of diversity and biomass for the whole of the reservoir. Although this strategy is desirable, Brazilian fishermen also fish in this region so intervention by the Brazilian authorities is necessary.

(4)    Presently, it is very easy to obtain a licence for fishing in the reservoir. To decrease fishing effort, restrictions which prevent the entrance of newcomers should be adopted.

(5)    Environmental education targeted at professional fishermen, teaching them about responsible fishing, has been a successful activity in recent years. It is now unavailable due to lack of funds. NUPELIA the organization responsible for this programme had a tri-monthly newsletter and radio programmes broadcast on eight radio stations; fishermen and their children were contacted at their home, in their villages and at school, and their response was enthusiastic. This activity, therefore, needs resurrecting.

## References


Agostinho Â.A. & Júlio H.F. (in press) Peixes da bacia do alto rio Paraná. In R.L. McConnell (ed.) *Ecologia de Comunidade de Peixes Tropicais*. Tradução de A.E.A.M. Vazzoler, A.A. Agostinho e P. Cunnighan. Editora da Universidade de São Paulo.

Agostinho Â.A., Júlio H.F. & Petrere M. (1994) Itaipu reservoir (Brazil): impacts of the impoundment on the fish fauna and fisheries. In: I.G. Cowx (ed.) *Rehabilitation of freshwater fisheries*. Oxford: Fishing News Books, pp 120–129.

Balon E.K. (1974) Fish production of a tropical ecosystem. In: E.K. Balon & A.G. Coche (eds) *Lake Kariba: A man-made tropical ecosystem in Central Africa*. The Hague: Dr W Junk, pp 257–573.

Hilborn R. & Walters C.J. (1992) *Quantitative fisheries stock assessment*. New York: Chapman and Hall.

Kimmel B.L., Lind O.T. & Paulson L.J. (1990) Reservoir primary production. In: K.W. Thornton, B.L. Kimmel & F.E. Payne (eds) *Reservoir limnology: Ecological perspectives*. New York: Wiley, pp 133–194.

Paiva M.P. (1982) *Grandes Represas do Brasil*. Brasília: Editerra.

Petrere M. (1986a) Amazon fisheries: I – Variations in the relative abundance of Tambaqui (*Colossoma macropomum* Cuvier, 1818) based on catch and effort data of the gill-net fisheries. *Amazoniana* **9**, 527–547.

Petrere M. (1986b) Amazon fisheries: II – Variations in the relative abundance of Tucunaré (*Cichla ocellaris, C. temensis*) based on catch and effort data of trident fisheries. *Amazoniana* **10**, 1–13.

Petrere M. & Agostinho A.A. (1993) The fisheries in the Brazilian portion of the Paraná river. Document presented at the UN/COPESCAL meeting 'Consulta de Expertos sobre los Recursos Pesqueros de la Cuenca del Plata', Montevideo, Uruguay, 5–7 May, 1993.

Ribeiro M.C.L.B. & Petrere M. (1990) Fisheries ecology and management of the jaraqui (*Semaprochilodus taeniurus, S. isignis*) in Central Amazonia. *Regulated Rivers: Research & Management* **5**, 195–215.

# III HYDROACOUSTIC ASSESSMENT

# Chapter 13
# Use of horizontal dual-beam sonar for fish surveys in shallow waters

J. KUBECKA *Royal Holloway, University of London, Egham, TW20 0EX Surrey, UK*
*Address for correspondence: Hydrobiological Institute, Czech Academy of Sciences, Na sadkach 7,*
*370 05 Ceske Budejovice, Czech Republic*

**Abstract**   The main problems associated with the use of horizontal sonar in shallow waters are conditions of low signal-to-noise ratios (SNR) and uncertainty as to the aspect of the fish body being insonified. The influence of noise can be reduced by proper positioning of the acoustic beam in the water column and by proper calibration *in situ*. Methods for processing (manual bottom tracking, setting and interpretation of noise thresholds, setting of pulse width and fish tracking criteria) and post-processing (sampling volume definition and density calculation, fish size identification) are described.

KEYWORDS: Fish density, fish size, horizontal sonar, noise

## 13.1   Introduction

Fisheries sonar is a well-established quantitative tool for fish studies in large waters (Mitson 1983; MacLennan & Simmonds 1992). Most fisheries sonar applications use vertically-oriented sound beams from the surface to the bottom. Following the recent development of narrow-beamed transducers with negligible side lobes it has become possible to use sonar horizontally in shallow waters with depths between 1.5 m and 5 m.

Horizontal sonar was used on a number of occasions for fixed location surveys monitoring migratory behaviour of mostly salmonid fish (Gaudet 1990; Mesiar *et al.* 1990; Johnston & Hopelain 1990; Johnson *et al.* 1992). There is little or no information about the use of horizontal sonar for quantitative fixed-location and mobile surveys of shallow waters. This contribution summarizes a three-year project to establish methods for quantitative fish stock assessment in British rivers and other shallow waters. The chapter concentrates only on the main features of horizontal beaming different from conventional vertical applications. In order to understand the chapter fully readers require an elementary background in sonar methods from textbooks such as Urick (1975), Mitson (1983) or MacLennan and Simmonds (1992). Most of the information was collected with BioSonics dual-beam sonars operating on 420 and 200 kHz frequency in standard configuration (Fig. 13.1). This configuration enables both real time processing by PC-based processors and storing data on digital audio tape recorder (DAT) for eventual reprocessing from playback.

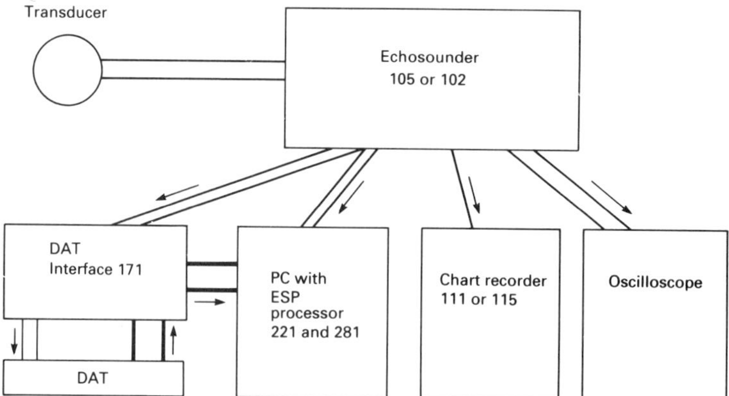


**Fig. 13.1**   Scheme of normal configuration of BioSonics dual beam sonar. Numbers indicate models of BioSonics products. Thin lines show signal trajectories during data collection, thick lines during playback.

## 13.2    Deployment of the transducer and the usable range

The transducer is the under-water component of the sonar system producing the sound cone beam. Various stands with pan-and-tilt rotators are used for fixed location surveys (Gaudet 1990; Mesiar *et al.* 1990; Johnston & Hopelain 1990; Johnson *et al.* 1992). A simple rod with the transducer attached to the bottom can be used in modest conditions. However, the presence of a remotely controlled rotator is highly desirable for adjusting the tilt angle of the beam.

Acoustic fish counters require special beam orientations (Gaudet 1990; Johnson *et al.* 1992), while fixed location and mobile studies for fish density and biomass estimations should cover maximum usable range (MUR). Figure 13.2 defines MUR as the distance from transducer where the acoustic beam fills all the water column between the bottom and surface (Figs 13.2c,d). The beam touches the surface or bottom at the end point of MUR and the oscilloscope display of echoes shows bottom or surface reverberation (BSR, Fig. 13.2c). BSR is usually of a rather diffuse type composed from a number of wide overlapping peaks of noise. This area should be eliminated from the processing by manual bottom tracking. Bottom marker (BM) on Fig. 13.2 shows the position of a manually imposed bottom which marks the end of processed data. Usable range usually decreases if the transducer is tilted down (Fig. 13.2b) or up (Fig. 13.2f).

When the transducer is tilted down, the bottom echo becomes 'stronger' (shorter distinct peak of very high amplitude) and the noise in the usable range is always low (less than 0.1 V apart from the side lobe area in Fig. 13.2a). The distribution of reflected echoes when the transducer is tilted up depends on the state of the water surface. Rough surface generates apparent surface reverberation similar to Fig. 13.2c. With a very calm surface the point where the beam touches the surface need not be apparent as whole beam is being reflected by the surface. This is erroneous because many of the targets behind this point are biased. False MUR of such surface-reflected

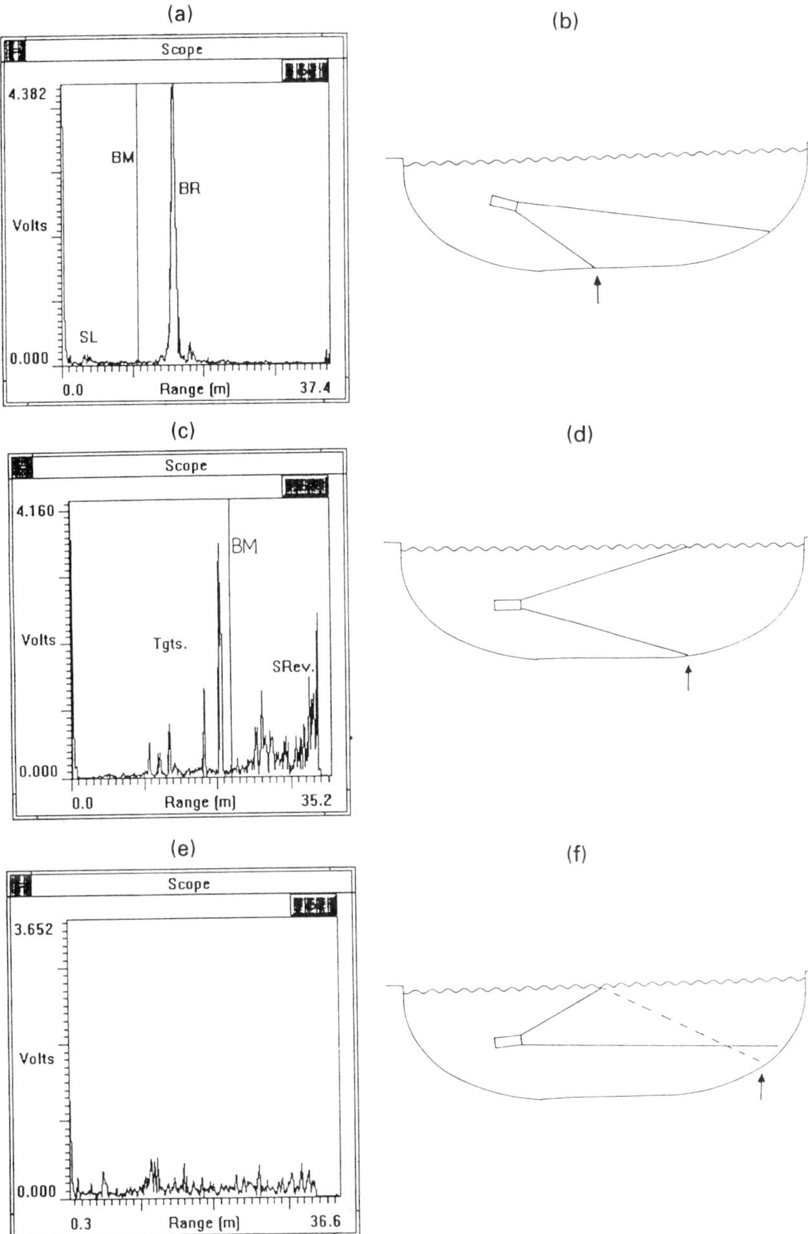



**Fig. 13.2**  Three positions of horizontal beam in shallow water and three equivalent oscilloscope narrow beam displays. BM = Bottom marker, indicates the position of manually imposed bottom; BR = bottom reverberation; SR = surface reverberation; Tgts = fish targets. a,b – beam is tilted down; c,d – beam uses maximum usable range; e,f – beam is tilted up and is being reflected by calm surface. Arrows show the range where the sound beam touches the bottom.

beams can be much bigger (Figs 13.2e,f) but the noise level is always higher compared with the true MUR (Figs 13.1c,d).

During mobile surveys in calm weather, the desirable state (Figs 13.2c,d) can easily be shifted to the surface-reflected type (Figs 13.2e,f). If noise level suddenly increases it is necessary to tilt the transducer down (to the position equivalent to the Figs 13.2a,b). MUR is reached by careful tilting up and the background noise level should not increase significantly.

Requirements for exact horizontal beam orientation in the water support the use of rigid transducer deployment, as described by Kubecka *et al.* (1994) for mobile surveys. The use of towed bodies is very sensitive to beam position changes resulting in drastic reverberation changes. Rigid fixation of the transducer enables changes in the boat speed and immediate shift to fixed location survey after stopping the boat.

## 13.3    Hardware setting

### 13.3.1    *Transducers*

Relatively high frequency systems (120–1000 kHz) are usually considered for horizontal use because of low cost, size and near field area. Further requirements are: narrow angle (especially in the vertical plane), concentricity of both beams and low side lobes (sensitivity to structures beyond the nominal beam angle). The last requirement can be underestimated; it is desirable for side lobes to be about 30 dB down compared with on-axis sensitivity. Common transducers with about −18 to −20 dB side lobes usually cause trouble at close range (Fig. 13.2a).

### 13.3.2    *Sonar setting*

Rules for sonar setting are similar to vertical use. Short pulse duration of 0.3–0.5 ms (For BioSonics 101–105 sounders) with equivalent frequency bandwidth is highly advisable for the resolution of single targets.

Pulse repetition rate (frequency of sound emissions) depends on the range of targets of interest. Theoretical pulse repetition rate can be calculated as

$$PRR = c/2R \tag{1}$$

where $c$ is the speed of sound (m s$^{-1}$) and $R$ is the range of targets of interest (m). A pulse repetition rate of about half the theoretical value, usually 10–30 Hz is recommended.

For mobile quantitative surveys it is recommended that data are collected with both time varied gains (TVG, i.e. corrections for spreading loss of the sound) 20 and 40 log $R$ simultaneously for echo integration (EI, integration of overall reflected energy without respect to individual targets) and dual-beam (DB) echo-counting of every single target. Analogue systems like BioSonics 101–105 sounders always require a backup tape record for eventual reprocessing.

**13.4     On-axis calibration**

On-axis calibration must accompany every quantitative survey (Johanesson & Mitson 1983; Foote *et al.* 1987). Tungsten carbide standard spheres (BioSonics 1989; MacLennan & Simmonds 1992) are especially suitable. According to Foote *et al.* (1987), on-axis calibration of BioSonics dual-beam system can be with several specifics.

(1)     All parameters of the sonar equation (Urick 1975) are for the BioSonics system defined on the basis of detected voltage (not root-mean-square voltage).

(2)     The use of the source level (*SL*) supplied by the manufacturer is recommended and receiver sensitivity (*RS*) is calculated from the sonar equation (Urick 1975):

$$RS = 20 \log Vst - SL - RG - TSst \tag{2}$$

where *Vst* is the detected voltage of the standard target, *RG* is the receiver gain set on sonar and *TSst* is the target strength of the standard target.

(3)     Procedure (2) must be repeated for both narrow and wide beams of dual beam system for 40 log *R* data and at least for the narrow beam for 20 log *R* data.

(4)     It is necessary to ensure a rigid on-axis position of the standard target because even small absolute deviation from the axis can influence calibration at short ranges.

(5)     Voltages of the standard target from Equation (2) should be determined by the same echo signal processor (ESP) which will be used for subsequent data processing. Dual beam processors measure exact voltage peak values which can be used in the equation. With EI it is necessary either to read peak voltages from the computer screen oscilloscope or to use the square root of the echo-integrated value mean $V^2$ for a standard target peak. This ensures that calibration compensates eventual gain differences at any level of the data handling and processing. In DB calibration, it is sensible to process the standard target record with calibrated *RS* values. Estimated target strength of the standard target should be the same as the known value of *TSst*.

(6)     With 420 kHz frequency it was impossible to achieve 2% coefficient of variation of squared voltage measurements of on-axis standard target records as recommended by Foote *et al.* (1987). This was caused by higher noise level in shallow waters and also by higher variability of noise reflection at 420 kHz frequency. For calibration using horizontal beaming, a coefficient of variation of about 7% represents achievable acceptance level for maintaining calibration error within $\pm 0.6$ dB ($\pm 2$ SD).

**13.5     Software processing**

The PC-based BioSonics model 281 dual beam processor and model 221 Echo Integrator are described by Dawson *et al.* (1989) and BioSonics (1990). This section is only intended to add specifics for shallow water use.

Processing represents the first level of analysis during which fish targets are separated from noise. DB procedure usually precedes EI because average size of fish targets is necessary for scaling of EI. With DB there is a possibility of reapplying various filtering criteria (like thresholds, pulse width criteria, BPF criteria, etc.) during post-processing and it is therefore generally advisable to be less restrictive during this first level processing.

### 13.5.1     *Manual bottom tracking (for DB and EI)*

Manual bottom tracking using a mouse and a bottom positioning ruler on the PC screen is essential for manual adjustment of MUR, as shown by the bottom marker on Fig. 13.2. Bottom voltage shape varies drastically during horizontal surveys and there is no automatic way of bottom tracking available for such surveys. Manual bottom position must be in a range shorter than SBR in diffuse (Fig. 13.2c) or strong bottom (Fig. 13.2a) form (Section 13.2). Manual bottom should be set to a low range before the start of every processing run and should be extended further according to the screen oscilloscope.

### 13.5.2     *Strata definition and thresholding (DB and EI)*

Noise thresholds should be set three times higher than the average noise level seen on the screen oscilloscope. As the noise level increases with range (with 40 log $R$ TVG, Fig. 13.2), noise thresholds should also increase. Minimum detectable size of fish therefore increases with range from the transducer. The probability $P$ of recording targets of a certain size with certain thresholds applied can be calculated as:

$$P = A(V_{t,TS,R}) \,/\, A(_{B,R}) \tag{3}$$

where $A(_{B,R})$ is the cross-section area of the nominal beam at range $R$ and $A(_{Vt,TS,R})$, the beam cross-section area around the axis at the range $R$ where a target of strength $TS$ generates a voltage bigger than the threshold voltage $Vt$. It can be calculated as:

$$A(_{Vt,TS,R}) = \pi* \,(Tg \; a_{0.5} * R)^2 \tag{4}$$

where $\alpha_{0.5}$ is the beam half angle equivalent to the area $A(_{Vt,TS,R})$.

$$\alpha_{0.5} = a*|BPF|^{\,b} \tag{5}$$

$a$ and $b$ are regression parameters of the parabolic equation usually supplied by the manufacturer and the beam pattern factor (*BPF*) can be calculated from the sonar equation:

$$BPF = 0.5 * (20 \log Vt - SL - RG - RS - TS) \tag{6}$$

for 40 log $R$ TVG or

$$BPF = 0.5 * (20 \log Vt - SL - RG - RS - TS - 20 \log R) \tag{7}$$

for 20 log $R$ TVG. *BPF* must be negative; if *BPF* is bigger than 0, there is no possibility of recording a target of given *TS*.

Cross-section area of the nominal beam $A(_{B,R})$ can be calculated by substituting the nominal value *BPF* = –3 in Equation (5).

As a rule, thresholds of the same restriction towards the same *TS* should be used for both DB and EI processing (Kubecka 1994a). Table 13.1 gives an example of a threshold restriction table at different ranges from the transducer for a horizontal survey. Short ranges from the transducer (usually the first 4 m) should be excluded from the processing in mobile surveys because of disturbance of fish.

**Table 13.1** Probability of recording targets of different size in ranges with different noise levels to which appropriate noise thresholds have been applied. $SL$ = 218.5; $RG$ = –12; $a$ = 1.6681; $b$ = 0.5016; *BPF* set = 3.29; *TNR* = threshold-to-noise ratio.

| | Conditions | | | | | | |
|---|---|---|---|---|---|---|---|
| Range, m | 1–3 | 3–6 | 6–9 | 9–12 | 12–15 | 15–18 | 18–21 |
| Noise level, $V$ | 0.01 | 0.02 | 0.03 | 0.04 | 0.05 | 0.06 | 0.07 |
| Noise threshold $V$ | 0.03 | 0.06 | 0.09 | 0.12 | 0.15 | 0.18 | 0.20 |
| TNR | 3:1 | 3:1 | 3:1 | 3:1 | 3:1 | 3:1 | 2.9:1 |
| Noise-adjusted threshold voltage $V^2 + V^2$ model (Kubecka 1994a) | 0.028 | 0.057 | 0.085 | 0.110 | 0.141 | 0.170 | 0.190 |

| Target strength (dB) | Probability of recording targets of different target strength | | | | | | |
|---|---|---|---|---|---|---|---|
| –60.0 | 90 | < 1 | | | | | |
| –57.5 | 128 | 36 | | | | | |
| –55.0 | 167 | 74 | 21 | | | | |
| –52.5 | 205 | 113 | 59 | 21 | | | |
| –50.0 | 243 | 151 | 97 | 59 | 30 | 5 | |
| –47.5 | 282 | 189 | 135 | 97 | 68 | 43 | 30 |
| –45.0 | 321 | 228 | 174 | 135 | 106 | 81 | 68 |
| –40.0 | 398 | 305 | 251 | 212 | 182 | 158 | 145 |

### 13.5.3 *Pulse width criteria (DB only)*

Pulse width (the duration of the returned echo in ms) is usually used to distinguish between single and multiple targets. A specific feature of most horizontal sonar applications is the necessity not to use all $> 1/3$ amplitude pulse width criteria (i.e. echo duration measured at $1/3$ of the peak voltage amplitude) if the 3:1 threshold-to-noise ratio is used. With the BioSonics system only half amplitude pulse width filtering can be used (recommended values are given in Table 13.2). Quarter amplitude criteria can be used with a threshold-to-noise ratio of 5:1 or better.

Figure 13.3 shows the distribution of 15 333 single echoes from fish in front of the sonar operating with 0.4 ms pulse duration. The variability is caused by sampling error of the system and also by the noise which can prolong the pulse of weaker

**Table 13.2**  Recommended pulse width criteria for different pulse durations and with signal-to-noise ratios of 3:1 or better. Frequency bandwidth (kHz) multiplied by pulse duration (ms) should be >2.

| Pulse duration (ms) | Amplitude | Narrow beam (min) | Narrow beam (max) | Wide beam (min) | Wide beam (max) |
| --- | --- | --- | --- | --- | --- |
| 0.3 | 1/2 | 0.24 | 0.48 | 0.32 | 0.52 |
| 0.4 | 1/2 | 0.36 | 0.64 | 0.40 | 0.64 |
| 0.4 | 1/4* | 0.48 | 0.72 | — | — |
| 0.6 | 1/2 | 0.52 | 0.84 | 0.52 | 1.05 |
| 0.6 | 1/4* | 0.60 | 1.00 | — | — |
| 0.8 | 1/2 | 0.72 | 1.00 | 0.72 | 1.16 |
| 0.8 | 1/4* | 0.88 | 1.20 | — | — |
| 1.6 | 1/2 | 1.40 | 1.80 | 1.40 | 2.00 |
| 1.6 | 1/4* | 1.76 | 2.40 | — | — |

*To be used only when the signal-to-noise ratio is 5:1 or better.

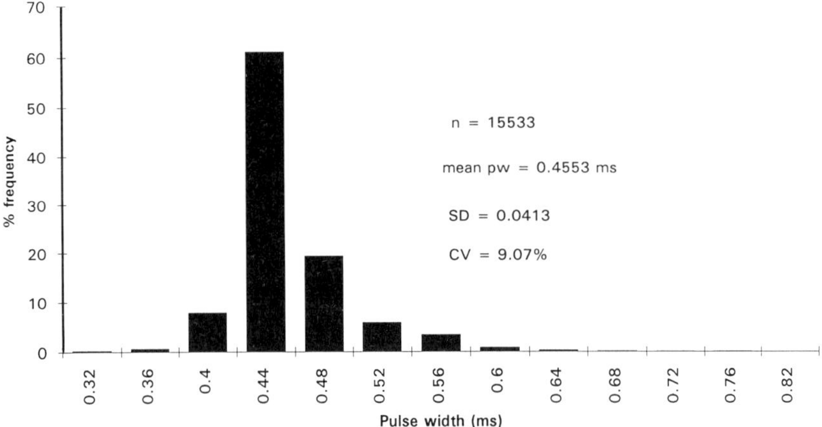


**Fig. 13.3**  Half amplitude pulse width frequency distribution of 15 533 fish echoes recorded from 43 known size fish mounted in front of the transducer (pulse duration 0.4 ms).

echoes (see also Robinson 1981). Figure 13.4 shows the relationship between average target voltage and pulse width showing longer pulses for smaller fish. Kubecka *et al.* (1992) gave an example of pulse width frequency distribution of natural fish population. Later experiments revealed that the percentage of multiple targets cannot be simply calculated from such distributions because the DB processor is not capable of processing some rather distant multiple targets (those with relative small echo overlap). There is therefore no mechanism to avoid EI if many multiple targets are present.

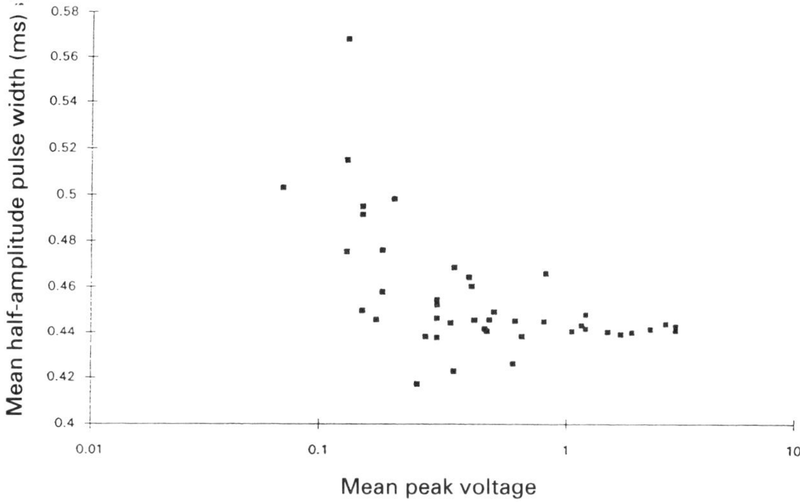


**Fig. 13.4**   The relationship between mean half amplitude pulse width and mean peak voltage of 43 fish mounted in front of the transducer (pulse duration 0.4 ms).

### 13.5.4   *Fish Tracking (DB only)*

Tracking of groups of targets belonging to one fish is essential for fish aspect or identification of direction of movement in fixed location surveys but can also be applied in quantitative mobile surveys. Minimal criteria are:

(1)   The fish tracking window (the range from the transducer interval in m, where the processor expects future echoes from individual fish) multiplied by pulse repetition rate (see Equation (1); in Hz) should not exceed 4 (2 in slowly moving dense populations).
(2)   The fish filter beam pattern threshold ($= BPF$ threshold) should be set at about –3 dB (nominal beam angle).
(3)   Every valid fish must contain at least one echo in the beam defined in (2), above.

Many other fish tracking filtering possibilities are available but it is usually more appropriate to apply them in post-processing.

### 13.6   **Post-processing**

Both DB and EI data need post-processing using either BioSonics commercial software (ESP_TS and ESP_CRUNCH) or spreadsheets.

DB echo counting can be carried out successfully using standard ESP_TS software. This data reduction programme enables application of noise-corrected receiver sensitivities in different ranges which is an effective way for the correction on noise interference (Kubecka 1994a). It is also possible to apply a second level of data filtering by higher threshold, narrower pulse width criteria and by a beam pattern factor (*BPF*). *BPF* defines the position of every target in relation to the acoustic axis of the beam. *BPF* is 0 dB on the axis and negative for targets further away from the acoustic axis (Figs 13.5, 13.6). It is therefore possible to define the sampling volume in terms of off-axis distance in dB or the angle from equations (4–6). At distances further from the axis (–5 and less dB), more and more targets are knocked down by the thresholds applied as their voltage decreases further from the axis (Fig. 13.6). The curve of volumes is only a function of the beam pattern of the transducer and in the example on Fig. 13.6 it does not fall with off-axis distance like the frequency of targets.

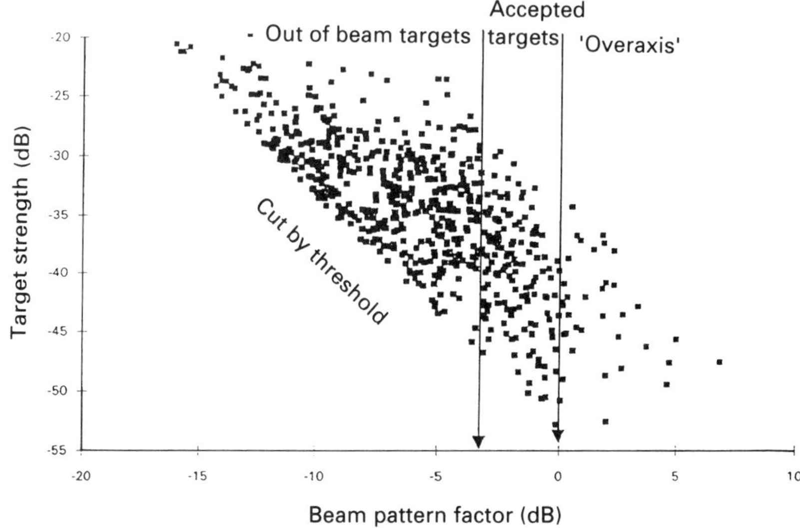


**Fig. 13.5** The relationship between target strength and the beam pattern in the population of targets in the River Thames. The two arrows define sampling volume. Most targets are out of nominal beam, some are 'overaxis'.

In horizontal applications of DB system some targets have positive *BPF*. These nonsense 'overaxis' targets have accidentally narrow beam echo voltage which is too high compared with their wide beam voltage. This may happen as a result of random variation of reflection or to noise interference. The smallest targets are prone to 'overaxis' position (Fig. 13.5). BioSonics usually handle these 'overaxis' targets as those just on axis (Dawson & Karp 1990) and accept targets with *BPF* –3 to +3 dB. This approach is justified with transducers having similar sampling volumes and therefore similar numbers of recorded echoes in all *BPF* bands around the imposed outer boundary of the sampling volume (usually –3 dB, Fig. 13.6). Figure 13.6 shows that about 6% of targets have positive *BPF* and that every *BPF* band between 0 and –4 dB contains 11–15% of targets. Similar to the case of randomly occurring positive *BPF* targets, there is likely to be an exchange of targets with randomly varying *BPF*

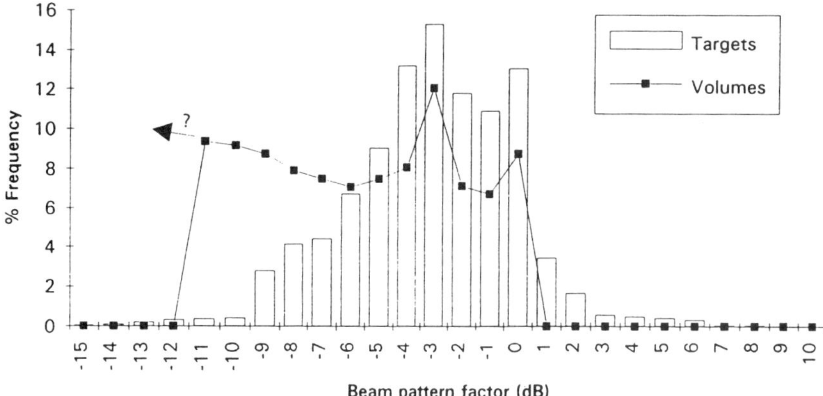



**Fig. 13.6**  Frequency distribution of recorded targets in various *BPF* bands compared with frequency distributions of sampled volumes (not calculated behind –11 dB). BioSonics 6/15 degrees transducer, River Thames near Abingdon, 9630 targets.

on the imposed nominal angle boundary of –3 dB. When the numbers of targets on both sides of the –3 dB boundary are similar (e.g. in Fig. 13.6) the random errors of increased and decreased *BPF* probably compensate for each other. Acceptance of positive *BPF* targets is problematic with poorer quality transducers where sampling volume may increase continually with off-axis distance.

Comparison of the *BPF* frequency distribution of target populations with the beam volume frequency distribution of the transducer (Fig. 13.6) is a useful test of the balance of DB system calibration.

Unlike DB processing, the transfer onto spreadsheets was found the most effective way to handle EI data. The so called 'area table' generated by BioSonics ESP_WIEW software gives EI fish densities per side scan area (*SSD* in fish m$^{-2}$) and average manual bottom range (*BR* in m). Fish density (*d* in fish.m$^{-3}$) is calculated for every EI period from:

$$d = SSD \ / \ BR \tag{8}$$

### 13.7    Data interpretation

Fish passage rate, size distribution and density are usual parameters required in horizontal sonar applications. For definitions of passage rates see the fixed location studies of Johnston and Hopelain (1990), Dawson *et al.* (1988), Mesiar *et al.* (1990), Gaudet (1990), Johnson *et al.* (1992). Fish acoustic TS vary dramatically in various aspects of the horizontal plane (Love 1977) and it is essential to know fish orientation for size reconstruction. In rivers, it appears that most fish move in relation to flow (i.e. upstream or downstream, Fig. 13.7). The transducer beaming perpendicularly across the river therefore records most fish from their side aspect and the length can be reconstructed from regressions (Love 1969; Kubecka 1994b).

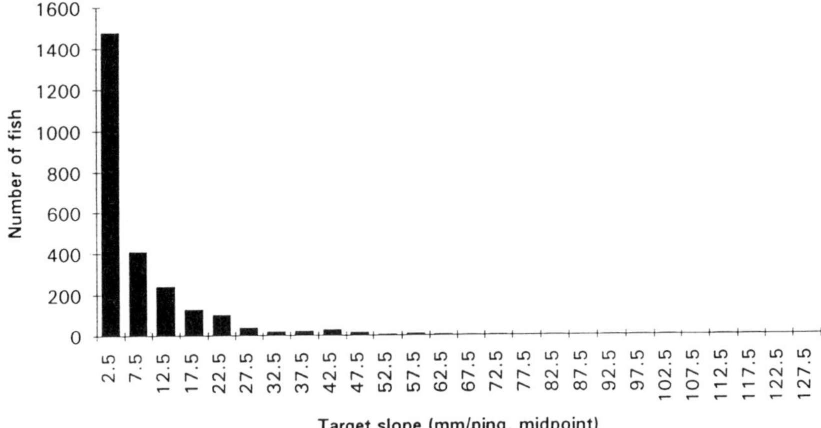



**Fig. 13.7**  Frequency distribution of the slope of fish targets as fish cross the horiziontal sound beam perpendicular to flow in River Thames near Chertsey. Even under low flow conditions most fish were moving parallel with the current, presenting their side aspect to a stationary transducer. About 2500 fish with four or more hits each.

The situation is much more complicated in shallow rivers without flow, where there are three options.

(1)  Numerous fixed location point samples where side-aspect fish can be filtered from the population using criteria called 'fish slope' (change of range in m per acoustic ping). Fish with the slope of $\pm 3$ mm per ping can be considered as recorded from the side aspect.

(2)  The deconvolution technique described in Kubecka *et al.* (1994) which enables size identification of fish recorded from an unknown aspect in mobile surveys.

(3)  An approximation of the average size of randomly-oriented fish can be obtained by applying average all-aspect TS vs length regression (Kubecka 1994b).

Fish density from DB counting is given by the number of accepted fish and the sampling volume. Unfortunately, DB counting provides only values of density per whole acoustic run, defined by the starting and stopping of the processor. EI reports can give much more detailed information about the distribution of fish density along the boat trajectory or sampling time. Density for mobile surveys can also be calculated by improved classical echo-counting, knowing the boat trajectory during acoustic run, vertical beam angle and the number of tracked fish with at least one echo in the beam.

### 13.8    Sampling strategy and season

Quantitative fish stock assessments by horizontal sonar should be carried out at night during warm periods of the year (May–September – Fig. 13.8). A good strategy for

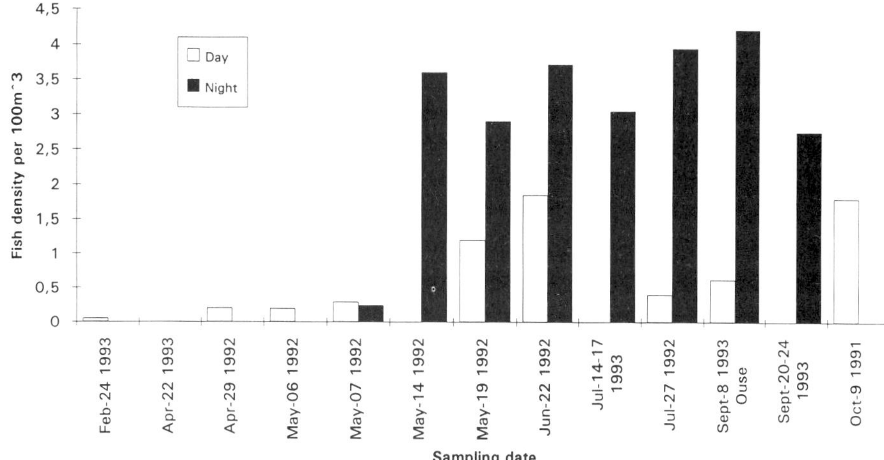


**Fig. 13.8**  Absolute densities of fish per 100 m³ estimated in the Rivers Thames and Ouse (one case) during day and night mobile surveys at various times of the year.

routine surveys is to combine mobile surveys of extensive areas of water which provide an overall picture of fish distribution with subsequent fixed-location observations of interesting areas determined during mobile surveying. Fixed-location observations with 'fish slope tracking' provide best acoustic size composition. The sound beam is usually oriented sideways from the boat (at one or both sides of the boat with multiplexing between two transducers), thus covering a 15–50 m (according to the set-up and conditions) band along the trajectory of the boat.

## Acknowledgements

Material for this chapter was collected during the project Hydroacoustic Methods of Fish Survey funded by the National Rivers Authority (UK). The help of Drs A. Duncan and A.J. Butterworth and an anonymous referee is highly appreciated.

## References

BioSonics (1989) BioSonics calibration spheres, User's manual, 23 pp.

BioSonics (1990) Echo Signal Processor, Operator's manual, 290 pp.

Dawson J.J., Brooks T.J. & Kuehl E.S. (1989) An innovative acoustic signal processor for fisheries science. *Proceedings of the Institute of Acoustics* **11**, 131–140.

Dawson J.J. & Karp W.A. (1990) In situ measures of target-strength variability of individual fish. *Rapport P.-v. Reun. Conseil international Exploration de la Mer* **189**, 264–273.

Dawson J.J., Karp W.A. & Raemhild G.A. (1988) Fixed-location hydroacoustics for quantitative fisheries studies. *BioSonics Memo*, 32 pp.

Foote K.G., Knudsen H.P., Vestnes G., MacLennan D.N. & Simmonds J. (1987) Calibration of acoustic instruments for fish density estimation: A Practical guide. *ICES Co-operative Research Report* **144**, 1–70.

Gaudet D.M. (1990) Enumeration of migrating salmon using fixed location sonar counters. *Rapport P.-v. Reun. Conseil international Exploration de la Mer* **189**, 197–209.

Johanesson K.A. & Mitson R.B. (1983) Fisheries acoustics: A practical manual for biomass estimation. *FAO Fisheries Technical Paper* **240**, 249 pp.

Johnson G.E., Sullivan C.M. & Erho M.W. (1992) Hydroacoustic studies for developing a smolt bypass system at Wells Dam. *Fisheries Research* **14**, 221–237.

Johnston S.V. & Hopelain J.S. (1990) The application of dual-beam target tracking and Doppler-shifted echo processing to assess upstream salmonid migration in the Klamath River, California. *Rapport P.-v. Reun. Conseil international Exploration de la Mer* **189**, 210–222.

Kubecka J. (1994a) Noise handling in fisheries surveys by horizontal sonar in shallow waters. In: L. Bjorno (ed.) *2nd European Conference on Underwater Acoustics*, European Commission, Luxembourg, **833–839**.

Kubecka J. (1994b) Simple model on the relationship between fish acoustic target strengths and aspect for high-frequency sonar in shallow waters. *Journal of Applied Ichthyology*, **10**, 75–81.

Kubecka J., Duncan A. & Butterworth A.J. (1992) Echo counting of echo integration for fish biomass assessment in shallow waters. In: M. Weydert (ed.) *European Conference on Underwater Acoustics*. London: Elsevier, pp 129–132.

Kubecka J., Duncan A., Duncan W.M., Sinclair D. & Butterworth A.J. (1994) Brown trout populations of three Scottish lochs estimated by horizontal sonar and multi-mesh gill nets. *Fisheries Research* **20**, 29–48.

Love R.H. (1969) Maximum side-aspect target strength of an individual fish. *Journal of Acoustic Society of America* **49**, 746–752.

Love R.H. (1977) Target strength of an individual fish at any aspect. *Journal of Acoustic Society of America* **62**, 1397–1403.

MacLennan D.N. & Simmonds E.J. (1992) *Fisheries Acoustics*. London: Chapman and Hall, 325 pp.

Mesiar D.C., Eggers D.M. & Gaudet D.M. (1990) Development of techniques for the application of hydroacoustics to counting migratory fish in large rivers. *Rapport P.-v. Reun. Conseil international Exploration de la Mer* **189**, 223–232.

Mitson R.B. (1983) *Fisheries sonar*. Oxford: Fishing News Books, 287 pp.

Robinson B.J. (1981) In situ target strength measurement; the problem of single fish echo recognition. In: J.B. Suomala (ed.) *Meeting on hydroacoustical methods for the estimation of marine fish population*. Ch. Stark Draper Laboratory, Cambridge Mass., 525–532.

Urick R.J. (1975) *Principles of underwater sound*. New York: McGraw-Hill, 384 pp.

# Chapter 14
# Application of a hydroacoustic sampling technique in a large wind-exposed shallow lake

P.J. MOUS *Wageningen Agricultural University, Dept. of Fish Culture and Fisheries, The Netherlands/Netherlands Institute for Fisheries Research (RIVO-DLO), P.O. Box 68, 1970 AB IJmuiden, The Netherlands*
J. KEMPER *Organisation for the Improvement of Inland Fisheries, P.O. Box 433, 3430 AK Nieuwegein, The Netherlands*

**Abstract**    A pilot survey was carried out to assess whether hydroacoustics is a suitable sampling tool to estimate the spatial distribution, size distribution and stock abundance of small pelagic fish in a large (180 000 ha), shallow (ca 4 m), wind-exposed lake. The targeted fish were perch (*Perca fluviatilis*) and smelt (*Osmerus eperlanus*) of 3–9 cm body length. A scientific echo-sounder (Biosonics® 2000ES) with a horizontally directed acoustic beam was used. The transducer was mounted in a towed body. Backscattered acoustic signals were identified as fish (*AIF*) by the echo signal processing software of Biosonics. Target strength estimation was facilitated by the dual beam technology of Biosonics. The swimming direction of *AIF* relative to the transducer could be estimated from the change in distance from the transducer of targets in consecutive pings. Detection of targeted fish was hampered by high noise levels and was dependent on the orientation of the fish in the beam. Most *AIF* avoided the research vessel. This is consistent with avoidance behaviour of fish. *AIF* avoiding the vessel had a higher target strength than fish not avoiding the vessel. This was caused by size-specific avoidance behaviour or by a difference in orientation between fish avoiding the vessel and fish not avoiding the vessel. Sampling conditions influenced the output of the hydroacoustic system: significantly less fish were detected with increasing windforce (1–5 Beaufort). Hydroacoustics are not recommended for estimating spatial distribution, size distribution and fish abundance on large, shallow lakes. Variable fish behaviour and sampling conditions are major sources of bias in the estimates.

KEYWORDS: hydroacoustics, sampling methodology, shallow lakes

## 14.1    Introduction

Lake IJsselmeer, a eutrophic lake in the Netherlands with an average depth of 4 m, is divided by a dike into a northern basin of 112 000 ha and a southern basin of 68 000 ha. It supports commercial fisheries for eel *Anguilla anguilla* (L.), perch *Perca fluviatilis* L., pikeperch *Stizostedion lucioperca* (L.) and smelt *Osmerus eperlanus* (L.), and it is an important foraging and breeding area of several species of fish-eating birds (van Eerden & bij de Vaate 1984). The overall objective of the study was to describe the interaction between three groups which utilize the smelt stock of Lake IJsselmeer: predatory fish (pikeperch, perch and eel), commercial fisheries and fish eating birds. To estimate whether the shared resource was scarce, the size and spatial distribution of the smelt stock needed to be determined. Hydroacoustics is potentially an effective tool for estimating stock size and spatial variation in abundance for three reasons. First, the sampled volume per unit time is high compared to other sampling methods (e.g. trawling and gill netting), which is an advantage in a large lake. Second, with hydroacoustics it is possible to obtain data on fish abundance with a high spatial resolution, even at a level where the distribution of individual fish can be studied. Third, with hydroacoustic equipment it is possible to study the swimming behaviour of insonified fish, a feature we used in this study to validate the results.

The use of hydroacoustics for the study of fish and other aquatic organisms is well established. Hydroacoustics are used for studies on vessel avoidance by fish (e.g. Gerlotto & Fréon 1992; Fréon *et al.* 1993; Misund & Aglen 1992), on diel vertical migration of fish (e.g. Eckmann 1991; Piersma *et al.* 1988; Hartmann *et al.* 1991), on large and small scale spatial distribution of fish and other aquatic organisms in relation to survey design or biological and physicochemical characteristics (e.g. Jolly & Hampton 1990; Kalikhman *et al.* 1992; Baroudy & Elliott 1993; Williamson 1982; Crawford *et al.* 1992; Baussant *et al.* 1993) on feeding ecology of fish (e.g. Janssen & Brandt 1980; Levy 1990) on fish migration in rivers (e.g. Mesiar *et al.* 1990; Johnston & Hopelain 1990; Gaudet 1990), and for studies on bioenergetics (Brandt & Kirsch 1993; Goyke & Brandt 1993). Both moving and stationary transducer platforms were used in these studies. Nearly all hydroacoustic studies deal with deeper waters ( > 10 m) using a vertically directed acoustic beam. The transducer is positioned just below the water surface, facing downwards. In shallow waters like Lake IJsselmeer, a vertical beam is not efficient because the maximum distance from the transducer where fish can be detected ($R_{max}$) is limited to the distance from transducer surface to bottom. $R_{max}$ can be increased by using a horizontal beam. In this case, the transducer is positioned in the water column, facing in a horizontal direction. Little is published on acoustic fish abundance estimation in shallow waters using a horizontal beam (e.g. Kubecka *et al.* 1992; Kubecka *et al.* 1994). This new technique seemed to be applicable to Lake IJsselmeer, but, to date, the technique has only been used in small water bodies under 'favourable sampling conditions', i.e. no wind or rain (Kubecka *et al.* 1994). Unfavourable sampling conditions generally result in high noise levels. In Lake IJsselmeer, favourable sampling conditions are rare, and for practical and logistic reasons the technique had to be used under the prevailing conditions during the sampling period. The hydroacoustic estimate of the fish biomass is biased if the backscattered sound energy of the targeted fish is low compared to the noise level. The backscattered sound energy of fish is related to its size, so it is important to know the size range of the targeted fish by some other sampling method to interpret the hydroacoustic data. In this study, the size range of the pelagic fish community was estimated with a pelagic beam trawl. A check was made to determine whether acoustically identified fish (*AIF*) showed vessel avoidance behaviour. Avoidance behaviour has consequences for the orientation of the fish in the acoustic beam, which has consequences for the detection and sizing of fish. The influence of varying wind speed on the output of the hydroacoustic system was studied.

## 14.2     Materials and methods

### 14.2.1     *Study area, survey design*

Two study areas, one in the northern part of Lake IJsselmeer (area A) and one in the southern part of the lake (area B), were selected. The depth varied between 3.5 and 5.5 m. Sampling took place between 14 and 18 June 1993. A partially randomized stratified sampling design was employed for the acoustic transect and the trawl haul locations, as recommended by Jolly and Hampton (1990). Randomization in time was not used for practical reasons.

Transects were positioned in an east–west direction. Areas A and B were allocated a number of transects proportional to their surface areas, i.e. three to area A and seven to area B. The shortest route between the end and the beginning of two adjacent transects was also sampled (Fig. 14.1). The pelagic fish community was sampled with a pelagic beam trawl. In area A, four pelagic beam trawl hauls were made and in area B 12 hauls were made. Haul duration was 10 min, during which about 0.26 ha were covered. The width of the pelagic beam trawl was 3 m, the codend had a stretched mesh size of 2 mm. The length–frequency distribution per species was established for each trawl catch. The length–frequency distributions from the pelagic beam trawl catches were averaged per area and per species.

### 14.2.2     *Hydroacoustic equipment and data collection*

A Biosonics ES2000 echo-sounder, with an operating frequency of 420 kHz in combination with an elliptical dual beam transducer with a $5.5° \times 17.5°$ wide beam and a $2.2° \times 7.5°$ narrow beam was used. The equipment generated a pulse width of $0.2 \ 10^{-3}$s and a ping rate of 20 s$^{-1}$. The system was calibrated by measurements on a tungsten carbide standard target. The unit was controlled by a 486 DX 66 MHz personal computer. An oscillograph of received signal voltage against distance from transducer was echoed on the screen of the computer. The equipment was used onboard a 20 m research vessel, powered by a 200 hp engine. The transducer was

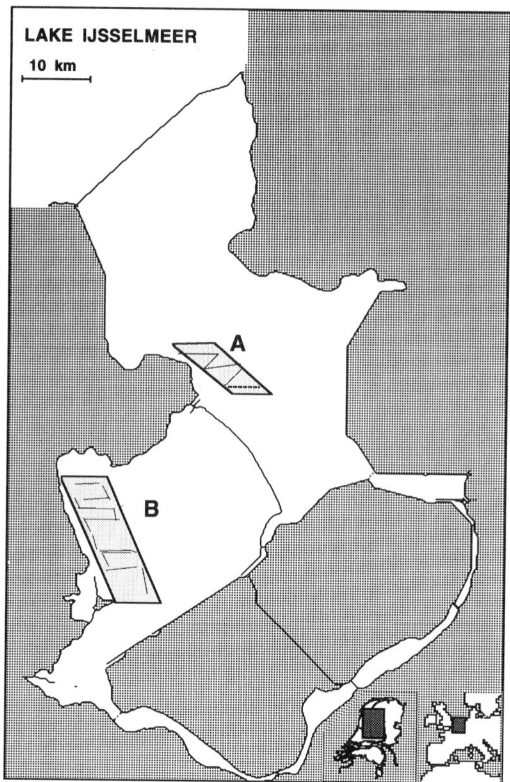


**Fig. 14.1**  Map of Lake IJsselmeer with sampling areas. The shaded parallelograms indicate the sampling areas A and B. The lines in the sampling areas represent acoustic transects. The dashed transect in area B was planned but not realized.

mounted on a towed body which was pulled alongside the vessel at a speed of 1.8 m s$^{-1}$. The beam was directed perpendicular to the towing direction (Fig. 14.2). Recordings were made at 2–20 m distance from the transducer. At 0–2 m from the transducer no recordings were made because of near field effects (MacLennan & Simmonds 1992). No recordings were made at distances >20 m because of permanently high noise levels.

Signal processing was performed by the ESP Model 281 Dual-Beam Processor using 40 log R time varied gain (echo counting). Dual beam technology is described in MacLennan & Simmonds (1992). The advantage of echo counting over echo integration is that it is possible to filter out backscattered signals of unwanted targets. Furthermore, echo counting makes it possible to study fish behaviour at the individual level. Processing took place during recording. The received signal was also recorded on a DAT recorder. Processing consisted of identification of single targets, target strength (*TS*) measurements and combination of consecutively recorded targets into acoustically identified fish (*AIF*). Single targets and *AIF* were identified

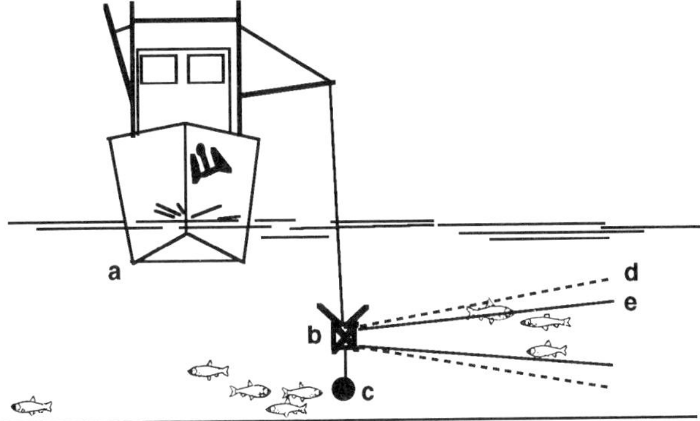


**Fig. 14.2** Front view of research vessel with towed body as used in this study. a = research vessel, b = towed body with transducer, c = depressor (to stabilize towed body), d = wide beam, e = narrow beam.

according to pre-set criteria. For each *AIF* the following variables were measured (Fig. 14.3):

- the number of single targets which make up the *AIF*;
- *TS*, estimated as the mean of the *TS* of *n* single targets which make up the *AIF* (dB);
- slope, the change in distance from the transducer in *n* single targets which make up an *AIF* (m.ping$^{-1}$);
- midrange, the mean of the distance from the transducer of the single targets which make up an *AIF* (m).

Each *AIF* was written as a record to a file. Data were exported to the SAS$^{©}$ software package for statistical processing.

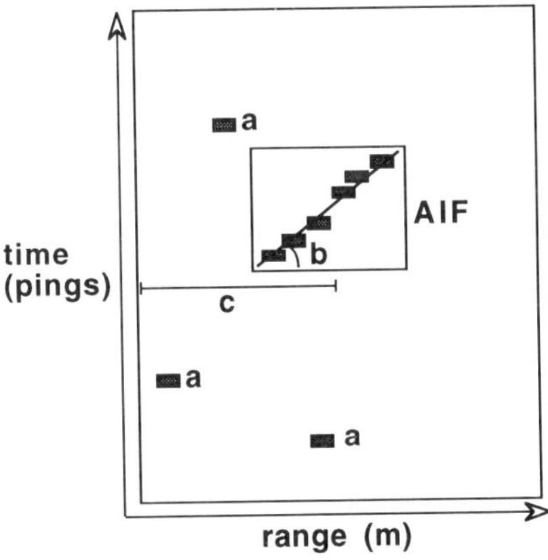


**Fig. 14.3** Schematic representation of an echogram. a = single targets (not interpreted as fish), *AIF* = acoustically identified fish, b = slope, c = mid-range. The *AIF* is made up of six echoes.

### 14.2.3   *Target strength measurements*

The *TS* of a fish is related to the size of the swimming bladder, to the shape of the swimming bladder and to the orientation of the swimming bladder in the acoustic beam. The other backscattering body tissues have minor influence on the *TS*. Love (1977) developed TS-length relations for 14 families of fish, insonified from every aspect. For each aspect, defined as an area on an imaginary spherical surface centred about the fish, he established a function which relates the backscattering cross-section to fish length, operating frequency of the echo-sounder and sound speed:

$$\sigma = a \ (L \ [f \ / \ c])^b \ (c \ / \ f)^2 \tag{1}$$

where $\sigma$ is the backscattering cross-section (m$^2$), $f$ the operating frequency of the echo-sounder (420 000 Hz), $c$ the speed of sound in water (1500 m.s$^{-1}$), $L$ the total length of the fish (m) and $a$ and $b$ are constants specific for the aspect under study.

The *TS* is related to $\sigma$ as:

$$TS = 10. \log \ (\sigma \ / \ 4\pi) \tag{2}$$

where *TS* is the target strength (in dB) and $\sigma$ the backscattering cross-section (in m$^2$).

When using a horizontally directed acoustic beam, variation in the orientation in the horizontal plane (yaw) is the main source of variation in *TS* of a fish of given length. It was assumed that the caudo-cranial axis of all insonified fish was horizontally oriented and that the ventral-dorsal axis was vertically oriented.

To make sure that no noise is interpreted as fish, it is recommended to set the noise threshold at three times the observed noise level on the oscilloscope (J. Kubecka, pers. comm.). During this survey, the threshold was set at approximately 1.5–2 times the observed noise level because there was a need to include small fish in the measurements. However, it was recognized that some noise would be interpreted as fish. The noise threshold of the acoustic equipment was set at the voltage corresponding to a *TS* of –56 dB. This value can be related to a range of fish lengths at several aspects according to the formulae of Love (1977). It was assumed that the acoustic equipment and the pelagic beam trawl sampled the same depth stratum and that the length-frequency distributions of the trawl catches reflected the size range and species composition of the targeted fish. The length–frequency distribution of the pelagic beam trawl and the *TS*-length relations of Love (1977) were used to evaluate roughly which part of the population could be detected by the hydroacoustic equipment.

### 14.2.4   *Avoidance reaction*

The avoidance reaction of fish was studied by analysis of the slope of an identified fish. A positive slope indicated a fish swimming away from the transducer, a negative slope indicated a fish swimming towards the transducer. If insonified fish did not react to the presence of the research vessel, the average slope was expected to equal zero. Since the slopes were not normally distributed, the null-hypothesis 'mean slope equals zero' was tested by calculating the signed rank statistic (SRS). In this analysis identified fish from all transects were pooled.

Fish which showed an avoidance reaction (*FA*) when the research vessel approached were expected to orientate themselves differently to the acoustical axis from fish which do not avoid the research vessel (*FNA*). Consequently, *FA* should have a different *TS* from *FNA*. First, a parametric model (ANOVA), was used to test whether *TA* of *FA* was different from *FNA* but residual analysis showed that the conditions of ANOVA were not met. Consequently, differences in shape of the *TS*-frequency distributions of *FA* and of *FNA* were tested with a non-parametric test (Kolmogorov–Smirnov).

### 14.2.5   *Influence of wind on fish abundance estimates*

During the survey, noise was predominantly caused by the backscattering properties of air bubbles in the water column and by backscattering of sound energy by the surface and bottom. Wind caused noise by increasing the number of air bubbles in the water column. The influence of wind increased from the bottom to surface, resulting in higher noise levels near the surface. Because targets are more difficult to detect at high noise levels, wind force has an effect on the number of detected targets. This effect is especially noticeable at higher distance from the transducer, because there the sampled volume extends to the surface layer of the water column. Consequently the number of detected fish is related to wind speed and to distance from the transducer. This source of bias was studied to estimate fish abundance by relating the median of the midrange of *AIF* of each transect to the wind speed. It was assumed that the influence of wind speed on noise level within a transect was constant.

### 14.3   **Results**

### 14.3.1   *The target strength of the insonified fish*

Hydroacoustic data were processed using a noise threshold corresponding to a *TS* of –56 dB. According to Love's *TS*–length relations, a *TS* of –56 dB corresponds to fish in the length range 3 cm (side aspect) to 16 cm (tail aspect) (Fig. 14.4). Most fish caught with the pelagic beam trawl ranged from 2 to 9 cm total length (Fig. 14.5). It is obvious that the signal-to-noise ratio was extremely unfavourable in this case. The smallest fish were recorded only if they were insonified in side aspect. Even the largest fish (9 cm total length) were not detected if they were insonified in tail aspect. Of all *AIF*, 95% had a *TS* of less than –45 dB; this corresponded to a fish with a total length of 9 cm, insonified in side aspect. This was in agreement with the length–frequency distribution of the fish caught with the pelagic beam trawl.

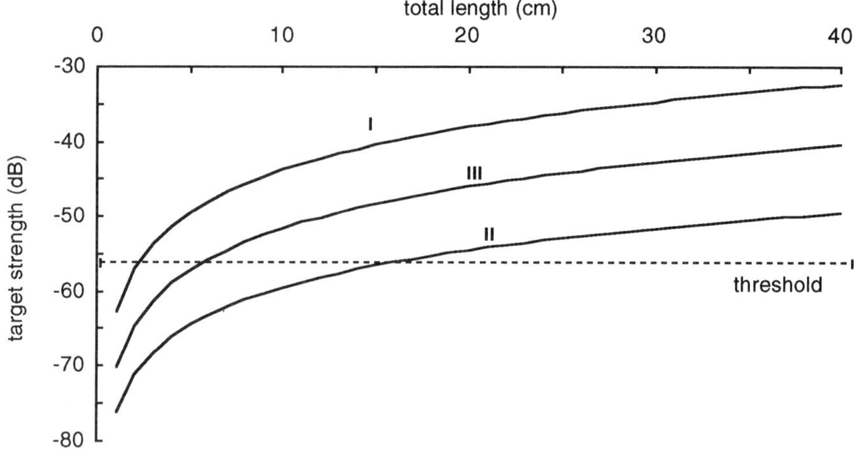

**Fig. 14.4**   Target Strength–length (*TS-L*) relationships for fish oriented with varying yaw angle in the acoustic beam. Curve I represents the *TS-L* relation for fish insonified in side aspect (the maximal *TS* for a fish of a given length), curve II represents the *TS-L* relation for fish insonified in tail-head aspect (the minimal *TS* for a fish of a given length). Curve III represents the average *TS-L* relation for fish insonified in all aspects between head-tail and tail-head aspect. Relations are calculated from formula 1 and 2 (see text) with a = $74 \times 10^3$ and b = 1.90 for curve I, a = $4.4 \times 10^3$ and b = 1.66 for curve II, a = $14 \times 10^3$ and b = 1.86 for curve III. The threshold value (–56 dB) as used during this study is indicated by the dashed line.

### 14.3.2   *Avoidance behaviour*

Most *AIF* had a slope larger than zero (Fig. 14.6) (*SRS* = 6708402, $P_{H0}(SRS > 6708402) = 0.001$) and were thus moving away from the vessel. At least part of the *AIF* recorded must have been fish since the 'behaviour' of *AIF* was consistent with vessel avoidance behaviour. Assuming that all *FNA* and an equal number of *FA* was noise, 31% of all *AIF* was fish and 69% was noise. Near the transducer, the proportion *FA* of all *AIF* increases with increasing midrange. The data suggest that at a mid-range of more than 12.5 m the proportion *FA* was relatively constant (Fig. 14.7). The increase over the full range (2–20 m from the transducer) was statistically significant (linear regression analysis, adjusted $r^2 = 0.75$, $P_{H0} << 0.01$). The increase of the proportion of *FA* was not consistent with vessel avoidance behaviour. Fish closer to the research vessel should have been more likely to show avoidance behaviour than fish which were further away. However, at a higher mid-range *AIF* consisted of more echoes (Fig. 14.8), and probably the swimming direction of fish relative to the transducer was more accurately estimated if the number of echoes which make up an *AIF* is higher. Thus, the estimate of the swimming direction relative to the transducer of fish located at higher midrange could have been more accurate. Avoidance behaviour seemed to be present at distances of at least 20 m from the research vessel.

The *TS*–frequency distribution of *FA* was different from the *TS*–frequency of *FNA* (Kolmogorov-Smirnov, *P* << 0.01) (Fig. 14.9). The difference in *TS*–frequency

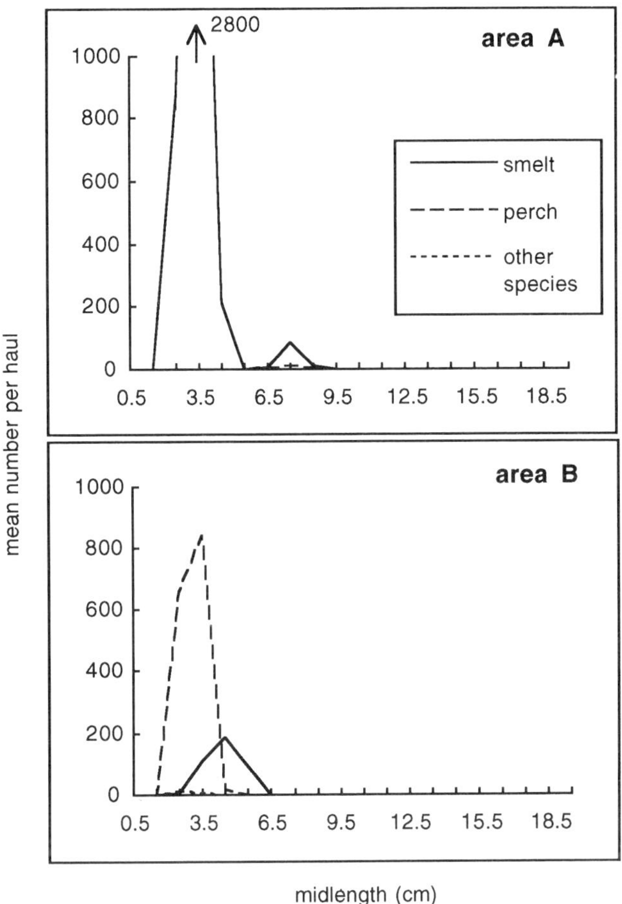


**Fig. 14.5**   Average length–frequency distributions of pelagic beam trawl catches in area A and B. In area A mainly 0-group perch (78.4% of the average catch in numbers) and 0-group smelt (20.5% of the average catch in numbers) were caught. Other species categories, including older perch and smelt, contributed 1.1% in numbers to the average catch. In area B 0-group smelt contributed 96.8% and 1 + smelt contributed 2.7% to the average catch in numbers.

distributions could be explained by bigger fish avoiding the vessel to a larger extent than smaller fish, by the difference in orientation in the beam between *FA* and *FNA* or by a combination of both effects.

### 14.3.3   *Influence of wind on fish abundance estimates*

At greater distance from the transducer fewer fish were detected than closer to the transducer (Fig. 14.10). Apparently, this was caused by the high noise levels at the end of the detection range, where the acoustic beam approached the water surface and the bottom. The median value of the midrange of *AIF* over all transects was 4.9 m. During the field work, the areas with high noise levels at the end of the detection range

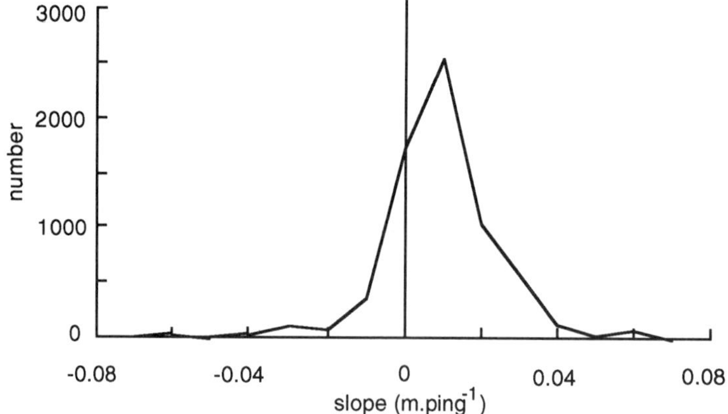

**Fig. 14.6**  Slope–frequency distribution of all identified fish. Positive slopes indicate fish which move away from the transducer. Values smaller than –0.08 and larger than 0.08 represent 0.6% of the total observations.

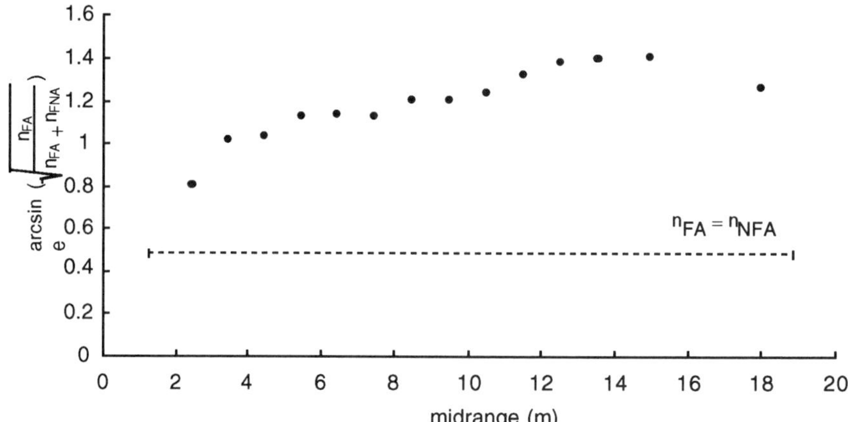

**Fig. 14.7**  Scatter plot of the proportion *FA* of all *AIF* against the distance from the transducer (mid-range). The number of observations per class was at least 20. The width of the mid-range classes is 1 m between 2 and 13 m. Mid-range classes 14–15 and 16–19 were combined. The proportion was arcsine transformed as recommended by Sokal and Rohlf (1981). The dashed line represents the level where *FA* equals *FNA*.

were observed to fluctuate between about 10 and 20 m distance from the transducer. When there were higher wind-induced waves, the noise area became larger and consequently, less fish were detected. This was demonstrated by the relation between median value of mid-range of *AIF* and the prevailing wind condition of each transect (Fig. 14.11). During the study period, sampling was performed at wind speeds varying between 1 and 5 Beaufort. At higher wind speed, the median value of the mid-range of *AIF* was lower, indicating that relatively fewer fish in relation to the population present were detected (linear regression analysis, adjusted $r^2 = 0.52$, $P_{H0} << 0.01$).

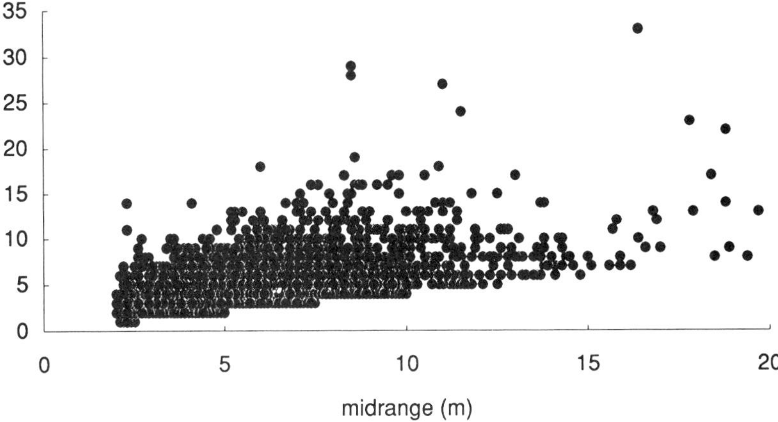


**Fig. 14.8** Scatter plot of the number of echoes which make up an *AIF* (*n*) against the distance from the transducer (mid-range).

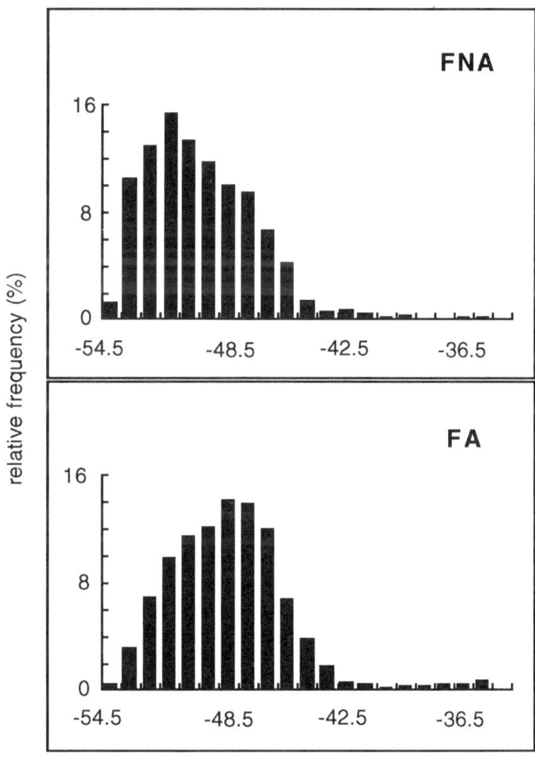


**Fig. 14.9** *TS*–relative frequency distribution of *AIF* with slope $\leq 0$ (*FNA*) and *AIF* with slope $> 0$ (*FA*). *FA* had higher *TS*.

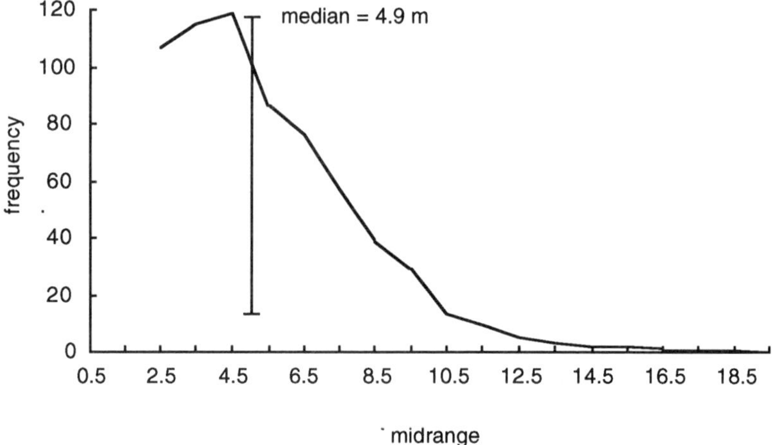


**Fig. 14.10**  Mid-range-frequency distribution *AIF* from all transects. The vertical line at mid-range = 4.9 m represents the median value of the mid-range of all *AIF*.

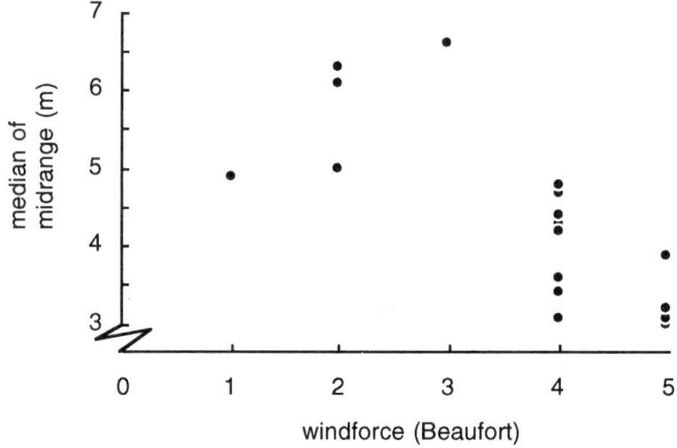


**Fig. 14.11**  Relation between wind force (Beaufort) and median of mid-range (m). Each data point represents *AIF* from one transect. At high wind speed, the median value of mid-range is lower than at low wind speed.

## 14.4    Discussion

It is possible to detect fish in large shallow lakes using dual-beam hydroacoustics with a horizontal beam employed in a mobile survey. However, results from the survey should be interpreted with care because of:

- the relatively high noise levels compared to the *TS* of the targeted fish;
- avoidance behaviour of insonified fish and possible consequences of avoidance behaviour for the *TS*-length relation; and
- variable sampling conditions (wind speed).

The use of a *TS*-length relation without taking into account the orientation of the fish (e.g. Braband 1991; Brandt *et al.* 1991; Brandt & Kirsch 1993; Burczynski & Johnson 1986; Burczynski *et al.* 1987; Hansson 1993; Jurvelius 1991; Luecke & Wurtsbaugh 1993) was inappropriate for Lake IJsselmeer. The orientation of the fish in the beam was a major source of variance in *TS* (Fig. 14.4). During this survey, the orientation not only affected the estimate of the size distribution of the insonified fish, but also the abundance estimate. Even the largest pelagic fish, smelt of 9 cm total length, could not be distinguished from noise if they were insonified in tail aspect.

Furthermore, because of the avoidance behaviour of the insonified fish, it could not be assumed that the yaw angle was randomly distributed over 360°. Avoidance behaviour of herring (*Clupea harengus*) and sprat (*Sprattus sprattus*) was demonstrated at a distance of up to ca 400 m from sea-going research vessels (Misund & Aglen 1992). *Sardinella aurita* avoided the research vessel at a distance up to 20 m (Gerlotto & Fréon 1992) and Ona & Godø (1990) noticed strong avoidance reactions to a trawling vessel in haddock (*Melanogrammus aeglefinus*) up to depths of 100 m. In this study, avoidance behaviour was observed in the total detection range (2–20 m from the research vessel), and it was probable that avoidance behaviour influenced the *TS*–length relation (Fig. 14.9). Other vessel-induced factors may also influence *TS* measurements: e.g. ships' lights caused a geometric reorientation of zooplankton which was responsible for a drop of 20 dB in backscattering strength in the area 60 m below the transducer (Sameoto *et al.* 1985). The importance of fish behaviour in relation to *TS* measurements has also been stressed by other authors (MacLennan *et al.* 1990; Blaxter & Batty 1990; Dawson & Karp 1990; Olsen 1990). The comparison of length–frequency distributions of the trawl catches and the *TS*–frequency distribution can be made by either peak identification (Bjerkeng *et al.* 1991; Guillard & Gerdeaux 1993) or maximum likelihood procedures (Bjerkeng *et al.* 1991). In this study, peak identification and maximum likelihood analysis were hampered by the absence of more than one peak in the *TS*–frequency distribution and in the length–frequency distributions of the trawl catches. Furthermore, the fraction of *AIF* which was actually noise was unknown. However, the length range of the fish caught with the trawl (3–9 cm total length) was approximately in agreement with the *TS*–frequency distribution.

Wind force had a major effect on the number of fish detected (Fig. 14.11). The effect of wind was reflected by a decrease in the median value of the distance from the transducer of insonified fish. Assuming that fish were distributed symmetrically above and below the acoustic axis it was possible to correct for the effect of wind speed. The median of the midrange of *AIF* could have been used as a correction factor. Correction for wind speed is not possible if fish are concentrated near the bottom or the surface. Furthermore, interaction between wind speed and vertical distribution of fish may lead to erroneous extrapolation of results from favourable to less favourable survey conditions. It is obvious that hydroacoustic abundance estimates can only be corrected for variable sampling conditions if the response of the fish to sampling conditions is known. Another method to account for wind conditions is abortion of the sampling programme at high wind speed (Pedersen & Boettner 1992). Experience during other hydroacoustic surveys suggests that if the weather situation remains unfavourable for hydroacoustic sampling during a longer period, not enough time may remain to complete the sampling programme.

Factors other than vessel-induced behaviour of fish or sampling conditions may influence hydroacoustic stock estimates. Diel changes in schooling behaviour and spatial distribution are responsible for decreased stock size estimates during periods of higher light intensity (daylight, full moon) (Luecke & Wurtsbaugh 1993; Bethke 1993; Northcote & Rundberg 1970). This may be caused by acoustic shadowing (Appenzeller & Legget 1992) or by problems with detecting schools closer to the bottom (Burczynski *et al.* 1987). During this survey, which was conducted during daytime, no schools were detected. This might have been caused by absence of schooling behaviour or by the vertical distribution of schools. Schools located near the bottom are difficult to detect by horizontal beam hydroacoustics. Day and night recordings with single beam hydroacoustic equipment in the summer of 1992 suggested that smelt and 0-group perch were more evenly distributed over the water column at night than at daytime (P.J. Mous, unpublished results), so problems in detecting fish aggregations may have had some influence on the results. Changes in fish behaviour and resulting changes in hydroacoustic detection characteristics may also be seasonal, as shown by Goldspink (1990) for perch (*Perca fluviatilis*) in UK, by Eckman (1991) for whitefish (*Coregonus lavaretus*) in Lake Constance in Germany and by Northcote & Rundberg (1970) for cisco (*Coregonus albula*) and smelt (*Osmerus eperlanus*) in Sweden. A complex of sources of bias in hydroacoustic estimates of fish biomass may lead to considerable discrepancies when results from hydroacoustic surveys and results from other sampling methods are compared. For example, Godø & Wespestad (1993) found a weak correlation between estimates of gadoid abundance by trawl surveys and hydroacoustic surveys. They concluded that relying on either of the two sampling methods may lead to severe bias in the abundance estimate.

The use of hydroacoustics as a sampling tool for studying fish abundance, size distribution and spatial distribution on a lake-wide scale on a large shallow lake is not recommended. Variable sampling conditions, high noise levels and avoidance behaviour of fish were the main factors constraining hydroacoustic sampling. Irrespective, the possibility of studying avoidance behaviour with hydroacoustics can be used to validate the abundance estimates from trawl surveys, where avoidance behaviour also affects the fish abundance estimate. Net avoidance behaviour in relation to water clarity (Buijse *et al.* 1992) could be studied in more detail in a simultaneous trawling and hydroacoustic surveying programme. Thus, in combination with other sampling methods like trawling or gillnetting, hydroacoustic sampling will increase the accuracy of abundance estimates (c.f. Dahm *et al.* 1992; Kubecka *et al.* 1994). Furthermore, the high spatial resolution makes hydroacoustics a valuable tool for studying the distribution of pelagic fish in more detail than is possible with trawling.

## Acknowledgments

We thank Mr A. Schelvis (Dept. of Technical and Technological Research, RIVO-DLO) for the design and construction of the towed body and the crew of the research vessel Stern for their co-operation in this project. Discussions with Dr A. Duncan and Dr J. Kubecka during the symposium were of great value for the improvement of this chapter. We thank W.L.T. van Densen and W. Dekker for critically reviewing the manuscript. The financial support of the project 'Lachs 2000' is gratefully acknowledged.

## References

Appenzeller A.R. & Legget W.C. (1992) Bias in hydroacoustic estimates of fish abundance due to acoustic shadowing, evidence from day-night surveys of vertically migrating fish. *Canadian Journal of Fisheries and Aquatic Sciences* **49**, 2179–2189.

Baroudy E. & Elliott J.M. (1993) The effect of large-scale spatial variation of pelagic fish on hydroacoustic estimates of their population density in Windermere (northwest England). *Ecology of Freshwater Fish* **2**, 160–166.

Baussant T., Ibanez F. & Etienne M.(1993) Numeric analysis of planktonic spatial patterns revealed by echograms. *Aquatic Living Resources* **6**, 175–184.

Bethke E. (1993) Probleme bei hydroakustischen Untersuchungen von pelagischen Fischarten in den Flachwassergebieten der westlichen Ostsee durch das Tag-Nacht-Verhalten der Fische. *Informationen für die Fischwirtschaft* 40 Nr., **2**, 66–72.

Bjerkeng B., Borgstrom R., Braband A. & Faafeng B.(1991) Fish size distribution and total fish biomass estimated by hydroacoustic methods, a statistical approach. *Fisheries Research* **11**, 41–73.

Blaxter J.H.S. & Batty R.S. (1990) Swimbladder 'behaviour' and target-strength. In: W.A. Karp (ed.). Developments in Fisheries Acoustics. *ICES Rapports et Procès-Verbaux des Réunions* **189**, 233–244.

Braband A. (1991) The estimation of pelagic fish density, single fish size and fish biomass of Arctic charr (*Salvelinus alpinus* L.) by echo-sounding. *Nordic Journal of Freshwater Research* **66**, 44–49.

Brandt S.B. & Kirsch J. (1993) Spatially explicit models of striped bass growth potential in Chesapeake Bay. *Transactions of the American Fisheries Society* **122**, 845–869.

Brandt S.B., Mason D.M, Patrick E.V., Argyle R.L., Wells L., Unger P.A. & Stewart D.J.(1991) Acoustic measures of the abundance and size of pelagic planktivores in Lake Michigan. *Canadian Journal of Fisheries and Aquatic Sciences* **48**, 894–908.

Burczynski J.J., Michaletz P.H. & Marrone G.M. (1987) Hydroacoustic assessment of the abundance and distribution of rainbow smelt in lake Oahe. *North American Journal of Fisheries Management* **7**, 106–116.

Burczynski J.J. & Johnson R.L. (1986) Application of dual-beam acoustic survey techniques to limnetic populations of juvenile sockeye salmon (*Oncorhynchus nerka*). *Canadian Journal of Fisheries and Aquatic Sciences* **43**, 1776–1788.

Buijse A.D., Schaap L.A. & Bult T.P. (1992). Influence of water clarity on the catchability of six freshwater fish species in bottom trawls. *Canadian Journal of Fisheries and Aquatic Sciences* **49**, 885–893

Crawford R.E., Hudon C. & Parsons D.G. (1992) An acoustic study of shrimp (*Pandalus montagui*) distribution near Resolution Island (Eastern Hudson Strait). *Canadian Journal of Fisheries and Aquatic Sciences* **49**, 842–856.

Dahm E., Hartmann J., Jurvelius J., Löffler H. & Völzke V. (1992) Review of the European Inland Fisheries Advisory Commission (EIFAC) experiments on stock assessment in lakes. *Journal of Applied Ichthyology* **8**, 1–9.

Dawson J.J. & Karp W.A. (1990) In situ measures of target-strength variability of individual fish. In: W.A. Karp (ed.). Developments in Fisheries Acoustics. *ICES Rapports et Procès-Verbaux des Rèunions 189*, 264–273.

Eckmann R. (1991) A hydroacoustic study of the pelagic spawning behavior of whitefish (*Coregonus lavaretus*) in Lake Constance. *Canadian Journal of Fisheries and Aquatic Sciences* **48**, 995–1002.

Eerden M.R. van & Vaate A. bij de (1984) Natuurwaarden van het IJsselmeergebied. Een inventarisatie van natuurwaarden van het open water in het IJsselmeergebied. Ministerie van Verkeer en Waterstaat. Rijksdienst voor de IJsselmeerpolders. Flevobericht 242, 73 pp.

Fréon P., Soria M., Mullon C. & Gerlotto F. (1993) Diurnal variation in fish density estimate during acoustic surveys in relation to spatial distribution and avoidance reaction. *Aquatic Living Resources* **6**, 221–234.

Gaudet D.M. (1990) Enumeration of migrating salmon populations using fixed-location sonar counters. In: W.A. Karp (ed.). Developments in Fisheries Acoustics. *ICES Rapports et Procès-Verbaux des Rèunions* **189**, 197–209.

Gerlotto F. & Fréon P. (1992) Some elements on vertical avoidance of fish schools to a vessel during acoustic surveys. *Fisheries Research* **14**, 251–260.

Godø O.R. & Wespestad V.G. (1993) Monitoring changes in abundance of gadoids with varying availability to trawl and acoustic surveys. *ICES Journal of Marine Science* **50**, 39–52.

Goldspink C.R. (1990) The distribution and abundance of young (I+ – II+) perch, *Perca fluviatilis* L., in a deep eutrophic lake, England. *Journal of Fish Biology* **36**, 439–447.

Goyke A.P. & Brandt S.B. (1993) Spatial models of salmonine growth rates in Lake Ontario. *Transactions of the American Fisheries Society* **122**, 870–883.

Guillard J. & Gerdeaux D. (1993) In situ determination of the target strength of roach (*Rutilus rutilus*) in Lake Boourget with a single beam sounder. *Aquatic Living Resources* **6**, 285–289.

Hansson S. (1993) Variation in hydroacoustic abundance of pelagic fish. *Fisheries Research* **16**, 203–222.

Hartmann J., Knöpfler G., Löffler H. & Quoss H. (1991) Tages- und jahresperiodische Vertikalwanderung der Fische im Bodensee. *Fischoekologie* **4**, 67–74.

Janssen J. & Brandt S.B. (1980) Feeding ecology and vertical migration of adult alewifes (*Alosa pseudoharengus*) in Lake Michigan. *Canadian Journal of Fisheries and Aquatic Sciences* **37**, 177–184.

Johnston S.V. & Hopelain J.S. (1990) The application of dual-beam target tracking and Dopler-shifted echo processing to assess upstream salmonid migration in the Klamath River, California. In: W.A. Karp (ed.). Developments in Fisheries Acoustics. *ICES Rapports et Procès-Verbaux des Réunions* **189**, 210–222.

Jolly G.M. & Hampton I. (1990) A stratified random transect design for acoustic surveys of fish stocks. *Canandian Journal of Fisheries and Aquatic Sciences* **47**, 1282–1291.

Jurvelius J. (1991) Distribution and density of pelagic fish stocks, especially vendace (*Coregonus albula* L.) monitored by hydroacoustics in shallow and deep southern boreal lakes. *Finnish Fisheries Research* **12**, 45–63.

Kalikhman I., Walline P. & Gophen M. (1992) Simultaneous patterns of temperature, oxygen, zooplankton and fish distribution in Lake Kinneret, Israel. *Freshwater Biology* **28**, 337–347.

Kubecka J., Duncan A. & Butterworth A. (1992). Echo-counting or echo-integration for fish biomass assessment in shallow waters. In: M. Weydert (ed.). European Conference on Underwater Acoustics. London, New York: Elsevier Applied Science, pp 129–132.

Kubecka J., Duncan A., Duncan W.M., Sinclair D. & Butterworth A.J. (1994) Brown trout populations of three Scottish lochs estimated by horizontal sonar and multimesh gill nets. *Fisheries Research* **20**, 29–48.

Levy D.A. (1990) Reciprocal diel vertical migration behavior in planktivores and zooplankton in British Columbia lakes. *Canandian Journal of Fisheries and Aquatic Sciences* **47**, 1755–1764.

Love R.H. (1977) Target strength of an individual fish at any aspect. *Journal of the Acoustical Society of America* **62**, 1397–1403.

Luecke C. & Wurtsbaugh W.A. (1993) Effects of moonlight and daylight on hydroacoustic estimates of pelagic fish abundance. *Transactions of the American Fisheries Society* **122**, 112–120.

MacLennan D.N., Magurran A.E., Pitcher T.J. & Hollingworth C.E. (1990) Behavioural determinants of fish target-strength. In: W.A. Karp (ed.) Developments in Fisheries Acoustics. *ICES Rapports et Procès-Verbaux des Réunions* **189**, 233–244.

MacLennan D.N. & Simmonds E.J. (1992) *Fisheries acoustics*. London: Chapman & Hall, Fish and Fisheries Series 5, 325 pp.

Mesiar D.C., Eggers D.M. & Gaudet D.M. (1990) Development of tehniques for the application of hydroacoustics to counting migratory fish in large rivers. In: W.A. Karp (ed.). Developments in Fisheries Acoustics. *ICES Rapports et Procès-Verbaux des Réunions* **189**, 223–232.

Misund O.A. & Aglen A. (1992) Swimming behaviour of fish schools in the North Sea during acoustic surveying and pelagic trawl sampling. *ICES Journal of Marine Science* **49**, 325–334.

Northcote T.G. & Rundberg H. (1970) Spatial distribution of pelagic fishes in Lambarfjarden (Malaren, Sweden) with particular reference to interaction between *Coregonus albula* and *Osmerus eperlanus*. *Report of the Institute of Freshwater Research, Drottningholm* **50**, 133–166.

Olsen K. (1990) Fish behaviour and acoustic sampling. In: W.A. Karp (ed.) Developments in Fisheries Acoustics. *ICES Rapports et Procès-Verbaux des Réunions* **189**, 147–158.

Ona E. & O.R. Godø (1990). Fish reaction to trawling noise, the significance for trawl sampling. In: W.A. Karp (ed.) Developments in Fisheries Acoustics. *ICES Rapports et Procès-Verbaux des Réunions* **189**, 159–166.

Pedersen G.E. & Boettner J.F. (1992) Application of hydroacoustic technology to marine fishery management in Washington State. *Fisheries Research* **14**, 209–221.

Piersma T., Lindeboom R. & Eerden M.R. van (1988) Foraging rhythm of great crested grebes *Podiceps cristatus* adjusted to diel variations in the vertical distribution of their prey *Osmerus eperlanus* in a shallow eutrophic lake in The Netherlands. *Oecologia* **76**, 481–486.

Sameoto D., Cochrane N.A. & Herman A.W. (1985) Response of biological acoustic backscattering to ships' lights. *Canadian Journal of Fisheries and Aquatic Sciences* **42**, 1535–1543.

Sokal R.R. & Rohlf F.J. (1981) *Biometry*. The principles and practice of statistics in biological research. Second edition. San Francisco: W.H. Freeman and Company, 859 pp.

Williamson N.J. (1982) Cluster sampling estimation of the variance of abundance estimates from quantitative echo sounder surveys. *Canadian Journal of Fisheries and Aquatic Sciences* **39**, 229–231.

# Chapter 15
# A hydroacoustic assessment of a rainbow trout (*Oncorhynchus mykiss*) population in a deep oligotrophic lake

M. CRYER *Science and Research Directorate, Department of Conservation, Private Bag, Turangi, New Zealand*

*(Present address: MAF Fisheries, PO Box 3437, Auckland, New Zealand)*

**Abstract**   Lake Taupo is a large oligotrophic lake on New Zealand's central volcanic plateau. It supports a valuable sport fishery for rainbow trout. Ten acoustic surveys were conducted in 1988 and 1989 using a split beam sounder in a stratified random design of 70 stations in 14 strata. Estimates of trout abundance, by echo counting, had the high precision (relative standard error <8%) and unknown accuracy typical of acoustic surveys. The size of the lake-resident population was estimated to vary from 190 000 to 2 124 000 individuals, with remarkable consistency between seasons and years. Mean recruited biomass was estimated to be about 125 t. This estimate is not consistent with previous (pre-1985) estimates of harvest from the fishery (up to 900 t), but is consistent with empirical estimators of yield, and with a recent creel survey. The reasons for the discrepancy and the likely accuracy of acoustic surveys in this environment are briefly discussed.

KEYWORDS: Hydroacoustics, Lake Taupo, *Oncorhynchus mykiss*, rainbow trout, split beam

## 15.1     Introduction

Lake Taupo supports a valuable recreational fishery for rainbow trout, *Oncorhynchus mykiss* (Walbaum). The lake is large (616 km$^2$), deep (mean 100 m; max. 185 m), and oligotrophic. Its geological history is short, the current basin being the result of a spectacular caldera eruption 2000 years ago on New Zealand's central volcanic plateau.

Small, indigenous fish species in the lake (koaro, *Galaxias brevipinnis* Günther, and common bully, *Gobiomorphus cotidianus* McDowall) were not attractive as sport fish to European settlers, and trout were introduced late in the nineteenth century (Parsons 1979; Burstall 1983). New Zealand smelt, *Retropinna retropinna* Richardson, were successfully introduced as a forage species in the 1930s. Other species present are catfish, *Ictalurus nebulosus* Le Sueur, goldfish, *Carassius auratus* L., and mosquitofish, *Gambusia affinis* Baird & Girard. All fish species other than trout are of small body size ( < 100 mm) or littoral–benthic habit.

In the 1980s, a combination of apparently increasing effort and catch combined with decreasing catch rates led to the requirement for an acoustic assessment of trout in Lake Taupo as part of a wider study to estimate actual and potential yield. This chapter concentrates on methodology, whereas the author discusses the implications of the findings in more detail in Chapter 28.


**15.2     Methods**

**15.2.1     *Overall survey design***

The survey design was based on 'echo counting' as opposed to 'echo integration'. The situation in Taupo is ideally suited to echo counting as the size of target fish (trout) is very large compared with non-target species, there is little likelihood of dense shoaling of target fish, and targets are widely separated.

Ten surveys of the lake were conducted between March 1988 and January 1990. This study was designed to 'map' trout distribution and to estimate abundance; consequently each survey consisted of a large number of short transects rather than the more usual acoustic survey design of a smaller number of long transects (e.g. Dahm *et al.* 1985; Burczynski & Johnson 1986; Schultz 1988). These short 'point' samples also resolve statistical problems associated with spatial auto-correlation and stationarity of data sampled *post hoc* from long transects (Williamson 1982; MacLennan & MacKenzie 1988).

The lake was stratified into 14 strata of unequal area designed around highly visible landmarks to facilitate navigation. Five of these strata covered the 'offshore' areas of the lake (depth > 100 m: approximately 50% of total area), and nine strata covered the 'inshore' area (depth 20–100 m). The remaining 'littoral' area was not amenable to acoustic surveying and was sampled by SCUBA divers making visual counts on 'manta boards' (Cryer 1991). Within each stratum, five transects of 1 km length were located at computer-generated random positions and headings. Optimisation of site allocations among strata to reduce variance was not attempted because the variance in all estimates was satisfactory and the objective of 'mapping' distribution was best met by retaining a more even spread of sites across the lake.


**15.2.2     *Acoustic equipment***

Surveying was conducted using a SIMRAD ES470 scientific sounder operating at 70 kHz using split acoustic beams, 40 log $R$ (*single target*) time varied gain (*TVG*), and a nominal beamwidth of 11.5°. In addition to the transceiver and colour VDU, a colour printer and a portable computer were carried. The printer recorded returning echoes of different target strength (*TS*) on echograms (one per transect) in colours ranging from blue through green, yellow, and orange to red, with red being the greatest target strength. Surveying was conducted at a constant speed of $2.5\,\text{ms}^{-1}$. The quadruple transducer was housed in a towed body which was deployed at about 2 m depth with a slight (2.5 to 4.5°) aft-facing tilt during surveying.

### 15.2.3    *Electronic data recording*

Internal logic in the ES470 screens each returning echo and selects for further analysis only those that are clearly separate from other echoes above and below and are estimated to be within 5° of the acoustic axis. This reduces the beamwidth for data outputs from 11.5° to 10.0°. A serial data output in the form: 'ping number'; 'depth'; 'TS' was recorded on the portable computer and simple communications software. Ping number is a numeric label incrementing by one for each transmission; 'depth' is the estimated depth in decimetres (tenths of metres); and 'TS' is the target strength corrected for depth and position with respect to the acoustic axis. TS is coded from 1 to 80, each step representing an increase of 0.375 dB. The selected sensitivity range for use in Taupo was –44 to –14 dB, the effective range of TVG at this setting being from 3.6 to 283 m from the transducer face.

### 15.2.4    *Echo counting*

Trace counting was carried out manually from echograms, counting the number of traces likely to have come from trout in each 10 m depth shell on each transect. As trout are the only large pelagic species in Taupo, the criterion for selection of a target as 'likely to have come from trout' was that a distinct trace was formed. Other species are present in the lake, but those present in the pelagic zone rarely exceed 50 mm in length and would be undetectable at the sensitivity used in this study. Shoaling smelt and bully larvae were undoubtedly detected by the sounder, but their depiction on echograms was as 'bands' or 'scattering layers' rather than 'fingernail' traces.

While conducting counts, the data file for the transect under examination was displayed on a computer screen, and frequent comparisons were made between the echogram and electronic copies. Traces detected on echograms could be 'pinpointed' in the data file by reference to depth and position along the transect. There was normally some consistency of TS among the multiple return echoes within a trace, except where fish swimming rapidly up or down produced unusual traces. In these cases TS was highly variable.

The counts of echo traces in each depth shell were converted to estimates of density using the effective sampling area of the acoustic beam at the midpoint of each shell, and an estimate of overall density at the station was derived by summing its shell densities. This method follows the vertical pooling of data suggested by Nunnallee (1973), and has the advantages of condensing the data set and avoiding problems with vertical auto-correlation. Mean densities for each stratum, and for the whole lake were calculated, and their associated confidence intervals derived according to Snedecor and Cochran (1989; unequal stratum sizes):

$$s^2_{(y)} = \Sigma W_i^2 \cdot S_i^2 \cdot (1 - \phi_i)/n_i$$

where $s^2_{(y)}$ is the variance of the mean, $W_i$ is the relative size and $S_i^2$ the sample variance of stratum $i$. $\phi_i$ is the sampled fraction correction which was set to unity.

Echograms from each survey were counted and analysed shortly after completion of fieldwork, but, to ensure internal consistency and following discussions with a range of other workers in the field, all 700 echograms were recounted (in random survey order) after the completion of the last survey. Repeat counts differed substantially from initial counts only for low TS targets which were undercounted in early surveys.

### 15.2.5 *Target strength*

Target strength in fish is maximal when the fish is insonified dorsally, or close to dorsally. As TS can be highly variable with only minor changes in fish tilt angle, it was not considered appropriate to estimate the size of each fish to estimate length frequency in the population. Instead, a global approach was taken by counting the number of occurrences of any given TS value in each depth band in inshore and offshore areas. Ideally, the mean tilt angle and its standard deviation, and their impact on TS, should be estimated (Foote 1980; Buerkle 1987) and corrected for. However, such an assessment was not possible and an approximate 'deconvolution' was carried out using data collected on other swimbladdered fish of a similar size in Norwegian waters (data supplied by H. Bodholt, SIMRAD, personal communication).

A deconvoluted TS-depth matrix was extracted from each raw matrix and counts were modulated to densities (of echoes) per hectare using the effective beamwidth at depth. A pair (inshore and offshore) of TS frequency distributions for each survey were estimated by weighted addition of the depth shell TS distributions (weights proportional to the density of trout at given depths). The two TS distributions were combined to give a single TS distribution for each survey by weighted addition, weights being proportional to the number of trout within the inshore and offshore regions for that survey.

Progressing from a TS frequency distribution to a length frequency distribution requires a length–TS regression (e.g. Nakken & Olsen 1977; Dahm *et al.* 1985; Foote *et al.* 1986; Buerkle 1987). Regressions relating peak TS to length are appropriate only after correction for position in the acoustic beam and fish attitude (tilt). If off-axis location and attitude are not known, a regression of mean TS on length should be used. The present survey was conducted using a split beam sounder capable of correcting for the position in the beam of each target, so a mean TS regression could not be used as one source of variability in TS was already removed (reducing the proportion of small TS values). Following application of an approximate deconvolution function to account for tilt angle variability in TS, a peak TS regression can properly be applied.

Published regressions of TS on length for salmonids are scarce and most previous acoustic studies on trout and salmon were analysed using more general predictors (e.g. Mathisen *et al.* 1977; Thorne 1979; Burczinski & Karp 1985). Nakken and Olsen (1977) report regressions for a range of swimbladdered fish, including some data for trout. Trout data fall very close to the pooled regression (at 38 kHz) selected for use in this study:

$$TS_{max} = 24.5 \log L - 67.5$$

The predicted target strengths of this equation and that offered by Simrad differ by about 4 dB for the smallest size class of trout (100 mm), but only by about 1 dB for the largest (600 mm). Nakken and Olsen (1977) collected their data at 38 kHz and 120 kHz, and reported a difference between the two of about 1 dB. The difference between 38 kHz and the 70 kHz used in this study should therefore be less than 1 dB, and was ignored.

### 15.2.6    *Calibration*

Calibration was conducted before the first and during each subsequent survey using a standard copper sphere (TS = −39 dB at 70 kHz) suspended below the transducer at a depth of ∼20 m using nylon fishing line. Precise on-axis calibration was not attempted as the split beam system automatically compensated for slightly off-axis targets (Simrad 1986). In practice, after the first calibration, subsequent checks revealed very little change, and it was seldom necessary to alter the calibration.

The TVG system of the sounder was checked in deep water by lowering the standard target to the bottom on a nylon fishing line, and allowing insonification for approximately 30 s at each of a range of depths as the target was retrieved. Mean TS for the standard target did not vary significantly with depth, so the TVG compensation could be regarded as adequate over the depth range encountered in Taupo. The overall standard deviation of recorded values (after compensation for position in the beam as well as depth) was 0.89 dB: only 74% of recorded values were within 1 dB of the mean, and 2.5% were further than 2 dB.

### 15.2.7    *Beamwidth and sampling volume*

The effective beamwidth of the sounder in field conditions was checked for each survey using the duration-in-beam (d-i-b) technique of Crittenden *et al.* (1988). The number of repeat detections of individual fish at measured depths at constant speed, ping rate and gain allows direct calculation of the beam angle without the need for reliance on the theoretical sonar equation (Thorne 1988). The d-i-b method has the advantage that no additional sampling is required: sampling volume was estimated using data already collected to estimate trout density. The weighted regression method of Crittenden *et al.* (1988) was selected as it has been shown to be without bias and minimum variance (Crittenden 1989):

$$\phi_{BLUE} = 2S\,\beta_{BLUE}\,/\,(\pi p)$$

where $\phi$ is the sine of half the effective beamwidth, $S$ is the speed of the survey vessel, $p$ is the pulse repetition rate, and $\beta$ is the slope coefficient from a weighted regression. Subscripts $_{BLUE}$ indicate 'best linear unbiased estimate'.

The pooled d-i-b beamwidth was $11.7°$, but removing data for the third survey in May 1988 (which were significantly different from the rest) reduced this to $11.5° \pm 0.15°$ (the nominal beamwidth before software reduction to $10°$). An effective beam angle of $11.5°$ was therefore assumed for all calculations of density for all surveys except the third, where the measured value of $13.0 \pm 0.50°$ was substituted. The anomalous performance during this survey was traced to wear in the transducer cable which was repaired.

## 15.3    Results

The estimated mean density of trout in that part of the lake amenable to acoustic sampling varied from 3.09 to 37.04 ha$^{-1}$, or total abundances of 0.18 to 2.12 million fish (Fig. 15.1). Relative standard errors were approximately 7 to 9%, giving 95% confidence intervals of $\pm 14$ to 17% of the mean. Including data from visual counts from the littoral zone (Fig. 15.2) increased the estimate of total trout numbers by an average of less than 1% (maximum 5%) and inflated the confidence limits by a small amount to $\pm 15$ to 18% of the mean.

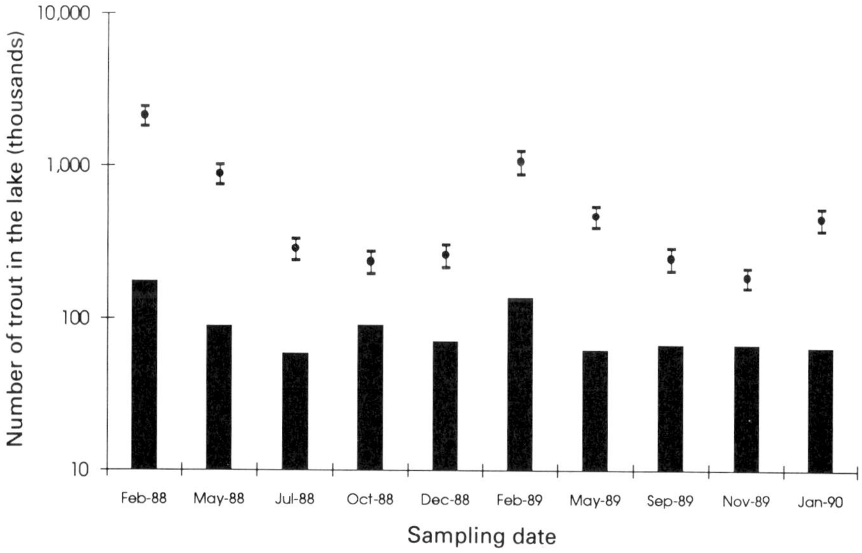


**Fig. 15.1**  Estimated total abundance of trout (with 95% confidence limits) in Lake Taupo. Estimated densities of legal-sized fish ( > 350 mm) are shown as histograms but confidence limits are not given because the variance in the conversion from TS to length is unknown.

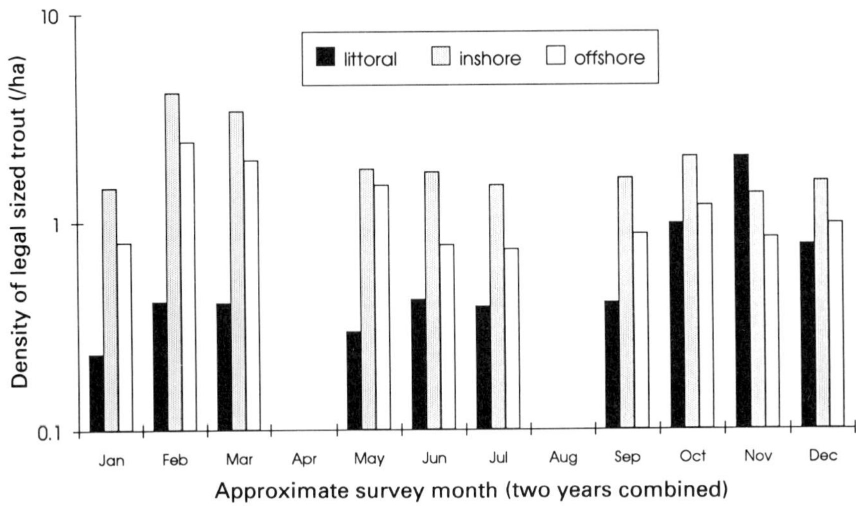


**Fig. 15.2**   Areal density of rainbow trout in three regions of Lake Taupo – two sampling years combined. Littoral areas ( < 20 m water depth) were surveyed by divers on manta boards whereas inshore (20–100 m) and offshore ( > 100 m) were surveyed acoustically.

For both annual cycles studied, trout density was greatest during the first survey (February–March), when the population consisted largely of small individuals high in the water column (Fig. 15.3). Abundance estimates were 2.12 million in 1988 and 1.07 million in 1989. Overall density and the proportion of small trout declined rapidly over the following months to a minimum during the fourth survey (October–November). Abundance estimates were 0.23 and 0.19 million trout for the two years. An increase in trout numbers was evident by the fifth survey for both years, but this increase was much greater for the second year when the last survey was undertaken in January as opposed to December. This pattern suggests a sizeable recruitment of juvenile individuals to the lake-resident population from tributary streams from December to March, followed by a period of mortality and/or migration into streams.

## 15.4   Discussion

The trout abundance estimates were considered of excellent precision. The consistency of population dynamics and spatial distribution between the two years is also encouraging as it suggests that surveys are likely to lead to good relative biomass estimates even if eventually found to be wanting with respect to estimating absolute abundance.

However, other studies carried out on Taupo suggest that biomass and production estimates (derived from abundance estimates) are probably fairly close to reality. The likely productivity of trout in Taupo's simple food chain can be estimated from production at lower trophic levels using theoretical predictors of transfer efficiency

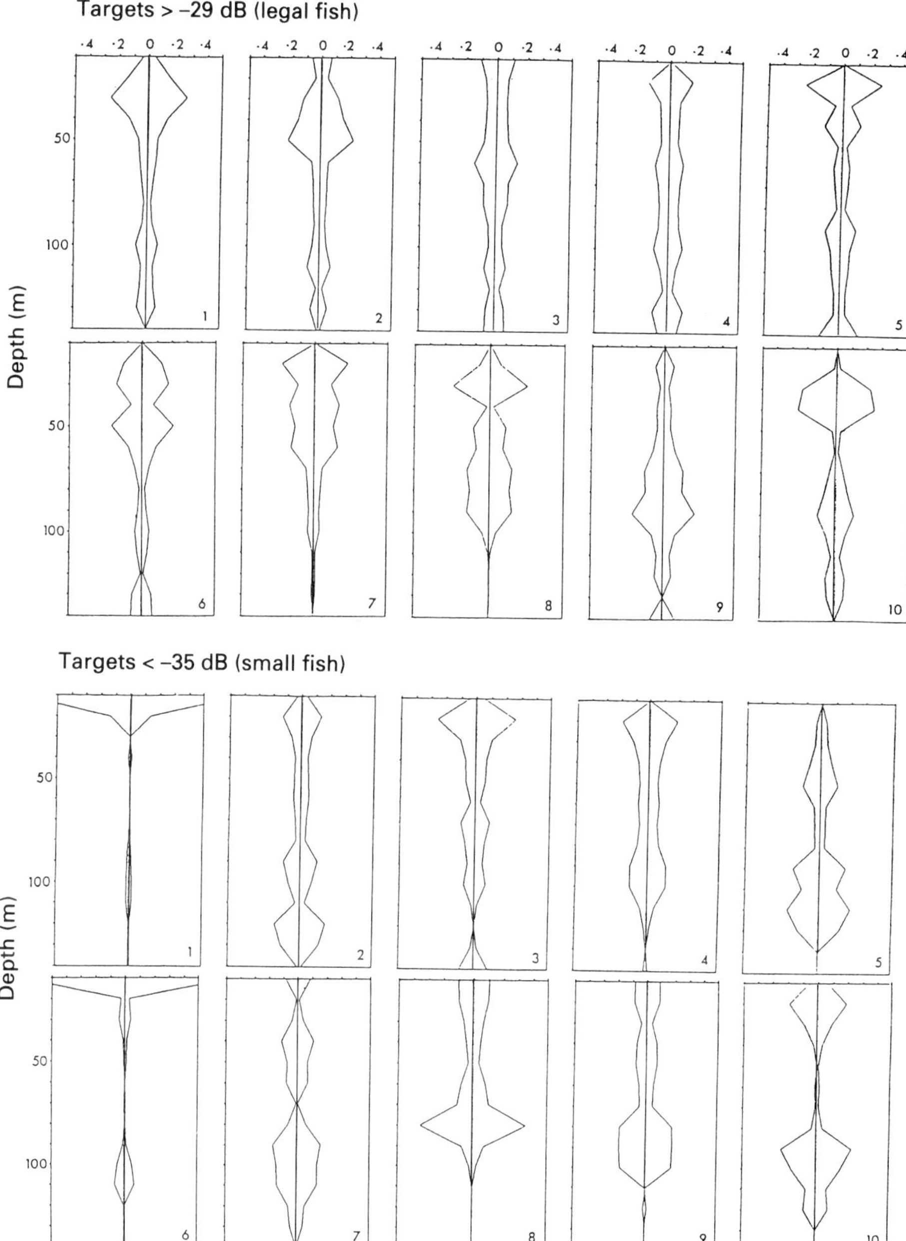



**Fig. 15.3** Vertical distribution of trout in Lake Taupo assessed by trace counting during 10 acoustic surveys. Sampling months for each survey are shown in Fig. 15.1.

based on particle size (e.g. Borgmann *et al.* 1984). Whilst primary and secondary production in the lake are only poorly understood, the production of smelt, which make up over 90% of the diet of trout, is well known. Annual smelt production of 5.1 to 6.3 g m$^{-2}$ and a predicted transfer efficiency of 11 to 23% lead to a likely range of trout production of 0.55 to 1.5 g m$^{-2}$. Calculated production was 0.55 and 0.88 g m$^{-2}$ in the two years of the study, at times when trout biomass was probably close to its historical minimum (Cryer 1991).

Estimates of likely yield for a lake of Taupo's size and trophic status were made from a range of published empirical models. The range of estimates was from 70 to 690 t per annum (Cryer, Chapter 28), but many of the models predicted yields close to 200 t, and the two extremes required assumptions regarded as unrealistic for Taupo. The measured catch in 1990 was approximately 175 t (DOC 1991). Both the likely long-term yield and the measured catch were compatible with the estimated mean recruited biomass in the lake of about 125 t, considering that a sizeable fraction of the adult population was absent from the lake during spawning periods, that recruitment of juveniles from streams continued year round, and that growth from recruiting juvenile (150 mm) to maturity (550 mm) took only 12 months (Cryer 1991).

The acoustic technique was developed for the circumstances prevalent in Lake Taupo. The lake is rather steep-sided and has only a narrow littoral zone. The target species seems to have a largely pelagic distribution in the lake, and tends to concentrate in or below the thermocline when present. Moreover, the target species was the only large pelagic species present, and its almost exclusive prey (also pelagic) were smaller than the lower limit of detection for the equipment. Finally, targets were widely dispersed, and trace counting was easily feasible. Under these special conditions, hydroacoustics seems to have produced a reasonable result.


## Acknowledgements

Thanks are due to innumerable people, but especially to all those hardy souls who assisted in the field: Errol Cudby, Brian Taylor, Dan Delaney, Glenn MacLean and John Gee.


## References

Borgmann U., Shear H. & Moore J. (1984) Zooplankton and potential fish production in Lake Ontario. *Canadian Journal of Fisheries and Aquatic Science* **41**, 1303–1309.

Buerkle U. (1987) Estimation of fish length from acoustic target strength. *Canadian Journal of Fisheries and Aquatic Science* **44**, 1782–1785.

Burczynski J.J. & Johnson R.L. (1986) Application of dual beam acoustic survey techniques to limnetic populations of juvenile sockeye salmon (*Oncorhynchus nerka*). *Canadian Journal of Fisheries and Aquatic Science* **43**, 1776–1788.

Burczynski J.J. & Karp W.A. (1985) The application of dual beam acoustic survey techniques for fish population estimation in lakes. *American Fisheries Society Conference on Small Hydropower & Fisheries*, Denver, USA.

Burstall P. (1983) Trout Fishery – History and Management. In: D.J. Forsyth & C. Howard-Williams (eds)

*Lake Taupo: Ecology of a New Zealand Lake.* Department of Scientific and Industrial Research Information Series No.158, Wellington, DSIR.

Crittenden R.N. (1989) Abundance Estimation Based on Echo Counts. PhD Thesis, University of Washington, Seattle, USA.

Crittenden R.N., Thomas G.L., Marino D.A. & Thorne R.E. (1988) A weighted duration-in-beam estimator for the volume sampled by a quantitative echo sounder. *Canadian Journal of Fisheries and Aquatic Science* **45**, 1249–1256.

Cryer M. (1991) Lake Taupo trout production. *New Zealand Department of Conservation Science and Research Series* No. 26, Wellington, Dept of Conservation, 191 pp.

Dahm E., Hartmann J., Lindem T. & Löffler H. (1985) EIFAC experiments on pelagic fish stock assessment by acoustic methods in Lake Constance. *EIFAC Occasional Paper* No. 15, Rome: Food and Agriculture Organisation.

DOC (1991) How many fish are caught at Taupo? *Target Taupo* Special Edition No. 1. 11 pp.

Foote K.G. (1980) Effect of fish behaviour on echo energy: the need for measurements of orientation distributions. *Journal du Conseil Internationale pour l'Exploration de la Mer* **39**, 193–201.

Foote K.G., Aglen A. & Nakken O. (1986) Measurement of fish target strength with a split-beam echo sounder. *Journal of the Acoustical Society of America* **80**, 612–621.

MacLennan D.N. & MacKenzie I.G. (1988) Precision of acoustic fish stock estimates. *Canadian Journal of Fisheries and Aquatic Science* **45**, 05–616.

Mathisen O.A., Croker T.R. & Nunnallee E.P. (1977). Acoustic estimation of juvenile sockeye salmon. *Rapports et proces-verbeaux de la Reunion de la Conseil Internationale pour l'Exploration de la Mer* **170**, 279–286.

Nakken O. & Olsen K. (1977) Target strength measurements of fish. *Rapports et proces-verbeaux de la Reunion de la Conseil Internationale pour l'Exploration de la Mer* **170**, 52–69.

Nunnallee E.P. (1973). A hydroacoustic data acquisition and digital data analysis system for the assessment of fish stock abundance. *University of Washington, Fisheries Research Institute Circular* No. 73–3, 48 pp.

Parsons J. (1979) *A Taupo Season.* Wellington, Collins.

Schultz H. (1988) An acoustic fish stock assessment in the Bautzen Reservoir. *Limnologica (Berlin)* **19**, 61–70.

Simrad (1986) *Simrad ES400 Scientific Sounder Instruction Manual P2092E.* Horten: Simrad.

Snedecor G.W. & Cochran W.C. (1989) *Statistical Methods*, 8th Ed. Ames: Iowa, Iowa State University Press.

Thorne R.E. (1979). Hydroacoustic estimates of adult sockeye salmon (*Oncorhynchus nerka*) in Lake Washington, 1972–75. *Journal of the Fisheries Research Board of Canada* **36**, 1145–1149.

Thorne R.E. (1988) An empirical evaluation of the duration-in-beam technique for acoustic estimation. *Canadian Journal of Fisheries and Aquatic Science* **45**, 1244–1248.

Williamson N.J. (1982) Cluster sampling estimation of the variance of abundance estimates derived from quantitative echo sounder surveys. *Canadian Journal of Fisheries and Aquatic Science* **39**, 229–231.

# Chapter 16
# Stock assessment of the Arctic charr (*Salvelinus alpinus*) population in Loch Ness, UK

C.W. BEAN *Freshwater Biological Association, The Ferry House, Far Sawrey, Ambleside, Cumbria LA22 0LP, UK*

I.J. WINFIELD and J.M. FLETCHER *NERC Institute of Freshwater Ecology, The Windermere Laboratory, Far Sawrey, Ambleside, Cumbria LA22 0LP, UK*

(All correspondence to I.J.W.)

**Abstract** Recent advances in echo-sounding technology, including data processing, provide a new technique to undertake stock assessment for open water fish species in fresh waters, although refinements are still required with respect to optimum survey design, data analysis and interpretation. The population of Arctic charr (*Salvelinus alpinus*) in Loch Ness, a large, deep and oligotrophic lake in the Highlands of Scotland, was assessed in July 1993 by a combination of quantitative echo sounding and trawling. Geometric mean densities (with 95% confidence limits) for the open water and inshore habitats were 36.7 (30.4, 44.3) and 8.8 (2.6, 25.8) individuals ha$^{-1}$, respectively, producing an overall population estimate of 181 500 individuals weighing 1260 kg. Maximum recorded age was 10 yr, with L$_\infty$ for males and females of 400 and 273 mm, respectively. Results of the assessment were compared with predictions from two production models and with published data on charr populations in other lakes.

KEYWORDS: Arctic charr, echo sounding, hydroacoustics, Loch Ness, stock assessment

## 16.1    Introduction

In large inland water bodies, stock assessments incorporating estimates of fish population densities pose considerable practical problems. While abundance in large but very shallow lakes can be determined using active netting techniques to sample known volumes of water (e.g. Jordan & Wortley 1985; van Donk *et al.* 1990), this approach is less reliable in deeper lakes where only a fraction of the entire water column can be fished at any one time. In such cases, trends in stock density can be followed using alternative approaches such as catch per unit effort (CPUE) (e.g. Mills & Hurley 1990; Dekker 1991) or virtual population analysis (Kipling 1984; Peltonen & Horppila 1992). However, such techniques usually require reliable data from commercial fisheries or studies of long duration involving extensive sample processing, and even then require further assumptions to be made before absolute values of density can be calculated. Not surprisingly, the need for rapid stock assessment has led to the development of simple predictive models of fish production or yield based on some measure of the nutrient status of the lake. A major problem of such models, which range from the original morphoedaphic index of Ryder (1965) to the phytoplankton production or total phosphorus models of Downing *et al.* (1990), is that they are empirical and the unknown principles which they incorporate may not apply to the water body under investigation. Consequently, there is still a need for new direct but rapid means of estimating fish density as part of wider stock assessments.

Recent advances in hydroacoustic technology, including the post-survey computer processing of data, offer one approach to the estimation of stock density in large, deepwater habitats. Work by the Geospace research group (Geospace Group 1993), for example, has demonstrated the application of such techniques in the marine environment, but also attests to their relative rarity in freshwater investigations. The latter is probably because of the high financial cost and, until recently, the requirement for relatively large survey vessels. Within Europe, the few freshwater studies which have been carried out have largely been on northern lakes (e.g. Lindem & Sandlund 1984; Jurvelius *et al.* 1988; Brabrand 1991; Elliott & Baroudy 1992; Sandlund *et al.* 1992; Snorrason *et al.* 1992) where the fish under investigation have included coregonids, cyprinids, osmerids and salmonids. Within this emerging spectrum of study systems, Brabrand (1991) noted few studies of ultra-oligotrophic lakes, which in northern European latitudes are typically dominated by Arctic charr (*Salvelinus alpinus* (L.)), and indicated the need for further work to determine the reliability of hydroacoustic surveys of sparse fish populations. Such investigations would also enable an examination of the performance of predictive production models at extremely low nutrient concentrations.

The oligotrophic Loch Ness, UK, possesses a simple fish community dominated by charr (Maitland 1981; Shine & Martin 1988) and so offers a particularly suitable site for the further study of fish stocks in low-nutrient systems. The aim of the study reported in this chapter was to undertake a comprehensive stock assessment of the charr population of Loch Ness using quantitative echo sounding to obtain information on within-loch distribution and density and open water trawling to obtain specimens for the determination of biological population parameters. Two fish production models were also run to investigate their performance in this oligotrophic lake.

## 16.2     Materials and methods

### 16.2.1     *Study site*

Loch Ness in the Highlands of Scotland (latitude 57° 15′, longitude 4° 30′) is the largest body of fresh water in the UK (volume $74.52 \times 10^8 \text{m}^3$, surface area 5700 ha). The *ca* 39 km long basin is morphometrically simple with steeply shelving sides (maximum depth 230 m, mean depth 132 m) and is elongated along a southwest–northeast axis (Fig. 16.1a). A deep (*ca* 40 m) seasonal thermocline is present and, although the loch is oligotrophic, the water is highly coloured by dissolved humic colloids. Several small rivers, including the Coiltie, Enrick, Foyers, Moriston, Oich and Tarff, flow into the loch, and the River Ness is the only outflow. Charr occur in the open water, although small numbers of Atlantic salmon (*Salmo salar* L.) also migrate through the loch and there are inshore populations of brown trout (*Salmo trutta* L.), three-spined stickleback (*Gasterosteus aculeatus* L.), eel (*Anguilla anguilla* L.), pike (*Esox lucius* L.) and minnow (*Phoxinus phoxinus* (L.)). No commercial fisheries operate on the loch, although the inflowing and outflowing rivers support valuable salmonid fisheries. Further information about Loch Ness can be found in Maitland (1981) and Shine & Martin (1988).

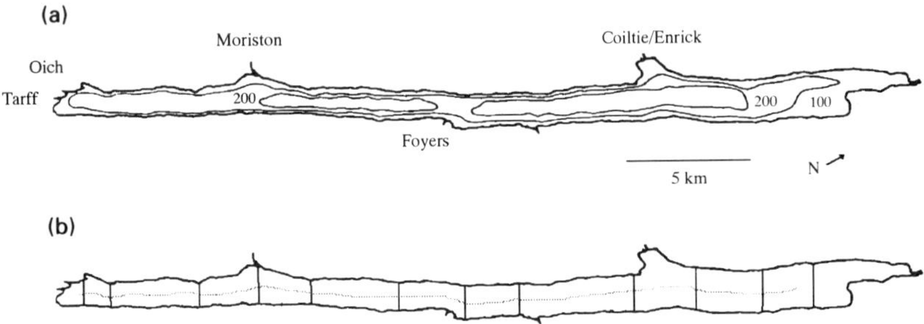



**Fig. 16.1**  Loch Ness showing (a) 100 and 200 m depth contours and the five river mouths covered by echo-sounding, (b) the mid-loch transect (broken line) used for trawling and echo-sounding, and 12 across-loch transects (continuous lines) used for echo-sounding only.

### 16.2.2    *Survey design*

An initial echo-sounding survey of the open water of Loch Ness was carried out in July 1992 (I.J.Winfield, unpublished data) as part of a study of fish–zooplankton spatial relationships. Throughout the loch, fish were mainly confined to the upper 50 m, with relatively higher densities in the southwest. The present stock assessment was carried out by a combination of trawling in the open water and quantitative echo-sounding in the offshore, inshore and river mouth areas of the loch. All of the echo sounding and almost all of the trawling (see below) was carried out in full daylight between 09.00 h and 18.30 h. Although Elliott & Baroudy (1992) recorded higher numbers of charr at night in Windermere, this effect was not observed in earlier surveys in Loch Ness.

### 16.2.3    *Trawling*

The fish community of the open water was sampled between 2 and 20 July 1993 by a pelagic trawl net (width 8 m, cod end mesh size 10 mm) pulled at a distance of *ca* 100 m behind a vessel moving at a speed of *ca* 1.3 m s$^{-1}$ (*ca* 2.5 knots). Trawls were made for durations of one hour at sites along a mid-loch line (Fig. 16.1b) and were initially carried out at depths of 15, 25, 50, 75 and 100 m, before 25 m was selected for all subsequent sampling as the most productive depth. In total, 35 trawls were made, only two of which were made after dark. All fish were killed by overdose with 2-phenoxyethanol and identified, measured (fork length, to the nearest mm), weighed (wet weight, to the nearest g) and frozen within minutes of capture. Subsequently, all fish were thawed and dissected to determine sex and to allow the removal and preservation of stomach contents for future analysis. Finally, the otoliths of charr were removed and viewed under a binocular microscope to determine their age. The latter was determined against a reference collection of charr otoliths from Windermere, England. A length-based von Bertalanffy growth curve was fitted using a nonlinear least squares technique (Saila *et al.* 1988).

### 16.2.4    *Echo sounding*

A Simrad EK 500 split-beam echo-sounder operating at 70 kHz, with a pulse duration of 0.6 ms and recording with a 40log$R$ Time-Varied Gain (TVG) to EP 500 software (Version 4.0) running on a personal computer (PC), was operated between 2 and 21 July 1993 from the trawl vessel cruising at *ca* 2.0 m s$^{-1}$ (*ca* 4.0 knots) along the same mid-loch line (see Fig. 16.1b), and produced echo absolute densities and target strengths. The transducer had a beam angle of 11° and was mounted 1.5 m below the water surface. During the transects, the target strength (*TS*) minimum threshold was set at $-70$ dB, although in subsequent analyses targets were classified as small fish (*TS* from $-61$ to $-58$ dB, length (*L*) from 1 to 5 cm), medium fish (*TS* from $-52$ to $-49$ dB, *L* from 5 to 10 cm), large fish (*TS* from $-46$ to $-34$ dB, *L* from 10 to 60 cm) and extra-large fish (*TS* greater than $-31$ dB, *L* greater than 60 cm) based on the relationship for physoclists of *TS* = 20log$_{10}$*L* $-$ 67.4 (Foote 1987). A total of 50 files from along the length of the loch was produced and subsequently analysed using EP 500 software with 10 m depth bands encompassing the upper 100 m to produce absolute densities (individuals ha$^{-1}$) for the four fish classes. Preliminary analysis of data from the single-beam system described below using both 2 m and 10 m depth bands showed a high degree of correlation ($r^2$ = 85%) and close agreement over the range of fish densities typically encountered in Loch Ness.

The above mid-loch echo-sounding was complemented by a programme of offshore and inshore echo-sounding using a Simrad EY 200P single-beam echo-sounder operating at 200 kHz, with a pulse duration of 1.0 ms and recording with a 40log$R$ TVG to analogue audio tape, deployed from a second vessel cruising at a speed of *ca* 2.4 m s$^{-1}$ (*ca* 4.7 knots). The transducer had a beam angle of 7° and was mounted 0.5 m below the water surface. Between 15 and 18 July 1993, 12 transects were made directly across the narrow axis of the loch from the northwest to southeast shores at *ca* 1, 2, 6, 8, 10, 14, 16, 19, 24, 26, 29 and 31 km from the southwest shore of the loch (Fig. 16.1b). Data collected were subsequently digitized and analysed for 10 m depth layers encompassing the upper 50 m using the PC-based hardware and software system HADAS (Version 4.02) (for further details see Walline *et al.* 1992) which produces echo absolute densities and target strengths, the latter being derived indirectly using a modification of the algorithm of Craig & Forbes (1969). Each transect was broken into 'inshore northwest' (start of transect until depth exceeded 120 m), 'inshore southeast' (latter part of transect beginning where depth had decreased to 120 m) and 'open water' (middle section of transect where depth exceeded 120 m). The latter component was also broken into five equal sections using the fish distribution facility of HADAS to determine distribution patterns across the main body of the loch.

The single-beam echo-sounding system was also used to investigate the abundance of fish near the mouths of six of the loch's inflowing rivers, i.e. the Rivers Coiltie and Enrick which share a common mouth, Foyers, Moriston, Oich and Tarff. Between 16 and 18 July 1993, single transects were made into the mouth of each tributary and the local abundance of fish was determined using the fish distribution facility of HADAS.

Overall in the echo-sounding survey, the ratio of coverage (length of surveys: square root of research area) was 6.5 : 1, which is comparable to those of other surveys (Jurvelius 1991).

In previous analyses of echo-sounding data, Jurvelius & Heikkinen (1988) and Baroudy & Elliott (1993) found that geometric rather than arithmetic means provided the best estimate of population density when some measure of precision was required. Consequently, both forms of mean (with 95% confidence limits) were calculated in the present study. Estimates of numerical densities were also converted to estimates of biomass densities using the weightings described below.


**16.2.5**    *Absolute numbers and biomass estimates*

Conversions of numerical densities to absolute values were made by multiplying by 4650 ha for the open water habitat and by 1050 ha for the inshore habitat, defined as areas of depth greater than and less than 100 m, respectively.

Estimates of numerical density from the split-beam echo-sounding were used to calculate estimates of biomass density using the median individual weights of small, medium and large charr (defined as earlier) sampled by trawling. A conservative estimate of the individual weight of extra large fish was generated using a log-transformed length–weight regression and a length of 60 cm (the minimum length of extra large fish). The individual weights used were thus 1, 5, 30 and 2000 g. For the single-beam echo sounding, for which estimates of target strength are less reliable, it was assumed that the fish recorded were an equal mixture of medium and large fish. Conversions of these biomass densities to absolute values were made using the multiplication factors given above.

**16.2.6**   *Fish production indices*

The original morphoedaphic index (*MEI*) (Ryder 1965):

*MEI = total dissolved solids* (ppm) / *mean depth* (feet)

was calculated using a total dissolved solids value of 31 ppm measured on 20 August 1993. This index was converted to fish production using the following equation derived from 34 North American water bodies (Ryder 1965):

*Fish production* = 2.094 $MEI^{0.4461}$

Fish production was also estimated using the relationship described by Downing *et al.* (1990):

$\log_{10}$ (*Fish production*, kg ha$^{-1}$) = 0.332 + 0.531 $\log_{10}$ (*total phosphorus*, µg l$^{-1}$)

using a total phosphorus value of 4 µg l$^{-1}$ measured on 19 July 1993 (D.G. George, Institute of Freshwater Ecology, unpublished data).


## 16.3   Results

**16.3.1**   *Trawling*

A total of 188 fish was collected, comprising 180 (95.8%) charr, 7 (3.7%) trout and 1 (0.5%) three-spined stickleback. All but 16 charr and 2 trout were taken during daytime trawls. Mean CPUEs at trawling depths of 15, 25, 50, 75 and 100 m were 0.50, 7.09, 1.20, 1.50 and 2.00 individuals h$^{-1}$, respectively.

The 180 charr were composed of 105 individuals of indeterminable sex, 38 males and 37 females. Length and age distributions (Fig. 16.2) ranged between 49 and 330 mm, and 1 and 10 yr, respectively, with both sexes being represented equally among the classes. Fig. 16.2 also shows growth curves for male and female charr, and the mean lengths at age for both sexes and fish of indeterminable sex. Good fits to the von Bertalanffy growth model were found for males, females and all fish combined (Table 16.1), and showed a considerably higher L$_\infty$ ( for males over females (400 and 273 mm, respectively).

Minimum and maximum individual weights were 1 and 216 g, respectively, and the length-weight relationship of charr (all fish combined) was described ($r^2$ = 98.9%) by

$\log_{10}$ (*weight*, g) = −4.903 + 2.949 $\log_{10}$ (*length*, mm)

There was no obvious phenotypic evidence for the presence of more than one 'morph' or 'race' of charr, nor was there any indication from examination of the raw data of the growth and length-weight relationships that more than one population was present in the sample.

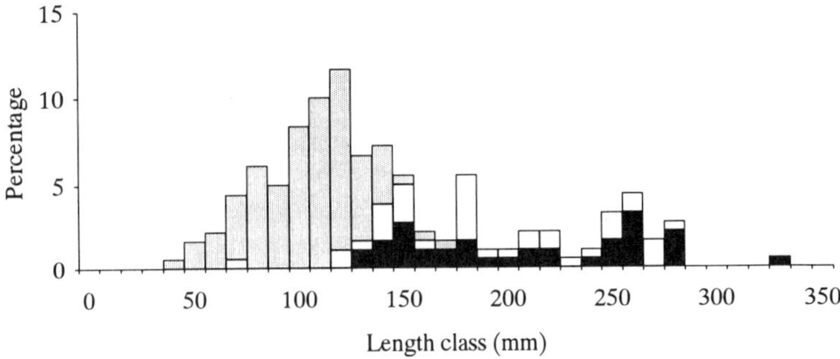

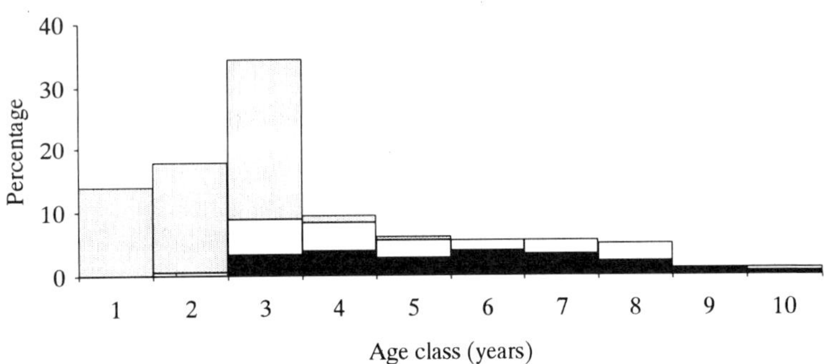

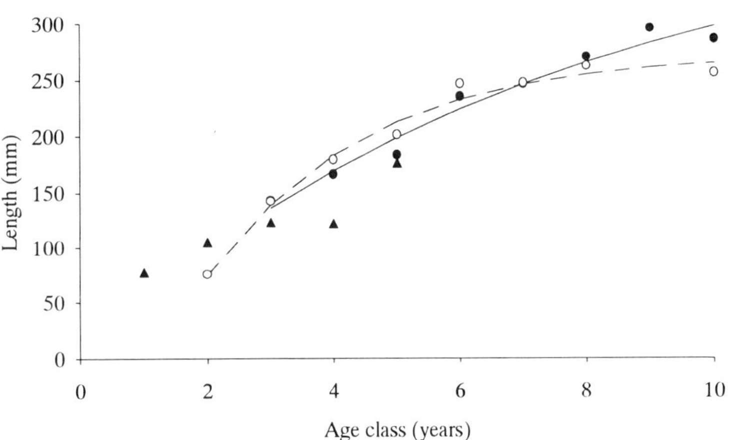

**Fig. 16.2**  Length and age frequency distributions and growth curves for charr. In the length and age distributions (upper and middle graphs, respectively), males, females and fish of indeterminable sex are shown as closed, open and stippled areas, respectively. For growth (lower graph), the mean lengths at age of males, females and fish of indeterminable sex are shown as closed circles, open circles and closed triangles respectively. Fitted von Bertalanffy growth curves for males and females are shown as continuous and broken lines, respectively.

**Table 16.1** von Bertalanffy growth curve parameters fitted to mean lengths at age of charr.

| Sex | L∞ (mm) | K | $t_0$ | $r^2$ (%) |
|---|---|---|---|---|
| Male | 400 | 0.135 | −0.081 | 97 |
| Female | 273 | 0.396 | +1.197 | 98 |
| Both* | 403 | 0.123 | −0.490 | 97 |

*Incorporates data from young fish of indeterminable sex.

### 16.3.2  *Echo sounding*

As suggested by the initial exploratory trawl catches, fish in all areas of the loch were concentrated in a depth band of 20 to 30 m. Target strengths ranged from −70 dB (the lower limit of detection) to −25 dB, which represents a fish of 132 cm in length. Along the long axis, densities of small, medium and large fish in the open water recorded by the split-beam system all tended to be higher in the southern parts of the loch (Fig. 16.3), while very few extra large fish were recorded anywhere. The total density peaked at 176 individuals ha$^{-1}$, although it was generally below 50 individuals ha$^{-1}$. Geometric and arithmetic mean densities, which were in close agreement, are given in Table 16.2.

Densities of fish recorded in the open water of the loch by the single-beam system (Fig. 16.4) were of similar levels and again showed a tendency for higher densities in the south. Fish distribution in the two inshore regions was more variable and while densities above 100 individuals ha$^{-1}$ were recorded seven times, many of the other inshore regions contained no fish. This patchy distribution led to a considerable divergence between geometric and arithmetic mean densities in the inshore habitat (Table 16.2). Taking geometric means as the more appropriate descriptor, the open water mean of 36.7 individuals ha$^{-1}$ produced by the split-beam system was slightly but significantly lower than that of 48.8 individuals ha$^{-1}$ produced by the single-beam system (two-way t test assuming unequal variances on $\log_{10}(x+1)$ transformed data, $t = 2.051$, df $= 31.901$, $P = 0.049$). Considering the single-beam results alone, the mean for the open water was significantly higher than that of 8.8 individuals ha$^{-1}$ for the inshore area (two-way t test assuming unequal variances on $\log_{10}(x+1)$ transformed data, $t = 3.276$, df $= 24.902$, $P < 0.01$). The numerical densities of fish among the five sections of the open water component of the single-beam transects, however, did not vary significantly (Kruskal-Wallis $H = 2.338$, $P > 0.10$).

The split-beam results showed that large fish dominated the total biomass density of the open water, while the single-beam data indicated that densities were lower in the inshore (Table 16.3). The latter is evident from a comparison of geometric means, although the less appropriate arithmetic means for these two habitats are very similar.

Numerical densities, and hence biomass densities, of fish near the river mouths were extremely variable and, given the limited nature of this aspect of the survey, are not reported in detail. However, an overall maximum of 5315 individuals ha$^{-1}$ (93 kg

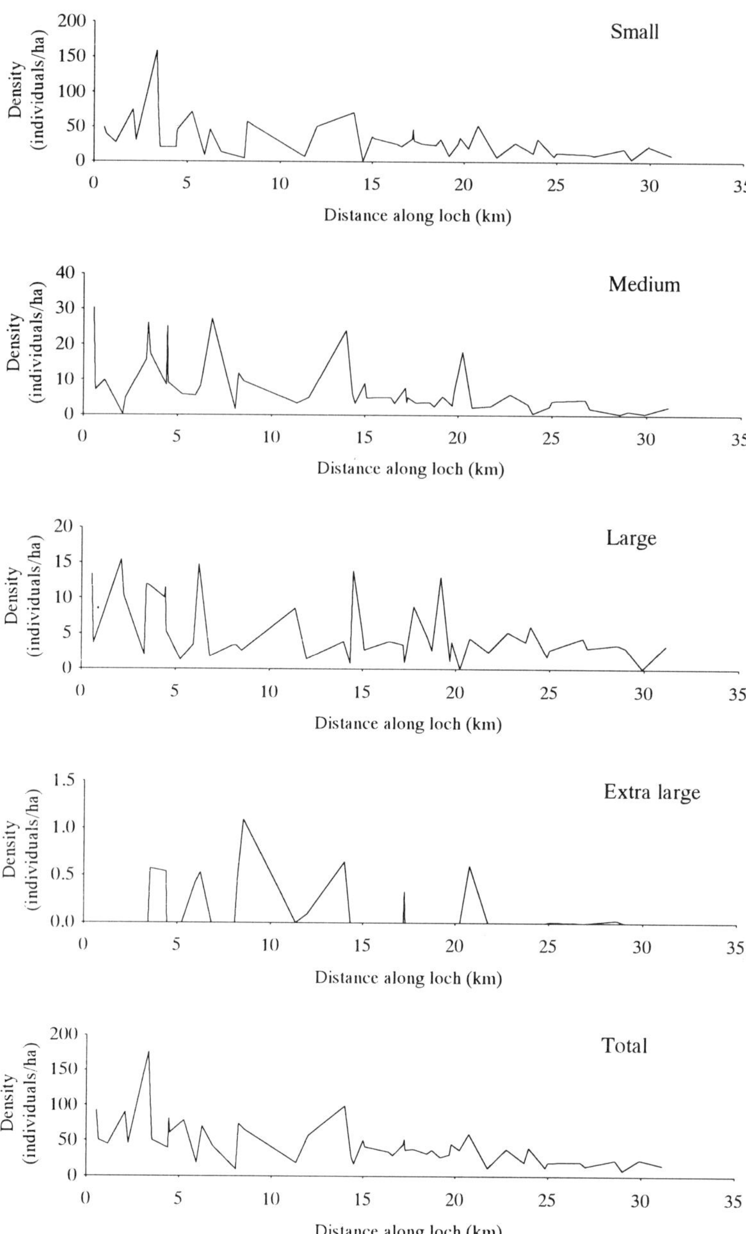



**Fig. 16.3** The density (individuals ha$^{-1}$) of small (1 to 5 cm), medium (5 to 10 cm), large (10 to 60 cm), extra large (greater than 60 cm) and total fish along the long axis of Loch Ness.

**Table 16.2**  Estimates of fish density in terms of numbers ha$^{-1}$.

| Sounder | Habitat | Fish class | Individuals ha$^{-1}$ (geometric mean with lower and upper 95% confidence limits in parentheses) | Individuals ha$^{-1}$ (arithmetic mean with lower and upper 95% confidence limits in parentheses) |
|---|---|---|---|---|
| EK 500 | Open water | Small | 23.5 (18.2, 30.3) | 32.1 (24.8, 39.5) |
| EK 500 | Open water | Medium | 5.2 (4.0, 6.8) | 7.6 (5.5, 9.6) |
| EK 500 | Open water | Large | 4.0 (3.1, 5.0) | 5.2 (4.0, 6.4) |
| EK 500 | Open water | Extra large | 0.1 (0.0, 0.1) | 0.1 (0.1, 0.2) |
| EK 500 | Open water | Total | 36.7 (30.4, 44.3) | 45.0 (36.4, 53.6) |
| EY 200P | Open water | Total | 48.8 (39.0, 61.1) | 51.8 (40.8, 62.7) |
| EY 200P | Inshore | Total | 8.8 (2.6, 25.8) | 57.3 (27.8, 86.8 |

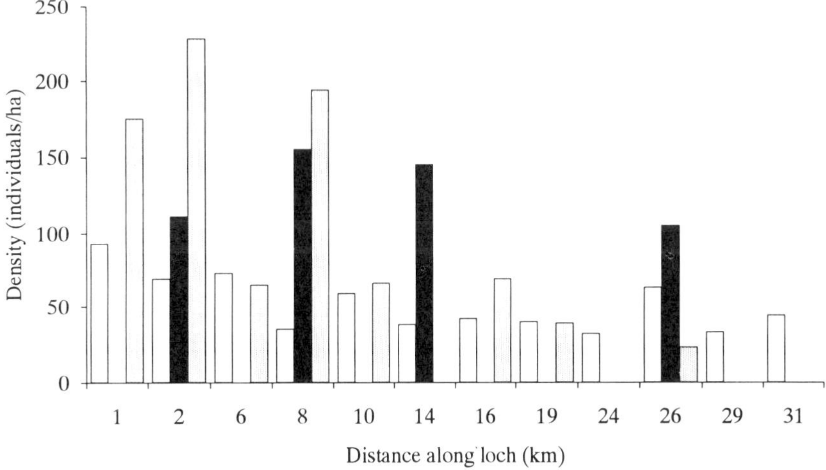

**Fig. 16.4**  The density (individuals ha$^{-1}$) of fish on 12 across-loch transects spaced along the long axis of Loch Ness. Each transect is itself broken into open water (open columns), inshore northwest (closed columns) and inshore southeast (stippled columns) components (see text for definitions).

**Table 16.3** Estimates of fish density in terms of biomass ha$^{-1}$.

| Sounder | Habitat | Fish class | Biomass (g) ha$^{-1}$ (geometric mean with lower and upper 95% confidence limits in parentheses) | Biomass (g) ha$^{-1}$ (arithmetic mean with lower and upper 95% confidence limits in parentheses) |
|---|---|---|---|---|
| EK 500 | Open water | Small | 24 (18, 30) | 32 (25, 40) |
| EK 500 | Open water | Medium | 24 (18, 32) | 38 (27, 48) |
| EK 500 | Open water | Large | 104 (76, 143) | 156 (121, 191) |
| EK 500 | Open water | Extra large | 4 (1, 9) | 214 (81, 347) |
| EK 500 | Open water | Total | 262 (197, 346) | 439 (295, 584) |
| EY 200P | Open water | Total | 853 (682, 1068) | 906 (714, 1098) |
| EY 200P | Inshore | Total | 40 (7, 204) | 1003 (486, 1519) |

ha$^{-1}$) recorded on a part of the River Tarff transect suggests that significant numbers of fish may exist in such areas.

### 16.3.3　*Absolute numbers and biomass estimates*

The total numbers and biomass of fish in Loch Ness, excluding concentrations near river mouths, are best derived by summing the open water data obtained by the split-beam system and the inshore data produced by the single-beam system (Tables 16.4 and 16.5). Using geometric means, this produces a population size of 181 500 individuals weighing 1260 kg. As noted above, the present study provided only preliminary estimates of fish densities near river mouths but such areas could contribute significantly to the total number of fish in the loch. Such calculations, however, would involve considerable assumptions which are considered further in the discussion.

### 16.3.4　*Fish production indices*

The *MEI* was 0.07, which produced a predicted fish production of 0.70 kg ha$^{-1}$ yr$^{-1}$, or 4.02 t yr$^{-1}$ for the entire loch. The total phosphorus model produced a predicted fish production of 4.48 kg ha$^{-1}$ yr$^{-1}$, or 25.53 t yr$^{-1}$ for the entire loch.

**Table 16.4**   Estimates of fish abundance in terms of numbers.

| Sounder | Habitat | Fish class | Number of individuals ($\times 10^3$) (geometric mean with lower and upper 95% confidence limits in parentheses) | Number of individuals ($\times 10^3$) (arithmetic mean with lower and upper 95% confidence limits in parentheses) |
| --- | --- | --- | --- | --- |
| EK 500 | Open water | Small | 111.6 (83.7, 139.5) | 148.8 (116.2, 186.0) |
| EK 500 | Open water | Medium | 23.2 (18.6, 32.6) | 37.2 (27.9, 46.5) |
| EK 500 | Open water | Large | 18.6 (14.0, 23.2) | 23.2 (18.6, 27.9) |
| EK 500 | Open water | Extra large | 0.5 (0.0, 0.5) | 0.5 (0.5, 0.5) |
| EK 500 | Open water | Total | 172.0 (139.5, 204.6) | 209.2 (167.4, 251.1) |
| EY 200P | Open water | Total | 226.9 (181.4, 284.1) | 54.6 (43.0, 66.2) |
| EY 200P | Inshore | Total | 9.5 (3.2, 27.3) | 59.8 (29.4, 91.4) |

**Table 16.5**   Estimates of fish abundance in terms of biomass.

| Sounder | Habitat | Fish class | Biomass (kg) (geometric mean with lower and upper 95% confidence limits in parentheses) | Biomass (kg) (arithmetic mean with lower and upper 95% confidence limits in parentheses) |
| --- | --- | --- | --- | --- |
| EK 500 | Open water | Small | 112 (84, 140) | 149 (116, 186) |
| EK 500 | Open water | Medium | 112 (84, 149) | 177 (126, 223) |
| EK 500 | Open water | Large | 484 (353, 665) | 725 (563, 888) |
| EK 500 | Open water | Extra large | 19 (5, 42) | 995 (377, 1614) |
| EK 500 | Open water | Total | 1218 (916, 1609) | 2041 (1372, 2716) |
| EY 200P | Open water | Total | 896 (716, 1121) | 951 (750, 1153) |
| EY 200P | Inshore | Total | 42 (7, 214) | 1053 (510, 1595) |

**16.4    Discussion**

The charr population of Loch Ness proved to be particularly suitable for a stock assessment in an oligotrophic lake and could be used in further development of this important branch of fisheries science. In practical terms, the loch is very accessible, including the opportunity of deploying large research vessels via the Caledonian Canal, and morphometrically simple, which eases the problems of devising systematic but practical survey designs. In scientific terms, the loch remains apparently unpolluted by increasing eutrophication or acidification which complicate charr stock assessments in many other parts of the UK (e.g. Elliott & Baroudy 1992; Maitland *et al.* 1991). In addition, the fish community of Loch Ness remains apparently free from recent species introductions which have occurred in many other large lakes (see review by Winfield 1992). The open water community is dominated by charr and the population biology of this species is not complicated by the effects of a fishery nor, apparently, by the presence of more than one 'morph' or 'race', as has been recorded elsewhere in Scotland (Walker *et al.* 1988) and England (Partington & Mills 1988). Loch Ness thus offers an increasingly rare opportunity of studying a pristine stock, and stock assessment methodology, in the comparatively simple open water of a near-pristine lake.

Interpretation of the echo-sounding data presented few difficulties when compared with the situations experienced in other charr lakes where species such as trout and perch (e.g. Elliott & Baroudy 1992) also make significant contributions to the open water fish community. The absences of multiple echoes and any cross-loch patterns of distribution in the open water of the present study were also notable. While hydro-acoustic surveys in other lakes are best undertaken at night when fish are most dispersed and thus amenable to counting (e.g. Walline *et al.* 1992), the charr in Loch Ness are also dispersed during the day. This feature is probably related to the high attenuation of light in the loch and the depth of 20 to 30 m at which the charr occur, where light levels remain extremely low at all times. The reason for this deep distribution of charr in Loch Ness, which was also indicated by the distribution of CPUE of the trawl, is unknown although it may be related to a similarly deep distribution of cladoceran zooplankton (D.G. George, Institute of Freshwater Ecology, unpublished data).

Whatever the biological explanation, this dispersion produces similar geometric and arithmetic mean estimates of charr density in the open water. In Loch Ness, these two measures diverge only in the inshore areas where the fish, which may include more trout or other species, have a more patchy distribution. The precision of the open water estimates was comparable with those reported for more dense fish populations elsewhere (e.g. Hansson 1993), showing that in this oligotrophic system at least, echo sounding offers a viable method for estimating fish abundance in the open water.

Much larger levels of variation were found when estimating the abundance of fish near the river mouths of Loch Ness, where localized areas held fish densities in excess of 1000 individuals ha$^{-1}$, typical of those observed in eutrophic lakes in the UK (I.J. Winfield, unpublished data). It is assumed that these concentrations included few charr which, when sympatric with trout, are restricted to open waters (Hindar *et al.* 1988). Furthermore, many multiple echoes were recorded near the river mouths, suggesting that they originated from tightly-shoaling species such as minnows or even inanimate objects such as gas bubbles or allochthonous material. Although river mouth concentrations of fish may contribute significantly to the total biomass and production of the entire fish community of Loch Ness, and so deserve further study, they probably make a negligible contribution to the charr population. The same is true for the depth layers greater than 100 metres which were excluded from the present analysis. Fish were rarely observed at these depths (always less than 1.1 individuals ha$^{-1}$) and tended to be large individuals with target strengths of $-25$ dB, equivalent to a length of 132 cm which is beyond the maximum size attained by charr in Loch Ness. It is more likely that these large individuals were ferox trout (*Salmo trutta* L.) which are known to exceed lengths of 90 cm in deep, oligotrophic Scottish lochs (Campbell 1979).

Before putting the results of the hydroacoustic component of the present stock assessment into a wider context, it is appropriate to consider the biological population parameters of the Loch Ness charr. Although this species comprises an important component of fish communities of Scottish highland lochs, reviews (Maitland *et al.* 1984; Maitland & Campbell 1992) show that little is known about its local ecology.

In a study of the threatened charr population of Loch Doon in southwest Scotland (Maitland *et al.* 1991), fish up to 8 years old with a sex ratio close to unity were found. A similar age structure, also encompassing a sex ratio close to unity, was found in the present study, although the maximum observed age was slightly greater at 10 years. Further examples of maximum ages of charr are given in Table 16.6 and show that the longevity of the Loch Ness population is comparable with those shown in other parts of Europe. In addition, the representation of all age classes in Loch Ness indicates stable recruitment over the last decade. A comparison with charr growth data collated by Maitland *et al.* (1984) indicates that in their early years Loch Ness fish grow more

**Table 16.6**  Maximum lengths and ages of European populations of Arctic charr. An asterisk indicates that variations were recorded in maximum lengths due to the study of more than one population (Pechlaner 1984) or the effects of marked changes in fishing pressure (Amundsen 1989). Note that no age data were given by Pechlaner (1984).

| Location | Length (mm) | Age (years) | Source |
| --- | --- | --- | --- |
| Lake Constance, Germany | 250 | 6 | Hartmann (1984) |
| Alpine lakes, Austria | 190–220* | — | Pechlaner (1984) |
| Takvatn, Norway | 220–250* | 11 | Amundsen (1989) |
| Konigsee, Germany | 370 | 11 | Bohl *et al.* (1989) |
| Windermere, England | 344–384 | 14 | Mills (1989) |
| Loch Doon, Scotland | 240 | 8 | Maitland *et al.* 1991 |

slowly than those of other Scottish waters, but achieve similar ultimate lengths. Maitland *et al.* (1991) reported a $L_\infty$ of *ca* 240 mm in the acidifying Loch Doon, which is below the figures of 400 and 272 mm found for males and females, respectively, in the present study. Maximum lengths of charr observed in other water bodies (Table 16.6), indicate that this aspect of charr growth in Loch Ness was similar to those reported for unstunted and unstressed populations elsewhere in Europe.

Comparisons with hydroacoustic, or any other, stock estimates of charr are hampered by a paucity of data from other sites. Long-term CPUE data from Windermere in England (Mills & Hurley 1990) suggested that the mature charr population amounted to 140 000 individuals weighing between 30 and 35 t, which is remarkably consistent with a whole population estimate of 768 000 charr (equivalent to 520 individuals ha$^{-1}$) made using a single-beam echo sounder and HADAS (Baroudy 1993). Similar charr densities were found in a hydroacoustic survey of Thingvallavatn in Iceland which recorded between 398 and 638 individuals ha$^{-1}$ (Snorrason *et al.* 1992), and mark-recapture methods in the Canadian High Arctic which estimated 362 individuals ha$^{-1}$ in excess of 130 mm in length (MacCallum & Regier 1984). These densities are considerably greater than the total of 36.7 individuals ha$^{-1}$ of charr in the open water of Loch Ness recorded by the split-beam system.

The relatively low abundance of charr in Loch Ness becomes more apparent when biomass is considered because the population was dominated by small fish. The estimate of the standing stock in Loch Ness (0.262 kg ha$^{-1}$) is less than 3% of the 9.23 kg ha$^{-1}$ reported by MacCallum & Regier (1984) in the Canadian High Arctic, and less than 1% of the 65 kg ha$^{-1}$ found in the productive Myvatn in Iceland by Kristjansson & Adalsteinsson (1984), who also noted that commercially-exploited Icelandic lakes typically yield 10 to 15 kg ha$^{-1}$ yr$^{-1}$.

There are no yield data available for the unexploited charr population of Loch Ness although some explorations may be made using the results of the yield indices. The morphoedaphic index predicted a yield of 0.70 kg ha$^{-1}$ yr$^{-1}$ or 4.02 t yr$^{-1}$ for the entire loch, while that produced by the total phosphorus model was considerably higher at 4.48 kg ha$^{-1}$ yr$^{-1}$ or 25.53 t yr$^{-1}$. A third approach can be made not through these indirect 'environment-driven' models, but by the model of Downing & Plante (1993) which predicts fish production as a function of fish standing biomass and maximum individual body weight. Using values of 0.262 kg ha$^{-1}$ and 216 g taken from the open water survey, the latter predicts a production of 0.238 kg ha$^{-1}$ yr$^{-1}$, which is much closer to that predicted by the morphoedaphic index. It should also be noted that both indirect indices predict the production of the entire fish community of a lake, which for Loch Ness would include the trout and other fish species associated with the inshore zone and river mouths.

The above exploration of the performance of the two indirect predictive models of fish production is speculative, but illustrates the dangers of their unconsidered application in a lake with peculiar characteristics such as Loch Ness. For example, the total phosphorus model implicitly assumes that lake, and hence fish, productivity is linked to the availability of phosphorus. However, this may not be the case in Loch

Ness, even though it is oligotrophic, because primary productivity may be limited not by nutrients but by light availability due to low light penetration which persists all year round.

Although models can be a valuable preliminary tool when assessing fish production in lakes and echo-sounding has been used to corroborate such predictions for the eutrophic Lake Kinneret in Israel (Walline *et al.* 1992), results from this study show the danger of their application to an atypical oligotrophic situation. By contrast, quantitative hydroacoustics deployed in this and other studies (Brabrand 1991; Sandlund *et al.* 1992; Snorrason *et al.* 1992) have produced reliable and direct biomass estimates. Such data still require the fishery manager to apply models, such as that of Downing & Plante (1993), which link standing stock with production before the yield of a stock can be predicted, but recent advances in hydroacoustics will themselves facilitate the further development of these models.

## Acknowledgements

We thank the crews of the *R.V. Calanus* and *R.V. Seol Mara* for their expertise and help in what were at times very trying conditions. We are also grateful to Adrian Shine and colleagues of the Loch Ness and Morar Project for their logistical support. The pelagic trawl was generously loaned by Dick Ferro of the SOAFD Laboratory in Aberdeen, while Simrad supplied the split-beam echo-sounding system which was expertly installed by Birnie Lees and Erik Stenersen. We also thank the Ness District Salmon Fishery Board and The Scottish Office for their permissions to undertake the trawling. This study was funded by Project Urquhart and we thank Nicholas Witchell in particular for his unstinting enthusiasm and support.

## References

Amundsen P. A. (1989) Effects of intensive fishing on food consumption and growth of stunted Arctic charr (*Salvelinus alpinus* L.) in Takvatn, northern Norway. *Physiology and Ecology, Japan* **Special Volume 1**, 265–278.

Baroudy E. (1993) *Some Factors Affecting Survival and Distribution of Arctic charr ( Salvelinus alpinus ( L.) ) in Windermere.* Unpublished PhD Thesis, University of Lancaster, 297 pp.

Baroudy E. & Elliott J. M. (1993) The effect of large-scale spatial variation of pelagic fish on hydroacoustic estimates of their population density in Windermere (northwest England). *Ecology of Freshwater Fish* **2**, 160–166.

Bohl M., Leuner E., Negele R.D. & Bohl E. (1989) Reduced growth of Arctic charr (*Salvelinus alpinus* L.) in the Lake 'Konigssee', Bavarian national park. *Physiology and Ecology, Japan* **Special Volume 1**, 639–646.

Brabrand A. (1991) The estimation of pelagic fish density, single fish size and fish biomass of Arctic charr (*Salvelinus alpinus* (L.)) by echosounding. *Nordic Journal of Freshwater Research* **66**, 44–49.

Campbell R.N. (1979) Ferox trout *Salmo trutta* L., and charr, *Salvelinus alpinus* (L.), in Scottish lochs. *Journal of Fish Biology* **14**, 1–29.

Craig R. E. & Forbes S. T. (1969) Design of a sonar for fish counting. *Fiskeriderektoratets Skrifter* **15**, 210–219.

Dekker W. (1991) Assessment of the historical downfall of the IJsselmeer fisheries using anonymous inquiries for effort data. In: I.G. Cowx (ed.) *Catch Effort Sampling Strategies: their Application in*

*Freshwater Fisheries Management*. Oxford: Fishing News Books, pp 233–240.

Downing J.A., Plante C. & Lalonde S. (1990) Fish production correlated with primary productivity, not the morphoedaphic index. *Canadian Journal of Fisheries and Aquatic Sciences* **47**, 1929–1936.

Downing J.A. & Plante C. (1993) Production of fish populations in lakes. *Canadian Journal of Fisheries and Aquatic Sciences* **50**, 110–120.

Elliott J.M. & Baroudy E. (1992) Long-term and short-term fluctuations in the numbers and catches of Arctic charr, *Salvelinus alpinus* (L.), in Windermere (northwest England). *Annales de Limnologie* **28**, 135–146.

Foote K.G. (1987) Fish target strengths for use in echo integrator surveys. *Journal of the Acoustical Society of America* **82**, 981–987.

Geospace Group (1993) The spatial organization of aquatic populations as observed using hydroacoustic methods. *Aquatic Living Resources* **6**, 171–174.

Hansson S. (1993) Variation in hydroacoustic abundance of pelagic fish. *Fisheries Research* **16**, 203–222.

Hartmann J. (1984) The charrs (*Salvelinus alpinus* L.) of Lake Constance, a lake undergoing cultural eutrophication. In: L. Johnson & B. Burns (eds) *Biology of the Arctic Charr: Proceedings of the International Symposium on Arctic Charr*. Winnipeg: University of Manitoba Press, pp 471–486.

Hindar K., Jonsson B., Andrew J.H. & Northcote T.G. (1988) Resource utilization of sympatric and experimentally allopatric cutthroat trout and Dolly Varden charr. *Oecologia* **74**, 481–491.

Jordan D.R. & Wortley J.S. (1985) Sampling strategy related to fish distribution, with particular reference to the Norfolk Broads. *Journal of Fish Biology* **27 (Supplement A)**, 163–173.

Jurvelius J. (1991) Distribution and density of pelagic fish stocks, especially vendace (*Coregonus albula* (L.)), monitored by hydroacoustics in shallow and deep southern boreal lakes. *Finnish Fisheries* **12**, 45–63.

Jurvelius J. & Heikkinen T. (1988) Seasonal migrations of vendace, *Coregonus albula* L. in a deep Finnish lake. *Finnish Fisheries* **9**, 205–212.

Jurvelius J., Lindem T. & Heikkinen T. (1988) The size of a vendace, *Coregonus albula* L., stock in a deep lake basin monitored by hydroacoustic methods. *Journal of Fish Biology* **32**, 679–687.

Kipling C. (1984) A study of perch (*Perca fluviatilis* L.) and pike (*Esox lucius* L.) in Windermere from 1941 to 1982. *Journal du Conseil International pour l'Exploration du Mer* **41**, 259–267.

Kristjansson J. & Adalsteinsson H. (1984) The ecology and management of the Arctic charr in Lake Myvatn, North Iceland. In: L. Johnson & B. Burns (eds) *Biology of the Arctic Charr: Proceedings of the International Symposium on Arctic Charr*. Winnipeg: University of Manitoba Press, pp 341–347.

Lindem T. & Sandlund O.T. (1984) New methods in assessment of pelagic freshwater fish stocks – coordinated use of echosounder, pelagic trawl and pelagic nets. *Fauna* **37**, 105–111.

MacCallum W.R. & Regier H.A. (1984) The biology and energetics of Arctic charr in Char Lake, N.W.T., Canada. In: L. Johnson & B. Burns (eds) *Biology of the Arctic Charr: Proceedings of the International Symposium on Arctic Charr*. Winnipeg: University of Manitoba Press, pp 329–340.

Maitland P.S. (1981) *The Ecology of Scotland's Largest Lochs: Lomond, Awe, Ness, Morar and Shiel*. Monographiae Biologicae **44**, J.W. Junk, The Hague, 297 pp.

Maitland P.S., Greer R.B., Campbell R.N. & Friend G.F. (1984) The status and biology of the Arctic charr *Salvelinus alpinus* L. in Scotland. In: L. Johnson & B. Burns (eds) *Biology of the Arctic Charr: Proceedings of the International Symposium on Arctic Charr*. Winnipeg: University of Manitoba Press, pp 193–215.

Maitland P.S., May L., Jones D.H. & Doughty C.R. (1991) Ecology and conservation of Arctic charr, *Salvelinus alpinus* (L.), in Loch Doon, an acidifying Loch in southwest Scotland. *Biological Conservation* **55**, 167–197.

Maitland P.S. & Campbell R.N. (1992) *Freshwater Fishes*. London: Harper Collins, 368 pp.

Mills C.A. (1989) The Windermere populations of Arctic charr, *Salvelinus alpinus*. *Physiology and Ecology, Japan* **Special Volume 1**, 371–382.

Mills C.A. & Hurley M.A. (1990) Long-term studies on the Windermere populations of perch (*Perca fluviatilis*), pike (*Esox lucius*) and Arctic charr (*Salvelinus alpinus*). *Freshwater Biology* **23**, 119–136.

Partington J. D. & Mills C.A. (1988) An electrophoretic and biometric study of Arctic charr, *Salvelinus alpinus* (L.), from ten British lakes. *Journal of Fish Biology* **33**, 791–814.

Pechlaner R. (1984) Dwarf populations of Arctic charr in high-mountain lakes of the Alps resulting from under-exploitation. In: L. Johnson & B. Burns (eds) *Biology of the Arctic Charr: Proceedings of the*

*International Symposium on Arctic Charr*. Winnipeg: University of Manitoba Press, pp 319–327.

Peltonen H. & Horppila J. (1992) The effects of mass removal on the roach *Rutilus rutilus* (L.) stock of Lake Vesijarvi estimated with VPA within one season. *Journal of Fish Biology* **40**, 293–301.

Ryder R.A. (1965) A method for estimating the potential fish production of north-temperate lakes. *Transactions of the American Fisheries Society* **94**, 214–218.

Saila S.B., Recksiek C.W. & Prager M.H. (1988) *Developments in Aquaculture and Fisheries Science* **18**. Basic Fishery Science Programs: A Compendium of Microcomputer Programs and Manual of Operation. Amsterdam: Elsevier, 230 pp.

Sandlund O.T., Gunnarson K., Jonasson P.M., Jonsson B., Lindem T., Magnusson K.P., Malmquist H.J., Sigurjonsdottir H., Skulason S. & Snorrason S.S. (1992) The Arctic charr *Salvelinus alpinus* in Thingvallavatn. *Oikos* **64**, 305–351.

Shine A.J. & Martin D.S. (1988) Loch Ness habitats observed by sonar and underwater television. *The Scottish Naturalist* **1988**, 111–119.

Snorrason S.S., Jonasson P.M., Jonsson B., Lindem T., Malmquist H.J., Sandlund, O.T. & Skulason S. (1992) Population dynamics of the planktivorous Arctic charr *Salvelinus alpinus* ('murta') in Thingvallavatn. *Oikos* **64**, 352–364.

van Donk E., Grimm M.P., Gulati R.D., Heuts P.G.M., de Kloet W.A. & van Liere E. (1990) First attempt to apply whole-lake food-web manipulation on a large scale in The Netherlands. *Hydrobiologia* **200/201**, 291–301.

Walker A.F., Greer R.B. & Gardiner A.S. (1988) Two ecologically distinct forms of Arctic charr (*Salvelinus alpinus* (L.)) in Loch Rannoch, Scotland. *Biological Conservation* **33**, 43–61.

Walline P.D., Pisanty S. & Lindem T. (1992) Acoustic assessment of the number of pelagic fish in Lake Kinneret, Israel. *Hydrobiologia* **231**, 153–163.

Winfield I.J. (1992) Threats to the lake fish communities of the U.K. arising from eutrophication and species introductions. *Netherlands Journal of Zoology* **42**, 233–242.

# Chapter 17
# Estimating fish production and biomass in the pelagic zone of Lake Malawi/Niassa: a comparison between acoustic observations and predictions based on biomass size-distribution theory

E.H. ALLISON *UK/SADC Pelagic Fish Resource Assessment Project, P.O. Box 311, Salima, Malawi. Address for correspondence: School of Development Studies, University of East Anglia, Norwich NR4 7TJ, UK*

**Abstract**   This study assesses the fishery potential of the entire pelagic zone of Lake Malawi/Niassa (approximately 24 000 km$^2$) and the trophic basis of pelagic fish production. Fish biomass estimates were obtained from five full-lake acoustic surveys, carried out between July 1992 and April 1993, in combination with trawling surveys. A lake-wide mean pelagic fish biomass of 6.6 g wet weight m$^{-2}$ was estimated. Size-spectrum models were used to predict fish biomass from size-spectra of phytoplankton, zooplankton (including larvae of the lakefly *Chaoborus edulis*) and fish larvae, all of which were sampled during the acoustic survey cruises. Observed plankton size-spectra showed good agreement, over broad trophic categories, with theoretical predictions of the structure of pelagic ecosystems, and provided good independent estimates of fish biomass (6.2–8.8 g m$^{-2}$).

Fish production, estimated from primary production using size-spectrum models, was of the order 8 to 13 g m$^{-2}$ yr$^{-1}$, and consisted mainly of zooplanktivores. By comparison, the production of clupeids in Lake Tanganyika was 65–70 g m$^{-2}$ yr$^{-1}$.

KEYWORDS: Lake Malawi, size-spectrum, fish biomass, pelagic production

## 17.1   Introduction

Lake Malawi/Niassa lies in the western arm of the East African Rift Valley between 9° 30′ S and 14° 30′ S. The lake is long and narrow (length 550 km, mean width 50 to 60 km), has a surface area of 28 800 km$^2$ and a volume of 8400 km$^3$, giving an average depth of 292 m; maximum depth is over 700 m. The lake has a three-layer structure, with a consistent barrier to vertical mixing occurring at approximately 230 m depth, which also approximates to the oxic-anoxic boundary layer, and a thermocline at a depth of 50 to 125 m that is strongly established in the hot, wet season but is weakened considerably during the cooler, windier season. Limited and variable rates of mixing between the vertical strata occur and this depends primarily on the density difference. Surface waters are super-saturated with oxygen, with $O_2$ concentrations decreasing to 6 mg l$^{-1}$ at 75 m depth and 3–4 mg l$^{-1}$ at 150 to 200 m. Epilimnion temperature is 24 to 29°C, metalimnion and hypolimnion temperatures are 22–24°C. Euphotic depth is 50 m. Detailed descriptions of the limnology are given by Eccles (1974) and Patterson and Kachinjika (1995).

The majority of the lake, comprising the offshore pelagic zone, is unexploited. The offshore pelagic zone is defined as all areas of > 50 m depth, or > 2 km offshore where the 50 m isobath is close to the shore. This study was part of a comprehensive survey of the pelagic ecosystem, the UK/SADC Pelagic Fish Resource Assessment Project, which aimed to quantify fish biomass, describe the trophic basis for fish production, and estimate potential fish yields.

Earlier surveys of the pelagic zone of the lake (Jackson *et al.* 1963; FAO 1982) were confined largely to the deep-water area within a few miles of Nkhata Bay, in the central part of the lake (Fig. 17.1). These studies indicated a relatively simple pelagic food web, comprising a speciose phytoplankton assemblage, a simple crustacean zooplankton community (four cyclopoid and two cladoceran species), larvae of the dipteran *Chaoborus edulis* Edwards, and a fish community consisting largely of cichlids (*Copadichromis* Eccles & Trewavas, *Rhamphochromis* Regan) and cyprinids (mainly *Engraulicypris sardella* Günther and *Opsaridium* spp. Günther) not considered to be truly pelagic and confined largely to the surface waters (0 to 50 m) within a few km of shore.

Trawl and gill-net surveys, which covered the whole lake (Fig. 17.1) indicated that the fish community comprised relatively evenly distributed stocks of zooplanktivorous cichlids of the genus *Diplotaxodon* Eccles & Trewavas, as well as smaller quantities of *E. sardella*, the mochokiid catfish *Synodontis njassae* Kielhack and piscivorous cichlids of the genus *Rhamphochromis* (Thompson *et al.* 1995a). Fish are distributed down to at least 220 m, with peak abundances between 50 and 150 m depth.

This chapter provides a preliminary synthesis of the structure of the pelagic ecosystem, based on a spatially extensive sampling programme, and provides a comparison of acoustic estimates of fish biomass with predictions obtained from analysis of the size-structure of the pelagic biota, using size-spectrum theory (Platt & Denman 1978; Borgmann 1982; 1987).

Sheldon *et al.* (1972) observed that, in the pelagic zones of oceans, the standing biomasses of phytoplankton, euphausids, tuna and whales showed only a slight (twofold) decrease between trophic levels. These authors suggested that this distribution of biomass could only arise if production was inversely related to size (Fenchel 1974; Calder 1984). Subsequent theoretical analyses have explained observed pelagic biomass distributions in terms of predator-prey size ratios and size-dependent rates of growth and metabolism (Kerr 1974; Sheldon *et al.* 1977; Platt & Denman 1978;

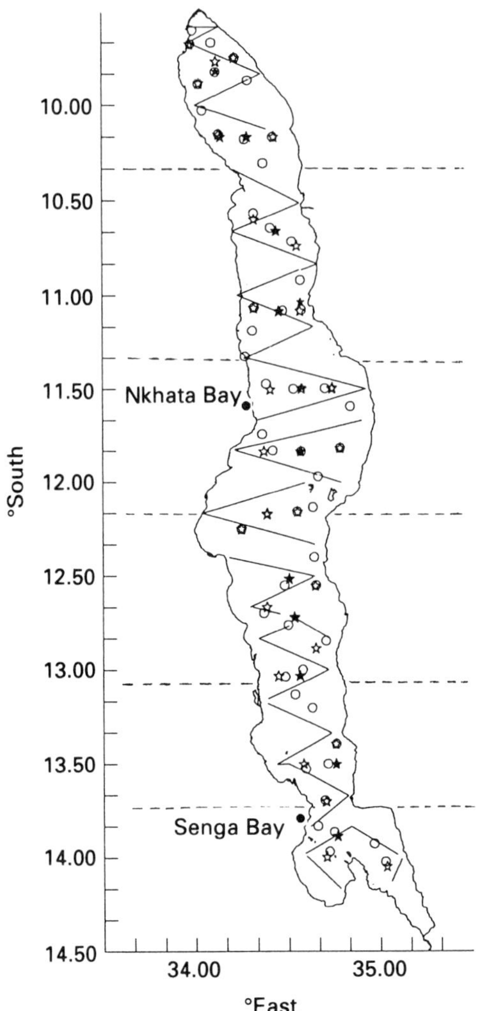



**Fig. 17.1**  Lake Malawi, showing acoustic transects (solid lines), and plankton-net (open circles), hydrographic probe (open star) and bottle (closed star) stations taken on a typical two-week research cruise (Sep–Oct 1992 cruise illustrated). Boundaries dividing the lake into sectors are also shown (dashed lines), as are the bases for previous pelagic surveys (Nkhata Bay) and this survey (Senga Bay).

Borgmann 1982, 1987; Dickie *et al.* 1987 for reviews). Size-spectrum theory allows biomass and production in one size category to be predicted from production and biomass estimates from another size category. Thus, primary production can be used to estimate potential fish production, or *vice versa* (Sheldon *et al.* 1977).

## 17.2    Materials and methods

### 17.2.1    *Data collection and processing*

Data for size-distributions of plankton and fish were collected between July 1992 and April 1993 on 10 research cruises of 8–14 days' duration covering the whole of the offshore pelagic zone of L. Malawi. Five of the cruises combined daytime sampling of phytoplankton and zooplankton with night-time acoustic surveys. The other five cruises sampled fish by mid-water trawl at six open water stations approximately 80 km apart on a N–S axis up the centre of the lake. These fishing stations were used to divide the lake into six arbitrary sectors (Fig. 17.1). Acoustic surveys used a systematic zig-zag grid of transects spaced at 10-nautical-mile (18.5 km) intervals over the whole lake (Fig. 17.1). Between 3 and 14 phytoplankton, zooplankton and fish-larvae samples were also taken in each lake sector on each cruise (Table 17.1).

**Table 17.1**  The main sampling cruises in the pelagic zone of L. Malawi, July 1992 – April 1993. Fish biomass estimates were calculated by combining acoustic echo-integrations with mid-water trawl catch-compositions from cruises listed in the adjacent columns.

| Cruise No. | Acoustics and plankton sampling | Cruise No. | Midwater trawling (and gillnetting) |
|---|---|---|---|
| 44 | 16 Jul–29 Jul 1992 | 45 | 26 Aug–2 Sep 1992 |
| 47 | 30 Sep–11 Oct 1992 | 49 | 10 Nov–18 Nov 1992 |
| 50 | 2 Dec–12 Dec 1992 | 49 | 10 Nov–18 Nov 1992 |
| 54 | 24 Feb–7 Mar 1993 | 53 | 3 Feb–18 Feb 1993 |
| 57 | 21 Apr–27 Apr 1993 | 58 | 18 Feb–25 Feb 1993 |

Phytoplankton were enumerated from water samples taken between 0 and 100 m depth using strings of 3.5 litre PWS bottles (Plastic Water Samplers – Hydrobios) and 2 litre Van Dorn bottles. Water samples were preserved with Lugol's iodine and stored in the dark. Algae were identified to taxa and counted under an inverted microscope using the Utermöhl method (Lund *et al.* 1958). The algal count values were converted to total cell volumes for each taxa, based on estimates of individual cell volumes for each taxa derived from geometric models of cell shape. The effective biovolume of colonial algae was calculated by multiplying individual cell volumes by the number of cells in the colony. For filamentous algae, an estimated mean number of cells per filament was used to compute effective biovolume. Volume was converted to mass (g wet weight) assuming a specific gravity of one (Munawar *et al.* 1974). All counts for one sampling occasion were integrated to give total wet weight (g) of individual taxa per $m^2$ lake surface area. Further sampling details are given in Patterson and Kachinjika (1995)

Zooplankton samples were collected using a GULF III high-speed plankton net (HSPN; Milligan & Riches 1983) consisting of a main (nose cone diameter 20 cm, mesh size 140 m) and pup net (nose cone diameter 4.5 cm, mesh size 53 m). Filtration efficiency of 0.93 was assumed in calculating the volume of water sampled (Nichols & Thompson 1991). Larval fish, consisting entirely of the cyprinid *E. sardella*, and larvae of the lake fly, *C. edulis*, were sampled from the main net; copepods and cladocerans were sampled from the pup net. Each sample consisted of a double-oblique haul to a maximum depth of 200 m at 1.5–2.0 m s$^{-1}$ tow speed. Zooplankton samples were preserved in 4% buffered formaldehyde solution.

All post-naupliar calanoid copepods and cladocerans in zooplankton samples were identified to species. The length-compositions of the zooplankton species in each sample were measured under a binocular microscope fitted with an eyepiece graticule. Dry weights of calanoid copepods and the cladoceran *Diaphanosoma excisum* (Sars) were obtained from length-dry weight regression equations established by weighing individuals or small batches of animals of uniform length (Irvine & Waya 1995). Dry weights of *Bosmina longirostris* Muller, which were not abundant in samples (Irvine 1995a), were obtained from previously published length-dry weight relationships (Bottrell *et al.* 1976).

*C. edulis* larvae in samples were identified to instar (I to IV) and the mean dry weight of each instar, derived from weighing replicate batches of animals, was used to calculate the weight-frequency distribution of the population (Irvine 1995b). Dry weights were converted to wet weights using an assumed conversion factor of 20% (median value from a range given in Bottrell *et al.* 1976) for crustacean zooplankton, and a measured conversion of 8% dry weight to wet weight for *C. edulis* larvae (Irvine & Waya 1995).

Larvae of *E. sardella* were sampled from HSPN hauls (larvae < 20 mm); length-frequencies were converted to weight frequencies using a length-dry weight relationship and a wet-dry weight conversion factor of 20% (Thompson 1995).

The species and size-composition of the pelagic fish community were determined using mid-water trawl catches from six deep water fishing stations located centrally within the six lake sectors. The trawl used was a 100 HP four-panel, semi-pelagic trawl with a 6 m headline length and a vertical opening of approximately 4 m when towed at 1.5 to 2.0 m s$^{-1}$. Meshes were graded from 150 mm to 38 mm in the main body of the net, with an 18 mm codend lined with 4 mm diamond-mesh to retain juvenile cichlids and *E. sardella*. Two or three hauls of 9 to 16 km, covering the whole depth range within which fish were found (0 to 220 m), were made during the night at each location on each fishing cruise.

Fish length-distributions were converted to weights using species-specific length-weight relationships (Thompson *et al.* 1995b). Unbiased estimation of the mean weight of species *j* in each 1 cm length category *i* was obtained from (after MacLennan & Simmonds 1992):

$$W_{ij} = a_j((L_i + \Delta L/2)^{bj+1} (L_t \, \Delta L/2)^{bj+1}) / (bj+1)\Delta L \tag{1}$$

where $\Delta L = 1$ cm and $a_j$, $b_j$ are the intercept and exponent of the length-weight relationship for species *j*. Relative biomass in each species/size category $W_{ij}$ was calculated by multiplying by the number of fish of length $L_{ij}$ in the catch.

Total fish numbers per lake sector were obtained from acoustic transects (Fig. 17.1). Average echo intensities were measured at 5 or 10 min intervals using a Bio-Sonics Model 102 echo-sounder operating at 120 kHz, and a BioSonics ESP Model 221 echo integrator. Integration was carried out to a maximum depth of 250 m in 10 m strata. Echo integrals were scaled to fish numbers using a mean target strength for each of the six lake sectors, estimated from a combination of *in situ* target strength measurements and trawl catch length-frequency data. *In situ* target strengths were measured by the dual beam method (Burczynski & Johnson 1986) using a BioSonics ESP model 281 dual beam processor. Biomasses were calculated by applying the estimated numbers of fish in each sector to the data on species and size compositions from trawl catches (Menz *et al.* 1995).

### 17.2.2    *Size-spectrum analysis*

The size spectrum is the distribution of biomass (g m$^{-2}$) by individual body size (g). Following Platt and Denman (1978), biomasses were summed within logarithmically equal size categories. Separate biomass size-spectra were generated for each cruise and lake sector. These individual spectra ($n = 30$) were averaged to give a whole-lake annual mean size spectrum. In calculating whole-lake arithmetic means, values from each sector were weighted by the lake surface area of the sector. Values from different cruises were given equal weighting.

To provide a general overview of the structure of the pelagic ecosystem, and to predict fish biomass and production from plankton biomass and production, two theoretical models were fitted to the observed whole-lake size spectrum.

Platt and Denman's (1978) 'normalized' size spectrum is a plot of $\log_{10}$ (*size category*) against $\log_{10}$ (*total biomass in each size category / width of the size category*). If the size spectrum conforms to theory, the data should lie on a straight line of slope approximating to $-1.0$ (equal 'number densities' in logarithmically equal size-categories).

The slope of the normalized plankton size spectrum was fitted by least squares regression, treating size categories with no detectable biomass as missing values, rather than zero values (Sprules & Munawar 1986). Biomass within any given size limits ($W_x$, $W_{x+i}$) can be predicted by integrating the area under the fitted line between $W_x$ and $W_{x+i}$. Extrapolation of the normalized size spectrum for plankton to size categories containing fish can therefore provide fish biomass predictions for comparison with acoustic estimates. Borgmann's (1982, 1987) biomass size spectrum theory was used to predict fish biomass from plankton biomass using:

$$B_{x,x+i} = b(W_{x+i}{}^{n-\epsilon} - W_x{}^{n-\epsilon}) \tag{2}$$

where $B_{x,x+i}$ is the biomass of organisms between sizes $W_x$ and $W_{x+i}$; $W$ is the individual size (g); $\epsilon$ is the particle-size conversion efficiency, a measure of the efficiency of biomass transfer scaled by relative sizes of predators and prey; $b$ is a fitted constant, calculated from the entire observed plankton size spectra; and $n$ is the exponent in the standard allometric production to biomass equation:

$$P/B = aW^{\,n} \tag{3}$$

In this case values for $n$ and $\epsilon$ were assumed from published values for other pelagic ecosystems ($n$ = 0.16–0.24 and $\epsilon$ = 0.20–0.26; Borgmann 1982, 1987; Minns *et al.* 1987). Examination of the data indicated a slight decrease in biomass with trophic level, so $n - \epsilon$, which is the slope of an idealized smooth spectra, was assumed to be negative. Sprules *et al.* (1991) used values of n – $\epsilon$ ranging from –0.02 to –0.04, median values in the observed range, for modelling the Lake Michigan size spectrum; the same values were adopted here for comparison. Biomasses were predicted for the following broad trophic groups, which have largely non-overlapping size distributions: phytoplankton, crustacean zooplankton and *C. edulis* first instar larvae, *C. edulis* II to IV instar larvae and *E. sardella* larvae < 20 mm *TL*, planktivorous fish, and piscivorous fish.

Potential net production of any size group of organisms can be predicted if actual production and body size are known for a reference group (Borgmann 1982). Net algal production in L. Malawi was calculated as 271 g C m$^{-2}$ yr$^{-1}$ (Degnbol & Mapila 1982). Assuming a carbon : wet weight conversion factor of 0.15 (Laws *et al.* 1984) production in terms of wet weight in 1980 was 1806 g m$^{-2}$ yr$^{-1}$. A more recent estimate (Hecky & Patterson, unpublished data 1994) derived from monthly estimates of Chlorophyll $a$ standing crop and a photosynthesis-irradiance-depth model (Fee 1990) gave an estimate of 2190 g m$^{-2}$ yr$^{-1}$ for 1992. Production in the size categories containing zooplankton, *E. sardella* larvae and fish were calculated from (Borgmann 1987):

$$P_{x,x+1} = c(W_{x+1}^{-\epsilon} - W_x^{-\epsilon}) \tag{4}$$

where $\epsilon$ is the particle size conversion efficiency, assumed to lie between 0.22 and 0.24, median values in the observed range (Borgmann 1987), and $c$ is a fitted constant, estimated from known phytoplankton production.

Observed biomasses, and either predicted or observed net production rates, were compared with estimates from the pelagic zones of two other great lakes: L. Tanganyika (Hecky 1991) and L. Michigan (Sprules *et al.* 1991). Trophic transfer efficiencies (e.g. net primary production/net secondary production) in the three lakes were computed for the major trophic steps.

### 17.3    Results

The size spectrum for the L. Malawi pelagic ecosystem, sampled in 1992–93, showed four major peaks (Fig. 17.2), corresponding to phytoplankton (PH), a mixture between large diatoms and zooplankton nauplii (PH/ZN), crustacean zooplankton copepodites and adults (ZO) and planktivorous fish (PLF). A minor peak between zooplankton and fish corresponded to *C. edulis* and *E. sardella* larvae (CH/ES) and a

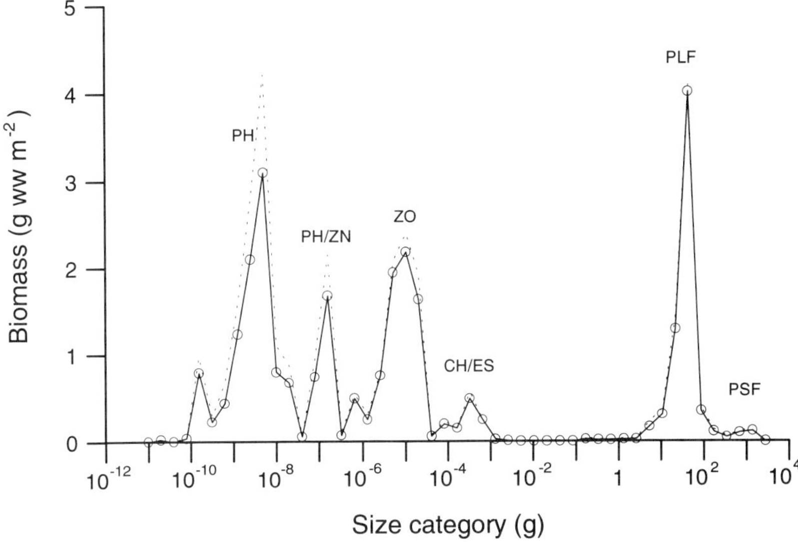



**Fig. 17.2**  Size spectrum of the L. Malawi pelagic ecosystem. Major biomass peaks are composed of phytoplankton (PH), a mixture of large diatoms and zooplankton nauplii (PH/ZN), crustacean zooplankton (ZO), *Chaoborus edulis* larvae and *Engraulicypris sardella* larvae (CH/ES), planktivorous fish (PLF), and piscivorous fish (PSF).

small peak at the larger end of the spectrum represented piscivorous fish (PSF). The major trophic groups were generally separated by size, although there was some overlap in individual biomass between large diatoms and the nauplii of crustacean zooplankton, between adult crustacean zooplankton and *C. edulis* larvae first instars, and between *C. edulis* larvae and *E. sardella* larvae. The biomasses of the major trophic groups decreased slightly from phytoplankton through to zooplanktivorous fish, but then fell off rapidly in the size categories occupied by the piscivorous fish. There was a large gap in the spectra, between sizes of $10^{-3}$ g and 3 g where biomasses were very low.

The normalized size spectrum (Fig. 17.3) indicated that pelagic biomass in L. Malawi was distributed approximately in accordance with size-distribution theory, in that the slopes of fitted regression equations approximate to –1; the slope of the plankton size spectra (including *C. edulis* larvae and *E. sardella* < 20 mm) is –1.03 ($\pm$ standard error 0.07). Although the error on the slope appears small, it leads to predictions of fish biomass of between 0.6 and 64 g m$^{-2}$ (95% CI), with a mean of 2.7 g m$^{-2}$. Clearly the normalized spectra was not useful for predicting fish biomass from this data set. The most serious departure from the theoretical prediction of equal biomass in logarithmically equal size categories occurred over the size-categories where juvenile fish would normally be found.

The equilibrium biomass size-spectrum equation of Borgmann (1982), when parameterized from the plankton spectra, provided estimates of fish biomass which fell within one standard error of acoustically observed biomasses (Table 17.2).

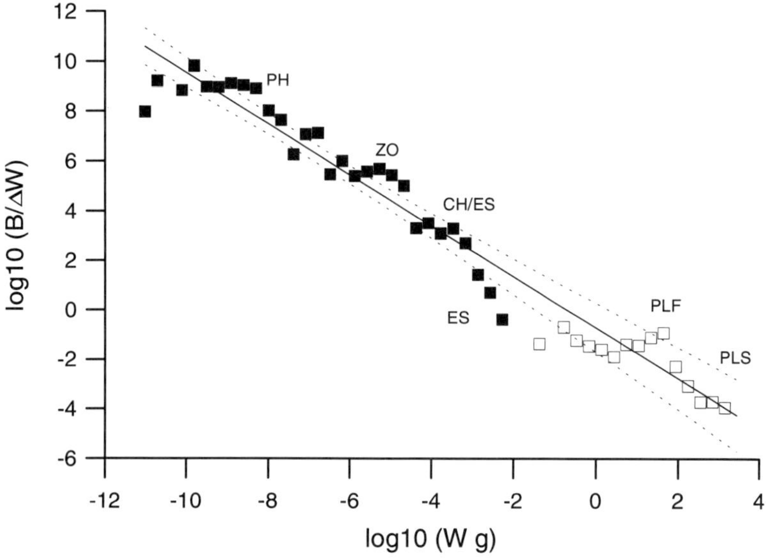

**Fig. 17.3**  Normalized size spectrum with regression equation ($\log_{10}(B/W) = -0.72-1.03.W$; $n = 29$, $r^2 = 0.89$) fitted to plankton spectra (closed squares) and extrapolated to predict fish biomass from the area under the line between the size limits of observed fish biomasses (open squares). The 95% confidence bands are also shown. (PH = phytoplankton, PH/ZN = large diatoms and zooplankton nauplii, ZO = crustacean zooplankton, CH/ES = *Chaoborus edulis* larvae and *Engraulicypris sardella* larvae, PLF = planktivorous fish, PSF = piscivorous fish).

**Table 17.2**  Observed and predicted biomasses for the major size-groups of pelagic organisms in L. Malawi; $W_{min}$ and $W_{max}$ are the limits of the size range of each group. Predictions are from Borgmann's (1982) size-spectrum model using literature values for $n$ and $\epsilon$, where $n$ is the exponent of the allometric production-biomass relationship (Equation (4)) and $\epsilon$ is the size-scaled trophic transfer efficiency. Observed values are whole-lake means from data collected between July 1992 and April 1993. Zooplankton includes *C. edulis* larvae and *E. sardella* larvae < 20 mm.

| | Individual body size (g) | | Biomass (g m$^{-2}$) Predicted | | Observed | |
|---|---|---|---|---|---|---|
| | $W_{min}$ | $W_{max}$ | $n{-}\epsilon = {}^-0.02$ | $n{-}\epsilon = {}^-0.04$ | mean | se |
| Phytoplankton | $7.33\ 10^{-12}$ | $4.80\ 10^{-07}$ | 11.94 | 12.96 | 11.73 | 0.486 |
| Zooplankton | $6.00\ 10^{-08}$ | $7.86\ 10^{-03}$ | 10.53 | 9.48 | 8.70 | 0.177 |
| Planktivorous fish | $1.04\ 10^{-01}$ | $1.29\ 10^{02}$ | 5.02 | 3.54 | 6.23 | 0.047 |
| Piscivorous fish | $1.29\ 10^{02}$ | $2.06\ 10^{03}$ | 1.76 | 1.12 | 0.38 | 0.011 |
| Total fish | $1.04\ 10^{-01}$ | $2.06\ 10^{03}$ | 6.78 | 4.66 | 6.61 | 0.050 |
| All biomass | $7.33\ 10^{-12}$ | $2.06\ 10^{03}$ | 29.17 | 26.61 | 27.04 | 0.434 |

Attempts to predict biomass on a finer scale, e.g. separating planktivorous and piscivorous fish, were less successful. Observed biomasses of planktivores were higher than predicted values, and observed biomass of piscivorous fish were lower than size-spectrum predictions (Table 17.2).

The concentration of biomass into relatively few size categories made fine-scale prediction using a smooth theoretical spectrum untenable. Predictions of crustacean zooplankton biomass, *C. edulis* and *E. sardella* larvae all diverge from observed values, although overall predicted zooplankton and phytoplankton biomasses were reasonably close to observed values (Table 17.2).

Production rates predicted from Borgmann's (1987) size-production relationship appear reasonable for all trophic levels (Table 17.3). As they were based on theoretical relationships between production, biomass and size, with assumed values for size-scaled trophic transfer efficiency, estimated values will tended to approach mean values derived from a wide range of ecosystems.

**Table 17.3**  Annual production of the major trophic groups (defined by their size ranges, $W_{min}$, $W_{max}$) in the L. Malawi pelagic ecosystem, predicted from Borgmann's (1987) size-spectrum model using observed annual net primary production values from: $PP_1$, Degnbol & Mapila (1982), and $PP_2$, Hecky & Patterson (unpublished data for 1992). Literature values are adopted for $n$, the exponent of the allotmetric production/biomass equation (Equation (4)).

| | Individual body size (g) | | Predicted production (g m$^{-2}$ yr$^{-1}$) | | | |
| | | | $PP_1 = 1806$ | | $PP_2 = 2190$ | |
| | $W_{min}$ | $W_{max}$ | $n = 0.22$ | $n = 0.24$ | $n = 0.22$ | $n = 0.24$ |
|---|---|---|---|---|---|---|
| Crustacean zoopl. + *C. edulis* I | 6.00 10$^{-08}$ | 3.07 10$^{-05}$ | 203.5 | 173.4 | 246.7 | 210.2 |
| *C. edulis* II–IV & *E. sard.* < 20 mm | 3.07 10$^{-05}$ | 1.00 10$^{-02}$ | 49.8 | 37.5 | 60.3 | 45.5 |
| Planktivorous fish | 1.00 | 1.29 10$^{02}$ | 4.6 | 2.8 | 5.6 | 3.5 |
| Piscivorous fish | 1.29 10$^{02}$ | 2.06 10$^{03}$ | 1.1 | 0.6 | 1.3 | 0.8 |

Production in the size range occupied by *C. edulis* and *E. sardella* larvae was fairly high, given the low observed biomasses. *E. sardella* biomass was about 10% of *C. edulis* biomass, so production of *E. sardella* was assumed to make up 10% of the production within those size categories. This gave a predicted production of *E. sardella* larvae of 4.6 to 6.0 g m$^{-2}$ yr$^{-1}$ which agrees well with calculated values of 5 g m$^{-2}$ yr$^{-1}$ for 1992 and 9.8 g m$^{-2}$ yr$^{-1}$ for 1993 (Thompson 1995).

Production/biomass ratios ($P/B$) are often reported in the literature (e.g. Banse & Mosher 1980) and are useful for evaluating model-predicted production values. $P/B$ values for L. Malawi crustacean zooplankton fall just below the lower limits of the known range for tropical zooplankton (Table 17.4), studies on which have mostly been conducted in shallow, highly productive systems (Mengestou & Fernando 1991, for review). Fish $P/B$ ratios (Table 17.4) appear to be plausible, given that the fish span a size range of 2–20 cm length (planktivores) and 20–50 cm (piscivores).

**Table 17.4**  Production/biomass ratios ($yr^{-1}$) for L. Malawi, calculated from observed or predicted production, and observed biomass, and comparisons with literature values for the same trophic groups.

| | min | max | Literature values |
|---|---|---|---|
| Pytoplankton | 186.8 | 186.8 | 299[1] |
| Crustacean zoopl. + *C. edulis* I | 28.0 | 32.9 | 40–70[2] |
| *C. edulis* II–IV & *E. sard.* <20 mm | 38.1 | 50.5 | No comparable value |
| All fish | 1.2 | 1.9 | 1.9–3.7[3] |
| Planktivorous fish | 0.5 | 0.9 | 0.9–2.9[4] |
| Piscivorous fish | 2.0 | 3.5 | No comparable value |

[1] calculated from data in Hecky (1984) for L. Tanganyika
[2] from a review of secondary production in tropical lakes in Mengestou & Fernando (1991)
[3] calculated from data in Hecky (1991) for L. Tanganyika fish (mainly clupeids)
[4] data from temperate fish reviewed in Banse & Mosher (1980), Table 4.

## 17.4    Discussion

The analysis of size spectra is a developing technique, in which a number of analytical approaches and theoretical explanations for observed distributions have been proposed. However, similar basic assumptions apply to the formulations of the size-spectrum cited in this chapter. These are that:

(1)    primary production is based on microscopic algae;
(2)    growth and metabolism are related to size, independently of taxonomic group;
(3)    predators are always larger than their prey, and the ratio of their sizes is constant at all trophic levels;
(4)    the conversion efficiency ($k$; predator food consumption/prey production) for each trophic transfer is constant;
(5)    all net production at one trophic level is consumed by the next trophic level;
(6)    the ecosystem is at equilibrium or 'steady state' over the time-scale being considered.

The two formulations of size-spectrum theory chosen for this study are those of Platt and Denman (1978) and Borgmann (1982, 1987). Both these models have the advantage of not requiring information on trophic level of all organisms in the ecosystem. Both formulations model the flow of biomass from smaller to larger organisms as a continuous process, rather than a series of discrete steps. This allows integration over any size range to obtain biomasses or production rate estimates for all organisms in that size range.

The two theories differ in one respect only: Borgmann's (1982, 1987) theory combines the assumptions of constant predator/prey size ratios and constant $k$ into the single assumption that $k$, when scaled by the relative size distribution of predators and prey, is approximately constant across whole ecosystems (the 'size conversion efficiency'; $\epsilon$). The efficiency of biomass transfer tends to be inversely related to the relative sizes of predator and prey (Paloheimo & Dickie 1966; Confer & Blades 1975). Thus, while predator/prey size ratios and $k$ may vary within an ecosystem, variations in the two parameters are compensatory, so that $\epsilon$ is likely to remain constant. $\epsilon$ shows remarkable stability across a variety of ecosystems and trophic levels ($\epsilon = 0.20$–$0.26$; Borgmann 1987, for review), although there are no estimates available from tropical freshwaters.

Are the assumptions of the biomass size spectrum applicable to the pelagic zone of L. Malawi? Analysis of ratios of stable carbon isotopes from the tissue of a variety of pelagic organisms from L. Malawi (R.E. Hecky & H. Bootsma, unpublished data) indicated that production originating from sources other than microscopic algae (e.g. anaerobic bacteria) is negligible.

There is good reason to believe, also, that unutilized production is low. The lake is fairly oligotrophic, and the phytoplankton support a long food chain (up to 5 trophic levels); these are usually indicators of high production/consumption ratios (Kerr & Martin 1970).

There is no reason to suppose that the lake is not at equilibrium over medium time scales (several years), although short-term fluctuations in biomass and production will occur. For the period from February 1992 to April 1993, the biomasses of the major components of the plankton were relatively constant (Allison *et al.* 1995a). The range of variability in the plankton is much less than that shown seasonally in temperate regions. This analysis covers the period July 1992 to April 1993, within a period of relative stability, but fish biomasses observed during this period may, of course, have been affected by differences in production which occurred prior to the start of this study.

Biomass distributions in lakes do not fit the smooth biomass size-distributions found in the pelagic zones of oceans (Sheldon *et al.* 1977) and predicted by theory. Obervations in temperate lakes suggest that biomass is distributed as a series of peaks separated by size categories containing little, or no biomass (Thiebaux & Dickie 1992). Averaged over broad trophic groups, however, the spectra from large lakes show a similar overall structure to those of oceanic systems (Sprules *et al.* 1983; Sprules & Munawar 1986; Boudreau & Dickie 1992) and enable examination of ecosystem-scale processes.

The size-distribution of biomass in the pelagic zone of L. Malawi conformed broadly to theoretical expectations of the regularity of pelagic ecosystem structure, although, as in temperate lakes, biomass was distributed as a series of peaks, rather than as a smooth continuum. It has recently been observed that the distributions of the biomass peaks are consistent between different pelagic systems, and existing size spectrum theory has been modified to take this secondary periodic structure into account (Thiebaux & Dickie 1992, 1993), although the developing theory has not yet been tested using field data.

The most serious departure from expected biomass distributions was the low observed biomass in the size range $10^{-3}$ to 3 g. This is the size range in which micronekton, including larval and juvenile fish would normally be found in pelagic ecosystems. In L. Malawi, the majority of pelagic fish biomass comprises mouth-brooding cichlids which produce few, large offspring. Fry and juveniles of the pelagic cichlids were present in low numbers in the offshore waters, and it is likely that they spend some of their life-cycle in areas further inshore than were sampled in the present study. Areas such as the south-east and south-west arms of the lake, and the far northern extremity, which are shallower ( < 100 m) and more productive (Patterson & Kachinjika 1995) may provide nursery grounds for the young of pelagic cichlids. Juvenile *E. sardella* (20–40 mm) which appeared to be absent from the offshore waters (Thompson 1995) were found in the littoral zone and are caught in beach seines. Phytoplankton, zooplankton and fish larval biomass estimates included coverage of the shallower areas, but sampling of post-larval fish in these areas was more limited. Fry and juvenile fish are pelagic feeders, so the assumption of a fully pelagic food web is not violated. The absence of small fish from the size-spectrum may thus be a sampling problem.

The assumption of a constant predator/prey size ratio over the whole ecosytem, which forms one of the central tenets of the earlier size spectrum theories (Kerr 1974; Sheldon *et al.* 1977; Platt & Denman 1978), appears to be untenable at the higher trophic levels in the L. Malawi pelagic zone. *Mesocyclops aequatorialis aequatorialis* (Say), *C. edulis* larvae and some of the planktivorous fish community all feed at the same trophic level (Allison *et al.* 1995b; Hecky & Bootsma, unpublished data) but differ in size by several orders of magnitude. Borgmann's (1982; 1987) theory that observed size-spectra in both lakes and oceans can be explained by constant size-scaled trophic transfer, rather than constant predatator/prey size ratios, appears more realistic, and a better predictor of fish biomasses in lakes (Leach *et al.* 1987; Sprules *et al.* 1991).

Predictions of fish biomass in L. Malawi from plankton size-spectra, using Borgmann's (1982, 1987) theory showed good agreement with biomass estimates derived from acoustic surveys. The agreement between the two independent estimates provides some validation for acoustic estimates, and also suggests that analysis of plankton size spectra may provide a method of generating independent estimates of pelagic fish biomasses in large lakes.

Size-spectrum data and production estimates allow preliminary comparison of the structure of the pelagic ecosystem of L. Malawi with those of L. Tanganyika and L. Michigan (Table 17.5). Phytoplankton biomasses and production rates are similar in L. Tanganyika and L. Malawi, while biomass is higher in L. Michigan, but production is lower. Zooplankton biomasses are also higher in L. Michigan, and include a high biomass of large, predatory zooplankton which occupy a similar position in the trophic structure of the pelagic ecosytem as *C. edulis* larvae do in L. Malawi. L. Tanganyika is thought not to contain a significant biomass of predatory zoo-plankton, although both *M. aequatorialis aequatorialis* and large coelenterates are present.

Fish biomasses in L. Malawi and L. Michigan, estimated using very similar methods (Brandt *et al.* 1991; Menz *et al.* 1995) are both between 6 and 7 g m$^{-2}$ and are

**Table 17.5**  Biomass (g m$^2$), production (g m$^{-2}$ yr$^{-1}$) and trophic transfer efficiencies (%) for the main trophic groups in the pelagic zones of three large lakes. Macroplankton includes large crustaceans (e.g. *Mysis* sp.) and fish and insect larvae. Data for L. Tanganyika are from Hecky (1991); data from L. Michigan are from Sprules *et al.* (1991).

|  | L. Malawi | L. Tanganyika | L. Michigan |
|---|---|---|---|
| **Biomasses** | | | |
| Phytoplankton | 11.73 | 10–20 | 35.0 |
| Herbivorous zooplankton | 7.50 | 10–25 | 20.5 |
| Macroplankton | 1.19 | — | 10.5 |
| Planktivorous fish | 6.23 | 19–35 | 6.1 |
| Piscivorous fish | 0.38 | — | 0.2 |
| **Net production rates** | | | |
| Phytoplankton | 2190 | 2000 | 1547 |
| Herbivorous zooplankton | 210–246* | 360 | 72–92* |
| Macroplankton | 46–60* | — | 10.6 |
| Planktivorous fish | 3.5–5.6* | 65–70 | 2.5–4.1* |
| Piscivorous fish | 0.8–1.3* | — | 0.3–0.5* |
| **Trophic transfer ratios (%)** | | | |
| Primary to secondary production | 9.6–11.2* | 18 | 4.7–5.9* |
| Secondary to tertiary production | 18.7–28.6* | 18–19 | 11.6–14.7* |
| Primary to fish production | 0.19–0.32* | 3.3–3.5 | 0.18–0.30* |

* prediction from size-spectrum model.

composed mainly of zooplanktivores of similar sizes. Estimated fish biomasses in the L. Malawi pelagic zone, which is largely unexploited, are between 17 and 30% of acoustic and trophic-model-based estimates of fish biomass in the L. Tanganyika pelagic zone (Coulter 1977, 1981; Hecky 1991). The L. Tanganyika estimates were made during the 1970s, when the pelagic fish were relatively lightly exploited and included high piscivore biomass.

Production rates at all trophic levels are higher in the two tropical lakes than in the temperate L. Michigan. The production rate of crustacean zooplankton in L. Malawi (including *C. edulis* first instar larvae), most of which are herbivorous (Irvine 1995a), is between 9.6 and 11.2% of observed phytoplankton production rate – a value-range which includes the generally accepted value of 10% for trophic transfer efficiencies (Slobodkin 1962) but is below the value of 15% considered more typical of tropical aquatic ecosystems (Hecky 1984). Similar predictions for L. Michigan (Sprules *et al.* 1991) indicate a 5–6% transfer efficiency. In the pelagic system of L. Tanganyika, a lake morphologically similar to L. Malawi, trophic transfer efficiency from producers to primary consumers is about 18–19% (Hecky 1991, for review).

The ratio of production of *C. edulis* and *E. sardella* larvae to production of their prey, the herbivorous zooplankton, is high (19–29%), indicative of high efficiency of trophic transfer, perhaps due to the small differences in their sizes (Paloheimo & Dickie 1966). The equivalent trophic step in L. Michigan has a transfer efficiency two or three times higher than other trophic steps.

Planktivorous fish in L. Malawi consist of a mixture of species, some of which feed on crustacean zooplankton, others feeding predominantly on *E. sardella* larvae and *C. edulis* larvae (Allison *et al.* 1994). The addition of this extra trophic step may reduce the efficiency with which primary production is converted to fish production. In L. Malawi planktivorous fish production is between 0.2 and 0.3% of primary production – similar to that in L. Michigan, which has a similar trophic structure. In L. Tanganyika, where primary production rates are comparable to L. Malawi ($2000 \text{ g m}^{-2} \text{ yr}^{-1}$; Hecky & Fee 1981) fish production is based on a shorter food chain, (phytoplankton – zooplankton – small clupeids) and is much more efficient (3.3–3.5% of primary production; calculated from Hecky 1991). Assuming the data from L. Tanganyika are representative, the pelagic system is exceptionally efficient at converting primary production into fish production – more so than some productive marine pelagic ecosystems (Hecky 1984; Lowe-McConnell 1987).

Overall, the size spectrum approach to ecosystem analysis looks promising. To date, complete size spectra have only been analysed for one lake (L. Michigan, Sprules *et al.* 1991) and partial spectra have only been published for temperate lakes (Sprules & Munawar 1986; Gaedke 1993) so that a complete spectrum from a tropical lake is of interest in the search for general laws describing the structure and function of aquatic ecosystems.

Analyses of size spectra are preferable to purely empirical models such as morphoedaphic indices (e.g. Ryder 1982) and fish yield/primary production models (e.g. Melack 1976; Downing *et al.* 1990) as they enable an examination of the structure of the ecosystem and an assessment of any likely departures from theoretical requirements. Estimates based on empirical models such as the morphoedaphic index are less amenable to critical scrutiny.

Although size spectra may only be useful for examination of ecosystem structure at a coarse scale, they have the advantage of not requiring detailed information on trophic relationships and biomass-transfer processes (production, consumption), making them more applicable in the early stages of an ecosystem study than even the simpler steady-state trophic models such as ECOPATH (Christensen & Pauly 1991), or the more complex dynamic simulation models (e.g. Anderson & Ursin 1977). These latter models often require a great deal of information to maintain realism, with age-structuring being required to accommodate changes in productivity and trophic position during different life-history stages, particularly for fish, which may increase in individual biomass by orders of magnitude and occupy two or more trophic levels during their lifetime. The biomass/size spectrum theory accommodates such changes, provided sampling is undertaken extensively in time and space; a decrease in biomass, production rate and an increase in trophic level are predicted with increase in size. R.E. Hecky (unpublished consultancy report to UK/SADC project 1994) suggested a combination of size spectra with information on trophic level, derived with precision from stable nitrogen-isotope ratios. This 'hybrid' model allows checking of size/ spectrum assumptions and more rigorous prediction of biomasses and production at different trophic levels, without the need for information on production and consumption rates. This may provide the best compromise between the simplicity of the size spectrum and the greater theoretical rigour (and more exacting data requirements) of a full trophic model.

This preliminary analysis of the L. Malawi pelagic ecosystem has shown that the structure does not differ markedly from that of large lakes in temperate regions (reviewed in Sprules & Munawar 1986). The pelagic food web is similar to that of L. Michigan, which also has five main trophic levels, and consists of organisms of similar size (Sprules *et al.* 1991). Fish biomasses in the L. Malawi pelagic (6.6 g m$^{-2}$, or approximately 160 000 t) are comparable with pelagic fish biomasses in the Laurentian Great Lakes (Leach *et al.* 1987; Sprules *et al.* 1991); they do not compare with estimates for L. Tanganyika (19–35 g m$^{-2}$ or 600 000 to 1 000 000 t), and it is unlikely that L. Malawi will ever support pelagic fisheries on the same scale.

## Acknowledgements

This chapter comprises a preliminary summary of some of the work of the UK/SADC Pelagic Fish Resource Assessment project (Lake Malawi/Niassa), funded by the UK Overseas Development Administration and carried out in co-operation with the Southern African Development Community states of Malawi, Mozambique and Tanzania. I would like to thank G. Patterson and O. Kachinjika (phytoplankton), K. Irvine and R. Waya (zooplankton), A. B. Thompson & B. P. Ngatunga (fish) and A. Menz and A. Bulirani (acoustics) for providing me with their data and for reviewing the manuscript. R. E. Hecky made many useful comments on an earlier draft, and allowed me to quote unpublished data, and I. J. Winfield's review of a later draft substantially improved the structure. W. G. Sprules and D. Tweddle provided helpful support and access to the recent literature. I am grateful for the assistance of the crew of the *R.V. Usipa*, the scientific technical staff, and the project support staff whose contributions made the collection of this comprehensive data set possible.

## References

Allison E.H., Patterson G., Irvine K., Thompson A.B. & Menz A. (1995b) The pelagic ecosystem. In: A. Menz (ed.) *Fisheries potential and the ecology of the pelagic zone of Lake Malawi/Niassa.* Natural Resources Institute, Chatham, Kent, UK, pp 351–68.

Allison E.H., Thompson A.B., Ngatunga B.P. & Irvine K. (1995a) The diet and food consumption rates of the offshore fish. In: A. Menz (ed.) *Fisheries potential and the ecology of the pelagic zone of Lake Malawi/Niassa.* Natural Resources Institute, Chatham, Kent, UK, pp 233–78.

Anderson K.P. & Ursin E. (1977) A multispecies extension to the Beverton and Holt theory, with accounts of phosphorus circulation and primary production. *Meddelelser Danmarks Fiskeri-og Havundersøgeleser* **7**, 319–435.

Banse K. & Mosher S. (1980) Adult body mass and annual production/biomass relationships of field populations. *Ecological Monographs* **50**, 355–379.

Borgmann U. (1982) Particle-size conversion efficiency and total animal production in pelagic ecosystems. *Canadian Journal of Fisheries and Aquatic Science* **39**, 668–674.

Borgmann U. (1987) Models of the slope of, and the biomass flow up, the biomass size spectrum. *Canadian Journal of Fisheries and Aquatic Science* **44 (Supplement 2)**, 136–140.

Bottrell H.H., Duncan A., Gliwitz Z.M., Grygierek E., Herzig A., Hillbricht-Ilkowska A., Kurasawa H., Larssen P. & Weglenska T. (1976) Review of some problems in zooplankton production studies. *Norwegian Journal of Zoology* **24**, 419–456.

Boudreau P.R. & Dickie L.M. (1992) Biomass spectra of aquatic ecosystems in relation to fisheries yield. *Canadian Journal of Fisheries and Aquatic Science* **49**, 1528–1538.

Brandt S.B., Mason D.M., Patrick E.V., Argyle R.L., Wells L., Unger P.A. & Stewart D. J. (1991) Acoustic measures of abundance and size of pelagic planktivores in Lake Michigan. *Canadian Journal of Fisheries and Aquatic Science* **48**, 894–904.

Burczynski J.J. & Johnson R.L. (1986) Application of dual-beam acoustic survey techniques to linmetic populations of juvenile sockeye salmon, *Oncorhynchus nerka*. *Canadian Journal of Fisheries and Aquatic Science* **43**, 1776–1788.

Calder W.A.III (1984) *Size, function, and life history*. Cambridge: Harvard University Press, Cambridge, MA, USA.

Christensen, V. & Pauly, D. (1991) A guide to the ECOPATH II software system (Version 2.0). *ICLARM Software* **6**, 71 pp.

Confer J.L. & Blades P.I. (1975) Omnivorous zooplankton and planktivorous fish. *Limnology and Oceanography* **20**, 571–579.

Coulter G.W. (1977) Approaches to estimating fish biomass and potential yield in Lake Tanganyika. *Journal of Fish Biology* **11**, 393–409.

Coulter G.W. (1981) Biomass, production and potential yield in the Lake Tanganyika pelagic fish community. *Transactions of the American Fisheries Society* **110**, 325–335.

Degnbol P. & S. Mapila S. (1982) Linmological observations on the pelagic zone of Lake Malawi from 1978 to 1981. In: *Biological studies on the pelagic ecosystem of Lake Malawi* FI:DP/MLWk/75/019, Technical Report 1 Rome: FAO, pp 5–47.

Dickie L.M., Kerr S.R. & Boudreau P.R. (1987) Size-dependent processes underlying regularities in ecosystem structure. *Ecological Monographs* **57**, 233–250.

Downing J.A., Plante C. & Lalonde S. (1990) Fish production correlated with primary productivity, not the morphoedaphic index. *Canadian Journal of Fisheries and Aquatic Science* **47**, 1929–1936.

Eccles D.H. (1974) An outline of the physical limnology of Lake Malawi (Lake Nyassa). *Limnology and Oceanography* **18**, 730–742.

FAO (1982) Biological studies on the pelagic ecosystem of Lake Malawi. *FAO FI:DP/MLWk/75/019, Technical Report* **1**, 182 pp.

Fee E.J. (1990) Computer programs for calculating *in situ* phytoplankton photosynthesis. *Canadian Technical Reports in Fisheries and Aquatic Science* **1740**, 32 pp.

Fenchel T, (1974) Intrinsic rate of natural increase: the relationship with body size. *Oecologia* **14**, 317–326.

Gaedke U. (1993) Ecosystem analysis based on biomass size distributions: a case study of a plankton community in a large lake. *Limnology and Oceanography* **38**, 112–127.

Hecky R.E. (1984) African lakes and their trophic efficiencies: a temporal perspective. In: D.G. Meyers & J.R. Strickler (eds) *Trophic interactions within aquatic ecosystems*. Washington DC: American Association for the Advancement of Science, pp 405–448.

Hecky R.E. (1991) The pelagic ecosystem. In: G. Coulter (ed.) *Lake Tanganyika and its Life*. Oxford University Press, UK, pp 90–110.

Hecky R.E. & Fee E.J. (1981) Primary production and rates of algal growth in Lake Tanganyika. *Limnology and Oceanography* **26**, 532–547.

Irvine K. (1995a) Standing biomasses, production and spatial and temporal distributions of the crustacean zooplankton. In: A. Menz (ed.) *Fisheries potential and the ecology of the pelagic zone of Lake Malawi/Niassa*. Natural Resources Institute, Chatham, Kent, UK, pp 85–108.

Irvine K. (1995b) Ecology of the lakefly, *Chaoborus edulis*. In: A. Menz (ed.) *Fisheries potential and the ecology of the pelagic zone of Lake Malawi/Niassa*. Natural Resources Institute, Chatham, Kent, UK, pp 109–40.

Irvine K. & Waya R. (1995) The Zooplankton: General sampling methods and estimation of biomass and development rates. In: A. Menz (ed.) *Fisheries potential and the ecology of the pelagic zone of Lake Malawi/Niassa*. Natural Resources Institute, Chatham, Kent, UK, pp 69–84.

Jackson P.B.N., Iles T.D., Harding D. & Fryer G. (1963) Report on the survey of Northern Lake Nyasa 1954-55. Government Printer, Zomba, Nyasaland, 171 pp.

Kerr S.R. (1974) Theory of size distributions in ecological communities. *Journal of the Fisheries Research Board of Canada* **31**, 1859–1862.

Kerr S.R. & Martin N.V. (1970) Trophic-dynamics of lake-trout production systems. In: J.H. Steele (ed.) *Marine Food Chains*. Berkeley: University of California Press, pp 365–376.

Laws E.A., Redalje D.G., Haas L.W., Bienfang P.K., Eppley R.W., Harrison W.G., Karl D.M. & Marra J. (1984) High phytoplankton growth and production rates in oligotrophic Hawaiian coastal waters. *Limnology and Oceanography* **29**, 1161–1169.

Leach J.H., Dickie L.M., Shuter B.J., Borgmann U., Hyman J. & Lysack W. (1987) A review of methods for prediction of potential fish production with application to the Great Lakes and Lake Winnipeg. *Canadian Journal of Fisheries and Aquatic Science* **44 (Supplement 2)**, 471–485.

Lowe-McConnell R.H. (1987) *Ecological studies in tropical fish communities*. Cambridge University Press, 382 pp.

Lund J.W.G., Kipling C. & LeCren E.D. (1958) The inverted microscope method of estimating algal numbers and the statistical basis of estimations by counting. *Hydrobiologia* **11**, 143–170.

MacLennan D.N. & Simmonds E.J. (1992) *Fisheries Acoustics*. London: Chapman & Hall, 325 pp.

Melack J.M. (1976) Primary productivity and fish yields in tropical lakes. *Transactions of the American Fisheries Society* **105**, 575–580.

Mengestou S. & Fernando C.H. (1991) Biomass and production of the major dominant crustacean zooplankton in a tropical Rift Valley lake, Awasa, Ethiopia. *Journal of Plankton Research* **13**, 831–851.

Menz A., Bulirani A., & Goodwin C.M. (1995) Acoustic estimation of fish biomass. In: A. Menz (ed.) *Fisheries potential and the ecology of the pelagic zone of Lake Malawi/Niassa*. Natural Resources Institute, Chatham, Kent, UK, pp. 307–50.

Milligan S.P. & B.F. Riches (1983) The new MAFF/Guildline high-speed plankton samplers. *ICES CM 1983/L7*, 6 pp (mimeo).

Minns C.K., Millard E.S., Cooley J.M., Johnson M.G., Hurley D.A., Nicholls K.H., Robinson G.W., Owen G.E. & Crowder A. (1987) Production and biomass size-spectra in the Bay of Quinte, a eutrophic ecosystem. *Canadian Journal of Fisheries and Aquatic Science* **44 (Supplement 2)**, 148–155.

Munawar M., Stadelmann P. & Munawar I. F. (1974) Phytoplankton biomass, its species composition and primary production at a nearshore and midlake station of Lake Ontario during IFYGL. In: *Proceedings of the 17th Conference on Great Lakes Research*, pp 629–652.

Nichols J.H. & Thompson A.B. (1991) Mesh selection of copepodite and nauplius stages of four calanoid copepod species. *Journal of Plankton Research* **13**, 661–671.

Paloheimo J.E. & Dickie L.M. (1966) Food and growth of fishes. III. Relations among food, body size, and growth efficiency. *Journal of the Fisheries Research Board of Canada* **23**, 1209–1248.

Patterson G. & Kachinjika O. (1995) Limnology and phytoplankton ecology. In: A. Menz (ed.) *Fisheries potential and the ecology of the pelagic zone of Lake Malawi/Niassa*. Natural Resources Institute, Chatham, Kent, UK, pp 1–68.

Platt T. & Denman K. (1978) The structure of pelagic marine ecosystems. *Rapports et Procès-verbaux. Reunion du Conseil International pour l'Exploration de la Mer* **173**, 60–65.

Ryder R.A. (1982) The morphoedaphic index – use, abuse, and fundamental concepts. *Transactions of the American Fisheries Society* **111**, 154–164.

Sheldon R.W., Prakash A. & Sutcliffe W.H.Jr. (1972) The size distribution of particles in the ocean. *Limnology and Oceanography* **17**, 327–340.

Sheldon R.W., Sutcliffe W.H.Jr. & Paranjape M.J. (1977) Structure of pelagic food chains and relationship between plankton and fish production. *Journal of the Fisheries Research Board of Canada* **34**, 2344–2353.

Slobodkin L.B. (1962) Energy in animal ecology. *Advances in Ecological Research* **1**, 69–101.

Sprules W.G., Brandt S.B., Stewart D.J., Munawar M., Jin E.H. & Love J. (1991) Biomass size-spectrum of the Lake Michigan pelagic food web. *Canadian Journal of Fisheries and Aquatic Science* **48**, 105–115.

Sprules W.G., Casselman J.M. & Shuter B.J. (1983) Size distribution of pelagic particles in lakes. *Canadian Journal of Fisheries and Aquatic Science* **40**, 1761–1769.

Sprules W.G. & Munawar M. (1986) Plankton size-spectra in relation to ecosystem productivity, size and perturbation. *Canadian Journal of Fisheries and Aquatic Science* **43**, 1789–1794.

Thiebaux M.L. & Dickie L. M. (1992) Models of aquatic biomass size spectra and the common structure of their solutions. *Journal of Theoretical Biology* **159**, 147–161.

Thiebaux M.L. & Dickie L.M. (1993) Structure of the body-size spectrum of the biomass in aquatic eco-systems: a consequence of allometry in predator-prey interactions. *Canadian Journal of Fisheries and Aquatic Science* **50**, 1308–1317.

Thompson A.B. (1995) Eggs and larvae of *Engraulicypris sardella*. In: A. Menz (ed.) *Fisheries potential and the ecology of the pelagic zone of Lake Malawi/Niassa*. Natural Resources Institute, Chatham, Kent, UK, pp 179–200.

Thompson A.B., Allison E.H. & Ngatunga B.P. (1995a) Spatial and temporal distribution of fish in the pelagic waters. In: A. Menz (ed.) *Fisheries potential and the ecology of the pelagic zone of Lake Malawi/Niassa*. Natural Resources Institute, Chatham, Kent, UK, pp. 201–32.

Thompson A.B., Allison E.H., Ngatunga B.P. & Bulirani A. (1995b) Fish growth and breeding biology. In: A. Menz (ed.) *Fisheries potential and the ecology of the pelagic zone of Lake Malawi/Niassa*. Natural Resources Institute, Chatham, Kent, UK.

# IV POPULATION MODELS

# Chapter 18
# A length-structured matrix population model used as a fish stock assessment tool

W. DEKKER *Netherlands Institute for Fisheries Research RIVO-DLO, P.O. Box 68, 1970 AB IJmuiden, The Netherlands*

**Abstract**   A simple matrix model for exploited fish stocks is presented. Unlike traditional fisheries' models, the model is completely structured by length. Its use for assessment of fisheries is discussed, and an application to three fisheries on Lake IJsselmeer (The Netherlands) is presented. Although the model is of limited relevance in its single species mode for the assessment of these intensively exploited fisheries, it will be a valuable starting point for more complex, length-structured multispecies models.

KEYWORDS: length structured, matrix model, population assessment

## 18.1   Introduction

Many classical stock assessment methods belong to the class of age-structured models (Getz & Haight 1989). Single species cohort analysis models and the more elaborate multispecies models (Andersen *et al.* 1973) can be traced back to Beverton and Holt (1957), who described an age-structured model for exploited fish populations. These models have been successfully applied to assess many fisheries. However, the most important processes affecting natural populations are not selective for age, but for length (Caswell 1989): either age groups overlap considerably (as in the eel, *Anguilla anguilla* (L.) data below), or processes affecting the stocks are very length selective (as in the perch, *Perca fluviatilis* L. and pikeperch, *Stizostedion lucioperca* (L.) data below). For example, the length frequency of the catches of pikeperch and perch (Fig. 18.1) split by age groups shows the wide overlap in size between age groups in the catch. In such cases, the need for length-structured stock assessment models has been stressed (Anon. 1992).

Length-structured stock assessment methods have been available for a long time (Pauly & Morgan 1987). Surprisingly, most of these models are built around the backbone of age-structured models: the length frequency is scanned to detect the underlying age groups; thereafter the traditional age-structured models are applied. Jones (1974) dropped the concept of underlying age groups completely, and analysed the interrelation of time and growth directly; however, he had to restrict his analysis to a steady state assumption.

Buijse *et al.* (1992) simulated a length selective fishery on the basis of a length- and

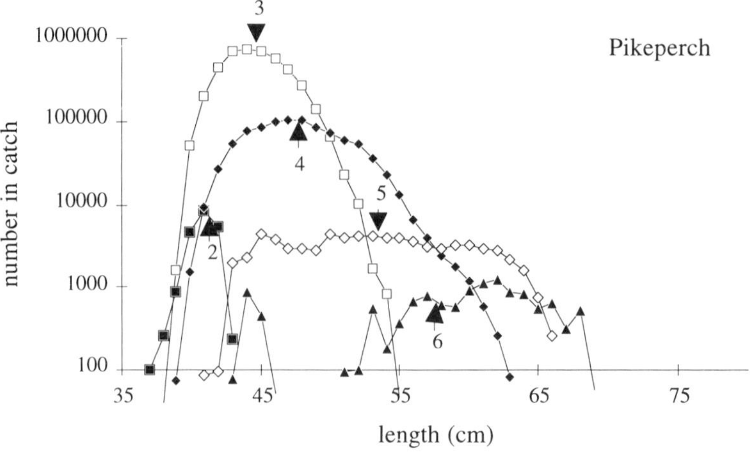


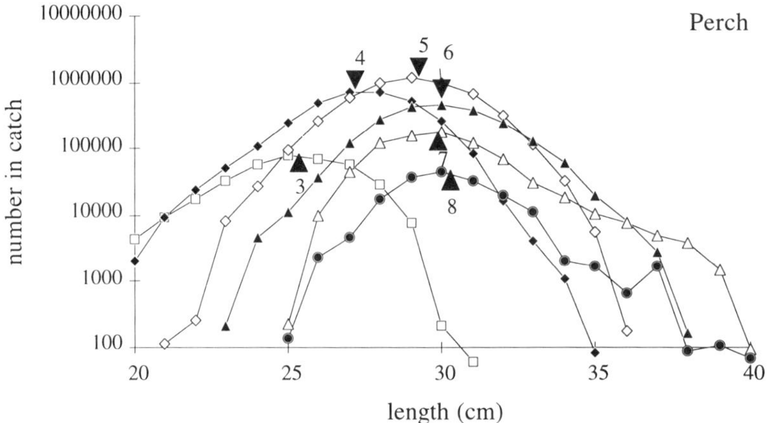


**Fig. 18.1**   Length frequency of the pikeperch and perch catches, summed over all seasons, per age group. Arrows indicate the average length per age group.

age-structured model, and showed an age-structured model differs from one with length and age structure. However, they did not consider the possibility of dropping age altogether, and therefore might have made the model needlessly complicated. Although their model provides insight into the behaviour of a (mixed) length selective fishery, it does not assess the actual history of the fishery, and therefore cannot be used as an assessment tool. Irrespective of this, discussions on their simulation model formed the onset to the work described here.

Dekker (1993) outlined a truly length-structured assessment model, based on an idea by Shepherd (1987), and applied that model to the eel fisheries on lake IJsselmeer, The Netherlands. In this chapter, the model is reformulated as a straightforward length-structured matrix model for single species stock assessment. An application to the mixed gill net fishery for perch and pikeperch on Lake IJsselmeer is added. Some of the unsolved problems of Dekker (1993) are tackled. In the future it is intended to extend the model to include a multispecies aspect, along the lines of the age-structured model of Helgason & Gislason (1979).

## 18.2    Material

The IJsselmeer fisheries have been described by Dekker (1986b, 1991) and Buijse *et al.* (1992). Since the main aim of this study is to present a new model, the data will be described only briefly.

Landings per species per year and lengthclass are derived from:

- auction statistics: the manager of the fisheries (the Ministry of Agriculture and Fisheries) collects information at the auctions on the amount of fish landed, per species and month;
- market sampling: during the fishing season samples of the landed fish are analysed for length, weight, sex, maturity, age (eel excluded), etc.

On the basis of of this information, the total landings per length class are calculated. For eel, summer totals are derived. It has been shown (Dekker 1987) that length frequencies are fairly constant over the fishing season. For perch and pikeperch the fishing season lasts from July through March 15 and listed totals relate to season totals. The most intensive fishing occurs in October and November, and therefore growth during the fishing season has been ignored. Fishing totals are labelled as referring to years, but represent fishing seasons and thus incorporate catches made after the first of January of the next year. Unfortunately, in 1985 no market sampling was carried out. The length frequencies in that year were based on a limited amount of data from research vessel catches.

Growth estimates are based on scale reading (perch and pikeperch, Willemsen 1977) and tetracycline marking experiments (eel, Dekker 1986a). In the present analysis, growth is assumed to be constant and all available age readings over the past 20 years have been combined. Estimates for empty cells were derived from an assumed linear relationship between length and growth, and an assumed linear relationship between length and the spread around the mean growth. Since few age reading measurements are available for eel, the growth function applied in this case is exclusively based on the assumed linear relationships.

## 18.3    Model description

The basic model is derived from the usual (age-structured) cohort analysis model, incorporating state-dependent features of matrix population models. The following is a brief recapitulation of the usual age-structured model:

Let $N_t$ represent the number of animals present in the population at time $t$, and $N_{t+1}$ the number one time interval later. $N_t$ and $N_{t+1}$ are related by the following equation:

$$N_{t+1} = \exp(-F_t - M) \times N_t \tag{1}$$

where $F_t$ denotes the instantaneous mortality induced by the fishery, and $M$ the instantaneous mortality caused by other factors. $F_t$ is usually interpreted as the product of the fishing effort and the catchability of the fish, while $M$ is taken as a constant. In the following, $M$ is assumed to be 0.2, although non-constant assumptions might be incorporated without much change to the model.

The catch of the fishery, during time interval $< t, t+1 >$ equals

$$C_t = F_t \,/\, (F_t + M) \times (1 - \exp(- F_t - M\,)) \times N_t \tag{2}$$

$$= F_t \,/\, (F_t + M) \times (\exp\,(F_t + M\,) - 1) \times N_{t+1}$$

The cohort of animals thus described, is recruiting at $t = 0$, with

$$N_0 = R \tag{3}$$

where $R$ represents the number of recruiting animals.

Virtual Population Analysis (VPA, Gulland 1965) solves Equation (1) and (2) backwards in time, starting with an assumed value for the terminal $N_t$. The elegancy of the method lies in the reasoning that errors in the assumed value for the latest $N_t$ diminish when going back in time (Pope 1972), i.e. the $N_t$ of the earlier years converge to their true value, irrespective of the assumed value for the terminal $N_t$.

Standard age-structured models use a separate set of Equations (1) and (2) for each cohort in a population, i.e. each year class is analysed separately. The model developed here diverges in the sense that the whole population is considered simultaneously, even though the population is subdivided into length classes. The processes affecting the population are considered to be state, i.e. length, dependent.

In the absence of any mortality and recruitment, the total number in the population remains constant, i.e. $N_t = N_{t+1}$. However, the structuring by length introduces the possibility that animals grow during the time interval, and thus change their state.

Introducing growth as a transition matrix $[G_{i,j}]$, it follows that in the absence of any mortality or recruitment

$$(N_{t+1,i}) = [G_{i,j}] \times (N_{t,j}) \tag{4}$$

Where $G_{i,j}$ denotes the probability that an animal will grow from lengthclass $j$ to lengthclass $i$ within one time interval. The notation used here adheres to the standard symbols in fish cohort analysis models. Thus capital versus lower case characters do *not* indicate matrices resp. vectors. Instead, matrices and vectors are given as indexed cells, enclosed in round (vectors) or square [matrices] brackets.) In the applications of the model below, $[G_{i,j}]$ is assumed to be constant over time, but this restriction might easily be lifted. Note that growth is assumed to have the Markov property, i.e. the growth of any individual is determined by its length only, and in particular not by its past growth history. Assuming that the maximum length included in the model is large enough, it will follow that

$$\sum_i G_{i,j} = 1 \quad \forall\, j \tag{5}$$

i.e. no animal gets lost by growing to another length class.

Since negative growth is not realistic, $[G_{i,j}]$ will be triangular, with non-zero entries mostly in a band below the main diagonal. It should be noted that adjacent columns of $[G_{i,j}]$ are nearly identical, i.e. animals in length class $i$ and $i+1$ at time $t$, will have almost the same probability of being in length class $j$ at time $t+1$. A graphical representation of the growth array for the data analysed below is given in Fig. 18.2 (a, b and d).

Mortality and recruitment is introduced as:

$$(N_{t+1,i}) = (R_{t,i}) + [A_{i,j}] \times (N_{t,j}) \tag{6}$$

where $[A_{i,j}]$ denotes the transition matrix describing the combined effect of growth and mortality, and $R_{t,i}$ denotes the recruitment. Note that recruitment is no longer treated as a new cohort, but merely adds to the existing population. Furthermore, recruitment is not included in the transition matrix, as in a classical Leslie matrix, since it will not be interpreted as a function of other system variables.

From Equations (1) and (4), the obvious form of $[A_{i,j}]$ is

$$[A_{i,j}] = [G_{i,j}] \times (\exp^{-F_{t,j}-M})^T \tag{7}$$

and from Equations (6) and (7)

$$(N_{t+1,i}) = (R_{t,i}) + [G_{i,j}] \times (\exp(-F_{t,j} - M))^T \times (N_{t,j}) \tag{8}$$

which can be rearranged to give

$$(N_{t,j})^T = (\exp(F_{t,j} + M))^T \times [G_{i,j}]^{-1} \times \{(N_{t+1,i}) - (R_{t,i})\} \tag{9}$$

i.e. the population at time $t$, derived from the population at time $t+1$. Since $(R_t)$ is represented by the lower length classes of $(N_{t+1})$, Equation (9) contains one time dependent variable: $F_{t,i}$, which – in analogy with Equation (2) – might be derived from

$$\begin{aligned}(C_{t,j})^T &= (F_{t,j} / (F_{t,j} + M) \times (1 - \exp(-F_{t,j} - M)))^T \times (N_{t,j}) \\ &= (F_{t,j} / (F_{t,j} + M) \times (\exp(F_{t,j} + M) - 1))^T \times [G_{i,j}]^{-1} \times \{(N_{t+1,i}) - (R_{t,i})\}\end{aligned} \tag{10}$$

Now, the usual reversal of time of the VPA leads to:

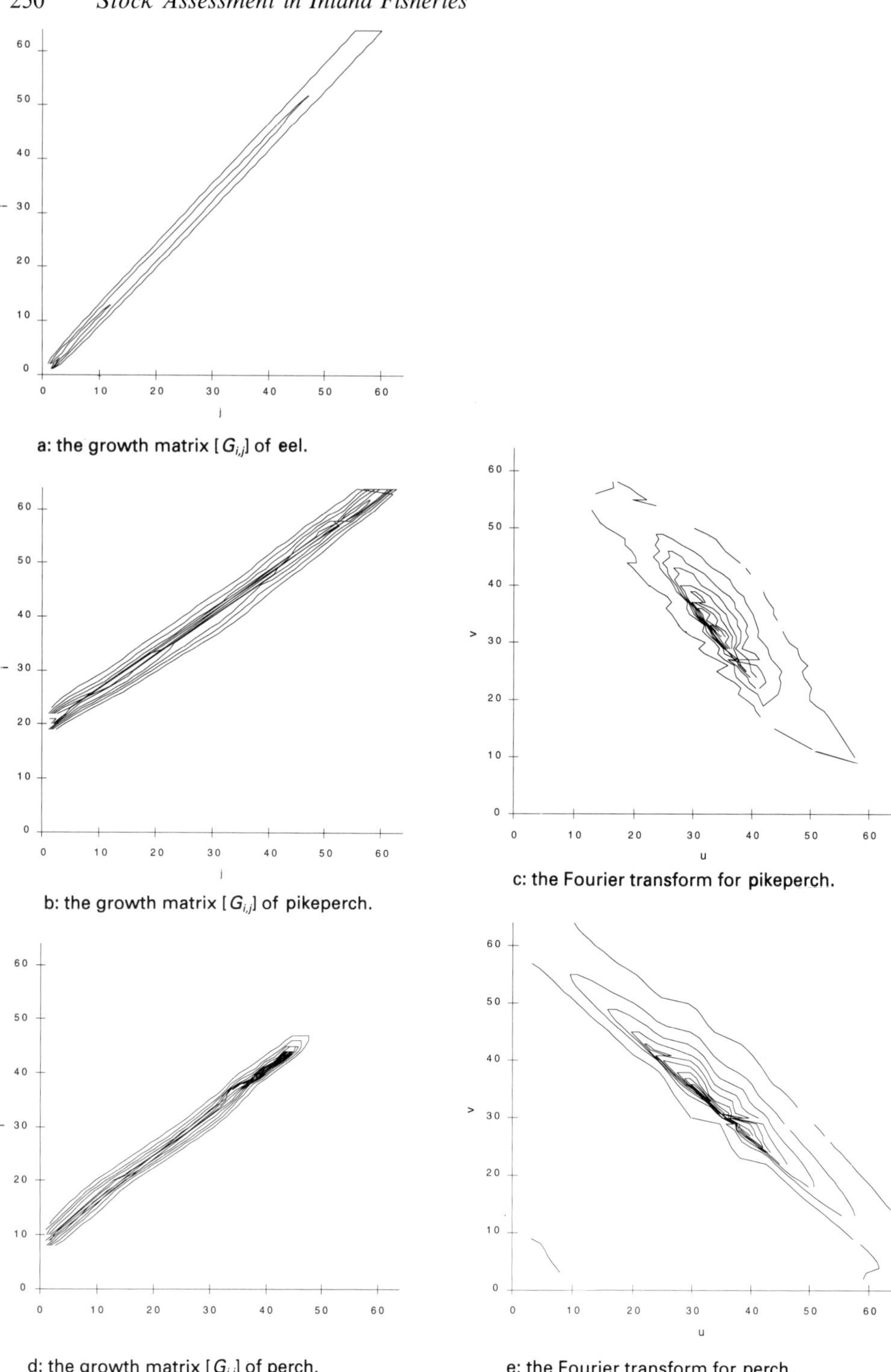



**Fig. 18.2**  Graphical representation of the growth matrices $[G_{i,j}]$, and the amplitude of their Fourier transforms $|F(u, v)|$. Ten equidistant contours from minimum to maximum are plotted.

(1)   Guess terminal values for $(N_{t+1,i})$; usually, this is the value resulting in a reasonable $(F_{t,j})$.
(2)   Estimate $(F_{t,j})$ from Equation (10), and the known catch $(C_{t,j})$
(3)   Solve $(N_{t,j})$ from Equation (9).
(4)   Repeat from step 1, for the previous time interval, with $(N_{t+1,i})$ taken from step 3.

### 18.4   Estimation procedures

The model described by Equations (9) and (10) was implemented in a computer program. As stated above, $[G_{i,j}]$ is – by its very nature – structurally ill-conditioned, and inversion will therefore be problematic. Therefore, inversion was avoided, and an approximation routine was developed to solve $(Y_i) = [G_{i,j}] \times (X_j)^T$ for $(X_j)$, based on a Newton-Raphson iteration, and the sums of squares of the difference between $(Y_i)$ and $[G_{i,j}] \times (X_j)^T$. Because of the structure of $[G_{i,j}]$, the Newton-Raphson approximation was restricted to the use of diag $(SS'')$, and used Seidel's updating procedure throughout.

Terminal values for $(N_{t+1,i})$ were chosen to give $(F_{t,j})$ reasonable terminal values, but no real tuning procedure has been applied. Thus, results in the convergence period might be unrealistic.

Dekker (1993) pointed out that the 'oldest age group' of age-structured models is replaced by a range of length classes containing animals that are in the population at time $t$, but have grown beyond the maximum length applied, at time $t+1$. Since no realistic estimates for these 'oldest length groups' were available, they were set to zero if no catch was found, and to 1, if a positive catch was found. This unrealistically low value of 1 will result in extremely high estimates of $(F_{t,j})$ for these length classes.

The computer program was able to recreate simulated test data to within rounding limits. However, introducing minor stochastic variation in the simulated data, resulted in aberrant behaviour: the stochastic variation was amplified strongly from one time step to another. Inspection of the results indicated that this amplification was the result of the structure of $[G_{i,j}]$:

Considering growth as a filtering process of the population number vector (a non-linear translation to greater length for the growth itself, followed by blurring, to account for the variance of the growth), degraded by measurement errors, the problem of finding the population vector in the previous year can be treated as an inverse filtering problem (see for instance: Gonzalez & Wintz 1977, p 199). In this context, it is well known that the behaviour of a filter (and its inverse) is largely determined by its Fourier transforms.

Figure 18.2 (c and e) presents the Fourier transform of the growth matrices $[G_{i,j}]$. The Fourier transform of $[G_{i,j}]$ drops off sharply even at medium frequencies. Thus, application of the inverse filter will not allow the higher frequencies of the signal to be distinguished from the noise. The results did indeed show a strong high frequency. It is likely that measurement errors will be most pronounced in the higher frequencies (so-called 'pink noise'), thereby intensifying the problem.

Gonzalez & Wintz (1977) suggest that higher frequencies in such cases are ignored completely. Since the cut-off point between high and low frequencies could not be fixed *a priori*, it was decided to suppress the higher frequencies gradually and not to leave them out completely. This was accomplished by a low pass filter. The low pass filter was implemented as a simple moving average over three length classes, giving equal weight to all three. Denoting this low pass filter by *LP*, Equations (9) and (10) were replaced by

$$(N_{i,j})^T = (\exp(F_{t,j} + M))^T \times [LP]^a \times [G_{i,j}]^{-1} \times \{(N_{t+1,i}) - (R_{t,i})\} \tag{11}$$

and

$$(C_{t,j})^T = (F_{t,j} / (F_{t,j} + M) \times (\exp(F_{t,j} + M) - 1))^T$$
$$\times [LP]^a \times [G_{i,j}]^{-1} \times \{(N_{t+1,i}) - (R_{t,i})\} \tag{12}$$

where

$$LP = \frac{1}{3} \times \begin{bmatrix} 1 & 1 & & & & & \\ 1 & 1 & 1 & & 0 & & \\ & 1 & 1 & & & & \\ & & 1 & & & & \\ & & & & 1 & & \\ & & & & 1 & 1 & \\ & 0 & & & 1 & 1 & 1 \\ & & & & & 1 & 1 \end{bmatrix} \tag{13}$$

The parameter *a* determines the degree to which high frequencies were suppressed. Figure 18.3 shows the population numbers for the three species in 1989, estimated with a value of $a = 0$ (no low pass filter) and $a = 2$. Higher values of *a* did not change the results markedly.

Note that in Dekker (1993), $[G_{i,j}]^{-1} \times [LP]^1$ was used, instead of $[LP]^a \times [G_{i,j}]^{-1}$, i.e. high frequencies were first suppressed, and then amplified; this gave rather unsatisfactory results.


### 18.5     Results

The model was applied to data from the eel, pikeperch and perch fisheries on Lake IJsselmeer. The natural mortality parameter *M* was fixed at a value of 0.2; it is known that bycatch and predation exert a much higher mortality on the stocks, but in current single species application, no alternative estimates are available. The low pass filter parameter *a* was fixed at a value of 2. The results are presented in Figs 18.4, 18.5 and 18.6. The number of data years was temporarily restricted to a maximum of ten, because of hardware limits. This restriction might easily be remedied in a future implementation. For eel, only seven years of data were available.

When applying age-structured stock assessment models, it is customary to present tables of catch numbers by age groups, as well as of the estimated population numbers and fishery mortalities by age groups. Although the same procedure might have been followed, the size of the tables would be very large, because of the extensive range of size classes (40 classes for eel and perch, 80 for pikeperch). In total, more than 4000

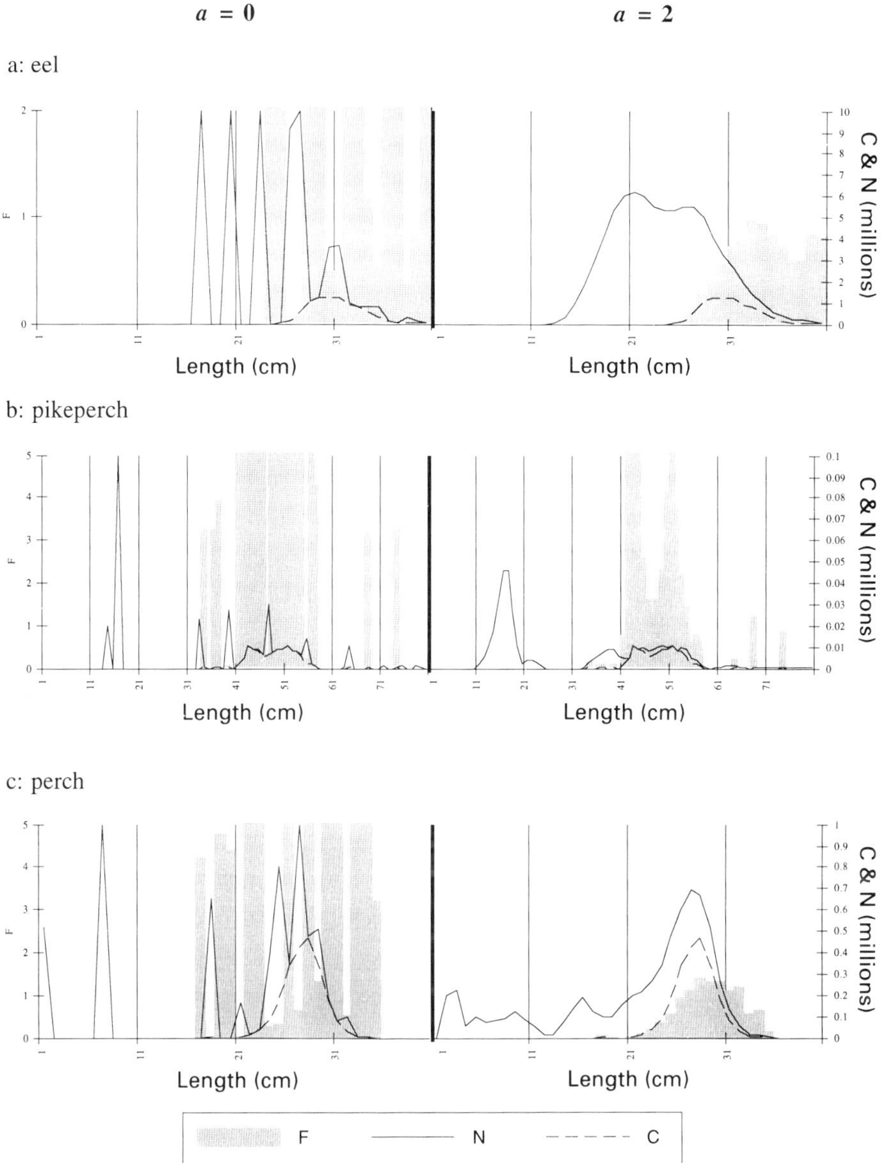



**Fig. 18.3**  Comparison of model results for the three species, for two values of the low pass filter parameter *a*. Results are presented for 1989 only.

table entries would have to be listed! Although appropriate summary statistics might be derived (averages over length intervals, cumulative statistics, etc.) in this methodological paper a full presentation of the results in graphical form has been preferred.

Figure 18.4 presents the results for the eel. Population numbers drop sharply over the entire time span of 5 years, while catches only start dropping in 1992. Fishing

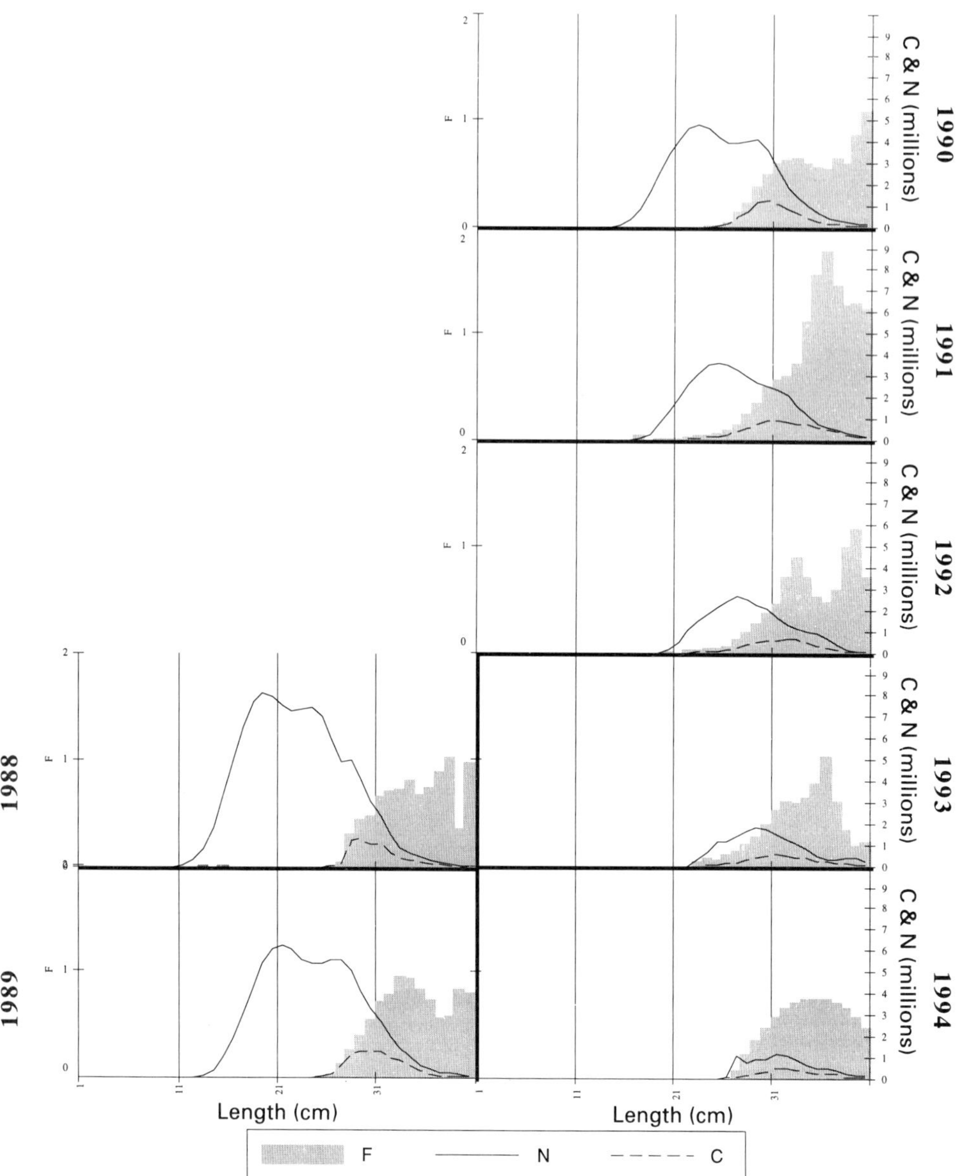

**Fig. 18.4**   Model results for eel, for all years with available data ($a = 2$).

mortality extends from ca 27 cm, peaking at 32 or 33 cm, at a level of $\pm 1.0$. The results for the larger eels vary considerably from lengthclass to lengthclass, and between years. This is evidently the consequence of estimates being based on a very small number in the catch, and the resulting estimates cannot be considered reliable. Since population numbers are reconstructed from the commercial catches, no

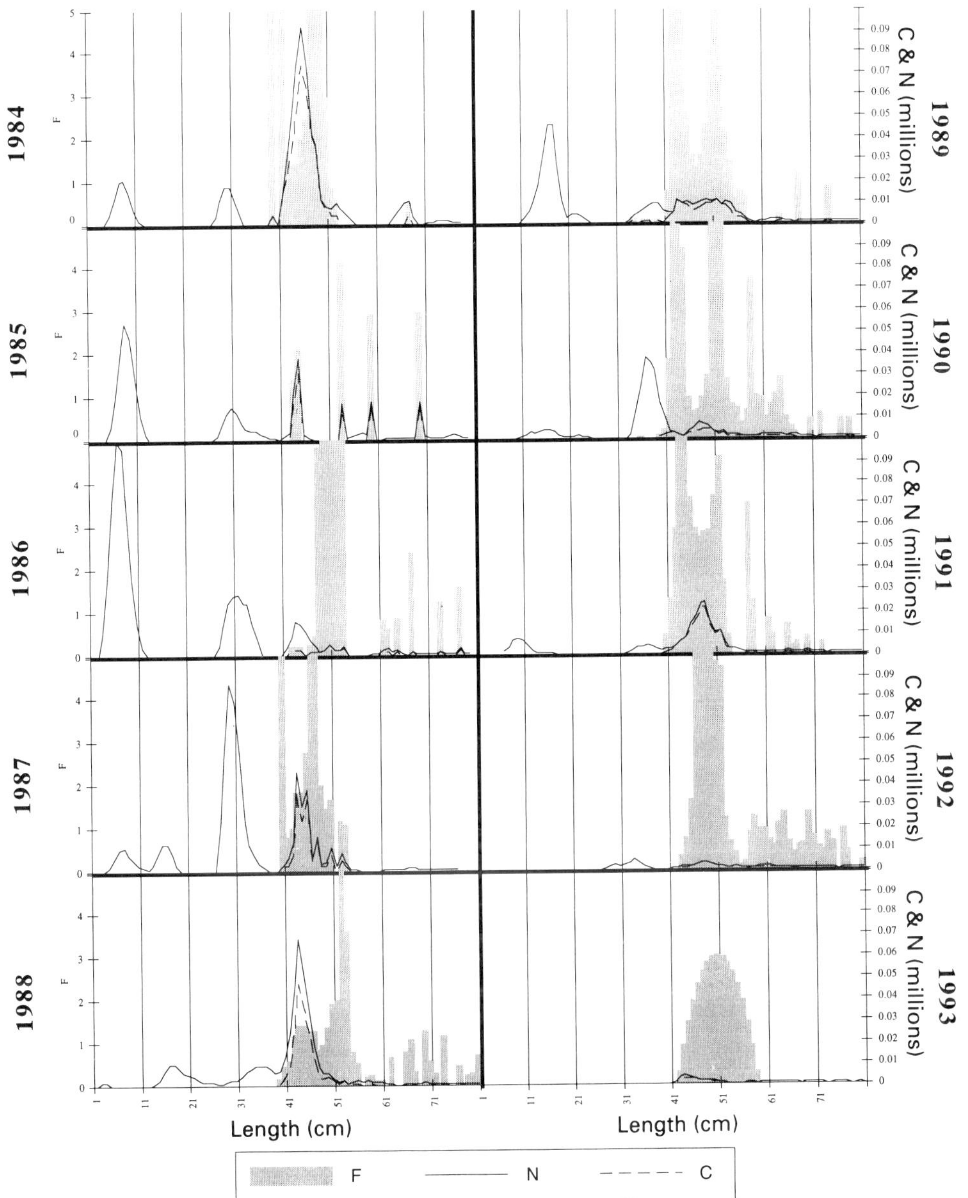

**Fig. 18.5** Model results for pikeperch, for the last ten years ($a = 2$).

estimates are available for the length classes below the official minimum length in the later years. In the earliest year, estimates are available down to a length of ca 20 cm.

Figure 18.5 presents the results for the pikeperch. During the last ten years pikeperch catches have dropped more or less continuously, because of a long period of weak year classes. Estimated population numbers vary strongly over the lengthclasses

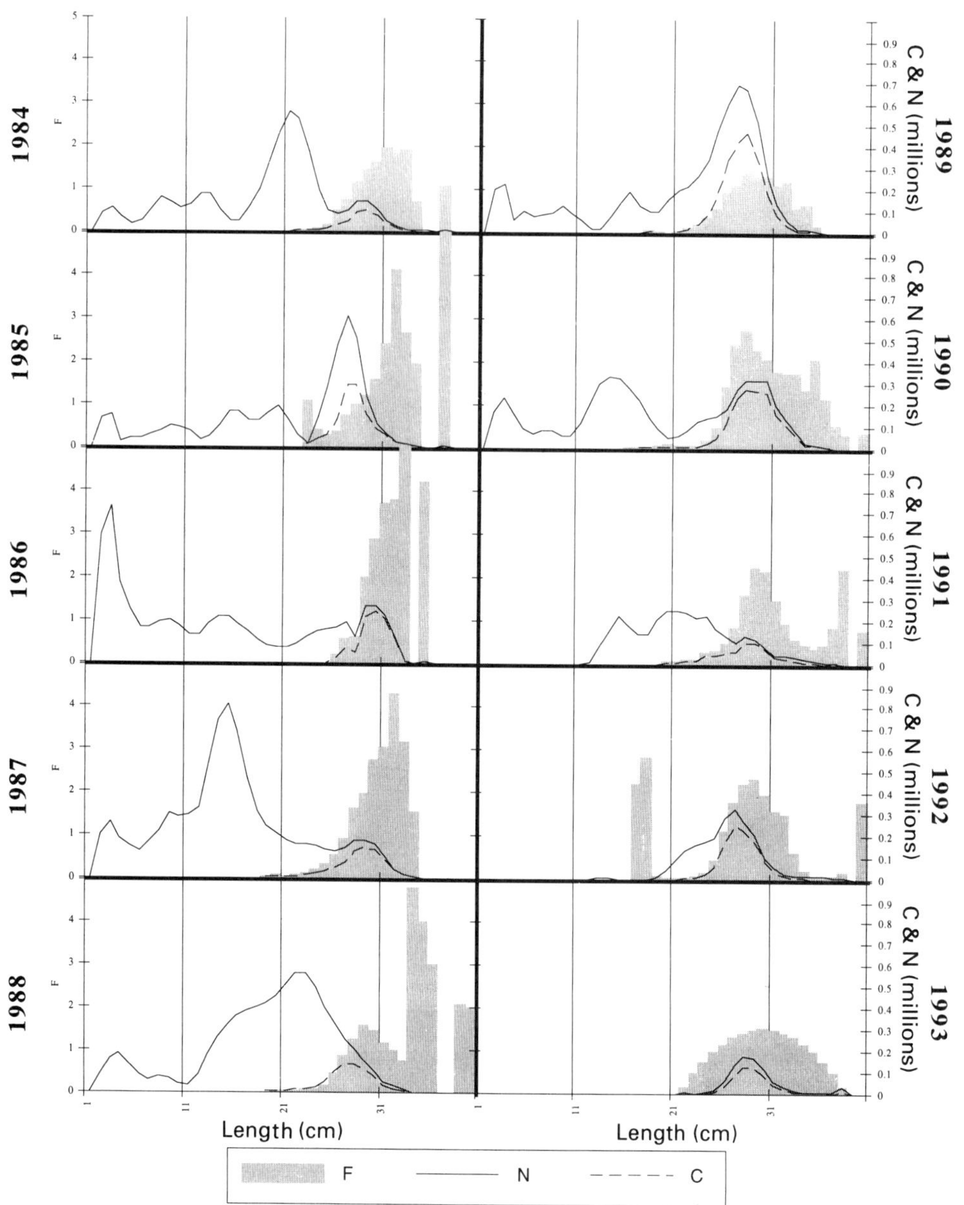



**Fig. 18.6**   Model results for perch, for the last ten years ($a = 2$).

and the years. Estimated fishing mortalities vary strongly over the lengthclasses, without a regular pattern over the years. Especially in the larger lengthclasses, the estimated fishing mortalities are either very high or very low and no intermediate values are observed. Estimates in 1985 (the year without sampling of commercial catches) differ from the other years, in the sense that catches and fishing mortalities seem to peak in exceptionally few lengthclasses.

Figure 18.6 presents the results for the perch. Population numbers vary over the years. Modes occur in the length frequencies, which seem to represent individual year classes, but strong overlap between adjacent modes occurs in some years (notably 1988, 1991 and 1984). Fishing mortality extends from ca 25 cm and peaks around 30 cm. As in eel, the fishing mortality in the larger lengthclasses varies between years, but the actual number of observations in this range is very small. Peak fishing mortalities are in the range 2 to 4. There is some indication that the peak in fishing intensity shifts to smaller lengthclasses in the recent years.

## 18.6     Discussion

A truly length-based assessment model has been presented, and applied to field data referring to the Lake IJsselmeer fisheries.

Most of the attractiveness of VPA-type models lies in the convergence property described by Pope (1972). Although the model presented here shares this convergence property with the age-structured cohort analysis models, it introduced a strong numerical instability caused by minor measurement errors. Indeed, the convergence is completely overruled by the amplification of measurement errors within the length frequency. However, the changes in the length frequencies are generated over very short length ranges (Fig. 18.3). Considering the raw results (i.e. without application of the low pass filter, as in Fig. 18.3, left), it is hardly conceivable that immense differences in abundance between neighbouring lengthclasses would be realistic. Interpretation of the results is only possible in terms of trends in abundance and fishing mortality over very broad length intervals. Thus, the use of a low pass filter seems reasonably justified. The choice of an appropriate low pass filter is rather subjective. In the current applications, a simple low pass filter was used, with only one parameter, but others might be developed.

In comparison with age-structured models, an extremely high number of parameters had to be fitted by the model. For instance, in the case of perch, approximately 6 age classes are observed in the length range from 20 to 40 cm. Therefore, more complex (multispecies) models might be based on the current model, but superimpose less overparameterized functional forms for various estimators, such as functional forms for net selectivities, or predator-prey size ratio functions. This superimposition of less parameterized models might in fact replace the low pass filter used in the current version of the model.

For the pikeperch data, even the application of the low pass filter did not smooth the results satisfactorily. However, it should be noted that the population of larger animals is extremely small, and the actual length distribution may be considered a small random sample from a large size range. Thus, the sampling intensity might be insufficient to cover the bigger animals, and the failure does not necessarily point to a potentially bad property of the model. Lumping of size classes, or *a priori* smoothing of the catch data might handle this problem.

Comparison of the model results to traditional age-structured models will be rather difficult, because they are bound to generate different results, while there is no independent evidence to judge between them. From looking at Figs 18.1 and 18.6, it is evident that perch age groups in the population do span a wide length interval. Therefore, even the older age groups are not completely recruited to the length selective fishery. Thus, the catch of older animals partly comprises animals newly recruited to the fishery, and not only survival from the previous season of previously recruited animals. In the model presented here, the presence of slow growing animals is explicitly included in the growth matrix $[G_{i,j}]$. One might develop an age-based model, taking into account the (varying) level of recruitment to the fishery, based on the length composition of each age group, but such a model would converge towards the model presented here.

Most so-called 'length-based methods' (Pauly & Morgan 1987) are age-based methods, because the length frequencies are partitioned into age groups, which process is aggravated by the cumbersome partitioning, involving many assumptions. The comparison of the current model to this group of methods will only duplicate the comparison with truly age-structured models, based on age readings.

Jones' (1974) truly length-based model represents an equilibrium model, and therefore cannot be used to detect year-to-year variation. From Figs 18.4, 18.5 and 18.6 it will be evident that year-to-year variation does occur, both in population numbers and in fishery mortality. A steady state model can only be used to predict long-term effects of changes in exploitation but cannot describe the history of a fishery.

Although the model presented is a truly length-structured model, it does include an ageing component: the growth of each size class should be specified beforehand. This is most likely done on the basis of field data, comprising age readings and back calculations. It should be noted that the model results are directly dependent on the growth estimates: increasing the assumed average growth rate increases the estimated mortality rates considerably. Thus, the current model cannot be applied where no growth estimates are present. In the case of the eel fishery, only rough growth estimates were available; analysis of the sensitivity of the results to the assumed growth rate is clearly needed.

As the bottom line, assessments were presented for three fisheries. For the eel fishery, no alternative age-structured model is possible, because ages can hardly be read. Partitioning of the length frequencies into age groups is practicably impossible, because age groups do overlap considerably. So the current model appears to be the only alternative. For the pikeperch and perch, the model did show the strong length selectivity in the gill net fishery, simulated by Buijse *et al.* (1992). Moreover, results showed some indication that the length selectivity has changed over the years.

All three fish species assessed appear to be so heavily over-exploited that the convergence property of cohort analysis models was not operative. Thus, the presented single species model is of limited relevance. However, it might serve as a building block for a more complex multispecies model, because the length selectivity of predation processes can be similarly treated in a straightforward way.

## References


Andersen K.P., Lassen H. & Ursin E. (1973) A multispecies extension to the Beverton & Holt assessment model with account of primary production. *ICES C.M. 1973/H:20.*

Anon. (1992) Report of the Multispecies Assessment Working Group, *ICES, Doc., C.M.* 1992/Assess:16.

Beverton R.J.H. & Holt S.J. (1957) On the dynamics of exploited fish populations. *Fisheries Investigations, London,* **19**.

Buijse A.D., Pet J.S., Densen W.L.T. van, Machiels M.A.M. & Rabbinge R. (1992) A size and age-structured simulation model for evaluating management strategies in a multispecies gill net fishery. *Fisheries Research* **13**, 95–117.

Caswell H. (1989) *Matrix Population Models.* Sunderland MA, Sinauer.

Dekker W. (1986a) Age reading of European eel using tetracycline labeled ototliths. *ICES C.M.* 1986/M:16.

Dekker W. (1986b) Regional variation in glass eel catches; an evaluation of multiple sampling sites. *Vie et Milieu* **36(4)**, 251–254.

Dekker W. (1987) Analysis of length frequency data by an ANOVA type model. *Paper presented at the EIFAC working party on eel.* Bristol, April 1987.

Dekker W. (1991) Assessment of the historical downfall of the IJsselmeer fisheries using anonymous inquiries for effort data. In: I.G. Cowx (ed.) *Catch Effort sampling strategies: their application in freshwater management.* Oxford: Fishing News Books, pp 233–240.

Dekker W. (1993) Assessment of eel fisheries using length-based cohort analysis; the IJsselmeer eel stock. *Paper presented at the EIFAC working party on eel*, Olsztyn, Poland, 24–27 May 1993.

Getz W.M. & Haight R.G. (1989) Population harvesting, Demographic models of Fish, Forest, and Animal Resources. Princeton University Press, Princeton, New Jersey.

Gonzalez R.C. & Wintz P. (1977) *Digital Image processing.* Massachusetts: Addison-Wesley Publishing Company, Reading.

Gulland J.A. (1965) Estimation of mortality rates. *Annex to Arctic Fisheries Working Group Report (meeting in Hamburg, January 1965). ICES C.M.* 1965, Doc. No 3.

Helgason T. & Gislason H. (1979) VPA-analysis with species interaction due to predation. *ICES C.M.* 1979/G:52.

Jones R. (1974) Assessing the long-term effects of changes in fishing effort and mesh size from length composition data. *ICES C.M.* 1974/F:33.

Pauly D. & Morgan G.R. (eds) (1987) Length based methods in fisheries research. *ICLARM Conference Proceedings,* **13**. ICLARM, Manilla, Philippines and KISR, Safat, Kuwait.

Pope J.G. (1972) An investigation of the accuracy of virtual population analysis using cohort analysis. *ICNAF Research Bulletin* **9**: 65–74.

Shepherd J.G. (1987) Towards a method for short-term forecasting of catch rates based on length compositions. *In*: D. Pauly & G.R. Morgan (eds) Length based methods in fisheries research. *ICLARM Conference Proceedings* 13. ICLARM, Manilla, Philippines and KISR, Safat, Kuwait, pp 167–176.

Willemsen J. (1977). Population Dynamics of Percids in Lake IJssel and some smaller Lakes in The Netherlands. *Journal of the Fisheries Research Board of Canada* **34**, 1710–1719.

# Chapter 19
# Uncertainty in fish stock assessment based on bottom trawl surveys in Lake IJsselmeer

A. D. BUIJSE *Rijkswaterstaat, Directorate IJsselmeergebied, P.O. Box 600, 8200 AP Lelystad, The Netherlands*

W. DEKKER *Netherlands Institute for Fisheries Research, P.O. Box 68, 1970 AB IJmuiden, The Netherlands.*

**Abstract**   Lake-wide bottom trawl surveys have been conducted in the 182 000 ha freshwater Lake IJsselmeer, The Netherlands, since 1966. The main species caught were smelt (*Osmerus eperlanus*), roach (*Rutilus rutilus*), bream (*Abramis brama*), ruffe (*Gymnocephalus cernua*), perch (*Perca fluviatilis*), pikeperch (*Stizostedion lucioperca*) and, recently, flounder (*Platichthys flesus*). The level of uncertainty in 0-group fish abundance and total stock biomass was evaluated. Since the historical data set did not provide any replicate sampling, this uncertainty was assessed by means of a GLM model (multiplicative model, Poisson error distribution). In the northern zone clear trends in the stocks of all species could be statistically traced. In the southern zone, however, catches were lower for all species associated with higher uncertainties which mask to a large extent annual variation. Uncertainty varied among the species and was lower for 0-group abundance than for total stock biomass estimates. Causes for year-class strength variations and developments in stock biomass are briefly discussed. The consequences for future monitoring programmes and for management are evaluated.

KEYWORDS: Cyprinids, flounder, monitoring, 0-group, percids, smelt

## 19.1    Introduction

*'Even under the best of circumstances, the estimation of fish abundance is an imprecise science. Fishery exploitation and management decisions should take account of this uncertainty.'* These are the concluding words of the extensive review by Sissenwine *et al.* (1983) on the use of bottom trawl surveys, ichthyoplankton surveys and acoustic surveys for determining the abundance of fish. In addition, they stated that

'Bottom trawl survey data are extremely useful for monitoring trends in abundance in spite of the fact that estimates are inherently inaccurate and imprecise. The results are inaccurate because bottom trawls are less than 100% efficient in capturing fish in their path, thus density is underestimated. The results are imprecise because of contagious distribution of the fish (spatial variability), variability in the efficiency of the bottom trawl as a sampler (measurement variability) and variability in environmental conditions between surveys (environmental variability).'

Byrne *et al.* (1981) give examples of factors affecting variability of research vessel trawl surveys (e.g. type of vessel, towing speed, wave height, water temperature and sampling period).

The monograph on bottom trawl surveys (Doubleday & Rivard 1981) addresses methodological aspects like sampling techniques and survey design, biological aspects in terms of distribution patterns and the impact of surveys on management advice.

The historical rationale for carrying out bottom trawl surveys in Lake IJsselmeer (Fig. 19.1) was for assessing the commercially exploited fish stocks. The surveys function merely as an indicator to keep a 'finger at the pulse'. Buijse (1992) gives a description of the history of the Lake IJsselmeer fishery with special attention to changes in the use of various types of commercial fishing gear. The development was characterized by a shift from active to passive gear.

At present, the fish stocks are exploited by means of passive fishing gear. The target species are eel (*Anguilla anguilla* (L.)), which are caught with fyke nets, eel boxes and long lines, perch (*Perca fluviatilis* L.) and pikeperch (*Stizostedion lucioperca* (L.)), caught with gill nets and spawning smelt (*Osmerus eperlanus* (L.)), caught with fyke nets. Other abundant species include roach (*Rutilus rutilus* (L.)), bream (*Abramis brama* (L.)), ruffe (*Gymnocephalus cernua* (L.)). Also, since a change in the operation of the sluices to the Wadden Sea in 1991, significant numbers of flounders (*Platichthys flesus* (L.)) migrate into the northern zone of the lake.

In recent years, more attention has been given to fish stocks as a component of complex ecosystems rather than solely as a renewable natural resource for maintaining a viable fishery. For example, in smaller lakes experiments have been conducted to establish whether manipulations of the fish stock are useful in integrated water management. Lake IJsselmeer is regarded as a wetland with a high international value. Bird species like smew (*Mergus albellus* L.), cormorant (*Phalacrocorax carbo* (Blumenbach), black tern (*Chlidonias niger* (L.)), and others exceed the 1% norm of the RAMSAR convention (Van Eerden & Bij De Vaate 1984), and are dependent on the size and composition of the fish stock in Lake IJsselmeer. Small planktivorous fish like smelt form the forage base for piscivorous birds (Buijse *et al.* 1993).

The management of Lake IJsselmeer is presently discussed among policy makers in terms of steering variables and target values of state variables (Iedema, *et al.* 1995). The commercial fishery as well as phosphate concentrations are regarded as potential steering variables, and biodiversity and 'naturalness' as state variables. The result of 'turning a steering variable' frequently does not lead to the expected change in the state variable, because steering variables are not the only factors involved, and due to the uncertainty in the interactions.

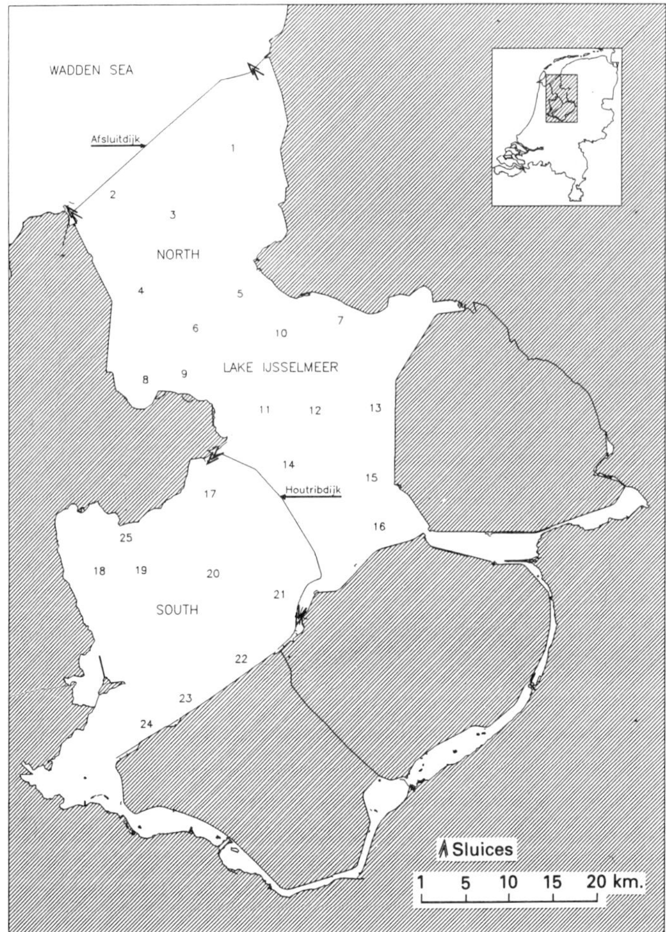


**Fig. 19.1**   Location of Lake IJsselmeer and of the sampling sites used during the trawl surveys in October or November.

Decisions to manipulate ecosystems are usually based on point estimations, disregarding uncertainty, derived from monitoring programmes. Often monitoring programmes cannot be expanded, because funding is limited. Therefore, scientists and managers have to make the best of the available data. Realizing this, Hickley & Starkie (1985) gave examples of effective sampling. However, they did not quantify the precision and accuracy of the survey data. As a start their approach is useful, but it is considered insufficient as a basis for management decisions, because it could also be based on 'best professional judgement' without further data analysis. If decision-making is to be based on data analysis, confidence limits of the point estimates are required, because these give insight in the probabilities that predicted changes will take place and also will be detectable.

Since 1966, fish stocks in Lake IJsselmeer have been assessed by means of fishery-independent surveys. The bottom trawl surveys in October/November have been consistently conducted in all years, although the number of stations varied. This chapter evaluates these surveys in terms of 'least significance difference', i.e. the minimal change in the system that could still be detected by the survey.


## 19.2    Material and methods

A description of the chemical and physical characteristics of Lake IJsselmeer, a shallow eutrophic 182 000 ha freshwater lake (52°45′N, 5°20′E), is given in Willemsen (1977) and Buijse *et al.* (1992). Since 1932, the lake has been separated from the Wadden Sea by a dam containing sluices ('Afsluitdijk'). The lake was divided in two zones by construction of a dam ('Houtribdijk'). Construction started as early as 1963; 22% was completed in 1966, 46% in 1968, 74% in 1972, and the project was finished in 1975. The two zones are connected by sluices. The lake-wide survey with a bottom-trawl was described by Buijse *et al.* (1992). Sixteen sampling sites were located in the northern zone and nine in the southern zone (Fig. 19.1). Not every site was sampled in every year.

The species composition by weight and the size compositions by species were determined by trawl. This gear was unsuitable for eels and only results for the other abundant species are considered. For bream, roach, pikeperch, perch and flounder an index of recruitment, i.e. 0-group abundance, was calculated for the northern and the southern zone separately. 0-group and older fish could be readily distinguished on the basis of the length frequency distributions (LF), because they showed no overlap. For smelt and ruffe, a separation into age-groups was not possible, because often only weights were recorded and an LF-distribution was not available for all sites.

The year-to-year variation in total stock biomass was assessed on the basis of the catches in the surveys. The total weight of the catch per species was estimated from the LF-distributions using length-weight relationships, again with the exception of smelt and ruffe, for which species the weight of the total catch was used. Haul duration varied between 10 and 45 minutes. Catches are expressed per hour trawling, which corresponded to an area fished of 4.8 ha (Buijse *et al.* 1992).

The sampling strategy was different in the years 1987–1989, which may have resulted in generally lower catches. With increasing water clarity, the efficiency of the bottom trawl decreased significantly for pikeperch and ruffe (Buijse *et al.* 1992). Prior to 1987 fishing was only conducted under favourable sampling conditions when water clarity was less than 0.8 m. From 1987 up to and including 1989 the time available for the surveys was severely restricted and therefore sampling conditions had to be taken for granted. Since 1990, a longer period was reserved to conduct the surveys so favourable conditions could again be awaited.

### 19.2.1    *Statistical analysis*

The catches were analysed by an Analysis of Variance-type model. Both numbers of 0-group and total weight of the catch were analysed. The ANOVA-model used was identical for both dependent variables and all species, although some terms in the model were not significant in some cases since varying the model would have made results incomparable. However, none of the species/dependent variables combinations showed marked departures from the assumptions underlying the common ANOVA-model.

The ANOVA model selected belongs to the class of Generalized Linear Models (McCullagh & Nelder, 1983), and was analysed using the GLIM-package (Baker & Nelder, 1978). Standard (Normal distribution) statistics were based on a number of assumptions which in practice were rarely met. McCullagh and Nelder (1983) gave the example of a dependent variable, being a natural count, which:

(1)   should be transformed to $\sqrt{y}$ to make the residual variance independent of the predicted value,
(2)   should be transformed to $y^{2/3}$ to normalize the residual variance, and
(3)   should be transformed to $\log(y)$ to produce an additive linear model.

Clearly, no single transformation can meet these three requirements at the same time. Using GLIM, the assumptions were uncoupled, in the sense that a link function took care of the additivity of the predictive model, while an independent variance model produced the appropriate error structure. The goodness of fit was measured by the Deviance (twice the difference of the log likelihood of the data themselves and the data given by the model), which replaces the sums of squares of the Normal distribution models.

In addition, zero observations were handled explicitly. The frequency of zero observations ranged from 12 (perch) to 60% (bream) in 0-group numbers, and 1 (perch) to 68% (flounder) in total catch weight. Inspection of the geographical and temporal distribution of the zero observations suggested that these zeroes were not produced by sampling outside the distribution area of a species, but were produced by chance when sampling a sparse density. Stefánsson (1991) analysed groundfish survey data by explicitly handling the zero observations in a binomial model, assuming a gamma error distribution for the positive catches. However, this procedure was based on the assumption that at least part of the zero observations were the result of limited distribution of the fish, which appeared not to be the case in this study. Thus, the analysis was limited to a multiplicative model with a Poisson error structure, and the error structure was checked *a posteriori*. The independent, explanatory variables included were: the year of sampling, the site of sampling, separate year effects for the two zones sampled, and the ratio of total water depth to Secchi depth (Buijse *et al.* 1992). Thus, the basic model reads for each observation $I$ was:

$$Y_i = \exp[a^*TS + zone(j) \, . \, year(k) + year(k) + site(l)] + \epsilon_I \tag{1}$$

where $Y$ is the catch, either in number of 0-group per hour, or in total weight per hour; $TS$ the ratio total water depth : Secchi depth; $a$ the regression coefficient of $TS$; $j$ the index of the zone (northern and southern respectively); $comp(j)$ the regression coefficient of the zone; $k$ the index of the year (1966–1993); $year(k)$ the regression coefficient of the year; $l$ the index of the sampling site (1–24); $site(l)$ the regression coefficient of the site and $\varepsilon_i$ the error term, having a Poisson distribution.

The results are presented in terms of the percentage variance explained. The explanatory variables are inorthogonal; the overlap in explained deviance is listed explicitly as 'multicollinearity'.

Model checking included graphical inspection of the residuals as function of the predicted values, of the normal probability plots of the residuals, and of the predicted values as a function of the observed values (link function); residuals were calculated as deviance residuals, being in line with the model fitting procedure. No obvious departure from the assumptions was detected. There was a slight tendency of the residuals to diminish at the larger predicted values, thus contradicting that a gamma error would exceed a Poisson error.

The error structure assumed was a Poisson distribution, characterized by a constant ratio of mean to variance, $\phi$. This scaling parameter $\phi$ was estimated from the residual deviance of the model of (1), as $\hat{\phi} = D / (n{-}r)$ (McCullagh & Nelder 1983), where $D$ is the deviance; $n$ is degrees of freedom of an empty model; and $r$ the degrees of freedom taken by the model of Equation (1).

The prime interest of the analysis was to assess the signal-to-noise-ratio of the survey data. The signal was estimated from (1), while the noise was assessed from the residual variance. Under the Normal distribution theory, the uncertainty of an estimate can be expressed as a Least Significant Difference LSD (see, for instance, Sokal & Rohlf 1981), this being the least difference between parameter estimates that is still significant (e.g. the least difference between year-classes that could still be detected). This LSD is only a function of the residual variance and the number of data points. However, using a GLIM model, the significance of a difference is no longer independent of the absolute value of the parameter estimates, nor of their ratio. Following the line of reasoning behind the LSD and the significance testing of GLIM models, it was derived that just significantly different parameter values should satisfy:

$$\mu_1 \times {}^e\!\log(\mu_1 / \mu_2) + (\mu_2 - \mu_1) = \hat{\phi} * F_{v_1,1:\infty} / v_1 \tag{2}$$

in which $\mu_1$ is the maximum likelihood estimator, and $\mu_2$ the just significantly different value. Since this equation cannot be solved analytically, Newton-Raphson approximation was used. The right hand side of Equation (2) only depends on $v_1$, i.e. the number of samples involved in the test. The average number of samples per year was 10 in the northern and 5 in the southern zone, respectively. Thus results are presented for these numbers of observations. The left hand side of Equation (2) can be interpreted as the relative difference between $\mu_1$ and $\mu_2$ (log of their ratio), plus their absolute difference, the first being weighted by the absolute value of $\mu_1$. Consequently, for large expected values, confidence limits will be approximately proportional to the expected value, while for small ones the confidence interval will have a more or less absolute width. In this context, 'large' and 'small' were determined by $\phi$ in the right hand side, which was estimated from Equation (1).

Since there was only one observation per year and site, no interaction term *year.site* could be tested. Consequently, the *year.site* interaction was interpreted as arising from stochastic variation only.

Data from years before 1976 (i.e. before the division into zones was completed) were given a prior weight of 0.001, so that station effects were estimated almost exclusively based on the later years, while year-class strengths before 1976 were conditional on these station effects.

### 19.3    Results

The total number of sites that was sampled varied considerably over the years (between 4 and 24; Fig. 19.2). In the northern zone the range was between 3 and 16 and in the southern zone between 1 and 9. The number of sampling sites increased over the years.

The model explained 50 to 84% of the variation in the catches (Table 19.1). Not surprisingly, year explained most of the variation in 0-group abundance (Fig. 19.3).

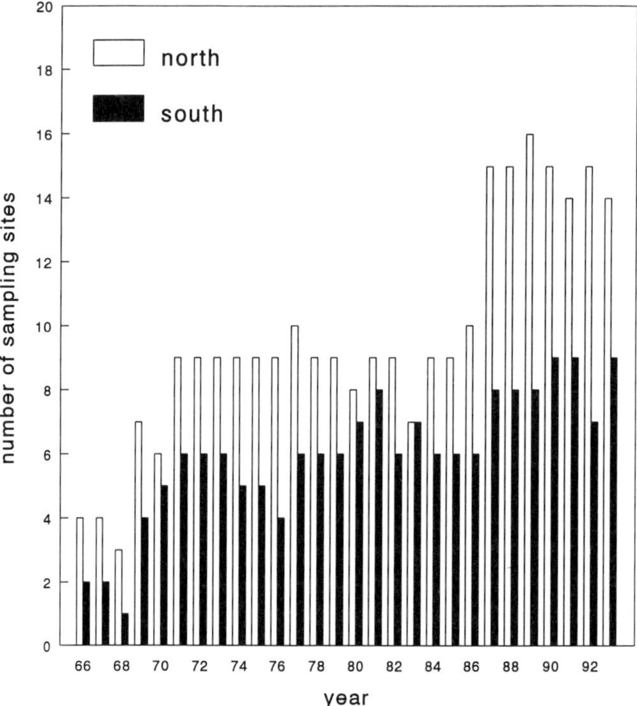


**Fig. 19.2**  Number of sampling sites per year for the northern and southern zone.

**Table 19.1**  Percentage explained variance of variables which contributed significantly to the model.

|  | *TS* | Zone.year | Site | Year | Multi-collinearity | Total explained | $\phi$ |
|---|---|---|---|---|---|---|---|
| *0-group* | | | | | | | |
| Bream | 1 | 2 | 13 | 56 | 11 | 84 | 75 |
| Roach | n.s. | 8 | 19 | 54 | 3 | 84 | 400 |
| Perch | 0 | 7 | 14 | 54 | 0 | 76 | 3488 |
| Pikeperch | 4 | 4 | 11 | 45 | 19 | 84 | 61 |
| Flounder | 1 | n.s. | 30 | 37 | 13 | 80 | 10 |
| *Stock biomass* | | | | | | | |
| Smelt | 2 | 7 | 19 | 21 | 6 | 55 | 41 |
| Bream | n.s. | n.s. | 13 | 15 | 19 | 50 | 109 |
| Roach | n.s. | n.s. | 16 | 11 | 32 | 61 | 53 |
| Ruffe | 0 | 8 | 9 | 28 | 10 | 56 | 48 |
| Perch | n.s. | 10 | 15 | 22 | 2 | 48 | 87 |
| Pikeperch | 2 | 4 | 21 | 34 | 20 | 81 | 8 |
| Flounder | 1 | n.s. | 16 | 30 | 23 | 70 | 2 |

*TS* = total water depth: Secchi depth ratio;
$\phi$ = scaling parameter (see text);
n.s. = not significant.

The total water depth: Secchi depth ratio (*TS*) was highly significant, but contributed little to the explanation of the variation. The high value of 81% explained by the model for pikeperch stock biomass may partly be due to the fact that 0-group contributes more than for the other species. Percentage contribution of 0-group, expressed as weight to the total biomass, was 2% for bream, 4% for roach, 12% for perch, 28% for pikeperch and 12% for flounder. A significant *comp.year* interaction indicated that the stocks in the two zones behaved differently, which could, for example, reflect that they do not produce strong year-classes in the same year. However, this chapter focuses on the ability to discriminate between years; no further attention is given to the interpretation of *comp.year* interaction, *TS* and site.

A short characterization of all six species is given by the trends observed in the development in stock biomass and the year-class strength variation (YCSV). The comments partly refer to significant differences, but also describe the observations at *prima facie*. The error bars depict 95% confidence limits based on the average number of sites sampled per year (northern zone: 10; southern zone: 5).

The development in the fish stocks in the northern and southern zones is described by the estimates made for sites 9 and 19 respectively (Fig. 19.1). As shown in Table 19.1, site is significant for all species, i.e. catches vary among sites between 0-group abundance and stock biomass, and between species. However, this bears no consequence for the trends given, although it influences the confidence limits: sites with lower catches have more uncertain estimates. In general, site 9 selected from the northern zone was rich for most fish species compared to other sites, whereas site 19 in the southern zone yielded moderate catches in comparison with other sites.

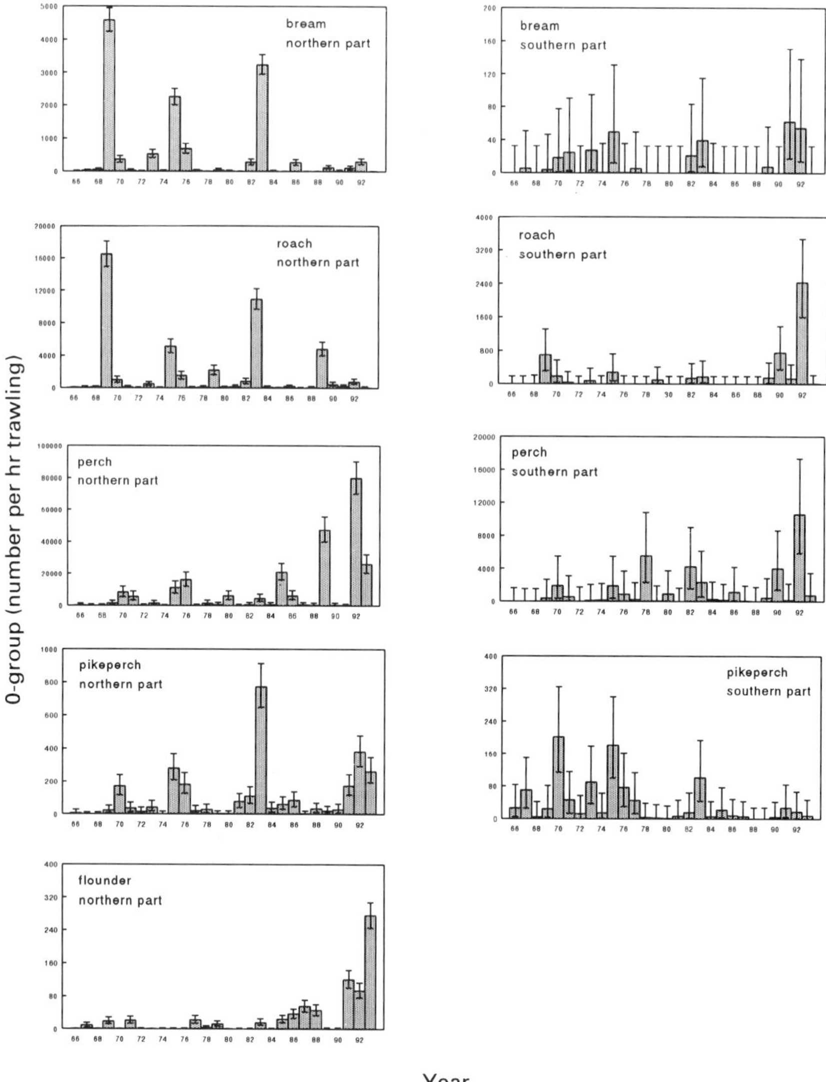



**Fig. 19.3** Fitted values of Equation (1) estimates of 0-group abundance per species, year, and lake zone caught per hour trawling. Error bars are 95% confidence limits. Note differing vertical scales between the northern and southern zone and between species.

### 19.3.1    *Smelt*

The smelt stock has fluctuated extensively over time, which is probably related to its short life; it comprises two year-classes only. In the northern zone, smelt seems to have recovered at the beginning of the 1970s, possibly in relation to the ban on commercial trawling in 1970, to reach its highest biomass in 1973–74 (Fig. 19.4). In recent years the smelt stock declined to an all time low in 1992, with a small recovery

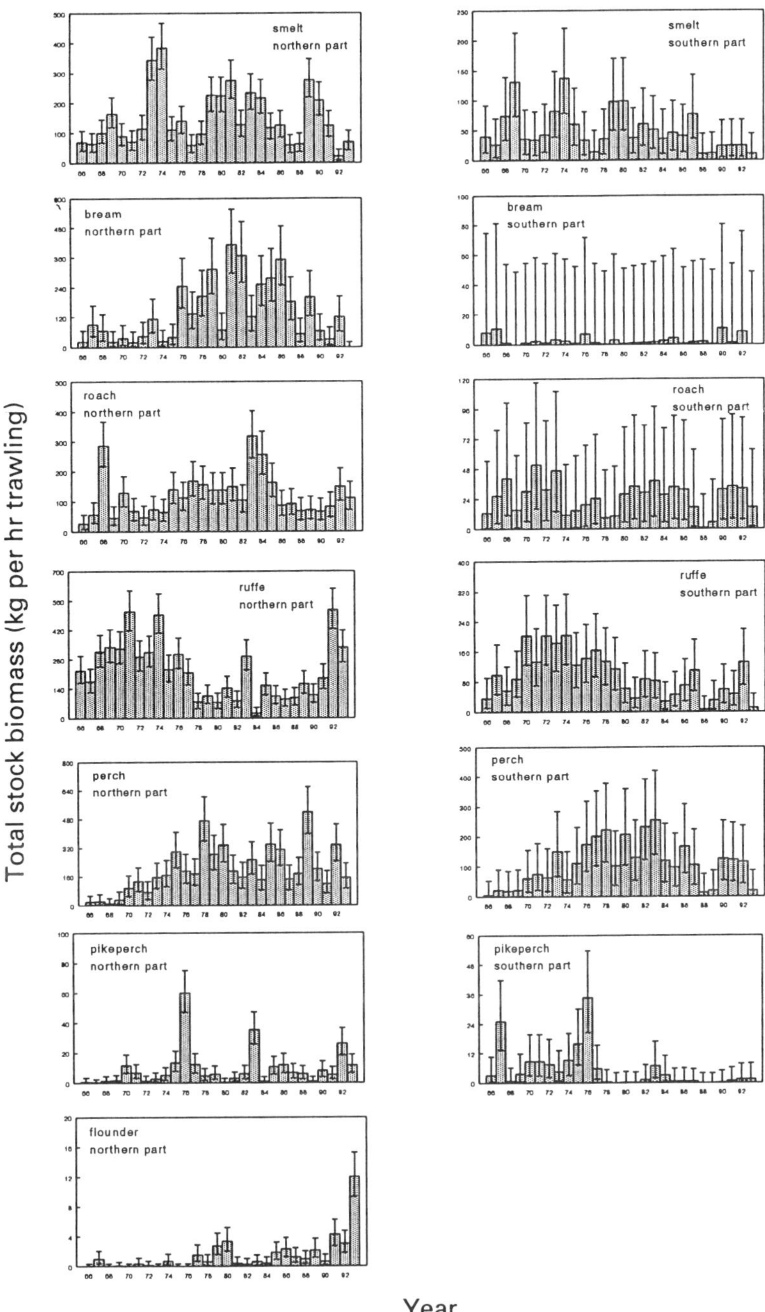

**Fig. 19.4**   Fitted values of Equation (1) estimates of mean biomass per species, year, and lake zone caught per hour trawling. Error bars are 95% confidence limits. Note differing vertical scales between the northern and southern zone and between species.

in 1993. Smelt biomass was also at its highest in 1974 in the southern zone, but peaked at the end of the 1960s, apparently not influenced by the commercial trawl fishery. Based on the analysis little can be said about the development of the smelt stock in the southern zone. After 1988 smelt biomass was lower than the long-term average.

### 19.3.2    *Bream*

The abundance of 0-group bream was much lower in the southern zone. In the northern zone all strong year-classes were significantly different from each other and recruitment was extremely variable. In the southern zone, however, year-classes did not differ much. Strong year-classes of bream were usually observed in years with warm summers, e.g. 1969, 1975 and 1983 (Figs 19.3 and 19.5), although high temperatures do not guarantee good recruitment. Of the 11 years with a sum of degree-days over 14°C significantly above average, only three produced very strong year-classes in the northern zone. The stock in the northern zone was relatively low before 1970. After the ban on commercial trawling in 1970 the stock started to increase, due

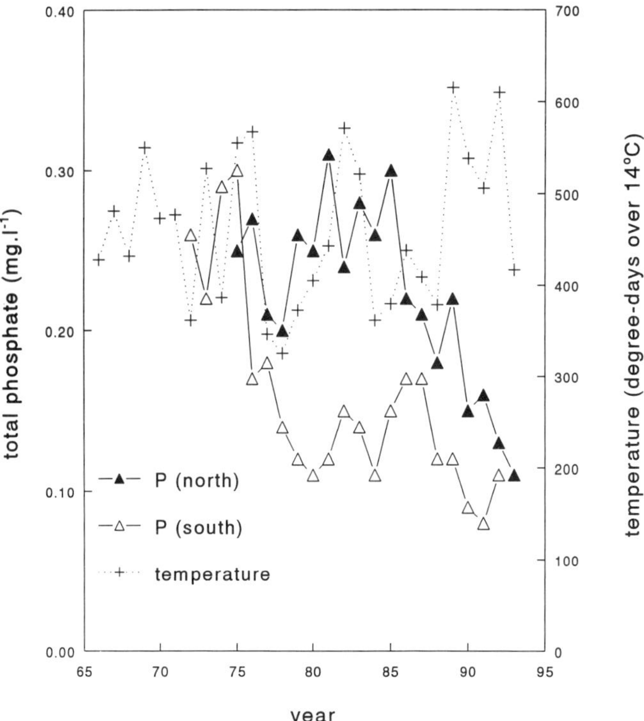


**Fig. 19.5**  Total phosphate level in the northern and southern zone and water temperature of Lake IJsselmeer. Water temperature is expressed as the sum of degree-days over 14°C. Phosphate levels are measured at station M108 for 1972–81, M111 for 1982–1993 in the southern zone and at station IJ23 in the northern zone. Station codes are from Rijkswaterstaat.

to the strong year-classes of 1969 and 1975, to reach its highest level in 1981. The low value in 1980 was probably erroneous because the stock increased as fast as it declined, which is unrealistic for a population consisting of several age-groups. After 1986 the stock gradually declined to the level prior to 1975. Possible explanations for the decline are the absence of strong year-classes since 1983, the intensification of the gill net fishery or the decrease of phosphate loading (Fig. 19.5). Similar to 0-group abundance, stock biomass of bream was much lower in the southern than in the northern zone. No trend was observed in this stock, since annual variation was masked by the noise in the observation.

### 19.3.3    *Roach*

Roach also produced strong year-classes during warm summers, and the pattern largely followed that of bream. Again, the YCSV was extreme in the northern zone (Fig. 19.3). In the southern zone only the three strongest year-classes differed significantly from all others, but these did not match the pattern observed in the northern zone. In the latter, the stock biomass increased gradually from the early 1970s to reach its highest biomass in 1983, followed by a decline to the original level. The high value in 1968 appears to be an outlier. The abundance in the southern zone is much lower and possible trends are masked by the very high confidence limits of the individual estimates.

### 19.3.4    *Ruffe*

The ruffe stock showed an increase in both zones during the first five years after the ban on commercial trawling in 1970 (Fig. 19.4). After the mid-1970s there was a gradual decline except for 1983. This decline might be due to the increase in the number of fyke nets, which yielded a high by-catch of ruffe (Dekker *et al.* 1993). The stock of ruffe increased again in the northern zone after 1990 for unknown reasons, although the nominal fishing effort of fyke nets has been reduced, the effective effort has not (Dekker 1994). Stock biomass in the southern zone was twice as high during the 1970s than during the 1980s. The stock has not recovered in recent years.

### 19.3.5    *Perch*

Factors affecting the YCSV of perch and pikeperch were analysed by Buijse (1992). For both species, strong year-classes were favoured by high summer temperatures and low temperatures in April. Stock-recruitment relationships could not be demonstrated. In recent years (1989, 1992) perch yielded exceptionally strong year-classes in the northern zone, which coincided with the warmest summers during the entire period (Figs 19.3, 19.5). In the southern zone only seven were significantly more abundant than the others. The perch stock showed a clear example of recovery after the ban on commercial trawling in 1970 (Fig. 19.4). The stock increased in the northern zone until 1975, then it varied without obvious trend. In the southern zone the stock initially followed the same pattern, but there was a decline after the mid-1980s.

For most species (smelt, roach, ruffe and perch), exceptional low catches were made in the southern zone in 1988 and 1989. This might be erroneous due to poor sampling conditions as a consequence of very clear water during a period of calm weather.

### 19.3.6　*Pikeperch*

[See also perch for causes of YCSV.] 0-group abundance was high in years with warm summers (Figs 19.3, 19.5). Strong year-classes were observed in recent years only in the northern zone. Only three moderate and no strong year-classes were produced in the southern zone after 1975. Since 1983 all YCS have been weak. It seems possible that the division into two zones, which was completed in 1975, resulted in unfavourable conditions for pikeperch in the southern zone. However, this explanation is in itself unexpected, because pikeperch appear to favour turbid waters and the southern zone is more turbid than the northern zone. Unlike other species, stock biomass showed rather similar variations as 0-group abundance (Fig. 19.4). This is probably related to the intensive commercial fishery which exploited the pikeperch from 2 years old. Therefore the stock consists almost exclusively of immature fish and a strong year-class is immediately reflected as an increase in stock biomass.

### 19.3.7　*Flounder*

Flounder is a catadromous species. The dam and its sluices form an artificial barrier between the freshwater lake and the marine Wadden Sea and therefore hamper migration of flounder. Since 1991 the management of the sluices has changed by opening the sluices earlier during ebbing tide for draining off fresh water from the lake to facilitate the migration of sea trout (*Salmo trutta* L.). This procedure facilitates the migration of 0-group flounder and must be held responsible for the increased abundance during recent years. In the southern zone flounder has not been caught.

### 19.3.8　*Uncertainty in abundance estimates as a function of the number of sites sampled*

The 95% confidence limits around the long-term arithmetic mean at the two sampling sites for 0-group abundance and estimates of stock biomass in the two zones are plotted as a function of the number of sampling sites in Figs 19.6 and 19.7 respectively. The confidence limits were wider for the site in the southern zone as a consequence of the lower abundance of fish. Confidence limits are in general also wider for stock biomass than for 0-group abundance. On the basis of these graphs it can be decided how many sampling sites should be visited to meet the maximum uncertainty required for management. However, it must be noted that uncertainty is a 'function' of species and abundance. To illustrate this, a section of Figs 19.6 and 19.7 is given in Table 19.2 listing the confidence limits around 0-group abundance and stock biomass if 10 sites are sampled in both zones.

In the northern zone 0-group abundance increased as follows: flounder < pike-perch < bream < roach < perch, whereas uncertainty decreased from pikeperch > perch > roach > flounder > bream. Thus, although numbers were lower than for roach and perch more certainty is known about YCSV of bream and immigrated

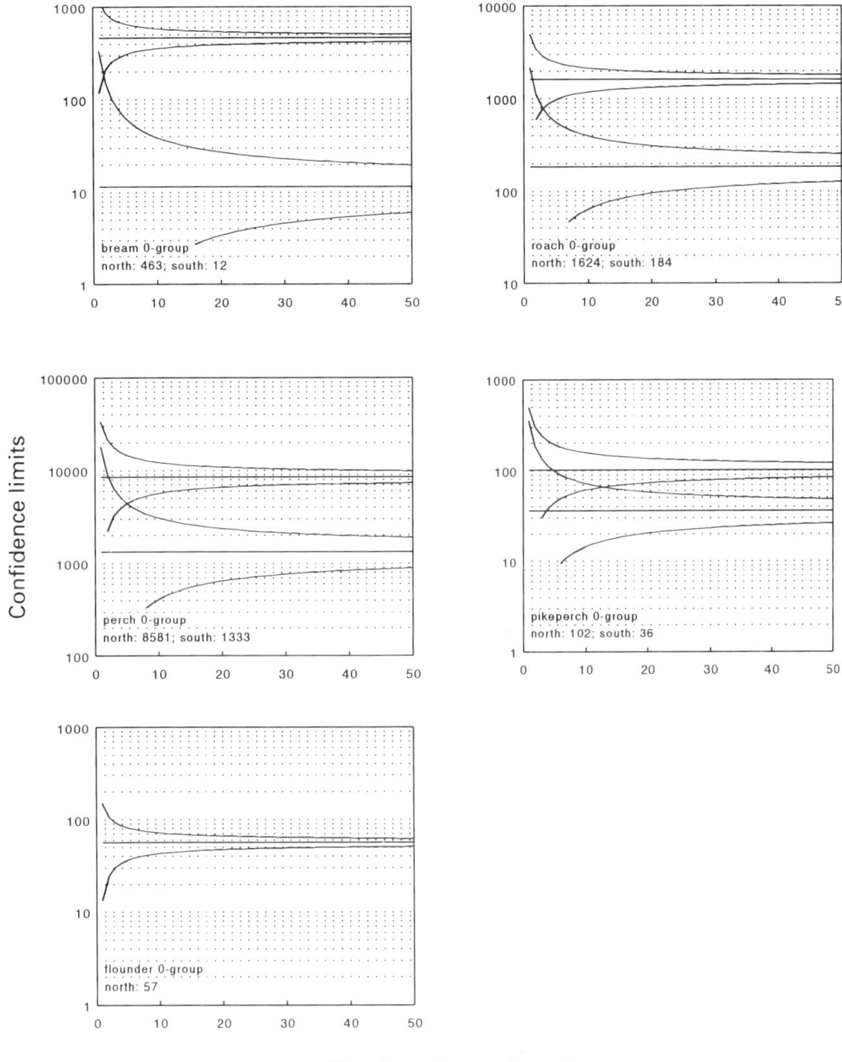


**Fig. 19.6** 95% confidence limits of the estimate of 0-group abundance per species as a function of the number of sampling sites. Confidence limits were calculated for the arithmetic mean over all years at site 9 (north) and at site 19 (south). Note differing vertical scales.

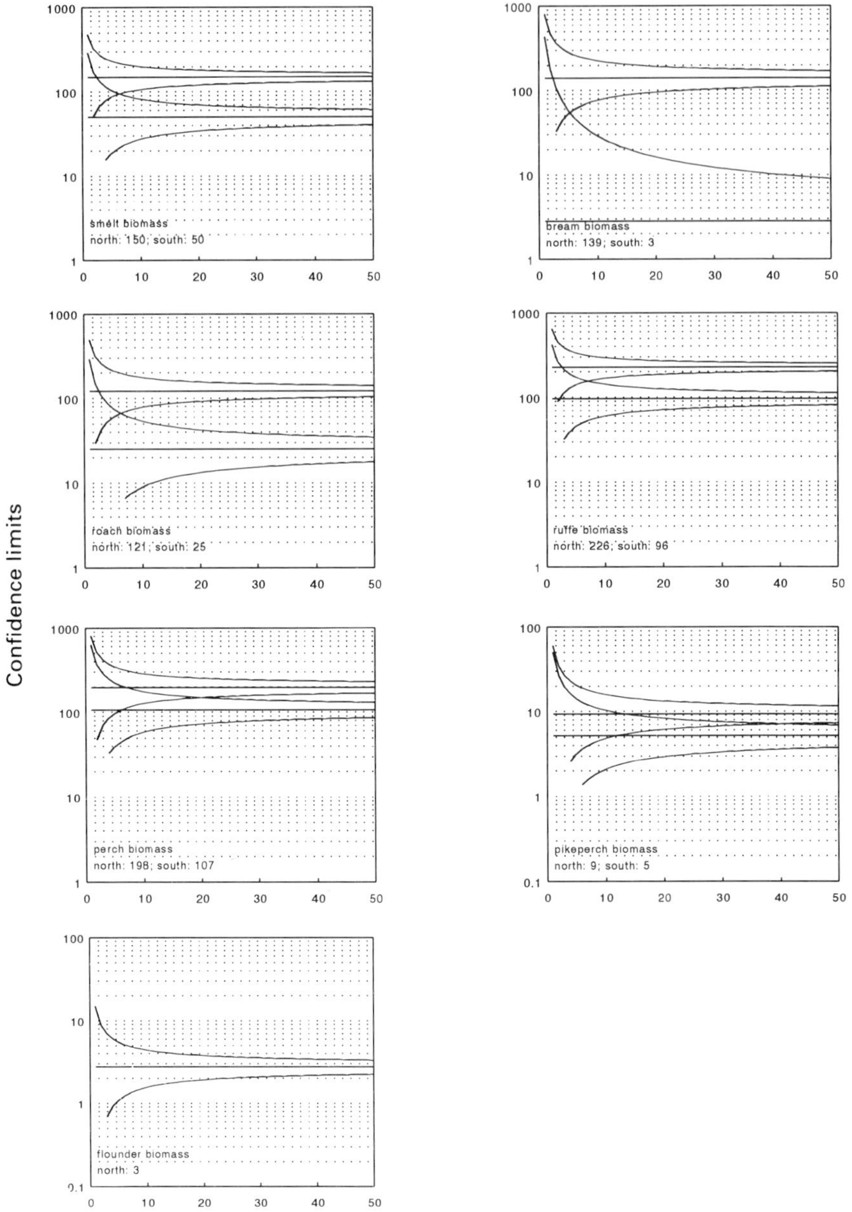



**Fig. 19.7**   95% confidence limits of the estimate of stock biomass per species as a function of the number of sampling sites. Confidence limits were calculated for the arithmetic mean over all years at site 9 (north) and at site 19 (south). Note differing vertical scales.

**Table 19.2**  0-group abundance and stock biomass per species and lake zone. Mean = arithmetic mean over all years. Percentages give lower (L%) and upper (U%) 95% confidence limits and were respectively calculated by 100 × (lower 95% confidence limit/mean) or 100 × (upper 95% confidence limit/mean). Confidence limits are based on 10 sampling sites.

| | North (site 9) | | | | | | South (site 19) | | | | | |
| | 0-group | | | Biomass | | | 0-group | | | Biomass | | |
| | L% | Mean | U% | L% | Mean | U% | L% | Mean | U% | L% | Mean | U% |
|---|---|---|---|---|---|---|---|---|---|---|---|---|
| Smelt | — | — | — | 72 | 150 | 135 | — | — | — | 55 | 50 | 165 |
| Bream | 78 | 463 | 126 | 55 | 139 | 164 | 0 | 12 | 342 | 0 | 3 | 1050 |
| Roach | 73 | 1624 | 133 | 65 | 121 | 145 | 35 | 184 | 218 | 36 | 25 | 215 |
| Ruffe | — | — | — | 75 | 226 | 131 | — | — | — | 63 | 96 | 149 |
| Perch | 66 | 8581 | 144 | 65 | 198 | 146 | 31 | 1333 | 232 | 55 | 107 | 165 |
| Pikeperch | 60 | 102 | 155 | 52 | 9 | 171 | 40 | 36 | 201 | 40 | 5 | 201 |
| Flounder | 77 | 57 | 128 | 57 | 3 | 160 | — | — | — | — | — | — |

numbers of 0-group flounder. In the southern zone 0-group abundance increased from bream < pikeperch < roach < perch. Uncertainty decreased from bream > perch > roach > pikeperch. Both lakes therefore showed that there was comparatively large uncertainty in the estimate of year-class strength of perch despite its being the most abundant species.

In the northern zone, stock biomass increased from flounder < pikeperch < roach < bream < smelt < perch < ruffe. The uncertainty decreased from pikeperch > bream > flounder > roach > perch > smelt > ruffe. Thus, where 0-group abundance of bream was determined with little uncertainty, there was large uncertainty around the estimate of stock biomass. In the southern zone stock biomass increased from bream < pikeperch < roach < smelt < ruffe < perch. The uncertainty decreased from bream > roach > pikeperch > perch, smelt > ruffe. Thus most certainty is known about the stocks of fish species which are in greatest abundance.

## 19.4   Discussion

The Lake IJsselmeer coarse fish surveys were set up in the middle of the 1960s. A detailed account of the rationale for the survey design was not handed down, but the catching of marketable fish (i.e. a direct monitoring of the prospects of the commercial fishery) was presumably the major aim. Such an approach would be in line with, e.g. Leaman (1981), who suggested no sampling at sites where there were few fish, and Scott (1981), who suggested that surveys should be optimized for certain species. At that time, statistical analysis of the data was minimal. Moreover, the originators were not aware of variances in survey results: catch rates were perceived as representing the absolute truth. No allowance was made for replicate samples.

In this chapter, interest focused on the uncertainty in the survey-based stock assessment. This means that the aim was to elucidate aspects which were explicitly not incorporated in the survey design. Since no replicates were available, no direct estimates of the variance could be derived. As an alternative, the variances in the catches were assumed to be generated by a single, Poisson-type stochastic process, and the expected catch was specified as a simple, multiplicative model. These two assumptions were (for the major part) explicitly tested. The most important conclusion of the analysis was that – despite the 'thoughtless' survey design – the data set exhibited a quite regular statistical behaviour, allowing for an analysis of uncertainty in the estimates. Thus, even without having access to data on replicate sampling, the careful compilation of a statistical model can, to a large extent, compensate for shortcomings in survey design.

The historical trawl survey data were not optimal. Leaman (1981) discussed survey design in relation to the collection of more precise and accurate estimates. He distinguished four types of survey: random, stratified random, systematic and encounter-response. The survey in Lake IJsselmeer can be regarded as a systematic survey. It was possible to discriminate between strong and weak year-classes of all fish species in the northern zone and between their fluctuations in stock biomass. However, hardly any discrimination between years regarding stock biomass was possible for the southern zone, which was sampled on average at only five stations per year. In particular, the bottom trawl surveys provided no information on stock fluctuations for the stocks of bream and roach. Since the northern zone showed significant changes in the stock biomass, changes in the southern zone would likewise be expected. It was considered a methodological problem and not a biological phenomenon that it was not possible to discern between years. To resolve the problem the number of sites in the southern zone should be expanded to at least 10 sites.

The analysis presented appears to give a reliable estimate of the order of magnitude of uncertainty in 0-group abundance and stock biomass for the period 1966–1993. An underlying assumption was that catches at different, lake-wide sites reflected the uncertainty in the abundance estimates. In this analysis the results were interpreted as if there were 10 independent catches. Distribution patterns were probably related to bottom types and depth (spatial variability). The fixed location of the sites may have resulted in an overestimation of the precision, since most of them were, at least in the northern zone, located in depressions which represent relics of former tidal streams. If the greater part of the fish stock was found in autumn in these depressions then the estimate of uncertainty can be considered representative for areas with high concentrations of fish and thus for the total stock. To remain on the conservative side no attempt was made to extrapolate the results to a total lake estimate but the analysis was restricted to one site with relatively high abundance in each zone.

Coles *et al.* (1985) analysed survey data originating from seining, electrofishing and trawling for their efficiency and uncertainty as a basis for fisheries management in Anglian Waters. They concluded that the highest uncertainty came from contagious distributions, especially during winter. Bream was the most clumped species. Therefore, Jordan & Wortley (1985) advised that sampling took place in summer when fish are maximally dispersed. Agreement, in principle, that surveys should be conducted when fish are most randomly distributed, resulting in low spatial variability, is an output of this study. However, in Lake IJsselmeer most fish migrate to deeper areas when temperature drops in autumn and it was these areas that were generally sampled. To extrapolate these results to estimates of total stock biomass a stratified sampling scheme would be required, but this would result in measurement variability, because the bottom trawl is less efficient in shallow areas due to earlier detection of the trawl and vessel by fish.

Scott (1981) addressed criteria for stratified sampling. Each species is characterized by a specific and distinctive distribution pattern, which is often thought to be related to depth, temperature and bottom type. He suggested sampling sites where fish are abundant, using information from commercial fishermen on distribution patterns, and to select for certain species for which the survey should be optimized. Koeller (1981) argued that surveys can never be optimized for all species simultaneously. The variance associated with spatial variability may be reduced by refined procedures for stratification and station allocation, but the scope for improvements is probably limited for multi-species surveys.

Measurement variability can be reduced by standardisation of sampling procedures. Unfortunately, the magnitude of the major source of environmental variability in survey results is difficult to estimate (Byrne *et al.* 1981). In Lake IJsselmeer, water temperature and water clarity in October and November appear to be major causes of year-to-year variations in fish distribution and catchability (Buijse *et al.* 1992).

Cyr *et al.* (1992) analysed the relationship between the variance and the expected value of larval fish surveys, based on recent literature data. However, their analysis did not take into account that the variance of survey data comprised aspects of the sampled fish population as well as aspects of the survey design. A high degree of aggregation, as found in schooling fish, and very small samples sizes will result in high variances. In this study, data refer to one survey only. The number per species varied, and consequently subsampling percentages differed among species. Therefore, the sampling procedure was not really constant. This bears the consequence that, in interpreting the results, one cannot distinguish between uncertainty generated by sampling design, and uncertainty inherent on the population sampled. Although it was not possible to discriminate between measurement and spatial variability, a large part of the observed differences between species was probably due to distribution patterns. The generalizations given in Cyr *et al.* (1992) neglected such species-specific behaviour and therefore should be applied with great care.

The present study showed that precision in estimates of 0-group abundance and stock biomass varied between species. Such information can be used in uncertainty analysis of ecosystems models. Rivard (1981) focused on age-related contagious distribution patterns for cod (*Gadus morhua* L.) and the consequences for the relative error in abundance estimates. The uncertainty was highest for age 1, relatively constant for age 2–9, but increased for older age-groups, due to low numbers caught. In this study, an analysis for abundance of 0-group bream based on the point estimate could suffice, because YCSV over-ruled all measurement variability, while an analysis of stock biomass clearly should also take the uncertainty into account.

The question of what level of uncertainty is acceptable must be answered by managers and user-groups (Hilborn *et al.* 1993). However, uncertainty must be taken into account by scientists before they embark on more complex models. Leaman (1981) drew attention to differences in interest between researchers and managers. Scientists try to optimize surveys to reduce uncertainty, while managers are often only interested in point estimates forgetting aspects of precision. However, the expected effectiveness of management measures depends strongly on the associated uncertainty of scientific information. Within the Netherlands, the ecological state of large water bodies is presently being characterized by the difference between the present stock size and the desired stock size of key species including pikeperch, bream and smelt (the so-called AMOEBE approach: Vanhemelrijk *et al.* 1993). In a target state, stock biomass is allowed to fluctuate between 75% and 200% of the desired level. The analysis showed that only the smelt survey estimates were sufficiently precise for such a conclusion to be drawn. On the basis of this uncertainty analysis, it becomes obvious to managers whether the data collection for fish is sufficiently precise to be used in an evaluation of the expected response of such key species to proposed management measures.

## Acknowledgements

The fieldwork of all those who were involved in the sampling programmes over the years is highly appreciated; especially that of the crew of research vessel *De Stern*, who were in charge of the actual trawling, and that of Leo Schaap and Jan van Willigen, who, for many years, were responsible for the co-ordination of the surveys. Prof. Dr Niels Daan critically reviewed the manuscript.

## References

Baker R.J. & Nelder J.A. (1978) The GLIM system, release 3. Royal Statistical Society, Oxford, mimeo.

Buijse A.D. (1992) Dynamics and exploitation of unstable percid populations. PhD thesis. Wageningen Agricultural University. Wageningen, The Netherlands. ISBN 90-5485-001-9.

Buijse A.D., Schaap L.A. & Bult T.P. (1992) Influence of water clarity on the catchability of six freshwater fish species in bottom trawls. *Canadian Journal of Fisheries and Aquatic Sciences* **49**, 885–893.

Buijse A.D., Van Eerden M.R., Dekker W. & Van Densen W.L.T. (1993) Elements of a trophic model for IJsselmeer (The Netherlands), a shallow eutrophic lake. In: V. Christensen & D. Pauly (eds) *Trophic models of aquatic ecosystems*. ICLARM Conference Proceedings **26**, pp 90–94.

Byrne C.J., Azarovitz T.R. & Sissenwine M.P. (1981) Factors affecting variability of research vessel trawl surveys. In: W.G. Doubleday & D. Rivard, (eds) Bottom trawl surveys. *Canadian Special Publications of Fisheries and Aquatic Sciences* **58**, 258–273.

Coles T.F., Wortley J.S. & Noble P. (1985) Survey methodology for fish population assessment within Anglian water. *Journal of Fish Biology* **27 (Supplement A)**, 175–186.

Cyr H., Downing J.A., Lalonde S., Baines S.B. & Pace M.L. (1992) Sampling larval fish populations: choice of sample number and size. *Transactions of the American Fisheries Society* **121**, 356–368.

Dekker W. (1994) Visstand en visserij op het IJsselmeer en Markermeer: de toestand in 1993 (Fish stock and fisheries in and on the northern and southern zone of Lake IJsselmeer in 1993). Netherlands Institute for Fisheries Research, IJmuiden. Report 94.001 (in Dutch).

Dekker W., Schaap L. & Van Willigen J. (1993) Bijvangsten in de fuiken-visserij op het IJsselmeer (By-catch in the fyke net fishery on Lake IJsselmeer). Netherlands Institute for Fisheries Research, IJmuiden. Report 93.021, 29 pp (in Dutch).

Doubleday W.G. & Rivard D. (eds) (1981) Bottom trawl surveys. *Canadian Special Publications of Fisheries and Aquatic Sciences* **58**, 273 pp.

Hickley P. & Starkie A. (1985) Cost effective sampling of fish populations in large water bodies. *Journal of Fish Biology* **27 (Supplement A)**, 151–161.

Hilborn R., Pikitch D.E. & Francis R.C. (1993) Current trends in including risk and uncertainty in stock assessment and harvest decisions. *Canadian Journal of Fisheries and Aquatic Sciences* **50**, 874–880.

Iedema C.W., Platteeuw M. & Rijsdorp A.A. (1995) Natuur in het natte hart: een verkenning van de kansen voor natuurontwikkeling in het IJsselmeergebied (An exploration of the prospects for nature development in the Lake IJsselmeer area). Rijkswaterstaat, Directorate Flevoland, Lelystad (in Dutch) (in press).

Jordan D.R. & Wortley J.S. (1985) Sampling strategy related to fish distribution, with particular reference to the Norfolk Broads. *Journal of Fish Biology* **27 (Supplement A)**, 163–173.

Koeller P.A. (1981) Distribution and sampling variability in the southern Gulf of St Lawrence groundfish surveys. In: W.G. Doubleday & D. Rivard (eds) Bottom trawl surveys. *Canadian Special Publications of Fisheries and Aquatic Sciences* **58**, 194–217.

Leaman B.M. (1981) A brief review of survey methodology with regard to groundfish stock assessment. In: W.G. Doubleday & D. Rivard (eds) Bottom trawl surveys. *Canadian Special Publications of Fisheries and Aquatic Sciences* **58**, 113–123.

McCullagh P. & Nelder J.A. (1983) *Generalized linear models.* London: Chapman & Hall. 261 pp.

Rivard D. (1981) Catch projections and their relations to sampling error of research surveys. In: W.G. Doubleday & D. Rivard (eds) Bottom trawl surveys. *Canadian Special Publications of Fisheries and Aquatic Sciences* **58**, 93–109.

Scott J.S. (1981) Summer distribution of groundfishes on the Scotian Shelf. In: W.G. Doubleday & D. Rivard (eds) Bottom trawl surveys. *Canadian Special Publications of Fisheries and Aquatic Sciences* **58**, 181–193.

Sissenwine M.P., Azarovitz T.R. & Suomala J.B. (1983) Determining the abundance of fish. In: A.G. MacDonald & I.G. Priede (eds) *Experimental biology at sea.* London: Academic Press, pp 51–101.

Sokal R.R. & Rohlf F.J. (1981) *Biometry*, 2nd ed. San Francisco: Freeman & Co., 859 pp.

Stefánsson G. (1991) Analysis of groundfish survey data: combining the GLM and delta approaches. *ICES C.M.; 1991/D:9.* mimeo, 16 pp.

Van Eerden M.R. & Bij De Vaate A. (1984) Natuurwaarden van het IJsselmeergebied (Nature values of the Lake IJsselmeer area). Rijksdienst voor de IJsselmeerpolders. Lelystad. Flevobericht nr, **242** 73 pp (in Dutch).

Vanhemelrijk J., Peters J., Butijn G., Vermij S., Lammens E., Laane W. & Wortel A. (1993) Amoebes IJsselmeergebied: studie naar ecologische ontwikkelingsrichtingen voor het IJsselmeergebied (Amoebes Lake IJsselmeer area: study of directions of ecological development for the Lake IJsselmeer area). General Report. Institute for Inland Water Management and Waste Water Treatment, Lelystad. Report 93.014. ISBN 90-3690-362-9, 250 pp (in Dutch).

Willemsen J. (1977) Population dynamics of percids in Lake IJssel and some smaller lakes in The Netherlands. *Journal of the Fisheries Research Board of Canada* **34**, 1710–1719.

# Chapter 20
# Technical interactions in tropical floodplain fisheries of south and south-east Asia

D.D. HOGGARTH *Marine Resources Assessment Group Ltd, 27 Campden Street, London W8 7EP, UK*

G.P. KIRKWOOD *Renewable Resources Assessment Group, Imperial College of Science, Technology and Medicine, 8 Prince's Gardens, London SW7 1NA, UK*

**Abstract**   The effect of seasonal fishing effort levels and gear utilization patterns on fish catches in floodplain fisheries in Bangladesh, Thailand and Indonesia were investigated. These artisanal fisheries are multispecies and multigear, but they have broad similarities arising from their underlying hydrological regimes. Exploitation levels, however, vary in line with human population densities and consequent demand for fish and employment. In Bangladesh, even the smallest fish are intensely exploited throughout the year, while in Indonesia, fishing is concentrated on larger fish mostly caught late in the growing season.

A one-year sampling programme provided comparative data from three field sites on catches, fishing efforts, demographic parameters of key fish species and social, economic and political factors. These data were used to compare the impacts of different exploitation strategies on fish yields relative to the range of lifespans and migratory behaviour shown by the various fish species. The implications of these results for the management of these fisheries and seasonal floodplains in general are discussed, bearing in mind the constraints arising from the need to achieve a balanced allocation of resources to different components of the fishing communities.

KEYWORDS: Asia, floodplain fisheries, population dynamics, technical interactions, multispecies, multigear modelling, fisheries management

## 20.1   Introduction

Tropical floodplain rivers are ecologically diverse and highly seasonal, with habitats and resources appearing and disappearing according to a regular pattern of flooding and drought. Consequently, they are populated by many types of fish, from air-breathing 'blackfish' in floodplain pools to more migratory 'whitefish' in the rivers, ranging in size from the smallest school fish to the largest predators and major carps (see Welcomme 1985 for a review). The fisheries that have evolved in these river systems are equally diverse with up to 20 different gear types, each one adapted to a particular time or place in the flood cycle.

This chapter describes the 'technical interactions' which occur between these various fish species and fishing gears, and the importance of these effects for the management of such fisheries. Technical interaction may be defined as 'competition between two fishing gears over a unit stock of one or more species of fish'. Depending on the mesh sizes and selectivities, two gears may either compete directly, catching fish at the same time and place, or sequentially, with one gear reducing the potential catches taken later by the other gear. Such interactions are associated with fishing gears, and contrast with purely 'biological interactions' between fish species, such as competition for food or predation. Technical interactions are far easier to study than biological ones (Hilborn & Walters 1992) and multigear fishery models are now common stock assessment tools (e.g. Daan & Sissenwine 1991).

During this research, three inland fisheries in Bangladesh, South Sumatra in Indonesia and Thailand were examined to give a full bio-socio-economic understanding of their fish stocks and fishing communities. The three sites vary in their hydrology and morphology, and in community structures and human populations from the highest levels in Bangladesh to the lowest in Sumatra. This chapter compares the fishing gears and fish resources of these sites, and draws conclusions on the potential fish catches from different management strategies.

## 20.2    Materials and methods

The three study sites were: Hail Haor, a natural 150 km$^2$ depression around the Gopla river in north eastern Bangladesh; a 20 km section of the Lempuing River in South Sumatra along with its associated floodplains, pools and swamps; and the 280 km$^2$ Thale Noi lake and savanna floodplain in southern Thailand. Each of these sites was intensively monitored for a full 12-month cycle between December 1992 and February 1994. Catch and effort data, stratified by gear type and fish species, were sampled monthly or fortnightly from a selection of between 30 and 90 fishing households in the three communities. Approximately bimonthly time series of length frequency samples were taken for 3–7 key fish species at each site. Using these data, von Bertalanffy growth parameters were estimated where possible by the ELEFAN 1 methodology (Pauly & David 1981) and total mortality rates were estimated from the ratios of the numbers of fish sampled in identifiable age classes. Natural mortality rates were estimated from an empirical model modified by Trenkel (1993) from the original equation of Pauly (1980) using an enlarged data set for fish species living in water temperatures above 5°C.

The effects of different management strategies on total yields from the three fisheries were simulated using FAO's 'BEAM4' methodology (Sparre & Willmann 1991). BEAM4 is a cohort based, multispecies, multigear simulation model using standard equations for exponential mortalities and von Bertalanffy growth. Each species in the model was assigned its own biological parameters ($K$, $L_\infty$, $t_0$, $Z$ and $M$ in standard notation), and each gear type was assigned a specified selectivity ogive for each species, determined approximately from the length frequency data. The seasonality of the fisheries is modelled by inputting the actual monthly fishing efforts of each gear. The number of recruits and the gear/species catchability coefficients were tuned to give the best possible approximation to the observed catches. Such a model is capable of detailed outputs on the seasonal patterns of catches of each species by each gear type. In this chapter, the overall changes in total yields to the fishery were examined for four different types of management scenario: changes in overall effort levels, closed seasons over the high water growth period, increases in mesh sizes and total bans on single fishing gears.

### 20.3    Results

#### 20.3.1    *The multispecies fish communities*

Broadly similar fish communities were found at each of the three study sites, with around 30 different species commonly found in the catches. The fish species fall into six main taxonomic groups, as shown below with common examples:

**Carps (Cypriniformes, Cyprinoidei)**
  *Puntius* spp
  *Osteochilus* spp
  *Labeo rohita* (Hamilton)

**Catfish (Cypriniformes, Siluroidei)**
  *Clarias* spp
  *Mystus* spp

**Perches (Perciformes)**
  *Anabas testudineus* (Hamilton)
  *Trichogaster* spp
  *Colisa* spp

**Snakeheads (Channiformes)**
  *Channa* spp

**Other fish**
  *Notopterus* spp
  various eels

**Prawns**
  *Macrobrachium* spp
  *Palaemon* spp.

From these various species, the greatest total catches at each site were the air-breathing 'blackfish', capable of surviving the deoxygenated conditions of the dry season (up to over 70% by weight at the Indonesian and Thai sites). The non-air-breathing 'whitefish', which must emigrate from the floodplain to the well-oxygenated rivers to survive the dry season, made relatively smaller contributions to the catches.

The above species were highly variable in terms of their maximum sizes, reproductive behaviour, feeding preferences and migration patterns. The vulnerability of different fish to different gear types was mainly determined by the latter two characteristics. For the BEAM4 modelling, each species was assigned to one of three to five fish guilds on the basis of its biological characteristics and gear vulnerability.

Approximate biological parameters were assigned to each guild, based on the results from the length-frequency sampling programme (Table 20.1). A slightly different selection of guilds was used at each site, reflecting the dominance of different species in the catches. All the sites contained similar guilds of large predators (mostly snakeheads including *Channa striatus* (Bloch)), medium-sized blackfish (e.g. *Notopterus notopterus* (Pallas), *Anabas testudineus* and *Clarias* spp) and medium-sized or small whitefish (e.g. *Puntius* spp, *Rasbora* spp). The Bangladesh model included a separate guild for the large major carps including *Labeo rohita* and *Catla catla* (Hamilton), and the Indonesian model included a separate guild for riverine predators, particularly the catfish *Mystus nemurus* (Valenciennes) (for further details on this site see Hoggarth & Utomo 1994). Shrimps were common in catches at all three sites, but the Bangladesh fishery took small *Palaemon* species while the Indonesian fishery had access to the much larger *Macrobrachium rosenbergii* (De Man). Insufficient information was collected on shrimps at the Thai site to include them in the model. However, since shrimps at this site were only caught by specialized shrimp traps, their lack of interactions meant that their exclusion from the model did not affect the results or conclusions for the other species and gears.

**Table 20.1** Approximate Von Bertalanffy growth parameters and instantaneous mortality rates for the fish guilds used in the three BEAM4 models. Parameters estimated from the project length-frequency sampling programme, or based on comparisons with other species guilds (indicated by apostrophes).

| Study Sites | Von Bertalanffy growth rates | | Mortality rates ($yr^{-1}$) | |
| Fish species guilds | $K$ ($yr^{-1}$) | $L_\infty$ (cm) | Natural, $M$ | Fishing, $F^1$ |
| --- | --- | --- | --- | --- |
| Bangladesh | | | | |
| Large carps | 0.3 | 75 | 0.6 | 4.1 |
| Large predators | 0.7 | 70 | 0.9 | 2.1 |
| Medium-sized blackfish | 0.7 | 30 | 2.4 | 2.5 |
| Small whitefish | 0.7 | 15 | 3.0 | 2.4 |
| Small shrimps | 1.0' | 4' | 3.0' | 1.9' |
| Indonesia | | | | |
| Floodplain predators | 0.7 | 65 | 0.9 | 2.0 |
| River predators | 0.6 | 58 | 0.9 | 2.0 |
| Medium-sized blackfish | 0.65 | 28 | 1.2 | 2.0 |
| Medium/small whitefish | 0.7' | 20' | 1.5' | 2.0' |
| Large shrimps | 1.0' | 30' | 2.0' | 2.0' |
| Thailand | | | | |
| Snakehead predators | 0.5 | 65 | 0.9 | 2.0 |
| Medium-sized blackfish | 0.6 | 30 | 1.2 | 2.0 |
| Medium-sized whitefish | 0.6' | 30' | 1.2' | 2.0' |

[1] Maximum fishing mortality rate, for fish at lengths fully selected by all gear types.

**20.3.2**    *The multigear fisheries*

All three fish stocks were exploited by a wide range of fishing gears including seine nets, gill nets, individual fish traps, barrier traps, hooks and other less familiar approaches. The relative importance of the different fishing gears at each site depended on morphological and hydrological characteristics. At the most riverine Indonesian site, exploitation was notable by the extensive use of efficient barrier traps (Fig. 20.1), catching fish as they migrate around the floodplain and river system. In the more lacustrine Thai site, gill nets took more than 50% of the total catch, being most effective in the calm waters of the lake environment. In the dry seasons, fish in Bangladesh and Thailand were caught by dewatering isolated ponds with diesel pumps. In Indonesia, such fish were caught by seining enclosed ponds or river sections up to 1 km in length with nets and sections of bamboo fences.

Similar mesh sizes were used in both the Indonesian and Thai fisheries, but all gears at the Bangladesh site were mainly small meshed (Fig. 20.1). As a result of these gear characteristics, the catches in Bangladesh comprised many more small fish (Fig. 20.1) with even the largest species often being caught as fry of only 2–3 cm in length.

The seasonalities of the fisheries were also determined by flow regimes, with each gear being best adapted to a certain period in the flood cycle. The three study sites again displayed notable differences. In Indonesia, the main catches from the barrier traps were taken when fish were migrating in the ebb and flood seasons, while the seining methods were effective during the drought. Only a few fish were taken by fish traps during high water (Fig. 20.2; see also Hoggarth & Utomo 1994). In Thailand, in contrast, the greatest catches of the gill net fishermen were in the flood season when fish were actively foraging for food. In Bangladesh, different methods were used in each season, such as hooks in the flood and seines and dewatering in the drought: many fish, however, were caught at all times, and a fairly constant total catch was taken throughout the year (Fig. 20.2).

The technical interactions between the various gears were partly determined by their mesh sizes and seasonalities, and partly by the species caught. Certain gears were more specific than others, and many of the smaller gears, such as fish traps and hooks, were strongly targeted at individual guilds of fish (see e.g. Fig. 20.3 for the Indonesian site). Other gears such as the barriers and seines in Indonesia (Fig. 20.3) and the gill nets in Bangladesh and Thailand were less specific and caught virtually any species they intercepted. With the various gears catching the same fish both within and between the different seasons, it was clear that both direct and sequential technical interactions were likely to be pronounced in these fisheries.

**20.3.3**    *Multispecies, multigear BEAM4 simulation modelling*

In terms of the total yield from these fisheries, none of the management strategies investigated would either increase the catches by more than 6%, or alternatively decrease them by more than 22%. While total yields were thus insensitive to man-

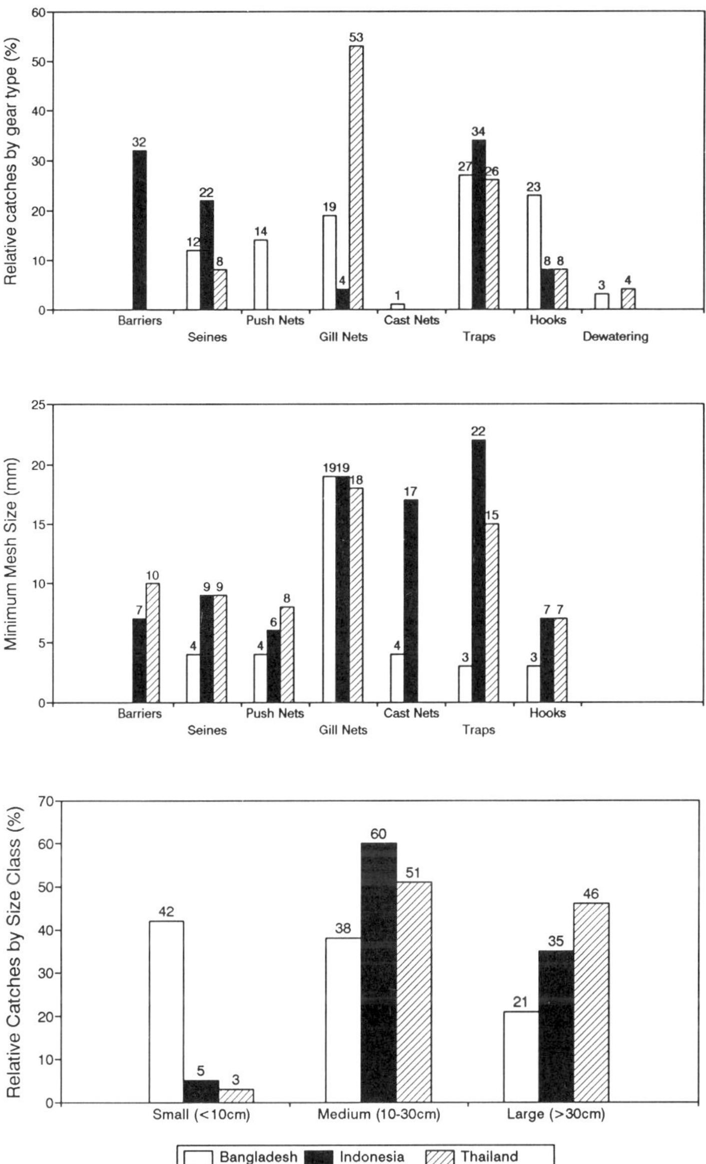



**Fig. 20.1**  Between-site comparisons of the relative catches and mesh sizes of the most important gear types, and the catches of fish species with maximum lengths in three different size classes. Total annual catches from the study sites were 1671 t in Bangladesh, 263 t in Indonesia, and 1475 t in Thailand.

agement, the impacts on the catches of individual gears could be much more pronounced: the *average* change to catches of individual gears was 22% while certain gears benefited from the different management strategies by up to 244% (Table 20.2). The greatest benefits were for those gears which exploited fish either late in the year,

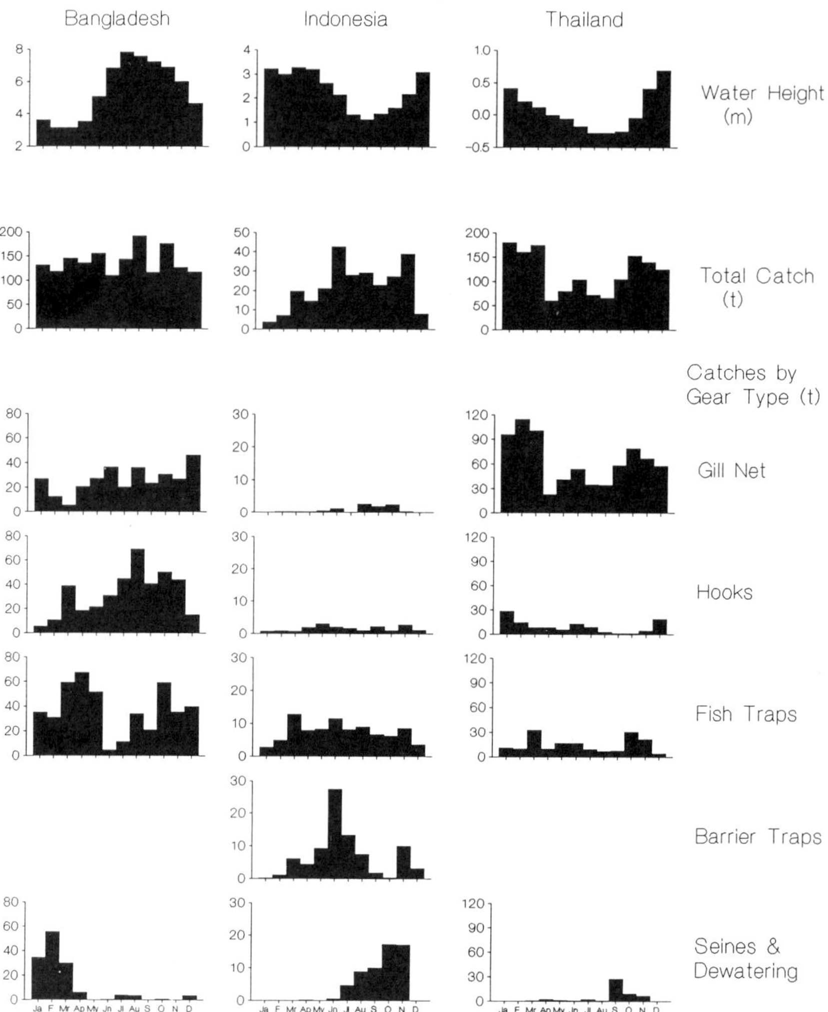


**Fig. 20.2** *Between-site comparisons of the seasonality of water levels, total fish catches and catches of selected, important gear types. Water height data sources; Bangladesh: Water Development Board, Moulvi Bazaar, October 1986 to August 1987; Indonesia: Research Institute for Freshwater Fisheries, Palembang, 1975 to 1991; Thailand: Royal Irrigation Department, Hat Yai, 1981 to 1991.*

or with large mesh sizes. The greatest management potential for these highly inter-active fisheries was thus in the *reallocation* of fish catches towards target fishing gears. The relative effects of the different management strategies at each study site depended on the mesh sizes and seasonalities of their most important gears, as shown in the following sections.

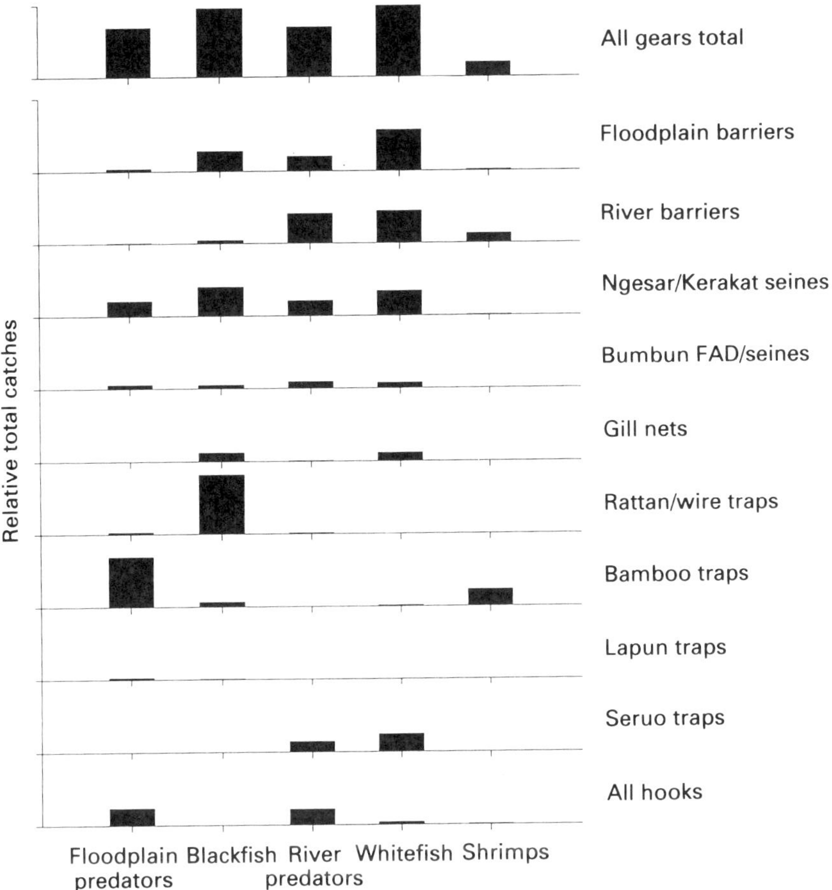



**Fig. 20.3** Comparison of the total catches of five fish species guilds by the ten most important gear types at the Indonesian study site. See Hoggarth & Utomo (1994) for further details on individual gear types.

### 20.3.4 *Changes in overall effort levels*

The relative effects of simultaneously changing the fishing efforts of all gears depended on the types of fish that each gear caught. The large, slow-growing species, such as carps and predators, were overexploited at the effort levels recorded, while the small species were under-exploited or more optimally exploited (Fig. 20.4). Increases in effort would thus increase the contribution of small fish to the catches while decreases would improve the contributions of large fish. The overall effect on total yield was minimal because the opposing results for different gears cancel out. Due to the extra effort placed on small, fast-growing fish, with small-meshed gears in Bangladesh, high fishing effort would be most rewarded at this site (Table 20.2). These results were supported by considering the relative catches taken at the three sites and the current fishing efforts as approximated by fishermen densities. Similar

**Table 20.2** The BEAM4-simulated % changes in the total yields of all species guilds from five different management scenarios for each study site.

| Study sites<br>Management scenarios | % Change to total yield of all gears | Mean absolute % change for single gears | Maximum % change for a single gear |
|---|---|---|---|
| **Bangladesh** | | | |
| 50% effort decrease | −17 | 26 | +49 |
| 50% effort increase | +6 | 12 | −28 |
| High water closed season[1] | −22 | 60 | +139 |
| Mesh size increase[2] | −14 | 23 | −50 |
| Single gear bans[3] | −3 | 10 | +83 |
| **Indonesia** | | | |
| 50% effort decrease | −4 | 14 | +60 |
| 50% effort increase | −3 | 11 | −39 |
| High water closed season[1] | 0 | 52 | +123 |
| Mesh size increase[2] | +2 | 27 | +97 |
| Single gear bans[3] | 0 | 15 | +244 |
| **Thailand** | | | |
| 50% effort decrease | −4 | 5 | −9 |
| 50% effort increase | −4 | 5 | −10 |
| High water closed season[1] | −11 | 49 | +113 |
| Mesh size increase[2] | +5 | 8 | −17 |
| Single gear bans[3] | −1 | 15 | +114 |

[1] June–December for Bangladesh; January–July for Indonesia and Thailand.
[2] Small-meshed gears increased to size of gill nets/fish traps.
[3] Mean and maximum results from simulations of total bans on each different gear type.

total catches per unit area were taken in both the Indonesian and the Bangladesh sites, but, as noted earlier (Fig. 20.1), the latter catches comprised far more small fish (Fig. 20.5).

### 20.3.5 *High water closed seasons*

Imposing a closed season over the high water period, to allow fish to grow to a reasonable size before capture, would mainly benefit the catches of those gears most effective in the later dry season. Such closed seasons would give a net gain to the Indonesian site where few fish were caught during the flood, but would give losses at the other sites due to the importance of their high water fisheries. High water closed seasons have the greatest reallocative potential of all strategies, with potential benefits for dry season gears of over 100% at all sites (Table 20.2).

### 20.3.6 *Mesh size increases*

Increasing the mesh sizes of all the small meshed gears to the larger sizes of either gill nets or fish traps would give small net gains to both the Indonesian and Thai fisheries,

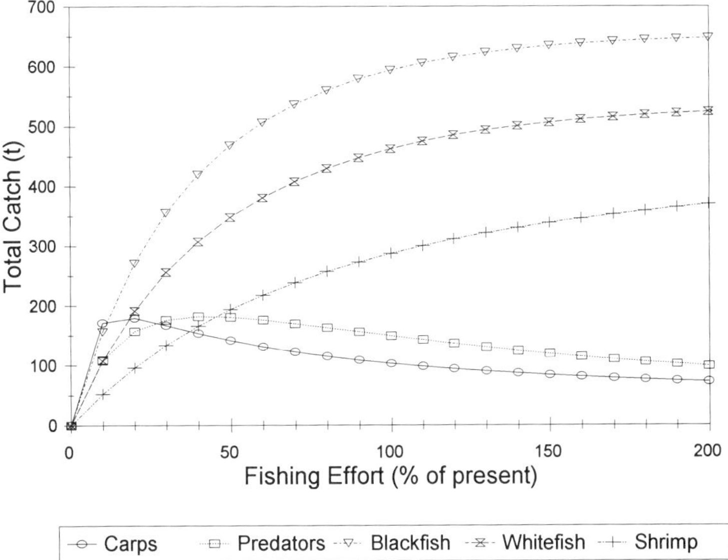



**Fig. 20.4**   BEAM4-simulated total catches to all gears, for the five fish species guilds at the Bangladesh study site, for a range of fishing effort levels.

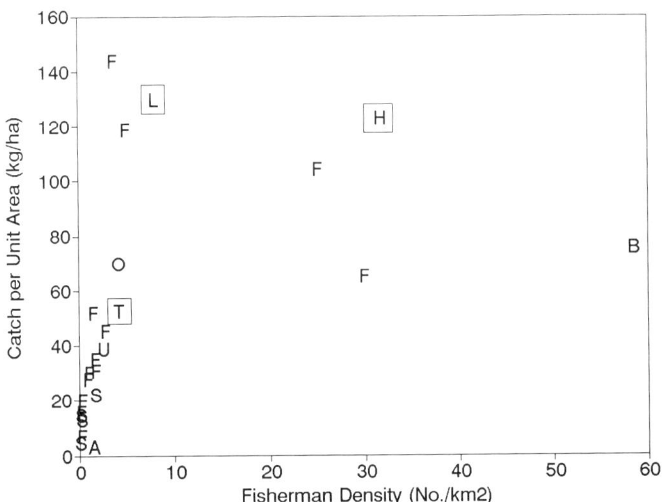



**Fig. 20.5**   Estimated total catches scaled by high water floodplain area, plotted against fisherman densities for the three study sites and other tropical floodplain fisheries in Asia, Africa and South America. H – Hail Haor, Bangladesh; L – Lempuing River, Indonesia; T – Thale Noi, Thailand; B – Western Bangladesh Beels (BCAS 1992); U – Lubuk Lampam, Indonesia (RIFF, Palembang, Sumatra, unpublished data); O – OKI regency, Indonesia (Department of Fisheries, Palembang, Sumatra); A – Kapuas River (MRAG 1993); S – South American rivers (MRAG 1993); F – African rivers (Bayley 1988).

but give a 14% overall reduction in catches in Bangladesh (Table 20.2) due to the importance of the many small-meshed gears presently in use at that site (Fig. 20.1).

### 20.3.7   *Single gear bans*

Total bans on single gear types would generally have the smallest overall impacts on fish catches, confirming the strength of the technical interactions in these fisheries. The losses from one gear can always be more or less made up by the increased catches to other gears. Certain individual gears would benefit by up to over 100% from the banning of other gears which presently take large catches of the same species (Table 20.2).

### 20.4   **Discussion**

The main potential of the management strategies considered for the interactive fisheries investigated lies in the reallocation of fish yields between different fishing gears. With additional information on the economic and social status of the fishermen and communities using the different gears, fishery managers should be able to identify target user groups and use the methodologies outlined to determine those management strategies giving the greatest benefits to those groups. In Bangladesh, for instance, the poorest fishermen were generally those taking the smallest fish with fine-meshed gears during the high water flood season, when there were no access restrictions on the waterbody (Alex Kremer, unpublished project documentation). Regulations on high water closed seasons or increases in mesh sizes at this site would thus be particularly disadvantageous for this most needy target group.

Due to the pronounced technical interactions between different fishing gears, it is clear that no one fishing gear or fish species should be managed in isolation from the others. Management should allow for the best overall balance of effects, depending on the interactions and seasonalities of the various gears and on the wide range of fish sizes and types available. This multispecies assessment has reaffirmed the effect of the 'fishing up' process on the composition of fish catches, previously discussed for river fisheries by Welcomme (1985) and observed elsewhere (e.g. Pope & Knights 1982; Turner 1981).

The BEAM4 results rely on several input parameters which were estimated with low precision, or simply by comparison with the results for other species. Due to this uncertainty, the sensitivities of the percentage changes due to the management scenarios were tested against variations in input parameters for mortality rates and growth rates of the fish guilds, and in the gear selectivities and estimated total catches of individual gears. In most cases, the results, particularly for the reallocation of benefits between gears, proved robust to these uncertainties. As expected, the relative magnitudes of natural and fishing mortality rates influence the exact fishing efforts giving the greatest yields; gear selectivities were also important to determine the optimal fishing season. In general, however, the percentage uncertainties were small compared to the percentage reallocations caused by the four management strategies (see Hoggarth 1994). Further uncertainty arises from the possibility of recruitment overfishing, since the BEAM4 models do not include any form of stock/recruit relationship. Although these multispecies fisheries have already proven resilient to high fishing pressure, the real upper thresholds to exploitation were difficult to determine, particularly in the face of environmental fluctuations and the rapidly changing conditions these systems were experiencing (Dudgeon 1992). In the face of such dangerous uncertainties, precautionary measures such as the use of closed areas or reserves should be recommended to protect these valuable fish resources.

## Acknowledgements

This research has been funded by the Overseas Development Administration of the British Government. The socioeconomic component of the project was based at the University of Bath Centre for Development Studies, with the biological components briefly outlined in this chapter provided by the Marine Resources Assessment Group. Local collaboration has been provided by the following institutes: the Bangladesh Institute for Development Studies, Dhaka, Bangladesh; the Coastal Resources Institute, Prince of Songkla University, Hat Yai, Thailand; and the Research Institute for Freshwater Fisheries, Palembang, Indonesia. Much of the field data were collected by associates from these institutes, whom the authors gratefully acknowledge. Thanks are also given to our colleagues at MRAG, for helpful discussions, particularly to Ms Julie Rossouw.

## References

Bayley P.B. (1988) Accounting for effort when comparing tropical fisheries in lakes, river floodplains, and lagoons. *Limnol. Oceanogr.* **33**, 963–972.

BCAS (Bangladesh Centre for Advanced Studies) (1992). *Third Fisheries Project, Support Technical Assistance, Floodplain Production Monitoring, Annual Report (August 1991–July 1992)*. Department of Fisheries, Government of Bangladesh, 22 pp plus figures and tables.

Daan N. & Sissenwine M.P. (eds) (1991) Multispecies models relevant to the management of living resources. *ICES mar. Sci. Symp.* **193**. 358 pp.

Dudgeon D. (1992) Endangered ecosystems: a review of the conservation status of tropical Asian rivers. *Hydrobiologia* **248**, 167–191.

Hilborn R. & Walters C.J. (1992) *Quantitative Fisheries Stock Assessment, Choice, Dynamics and Uncertainty*. New York: Chapman & Hall, 570 pp.

Hoggarth D.D. (1994) Management strategies and their sensitivity to parameter uncertainties in a model of artisanal river fisheries. *ICES C.M.*1994/T:30, 10 pp.

Hoggarth D.D. & Utomo A.D. (1994). The fisheries ecology of the Lubuk Lampam river floodplain in South Sumatra, Indonesia. *Fisheries Research* **20**, 191–213.

MRAG (1993) *Synthesis of simple predictive models for tropical river fisheries*. Report to ODA, London: MRAG, 85 pp.

Pauly D. (1980) On the interrelationships between natural mortality, growth parameters, and mean environmental temperature in 175 fish stocks. *J. Cons. CIEM.* **39**, 175–192.

Pauly D. & David N. (1981) ELEFAN 1, a BASIC program for the objective extraction of growth parameters from length frequency data. *Meeresforschung* **28**, 205–211.

Pope J.G. & Knights J. (1982) Comparison of length distributions of combined catches of all demersal fishes in surveys in the North Sea and at Faroe Bank. In: M.C. Mercer (ed.) Multispecies approaches to fisheries management advice. *Can. Spec. Publ. Fish. Aquat. Sci.* **59**, 116–118.

Sparre P. & Willmann R. (1991) Software for bio-economic analysis of fisheries: BEAM 4: Analytical bio-economic simulation of space-structured multispecies and multifleet fisheries. *FAO Fish. Tech. Pap.* **310.3**, Rome, 229 pp.

Trenkel V. (1993) *Multivariate analysis of fish population data.* Unpublished MSc. Thesis, University of Kent, UK.

Turner J.L. (1981) Changes in multispecies fisheries when many species are caught at the same time. *CIFA Tech. Pap./Doc. Tech. CPCA* **8**, 201–211.

Welcomme R.L. (1985) River Fisheries. *FAO Fish. Tech. Pap,* **262**. 330 pp.

# Chapter 21
# Estimation of Atlantic salmon (*Salmo salar* L.) and sea trout (*Salmo trutta* L.) run size and angling exploitation on the Welsh Dee, using mark-recapture and trap indices

I.C. DAVIDSON and R.J. COVE *National Rivers Authority (Welsh Region), Shire Hall, Mold, Clwyd CH7 6FA.*
N.J. MILNER *National Rivers Authority (Welsh Region), Priestly Road, Caernarvon, Gwynedd, LL55 1HR.*
W.K. PURVIS *National Rivers Authority (Welsh Region), Rivers House, St. Mellons Business Park, Cardiff CF3 0LT.*

**Abstract**   In 1991, NRA Welsh Region launched a programme to study stocks of salmon (*Salmo salar* L.) and sea trout (*Salmo trutta* L.) on the River Dee. The Dee Stock Assessment Programme (DSAP) is a long-term integrated programme to provide fisheries management with improved resource performance measures, a central component of which is a tag-recapture study designed to produce adult run and angling exploitation estimates.

Operation of a partial head-of-tide fish trap at Chester Weir provided large numbers of fish for tagging: 3162 salmon and 1962 sea trout in the first three years of the study. Sufficient numbers of salmon were recaptured by the rod fishery to produce statistically acceptable run estimates (95% CL within 30% of the estimate). By contrast, too few sea trout were recaptured by anglers for this purpose, but run estimates were produced from trap recaptures of previous spawners. Angling exploitation estimates for both species are also presented.

Results from the DSAP tag-recapture programme for 1991-93 are described, including the use of salmon radio-tracking data to examine underlying assumptions of run estimates (e.g. availability, tag mortality and behaviour effects, tag loss, susceptibility to capture and angler reporting rates). The use of the Chester trap catch as an indicator of salmon run size is also briefly examined.

KEYWORDS: Atlantic salmon, sea trout, trapping, mark-recapture, run size, exploitation

### 21.1     Introduction

Effective management of natural resources requires robust, objective measures of their performance. In fisheries, however, some of the more fundamental measures of resource performance – particularly stock size and related parameters (e.g. year class strength, spawning escapement) are rarely obtained. To improve on this situation for Atlantic salmon (*Salmo salar* L.) and sea trout (*Salmo trutta* L.) in Wales, the Welsh Region of the National Rivers Authority recently selected the River Dee for a long-term research and monitoring programme (the Dee Stock Assessment Programme or DSAP) designed to provide and interpret a number of resource performance measures for these species.

A central component of the DSAP is a mark-recapture programme to estimate salmon and sea trout run size and exploitation rates from angling catches. This requires trapping and tagging large numbers of adult fish at a partial, head-of-tide fish trap at Chester Weir (SJ 407 658) (Fig. 21.1) – the construction of which marked the formal start of the programme in 1991. This, coupled with a number of other activities – including radio-tracking, microtagging, rod and net fishery censuses, and juvenile monitoring, make the DSAP one of the most comprehensive studies of its type in England and Wales, and one which is already contributing to migratory salmonid management at a regional and national level.

The results of the first three years of the mark-recapture programme are described, utilizing radio-tracking data to examine underlying assumptions of the estimates used. The use of the Chester Weir trap catch as an indicator of run size is also briefly investigated.


### 21.2     The river and its fisheries

The Dee rises in the Bala region of Snowdonia and flows 160 km through rural, urban and industrial areas before entering the Irish Sea close to Flint (Fig. 21.1). It is one of the largest rivers in Wales (catchment area 2088 km$^2$; ADF 35 cumecs at Eccleston Ferry, 1991–93) with flows controlled by a series of headwater reservoirs for abstraction in the lower Dee close to Chester.

In addition to salmon and sea trout, the Dee supports a variety of non-salmonid species as well as brown trout (*Salmo trutta* L.) and grayling (*Thymallus thymallus* L.). A commercial net fishery for salmon and sea trout operates in the estuary from 1 March to 31 August, and consists of 30 draft and four trammel nets. Average net catches for 1982-91 (both gears combined) were 936 salmon and 130 sea trout; these compare with average rod catches of 538 and 115, respectively. The angling season for migratory salmonids runs from 26 January–17 October, with fishing activity concentrated in the main Dee between Farndon and Bala (Fig. 21.1).

**Fig. 21.1**   River Dee catchment showing main features and angling sections.

## 21.3    Materials and methods

### 21.3.1    *Trapping*

Chester Weir marks the upstream limit of tidal influence for most tides and lies between the net and rod fisheries (Fig. 21.1). The trap is a conventional inscale design located in a covered channel running from the top pool of the weir fish pass; oversails at the head of the pass block the normal route of ascent and direct most fish into the trap. The extent to which the weir acts as a barrier to migration varies with tidal state and freshwater discharge; around 20% of tides exceed the weir crest height (9.26 m A.O.D.) leading to flow reversal.

Trapping was carried out on a regular, but discontinuous, basis from May 1991. The trap was checked twice daily except when small catches were expected; a five in seven-day trapping cycle was adopted in 1991 and a six in nine-day cycle in subsequent years. Tagging and release of fish took place within the trap channel after it was isolated from the river flow by automated penstocks and the water level pumped down. When not in operation, movable inscales and doors within the trap were opened to allow fish free passage through the channel.

### 21.3.2    *Tagging*

Prior to tagging all fish were anaesthetized in 270 ppm 2-phenoxy ethanol. Scales (2–3) were removed for ageing purposes and fork lengths recorded to the nearest mm. Fish were sexed from external features and examined for parasites, and signs of damage and disease. A sub-sample of the catch was weighed to the nearest 25 g. After tagging, fish were passed to an aerated release pool within the trap channel to recover from the anaesthetic and await the opening of the penstocks. When large numbers of fish were present, only a proportion of the catch was tagged to avoid excessive delays and stress. Untagged fish were passed straight to the release pool after identifying the species.

Two types of individually coded external tag were used; salmon were exclusively tagged with 'Floy' brand T-bar anchor tags (Floy Tag Ltd, Seattle) – inserted just below the dorsal fin. The same tags were used for sea trout in 1991 and 1992, but were replaced by 'Visible Implant' or VI tags (North West Marine Technology, Washington) in 1993; the latter were inserted into transparent adipose tissue immediately behind the eye. Additional marks consisting of a single alcian blue Panjet mark (1992) and adipose fin hole punch (1993) were used to assess tag loss rates.

A number of salmon released at Chester Weir (as well as several fish released in the estuary) were also internally tagged with 'radio tags' or 'radio acoustic tags'; these tags were inserted into the stomach via the oesophagus using methods similar to those of Solomon and Storeton-West (1983). All radio tagged fish were externally marked with Floy tags; these fish are referred to in the following text as 'radio tagged' whereas fish which were Floy tagged only are termed 'Floy tagged'.

### 21.3.3    *Second catch and recaptures*

At the outset of the DSAP it was anticipated that the rod fishery would provide the main source of second catch and recapture for salmon (although sea trout remained a largely unknown resource on the Dee). To encourage tag returns, the programme was heavily publicized and tag rewards of £5.00 for Floy and £20.00 for radio-tags were offered from the start. Most tag returns were received within two weeks of capture and independently of catch returns – which were collected at the end of the angling season.

Angling catches were derived from two sources; a regional licence return targeting all anglers and considered to represent 76% of the total catch (Gregory *et al.* 1988), and a logbook scheme introduced to collect detailed catch-effort data and circulated among angling organizations and tag returnees. The former was used in 1991 only; in the following year an alternative licence structure resulted in a serious reduction in the quality of this return. Consequently, after 1991, catches reported from the logbook scheme were utilized, although no reliable estimate of the total rod catch could be made from this source.

In addition to the rod fishery, broodstock traps at Pont Barcer (SJ 044 437) and Celyn (SH 880 399) (Fig. 21.1), as well as the trap at Chester Weir, provided recaptures, the latter mainly of sea trout tagged in previous years.

### 21.4     **Results**

#### 21.4.1     *Tagging*

3162 salmon and 1962 sea trout were tagged at Chester Weir between the start of trapping in May 1991 and the end of 1993 (Table 21.1). Except for sea trout in 1993, tagging commenced at the beginning of each year and ceased on the 14 October – three days before the end of the angling season. This was to allow time for the last fish tagged to enter the rod fishery; tracking studies (1991–92) indicated that 93% (100/ 108) of salmon arrived in the lower reaches of the rod fishery (28 km upstream of Chester Weir) within 48 hours. Sea trout were tagged all year round in 1993 as it

**Table 21.1**  Numbers and sea age composition of salmon and sea trout tagged at Chester Weir, 1991–93 (SW = sea winter; PS = previous spawner).

| | Salmon | | | Sea trout | | |
|---|---|---|---|---|---|---|
| | 1991 | 1992 | 1993 | 1991 | 1992 | 1993 |
| No. tagged N (%) | | | | | | |
| Floy | 742 (94.3) | 1051 (94.6) | 1228 (97.2) | 422 (100.0) | 651 (100.0) | 0 (0.0) |
| Radio | 45 (5.7) | 60 (5.4) | 36 (2.8) | 0 (0.0) | 0 (0.0) | 0 (0.0) |
| VI | 0 (0.0) | 0 (0.0) | 0 (0.0) | 0 (0.0) | 0 (0.0) | 889 (100.0) |
| Total | 787 (100.0) | 1111 (100.0) | 1264 (100.0) | 422 (100.0) | 651 (100.0) | 889 (100.0) |
| Sea age composition N (%) | | | | | | |
| 0SW | 0 (0.0) | 0 (0.0) | 0 (0.0) | 155 (36.7) | 147 (22.6) | 310 (34.9) |
| 1SW | 659 (83.7) | 687 (61.8) | 868 (68.7) | 161 (38.1) | 223 (34.3) | 283 (31.8) |
| 2SW | 107 (13.6) | 375 (33.8) | 344 (27.2) | 15 (3.6) | 30 (4.6) | 32 (3.6) |
| 3SW | 0 (0.0) | 24 (2.2) | 20 (1.6) | 0 (0.0) | 0 (0.0) | 0 (0.0) |
| PS | 21 (2.7) | 25 (2.2) | 32 (2.5) | 91 (21.6) | 251 (38.5) | 264 (29.7) |
| All | 787 (100.0) | 1111 (100.0) | 1264 (100.0) | 422 (100.0) | 651 (100.0) | 889 (100.0) |

became evident from the first two years of the study that Chester trap, rather than angling, provided the greatest numbers of recaptures for this species (below). Accordingly, the run estimates described are based on trap recaptures for sea trout and angling recaptures for salmon.

Monthly fishing effort at Chester Weir (% available hours fished) remained fairly constant within each year (Fig. 21.2a). Most fish were trapped within the angling season – equivalent to more than 89% and 93% of the annual catch of salmon and sea trout, respectively, and the majority of these were tagged (at least 73% of salmon and 50% of sea trout in any one month) (Fig. 21.2b,c). Trapping was considered to provide an unbiased sample of fish larger than 350 mm, with smaller fish able to pass between the bars of the trap (spacing approximately 35 mm). This would have resulted in tagging bias among whitling (0-sea winter) sea trout, but for all other sea age groups, tagging was assumed to be representative of the population as a whole.

Tagged salmon populations in 1992-93 were dominated (62–69%) by 1SW (one-sea winter) fish (Table 21.1); 2SW fish made up most (27–34%) of the remainder with 3SW fish and previous spawners each less than 3% of the total. Early running 3SW salmon were absent from the part-season tagging programme in 1991, as this group usually enter the river before June; 2SW fish and previous spawners also contribute to the early season run. For sea trout, 0SW fish, 1SW fish and previous spawners accounted for similar proportions of the tagged populations in 1992–93 (23–39%), with the rest (< 5%) made up of 2SW fish (Table 21.1). Unlike salmon, the sea age composition of the 1991 sea trout population was not markedly different from that in

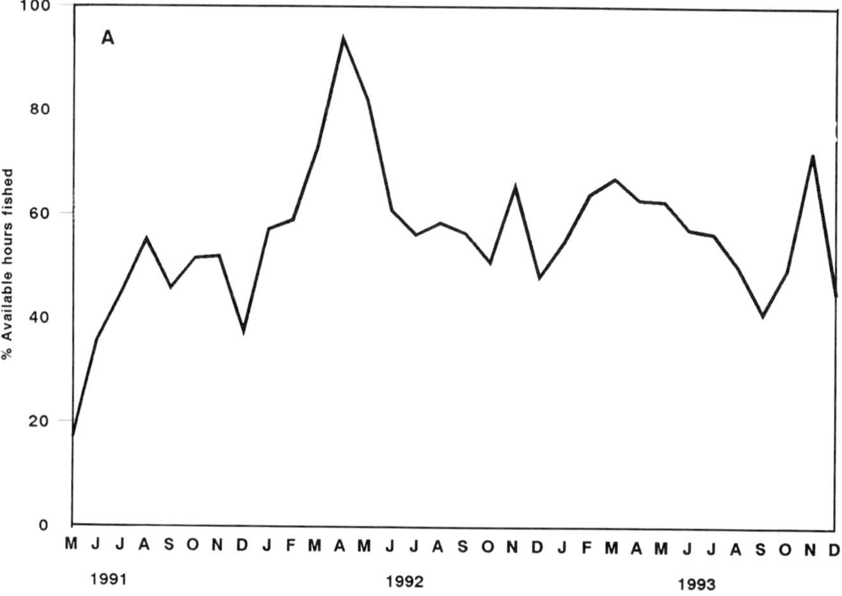


**Fig. 21.2** Trapping and tagging at Chester weir, 1991–3.
(a) fishing effort

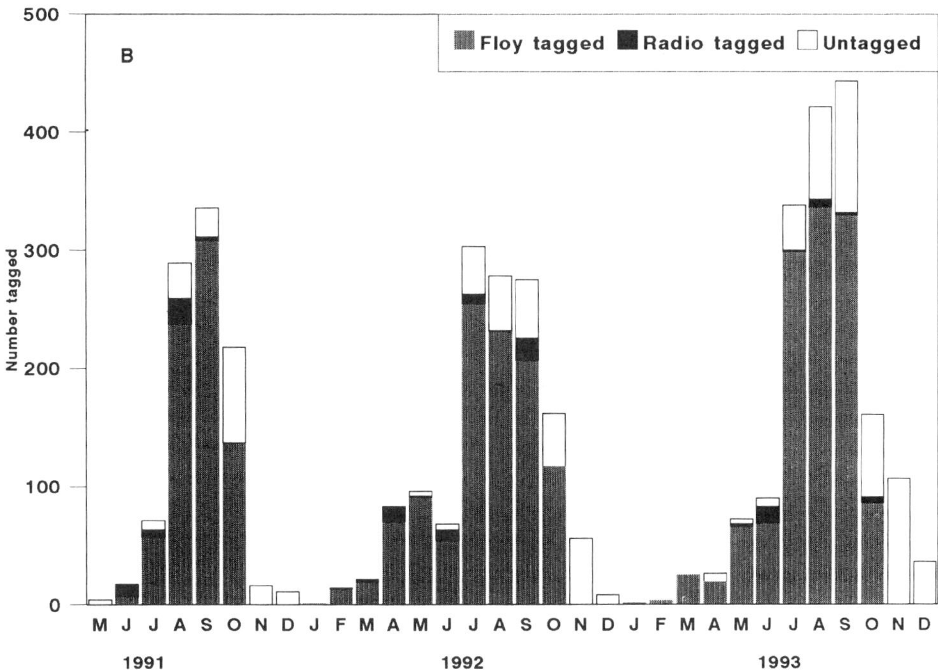


**Fig. 21.2**    (b) salmon catch

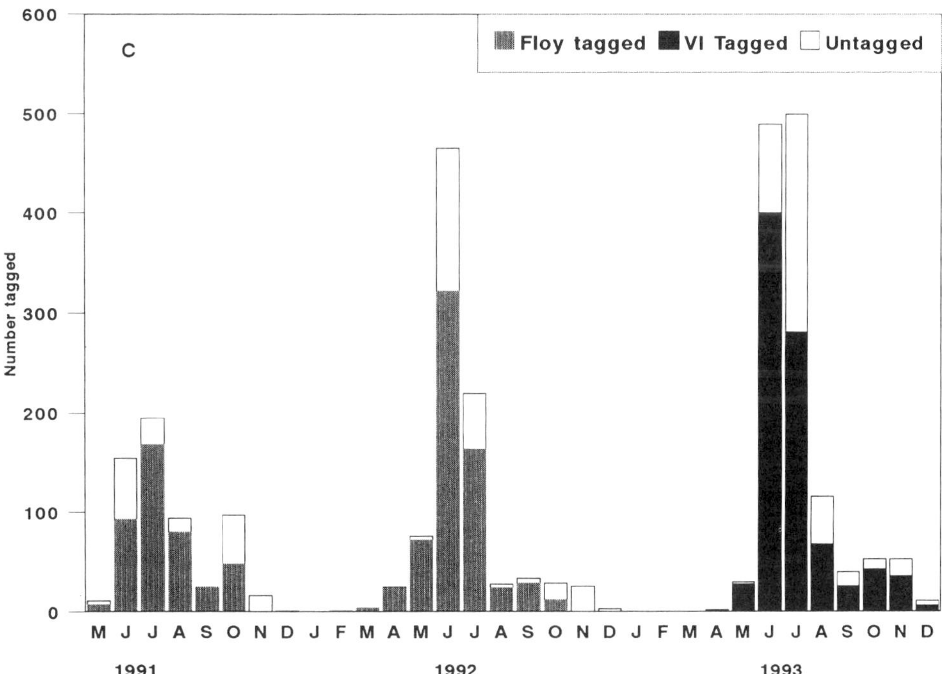


**Fig. 21.2**    (c) sea trout catch.

subsequent seasons, probably because few fish enter the river in early season, indeed 75% of the annual trap catch in 1992–93 was taken in June–July.

### 21.4.2    *Recaptures*

Angling recaptures of salmon and sea trout for the period 1991–93 (all returns) totalled 406 and 19, respectively (Table 21.2); 13% and 1% of all fish tagged. As indicated above, after 1991, larger numbers (78) of sea trout were recaptured at Chester Weir than by anglers. The majority of angling recaptures of salmon (at least 86%) and all sea trout were taken from the main Dee below the Alwen confluence, the remainder in the upper main river to Bala or the major tributaries Alwen and Tryweryn (Fig. 21.1). September and October were consistently the most prolific months for salmon, together accounting for 47–78% of recaptures.

**Table 21.2**   Angling and trap recaptures of salmon and sea trout tagged at Chester Weir, 1991–93.

| | Salmon | | | Sea trout | | |
|---|---|---|---|---|---|---|
| *Tagging year (n)* | | | | | | |
| | 1991 | 1992 | 1993 | 1991 | 1992 | 1993 |
| *All angling recaptures (year n)* | | | | | | |
| Floy | 48 | 173 | 165 | 3 | 14 | — |
| Radio | 3 | 8 | 9 | — | — | — |
| VI | — | — | — | — | — | 2(i) |
| Total | 51 | 181 | 174 | 3 | 14 | 2 |
| *Logbook angling recaptures (year n)* | | | | | | |
| Floy | 9 | 58 | 53 | 1 | 4 | — |
| Radio | 0 | 3 | 4 | — | — | — |
| VI | — | — | — | — | — | 2(i) |
| Total | 9 | 61 | 57 | 1 | 4 | 2 |
| *Chester trap recaptures (year n + 1)* | | | | | | |
| Floy | 0 | 5 | — | 19 | 59 | — |
| Radio | 0 | 1 | — | — | — | — |
| VI | — | — | — | — | — | — |
| Total | 0 | 6 | — | 19 | 59 | — |

(i) Few angling recaptures were anticipated for VI tagged fish because of the inconspicuous nature of the tag.

Logbook anglers recorded up to 35% of salmon recaptures (1992) and 33% of sea trout recaptures (1991 – Floy tag returns only) (Table 21.2). The distributions of salmon recaptures from this source (by month tagged and recaptured) were not significantly different (Chi square $P > 0.05$) from those for all reported angling recaptures in 1992 and 1993 – indicating that the returns from logbooks were representative of the angling population as a whole.

Angling recapture times ranged from 1-186 days after tagging in salmon and 4–114 days in sea trout. The mean recapture time for salmon tagged in June-October 1992 (31 days; 95% confidence limits 27–35 days) was significantly greater (one-tailed t-test; $P < 0.05$) than in 1991 (25 days; 95% CL 21-30), but similar to 1993 (28 days; 95% CL 24–32) (all data square root transformed).

### 21.4.3  Angling exploitation

Exploitation rates ($\mu$) were estimated from recaptures using $\mu = r / m$, where $r$ is the number recaptured and $m$ is the number of fish tagged, with 95% confidence limits based on Pearson's approximation to the Poisson distribution (Ricker 1975). Early running salmon tended to be the most heavily exploited, with rates declining as exposure to the fishery diminished (Fig. 21.3). Seasonal exploitation rates, like mean recapture times, were greatest (20%) in 1992 and least (8%) in 1991 (Table 21.3). (The same pattern was apparent when only June–October tagged fish were considered.) Similar trends in exploitation were evident among different sea age groups, where early running 3SW salmon were the most heavily exploited ($< 31\%$), and late running 1SW fish the least ($\leqslant 19\%$) (Table 21.3).

The exploitation estimates in Table 21.3 include corrections for Floy tag loss and unreported angling recaptures. Estimates of the former were 6.3% (1/16) in salmon and 4.2% (1/24) in sea trout – based on fish recaptured at Pont Barcer (mean 100 days after tagging; range 77–138 days) and Chester Weir (mean 362 days after tagging;

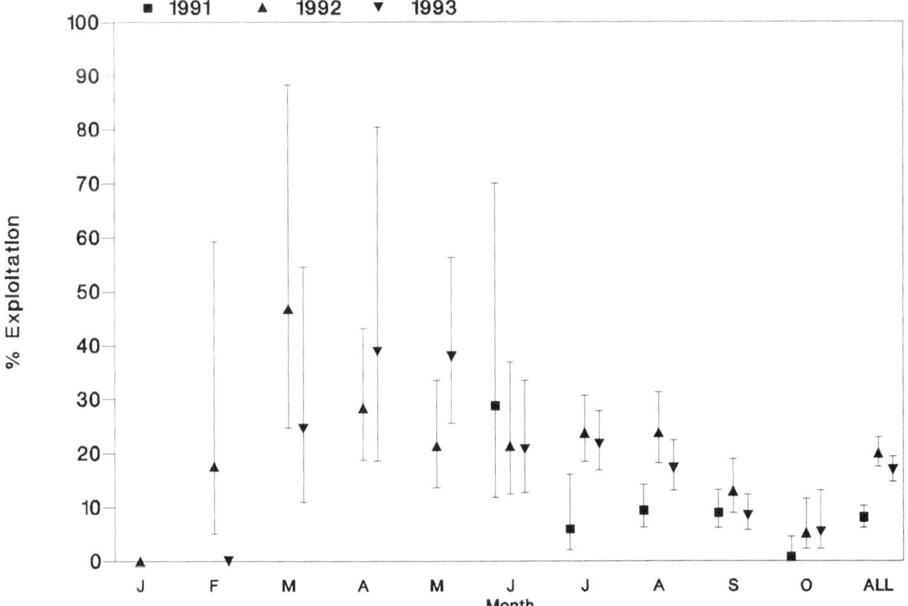


**Fig. 21.3**  Salmon angling exploitation rates for monthly entrant groups, 1991–3.
Note: 95% confidence limits shown.

**Table 21.3** Angling exploitation (% Exploitation and 95% CL) estimates for salmon and sea trout 1991–93.

| | Salmon | | | Sea trout | | |
|---|---|---|---|---|---|---|
| | 1991 (I) | 1992 | 1993 | 1991(i) | 1992 | 1993 |
| All ages | 8.0 | 20.1 | 16.9 | 0.9 | 2.6 | — |
| | (6.2–10.3) | (17.5–22.9) | (14.7–19.4) | (0.3–2.4) | (1.6–4.2) | |
| 1SW | 7.1 | 18.5 | 15.2 | — | — | — |
| | (5.3–9.5) | (15.4–22.1) | (12.7–18.1) | | | |
| 2SW | 13.8 | 22.7 | 21.5 | — | — | — |
| | (8.2–23.3) | (18.2–28.2) | (17.0–27.2) | | | |
| 3SW | — | 30.8 | 30.8 | — | — | — |
| | | (14.8–63.8) | (13.8–68.2) | | | |
| PS | 5.9 | 4.9 | 7.7 | — | — | — |
| | (1.0–30.2) | (0.8–25.4) | (2.2–26.0) | | | |

SW = sea winter;

PS = previous spawner; (i) 1991 estimate applies to part-season (14 Jun–Oct) tagging programme only.

range 282–476 days), respectively. Provisional radio-tracking estimates of unreported salmon recaptures in 1991–92 were 13.3% (2/15); this value was also assumed for sea trout.

### 21.4.4 *Population estimates*

The models of Chapman (1951) and Schaefer (1951) were used to estimate run size; the latter includes confidence limits based on Pearson's approximation to the Poisson distribution (Ricker 1975). Estimates over the three years ranged from 3772 to 8610 in salmon, and from 4304 to 7978 in sea trout (combined 0SW and >0SW estimates (Table 21.4). The Schaefer model utilized angling recapture data stratified by month and was therefore suitable for salmon only. Salmon estimates were not produced for separate sea age groups because insufficient information was available on the age composition of the rod catch. However, for sea trout separate estimates of 0SW and >0SW fish were derived from trap recaptures of 0 + SM + and other previous spawners, respectively (see Walker 1984).

The run estimates in Table 21.4 also include corrections for Floy tag loss, and unreported catches (Gregory *et al.* 1988) and recaptures described earlier. No adjustments for unreported catches or recaptures were applied to the 1992 and 1993 salmon estimates. These were based on logbook returns where reporting was considered complete due to the voluntary nature of the scheme.

Salmon run estimates derived from angling recaptures applied to the in-season period only. The out-of-season run was derived from the reverse prediction of $X$ from $Y$ (Sokal & Rohlf 1981) based on the regression relationship between estimated total

**Table 21.4**  Schaefer and Chapman population estimates for salmon and sea trout, 1991–93.

| | Salmon (i) | | | Sea trout (ii) | | |
|---|---|---|---|---|---|---|
| | 1991 | 1992 | 1993 | 1991 | 1992 | 1993 |
| **Schaefer** | | | | | | |
| All ages | 3772 | 4292 | 8610 | — | — | — |
| **Chapman (95% CL)** | | | | | | |
| All ages | 4528 | 4221 | 7157 | — | — | — |
| | (3522–5814) | (3298–5397) | (5546–9226) | | | |
| 0SW | — | — | — | 5202 | 2537 | — |
| | — | — | — | (2301–10252) | (1359–4531) | — |
| >0SW | — | — | — | 2776 | 1767 | — |
| | — | — | — | (1725–4407) | (1328–2347) | — |

SW = sea winter; (i) Salmon estimates based on all angling recaptures (1991) and those of logbook anglers only (1992 and 1993); (ii) Sea trout estimates based on recaptures at Chester Weir.

trap catch in period *i* (i.e. *observed catch per hour in period i* × *total hours available for fishing in period i*) and the equivalent run estimate from the Schaefer model (both variables $\log_{10}$ transformed; $r^2 = 0.942$; $P < 0.001$) (Fig. 21.4). Out-of-season estimates for the full season tagging years of 1992 and 1993 were 347 (95% CL 169–707) and 773 (95% CL 381–1597), indicating annual runs of 4639 and 9383, respectively (Table 21.4).

Mean monthly flows outside the season in 1992–93 (overall mean 68.5 cumec) were generally greater than those in-season (overall mean 21.6 cumecs). If higher flows resulted in a reduction in trap efficiency, the above regression model could prove unreliable in predicting the out-of-season run. However, there was no evidence of any significant correlation between trap efficiency (*total trap catch in period i* / *run estimate in period i*) and freshwater flow ($P > 0.05$), or between trap efficiency and river level downstream of Chester Weir ($P > 0.05$; Fig 21.5) – the latter influenced by both freshwater flow and tide height. Consequently, it would seem reasonable to use the regression model to predict out-of-season run size. Indeed, provisional results from the radio-tracking programme indicated that weir crossings by salmon were not confined to the higher river levels, with only 62% (25/45) of salmon crossing Chester Weir at levels within 1.0 m or less of the weir crest.

### 21.4.5  *Tagging mortality and behaviour effects*

Very few salmon (4) or sea trout (1) died at Chester trap prior to, or immediately following, Floy tagging, or were found dead subsequently in the pre-spawning period (9 and 2, respectively). Provisional results from the radio-tracking programme indicated that 10.3% (15/145) of salmon entering the Dee at Chester (1991–92) were lost prior to spawning due to unknown causes of mortality (i.e. not associated with the legal or illegal fisheries).

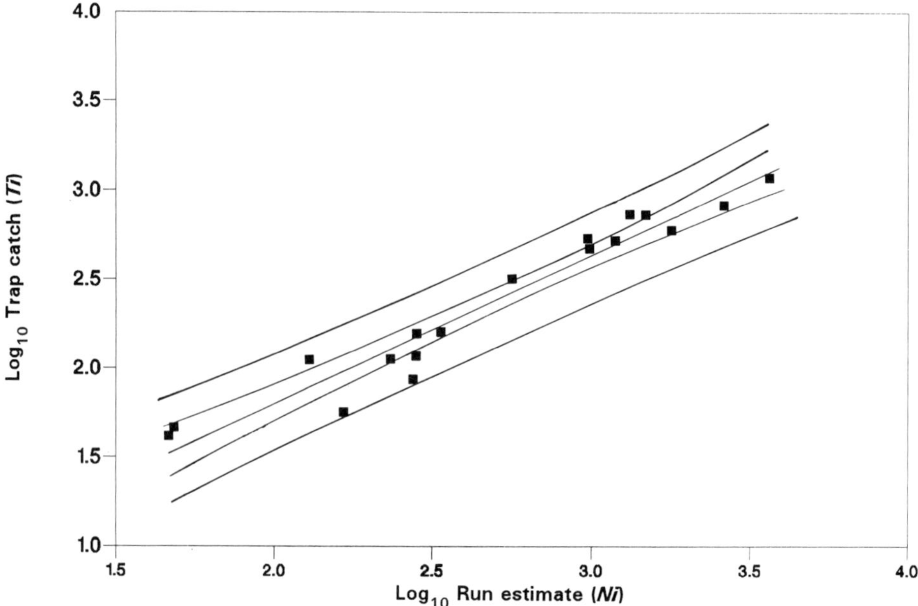

**Fig. 21.4**  Relationship between salmon run estimate (*Ni*) and trap catch (*Ti*) at Chester weir, 1991–3. Note: $\text{Log}_{10}\ Ti = 0.167 + 0.823\ (\text{Log}_{10}\ Ni)$; $R2 = 94.2\%$; $P < 0.001$. 95% confidence and prediction intervals shown.

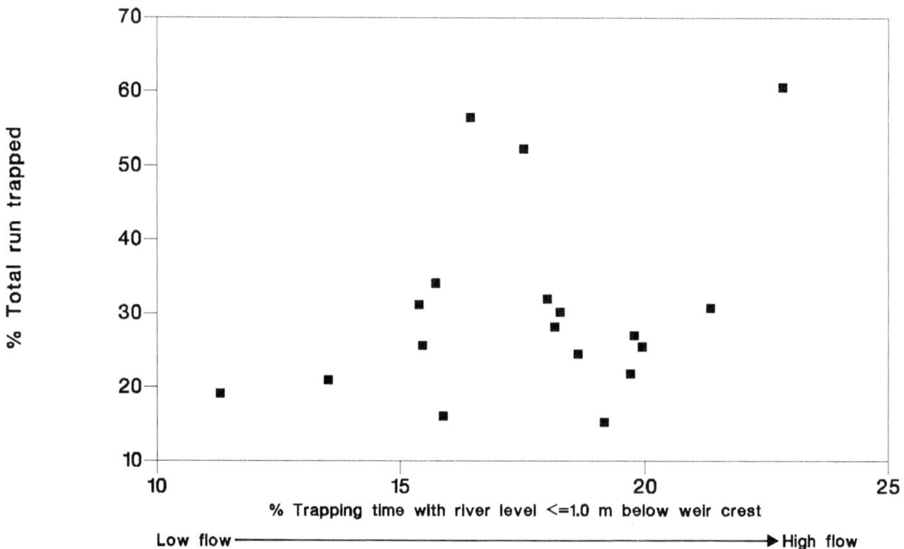

**Fig. 21.5**  Relationship between river level downstream of Chester weir and trap efficiency for salmon, 1991–3.

Enlargement and infection of the tag wound was a rare occurrence in both species. Of greater concern was the presence of filamentous weed growth (with strands up to 30 cm) on the tags of sea trout recaptured at Chester Weir. This may have impeded swimming ability in smaller sea trout, adversely affecting survival and biasing population estimates. For this reason, VI tagging was introduced for sea trout in 1993. This small subcutaneous tag should overcome the problems associated with Floy tags and although inconspicuous to anglers, the tag can be effectively screened at Chester Weir.

There was no evidence among radio-tagged salmon of unusual behaviour which might indicate a tagging effect. It is reasonable to assume that Floy tagging, which required a lesser degree of anaesthesia and handling, would not affect behaviour or survival. Indeed, there were no significant differences (Chi square $P > 0.05$) in the angling recapture rates of Floy and radio-tagged salmon (including radio-tagged salmon released in the estuary), indicating that the two groups of tagged fish behave similarly and could justifiably be treated as one for the purposes of mark-recapture.

### 21.5     Discussion

Mark-recapture methods have been used in a number of studies to estimate in-river populations of adult migratory salmonids (e.g. Swain & Champion 1978; Pratten & Shearer 1981; Walker 1984). Usually these rely on trapping to provide the second catch, but occasionally are based on fishery returns. For example, Pratten and Shearer (1981) used the net fishery in salmon tagging studies on the North Esk, but had the advantage of being able to screen the entire catch. Use of rod recaptures, as on the Dee, brings potential difficulties because of the need to secure acceptable recapture and reporting rates to support the statistical objectives of the study. However, this worked on the Dee due to efforts to ensure good communication with anglers, their enthusiasm for the scheme, and the fairly small number of individual fisheries involved.

As a result of a high level of angler support for the programme, it appears (from tracking studies) that acceptable tag reporting rates were achieved (87%) for salmon, and that in-season run estimates for this species had 95% confidence limits within $\pm 30\%$ of the estimate and approaching the target level of $\pm 20\%$ identified as most appropriate for stock assessment purposes (NRA, unpubl.). There was also a strong possibility of improvement in both these areas; support for the anglers' logbook scheme has grown considerably in the last three years, from 143 returns in 1991 to 243 in 1993, and tag reporting rates are likely to follow this trend. Similarly, the reintroduction of a tiered angling licence structure in 1994 should provide a reliable catch return and so allow all angling recaptures to be utilized – resulting in more precise estimates of salmon run size.

In addition to tag reporting rates, other assumptions associated with mark-recapture were examined and appear satisfactory, or appropriate adjustments were made. For example, there was no evidence that Floy tagging adversely affected survival or behaviour in salmon and although problems were encountered with sea trout, alternative (VI) tagging techniques were introduced to overcome these. Floy tag loss rates were also acceptable for both species ($< 7\%$). Dunkley (1992) estimated Floy tag loss rates of less than 1% in Panjet-marked salmon on the North Esk, but these data applied to fish recaptured much sooner after release (most within 15 days) than on the Dee (77–476 days).

Overall angling exploitation rates observed on the Dee for salmon (8–20%) and sea trout (1–3%) were comparable with other studies (see the review of Mills 1992), although rates reported in the Dee for individual sea age groups of salmon have few precedents elsewhere in the UK (Gee & Milner 1980; Beaumont *et al.* 1991). Exploitation rates among monthly salmon entrants and sea age groups appear positively correlated with the time fish are available to the fishery. However, there was no evidence that salmon tagged in the same months experienced any differences in angling exploitation rate associated with sea age (Chi square $P > 0.05$). This is an important observation implicit in the run estimates which make no distinction between sea age groups. Exploitation rates also appeared to be influenced by flow, with the higher rates in 1992 (20%) and 1993 (17%) occurring at greater mean monthly flows (23.2 and 22.6 cumecs) than in 1991 (14.5 cumecs; exploitation rate 8%). Times to recapture were also greater in the high flow years, indicating that fish remained susceptible to capture for longer – a behavioural response which may be one cause of increased exploitation.

Highly variable exploitation rates (i.e. recapture rates) among monthly entrants were an unavoidable source of bias in the recaptured sample. However, trapping was considered to provide an unbiased sample of marked fish, ensuring that resulting run estimates were unaffected (Ricker 1975). Of the two models used to estimate run size in this study, Ricker (1975) considered the Schaefer model the most reliable in the less than ideal (but usual) situation where individual groups of tagged fish (i.e. monthly entrants) neither remain separate nor mix randomly, but undergo some degree of partial mixing before recovery. The Schaefer model also confers the additional advantage of providing separate estimates at both time of sampling and recovery. Estimates from the two models did not differ greatly in the present study, although the Chapman model tended to produce a lower estimate than the Schaefer (Table 21.4). Ricker (1975) noted that this was usually the case except when the ratio of marked to recaptured fish was fairly constant between strata when the two models produced near identical estimates.

The production of annual run estimates for salmon and sea trout was one of the key objectives of the DSAP and this has been achieved. This information has already been used in a regional context to calculate egg deposition densities and define spawning targets as part of a net limitation order review. Run estimates for sea trout in 1991 and 1992, although based on part-year tagging programmes, were considered to provide a reasonable measure of the annual run. This was because most sea trout would have been available for tagging (even in 1991), and recaptured fish were well mixed within the second catch (i.e. they returned outside the month in which they were tagged). For salmon, annual run estimates required the use of trapping indices to estimate the out-of-season component, as well as in-season mark-recapture estimates. The regression model used to derive the latter may need to be refined as data from radio-tagged salmon are further examined to clarify the factors affecting trap efficiency.


## Acknowledgements

We would like to thank all those who have contributed to the work of the DSAP in the last three years – in particular the staff of the radio-tracking programme working under Bill Purvis. The views expressed in this chapter are those of the authors and do not necessarily reflect those of the National Rivers Authority.


## References

Beaumont W.R.C., Welton J.S. & Ladle M. (1991) The exploitation by rod and line of Atlantic salmon (*Salmo salar* L.) in a southern chalk stream. Joint Atlantic Salmon Trust/Royal Academy Workshop on: The Measurement and Evaluation of the Exploitation of Atlantic Salmon, 10 pp.

Chapman D.G. (1951) Some properties of the hypergeometric distribution with applications to zoological sample censuses. *University of California Publication on Statistics* **1**, 131–160.

Dunkley D.A. (1992) Measurement of exploitation levels in rod and net fisheries. Joint Atlantic Salmon Trust/Royal Irish Academy Workshop on: The Measurement and Evaluation of Exploitation in Atlantic Salmon, 96 pp.

Gee A.S. & Milner N.J. (1980) Analysis of 70-year catch statistics for Atlantic salmon (*Salmo salar* L.) in the River Wye and implications for the management of stocks. *Journal of Applied Ecology* **17**, 41–57.

Gregory J., Johnson P., Mawle G.M., Milner N.J. & Winstone A.J. (1988) Collection of catch data from salmon and sea trout rod fisheries in Wales. In: W.L.T. van Densen, B. Steinmetz and R.H. Hughes (eds) *Management of freshwater fisheries*. Pudoc: Wageningen, pp 540–546.

Mills C.P.R (1992) Estimates of exploitation rates by rod and line of Atlantic salmon (*Salmo salar* L.) and sea trout (*Salmo trutta* L.) in the British Isles. Joint Atlantic Salmon Trust/Royal Irish Academy Workshop on: The Measurement and Evaluation of Exploitation in Atlantic Salmon, 9 pp.

Pratten D.J. & Shearer W.M. (1981) Fishing mortality of North Esk salmon. CM 1981/M:26. Copenhagen: International Council for the Exploitation of the Sea, 10 pp.

Ricker W.E. (1975) Computation and interpretation of biological statistics of fish populations. *Bulletin of the Fisheries Research Board of Canada* **191**, 382 pp.

Schaefer M.B. (1951) Estimation of size of animal populations by marking experiments. *U.S. Fish and Wildlife Service Fisheries Bulletin* **52**, 191–203.

Sokal R. & Rohlf F.J. (1981) *Biometry*. New York: W.H. Freeman and Co, 859 pp.

Solomon D.J. & Storeton-West T.J. (1983) Radio-tracking of migratory salmonids in rivers, development of an effective system. Fisheries Research Technical Report, MAFF Directorate of Fisheries Research, Lowestoft. No. 75, 11 pp.

Swain A. & Champion A.S. (1978) The composition, exploitation and fecundity of adult salmon migrations on the River Axe, Devon. Salmon and Freshwater Fisheries Laboratory, MAFF, London. JC 122/1, 20 pp.

Walker A.F. (1984) The trout of the Findhu Glen Burn, Tayside. In: E.D. Le Cren (ed.) The biology of sea trout. Summary of a symposium held at Plas Menai; Atlantic Salmon Trust Publication, Pitlochry, 10 pp.

# Chapter 22
# Population dynamics of steelhead trout in a coastal stream, the Keogh River, British Columbia

B.R. WARD *Ministry of Environment, Lands and Parks, Fisheries Centre, 2204 Main Mall, University of British Columbia, Vancouver, B.C., Canada V6T 1Z4*

**Abstract**   The population dynamics of steelhead trout (*Oncorhynchus mykiss*) of the Keogh River, on northern Vancouver Island, British Columbia, Canada, have been studied since 1976. The estimated population of spawners ($P$) and the smolt recruitment provided information on the relation between spawners and recruits ($R$). $R$ as smolt yield and subsequent adult recruits were compared to spawner density using the relation between smolt size and smolt-to-adult survival. This required the number, age structure and size-at-age of smolts. The Beverton-Holt recruitment model was used with and without hatchery spawners included, as well as from years with and without whole-river nutrient enrichment. Results suggested a decline in recruitment when hatchery returns were high, and near doubling of smolt yield with nutrient addition. Broad variation in steelhead life history characteristics, such as adult size and fecundity, smolt age and size, and climatic events can severely alter $R$, justifying conservative management.

KEYWORDS: steelhead trout, population dynamics, recruitment, smolt yield, enrichment

## 22.1   Introduction

Assessment of fish populations requires knowledge of population dynamics and trends in production over considerable time periods. The impacts of factors affecting yield, such as habitat loss or gain, competition and predation, pollution, nutrient level, enhancement techniques, or harvest, cannot be confidently assessed without recruitment information.

Knowledge of the population dynamics of salmonids requires years of study. Rarely is this knowledge available because the data required must be obtained over many generations of fish which reach maturity at ages that can be greater than that of the managing agency or, at least, its current administration. Therefore, continuity and commitment over long periods is difficult to maintain; management often becomes based on piecemeal information where population fluctuations are not only poorly understood, but also inaccurately documented and difficult to relate to stock status

308

(Kope & Botsford 1988). Stock-recruitment relationships have been examined based on angler catch (Chadwick & Randall 1986), but require refinement. Fisheries managers in the province of British Columbia, Canada, recognized the need for knowledge of stock-recruitment information, and conducted life history studies of steelhead trout (*Oncorhynchus mykiss*) at the Keogh River research station, on Vancouver Island, B.C.

Assessment of stock dynamics in an index population may be useful for steelhead management in general. Examination of a steelhead population which is mainly unaffected by commercial fisheries interceptions (except for small numbers of kelts, spawned emigrants; Evans 1979) and with low (near zero) harvest from recreational fishing (catch-and-release) may provide valuable insight on the generalized stock-recruit relationship in steelhead. Such information could improve the definition of the minimum required escapement for this species. Whereas interception of winter-run steelhead may be low, summer-run fish (Withler 1966) may be harvested at relatively high rates when their migration pattern matches the target species in commercial fisheries of the west coast. Thus, there is a need to examine allowable harvest rates in steelhead trout based on recruitment data. Detailed study at each life stage is required to discern the relation of one stage to the next and assess at which stage density dependent and density independent mortality occurs. Accordingly, the numbers of recruits produced at various levels of spawning population size were monitored, while perturbation and enhancement experiments were ongoing in the watershed.

Life history studies focused on freshwater and marine life stages were combined to examine overall stock-recruitment in steelhead trout. Earlier studies examined the marine phase of steelhead life history (Ward *et al.* 1989; Ward & Slaney 1988) and aspects of the freshwater phase (Ward & Slaney 1993a). This study uses those results and further work to estimate the Beverton-Holt parameters (Hilborn & Walters 1992; Ricker 1975). Smolt yield was asymptotic, thus the Beverton-Holt curve was appropriate (Ward & Slaney 1993a). In addition, the role of nutrient levels in streams in relation to habitat capacity and in the control of productivity of trout are discussed, and the potential impact of over-escapement of hatchery fish on wild escapement is considered.


**22.2    Methods**

The Keogh River, a third order coastal stream, is 31 km in length, in a drainage area of 129 $km^2$ with mean annual discharge of 5.6 $m^2 s^{-1}$ (max. 254 $m^2 s^{-1}$), situated on northern Vancouver Island, British Columbia, Canada (coastal western hemlock biogeoclimatic zone). The Keogh River watershed was described in detail by Ward and Slaney (1979), Johnston *et al.* (1986, 1990), and Irvine and Ward (1989).

The Keogh River fish fence is situated 300 m upstream from the river mouth, at approximately the high tide mark. Various experiments in the past have impacted salmonid smolts, including whole-river fertilization (Slaney & Ward 1993; Johnston *et al.* 1990; Perrin *et al.* 1987), instream habitat structures (Ward & Slaney 1979,

1993b), net-pen rearing of steelhead smolts (Ward *et al.* 1994 in press; Ward & Slaney 1990), and tagging studies of steelhead (Ward *et al.* 1990) and coho (Irvine & Ward 1989).

The procedure for deriving the population estimates and ages of adult steelhead, and methods of enumerating and sampling smolts in the Keogh River were described by Slaney *et al.* (1990), Ward and Slaney (1988, 1990, 1993a), Ward *et al.* (1989, 1990) and Irvine and Ward (1989). Similarly, survival estimates and variation in age structure were documented by Ward (1988, 1989) and Ward and Slaney (1988, 1993a). Methods of data collection were unchanged to include recent years in this analysis.

As described in Ward and Slaney (1988, 1990) and Slaney *et al.* (1990), the total numbers of adult males and females were estimated separately by mark-recapture using the adjusted Petersen estimate (Ricker 1975) by marking upstream migrant adults (December to May) and capturing kelts during their downstream migration (March to June). Kelts, captured as they emigrated, were examined for marks. The estimated population size was calculated as:

$$N = (M+1)\,(C+1)\,/\,(R+1) \tag{1}$$

where $N$ is the Petersen population estimate, $M$ is the number of marked adults, $C$ is the number of captured kelts and $R$ is the number of recaptured kelts.

Confidence limits (CL) on the male and female population estimates were calculated based on the Poisson frequency distribution where $R$ is substituted for $x$ in Appendix II of Ricker (1975). The CL for joint estimates of males and females were approximated based on the principal of variance additivity:

$$-95\%\ \mathrm{CL} = N - [(N_m - L_m)^2 + (N_f - L_f)^2]^{-2} \tag{2}$$

$$+95\%\ \mathrm{CL} = N + [(U_m - N_m)^2 + (U_f - N_f)^2]^{-2} \tag{3}$$

where $N_m$ is the number of males, $N_f$ the number of females, $L_m$ is the lower CL on the male $N$, $L_f$ is the lower CL on the female $N$, $U_m$ is the upper CL on male $N$ and $U_f$ is the upper CL on female $N$.

Fish captured in the smolt trap were processed daily according to procedures in Ward *et al.* (1990). All fish were counted and randomly sampled for lengths and weights throughout the trapping period, with the exceptions of steelhead smolts in 1990 and 1991, when all tagged smolts were sampled during coded-wire tagging (Irvine & Ward 1989) in the adipose fin (Ward *et al.* 1990). Scale samples from above the lateral line and behind the dorsal fin were taken from steelhead smolts in a sampling regime that was stratified according to fish size and migration time (Ward *et al.* 1989; Ward & Slaney 1988).

The relationship between smolt number and number of adults returning was examined, as in Ward and Slaney (1988), but included data up to 1989 for a sample size of 12 smolt years, versus 7 in the previous analysis. The linearity of this relationship was tested with least squares regression. It had previously been demonstrated that steelhead survival during the marine phase was highly size-dependent (Ward & Slaney 1988, Ward *et al.* 1989). In this chapter, the relationship between mean smolt size and smolt-to-adult survival is further examined to include recent results and confirm previous conclusions. Using non-linear regression, a line of best fit was determined for the relation between back-calculated smolt size and survival (Ward *et al.* 1989). The back-calculation procedure was broader in the range of lengths considered for predictive purposes than that based on mean smolt length.

Steelhead smolt yield was tabulated into the brood year of origin based on fish age and total smolt yield, and compared to the spawner density (Ward & Slaney 1993a). A Beverton-Holt curve (Hilborn & Walters 1992) was fitted to data on smolt recruitment from smolts derived from years of natural nutrient level. Four smolt broods (1982 to 1985) were wholly or partially affected by whole-river nutrient addition (Johnston *et al.* 1990; Slaney & Ward 1993). The Beverton-Holt curve was forced through only one of these data points (1984) since it represented the only smolt group which contained smolts of all ages affected by fertilizer additions. (A more comprehensive experiment would have allowed continued fertilizer addition for several additional years to more accurately define the curve shape.)

Adult recruits (*R*) were calculated from a brood-year's yield of smolts and smolt's survival rate based on size at age. Smolt numbers and length-at-age were tabulated according to their brood year of origin (1976 to 1989) and age, based on ageing of smolts and counts through the fish fence (Ward & Slaney 1988, 1993a). Recruits from spawners (*P*) were calculated from the highly correlated relationship between back-calculated smolt size and survival (Ward *et al.* 1989). The number of recruits calculated within each smolt-age category was summed across all ages to give totals from each brood year.

This procedure assisted in filtering variation possibly caused by environmental factors in the ocean, and allowed clearer focus on environmental factors in fresh water which may have created variation in the stock-recruit relationship. Recruits based on age analysis were compared to the recruits based on smolt size and survival. This method of calculating recruits was preferred to the alternative of basing numbers on scale ageing and the number of adult returns to determine their brood year of origin. Adult scales were frequently regenerated or unreadable in the freshwater growth zone, and sample sizes were at times too small (particularly in years of low return) to provide reliable estimates over the broad range in age structure (Ward & Slaney 1988). Furthermore, a method based on age would not permit the filtering out of the El Niño effects (1982) on four broods, and reduced the available brood years for analysis (*n* = 8 *v.* 14), which was already minimal. Thus, the method of size-biased survival was used.

The relationship between recruits (*R*) and spawners (*P*) was asymptotic, thus the Beverton-Holt model (Hilborn & Walters 1992) was used to estimate stock-recruit parameters:

$$\log(R) = \log(\alpha P \, / \, (\beta + P)) + w \tag{4}$$

where $\alpha$ is a constant related to the asymptote (maximum recruitment) and $\beta$ is a constant that describes the spawning stock size needed to produce (on average) recruitment equal to $\alpha/2$.

Parameters $\alpha$ and $\beta$ were derived from non-linear estimation using $R$ and $P$ in Equation (4). Given $P$ and $R$, it was then possible to calculate several other recruitment statistics (Hilborn & Walters 1992), including: maximum sustainable yield (MSY), the number of spawners needed $(P_s)$ for MSY, the recruitment $(R_s)$ at MSY, and the rate of exploitation at MSY $(\mu_s)$.

These estimates were generated with and without hatchery spawners included in the calculations. In both cases, data of 1986 and 1987 were omitted in the analysis since they were below the replacement line of the recruitment curve and possibly outliers due to climate.

## 22.3    Results

The steelhead runs of the Keogh River remained relatively high from 1987 to 1990 compared to previous years (1976 to 1986; Ward & Slaney 1988), but recent declines (1991 to 1993) were indicated (Fig. 22.1).

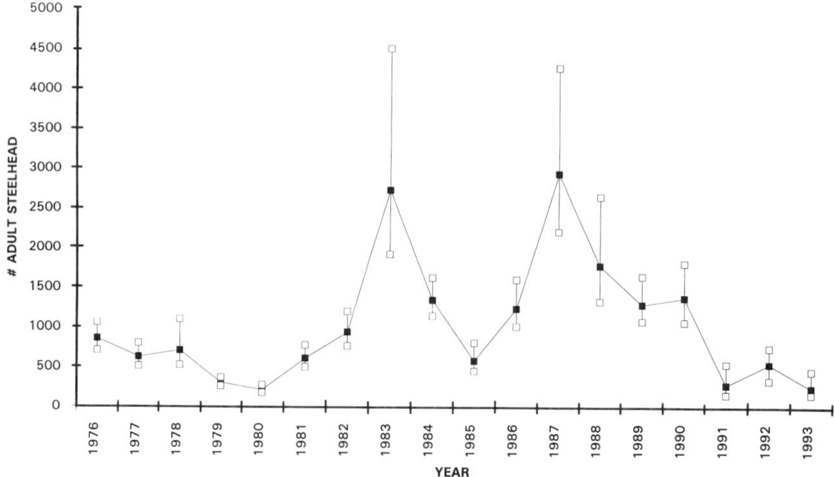


**Fig. 22.1**  Estimated numbers of the winter-run steelhead returning to spawn in the Keogh River, B.C., from 1976 to 1993 ($\pm 95\%$ confidence limits).

Ageing of the adults indicated broad variation in age at return (Ward 1989) and allowed determination of the smolt group from which they originated. Regression of 14 years of smolt output (1976 to 1989) on subsequent adult return suggested a linear relationship between smolts and adults (Fig. 22.2; $r^2 = 0.89$, $P < 0.05$) over a six-fold difference in smolt number, and can be described by the equation:

$$\ln(adults) = 1.31\ln(smolts) - 4.68 \tag{5}$$

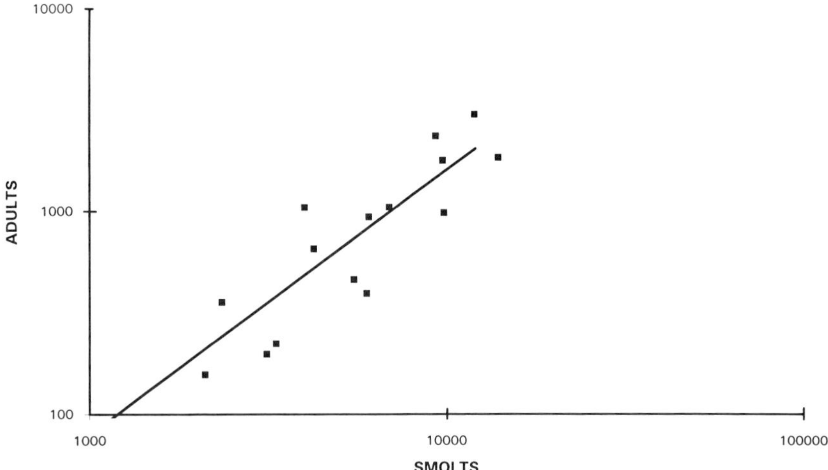



**Fig. 22.2**   The relationship between counts of smolts from 1977 to 1989 at the Keogh River fish fence and their subsequent return as adults (note natural logarithmic scale).

Most of the variation in the relation of smolts and adults was explained by smolt size (Ward & Slaney 1988). Inclusion of recent results ($n = 10$) improved on the relation of mean smolt length and survival from smolt-to-adult:

$$smolt\ survival\ (\%) = 5.158(smolt\ length_{cm}) - 73.996 \qquad (6)$$

The 1982 data point was, as in Ward and Slaney (1988), an outlier in this relationship ($r^2 = .85$, $P < 0.05$, Fig. 22.3), where survival was almost twice that expected.

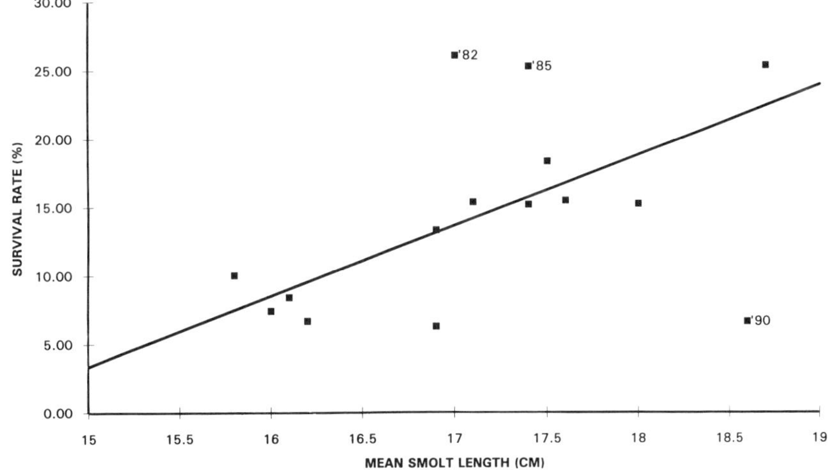



**Fig. 22.3**   The relationship between mean length (cm) of the steelhead smolts from 1977 to 1990 at the Keogh River, British Columbia, and their subsequent survival to return as maiden-run adults. The years 1982, 1985, and 1990 are suspected to be non-typical due to ocean conditions, as explained in the text.

However, an additional smolt year, 1985, was also warmer than normal, and likewise appeared to display above-average survival. Conversely, a recent and persistent El Niño on the west coast beginning in 1990 (Fig. 22.3) seemed to generate opposite results (lower than expected survivals), and needs further study.

For predictive purposes, the relation of back-calculated smolt size (*l*) and smolt-to-adult survival (Ward *et al.* 1989) was more useful, since the range in lengths examined was broader. A line of best fit for this relationship (Fig. 22.4) was a 3$^{rd}$ order polynomial:

$$smolt\ survival = 94.483 - 24.499l + 1.714l^2 - 0.033l^3 \tag{7}$$

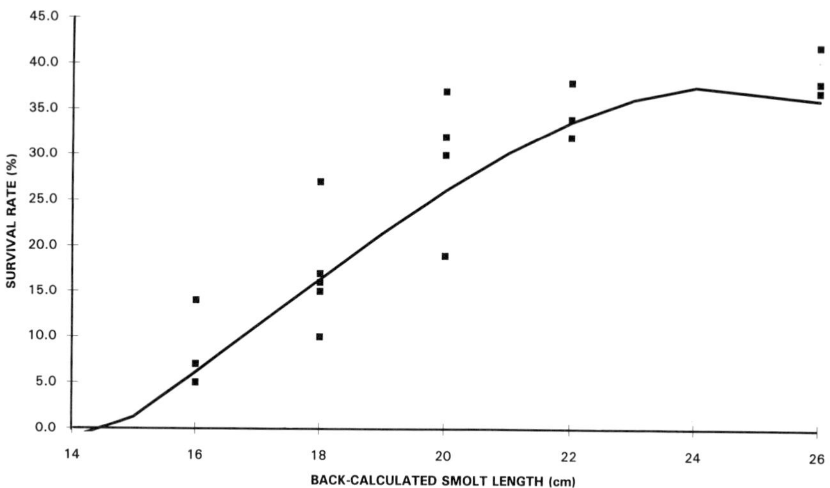

**Fig. 22.4**    The relationship between back-calculated smolt length and survival to adult return to the Keogh River, British Columbia. The solid line is the curve used to estimate adult recruits, the filled squares are the data from Ward *et al.* (1989).

Smolts were tabulated into their brood year of origin (1975 to 1989) based on ageing of smolt scales (Ward & Slaney 1988, 1993a). Smolt length at age was used to determine expected survival rate (Equation 7), and number per age was multiplied by this rate to give recruits per age. Summing across ages provided total recruits for 1975 to 1989 brood years (Table 22.1). Estimated recruits varied from 265 to 1251.

Comparison of mean escapement with mean recruitment can be an important indicator of population health. Mean escapement (1030; Table 22.1) was similar to estimates of carrying capacity (defined as the point where the recruitment curve crossed the line of replacement, or 1000 spawners; Figs 22.5, 22.6), and slightly higher than mean recruitment (849), suggesting the Keogh River steelhead population was stable. A population that was experiencing over-harvest might be expected to show mean spawning escapements that were substantially lower than mean recruitment estimates.

Smolt recruits varied from 2915 to 10 750 fish (Table 22.1). Asymptotic yields of

**Table 22.1** Steelhead smolt and adult recruitment to the Keogh River from brood years 1975 to 1989, based on the relationship between smolt size and survival.

| Brood Year | Spawners | | | Number of smolts by age | | | | | Total smolts | Smolt length by age | | | | | Recruits by smolt age | | | | | Total recruits |
|---|---|---|---|---|---|---|---|---|---|---|---|---|---|---|---|---|---|---|---|---|
| | Wild | Hatchery | Total | 1 | 2 | 3 | 4 | 5 | | 1 | 2 | 3 | 4 | 5 | 1 | 2 | 3 | 4 | 5 | |
| 1972 | | | | | | | | | | | | | | | | | | | | |
| 1973 | | | | | | | 227 | | | | | | 221 | | | | | | 77 | | |
| 1974 | | | | | | 1207 | 198 | | | | | 180 | 221 | | 0 | 0 | 198 | 67 | | 265 |
| 1975 | | | | 0 | 909 | 789 | 1173 | 44 | 2915 | | 152 | 176 | 225 | 252 | 0 | 20 | 113 | 411 | 16 | 561 |
| 1976 | 859 | | 859 | 0 | 1117 | 4892 | 561 | 118 | 6688 | | 149 | 179 | 191 | 281 | 0 | 9 | 776 | 123 | 32 | 940 |
| 1977 | 625 | | 625 | 0 | 835 | 3350 | 2713 | 16 | 6914 | | 151 | 164 | 226 | 272 | 0 | 15 | 273 | 958 | 5 | 1251 |
| 1978 | 706 | | 706 | 0 | 1546 | 5592 | 104 | 0 | 7242 | | 151 | 186 | 222 | | 0 | 27 | 1086 | 36 | 0 | 1149 |
| 1979 | 299 | | 299 | 0 | 3474 | 2376 | 401 | | 6251 | | 156 | 177 | 218 | | 0 | 144 | 353 | 133 | 0 | 629 |
| 1980 | 209 | | 209 | 0 | 1504 | 3701 | 866 | | 6071 | | 153 | 180 | 224 | | 0 | 40 | 606 | 302 | 0 | 948 |
| 1981 | 608 | 44 | 652 | 0 | 1951 | 3569 | 205 | | 5725 | | 160 | 184 | 209 | | 0 | 119 | 657 | 61 | 0 | 838 |
| 1982 | 938 | 556 | 1494 | 0 | 5291 | 2504 | 39 | | 7834 | | 163 | 188 | 206 | | 0 | 404 | 511 | 11 | 0 | 926 |
| 1983 | 2730 | 739 | 3469 | 0 | 6600 | 1257 | 167 | | 8024 | | 168 | 184 | 223 | | 0 | 672 | 232 | 58 | 0 | 961 |
| 1984 | 1344 | 2904 | 4248 | 0 | 8300 | 2443 | 7 | | 10750 | | 155 | 203 | 243 | | 0 | 303 | 670 | 3 | 0 | 975 |
| 1985 | 580 | 581 | 1161 | 226 | 9786 | 287 | 108 | | 10407 | 146 | 164 | 187 | 225 | | 3 | 796 | 57 | 38 | 0 | 894 |
| 1986 | 1237 | 2747 | 3984 | 1485 | 3030 | 1969 | 44 | | 6528 | 142 | 159 | 183 | 251 | | 0 | 170 | 353 | 17 | 0 | 539 |
| 1987 | 2939 | 1746 | 4685 | 0 | 2146 | 762 | 1119 | | 4027 | | 158 | 175 | 202 | | 0 | 110 | 105 | 302 | 0 | 517 |
| 1988 | 1778 | 283 | 2061 | 23 | 2324 | 4057 | 884 | | 7288 | 140 | 165 | 185 | 211 | | 0 | 201 | 768 | 271 | 0 | 1240 |
| 1989 | 1290 | 825 | 2115 | 14 | 836 | 4074 | | | 4924 | | 163 | 191 | | | | 64 | 892 | 0 | 0 | 956 |
| 1990 | 1374 | 542 | 1916 | 0 | 2185 | | | | | | 154 | | | | | | | | | |
| 1991 | 277 | 50 | 327 | | | | | | | | | | | | | | | | | |
| 1992 | 540 | 215 | 755 | | | | | | | | | | | | | | | | | |
| 1993 | 210 | 30 | 240 | | | | | | | | | | | | | | | | | |
| Mean | 1030 | 866 | 1656 | 117 | 3310 | 2677 | 573 | 45 | 6773 | 143 | 158 | 183 | 220 | 268 | 0 | 193 | 478 | 174 | 4 | 849 |

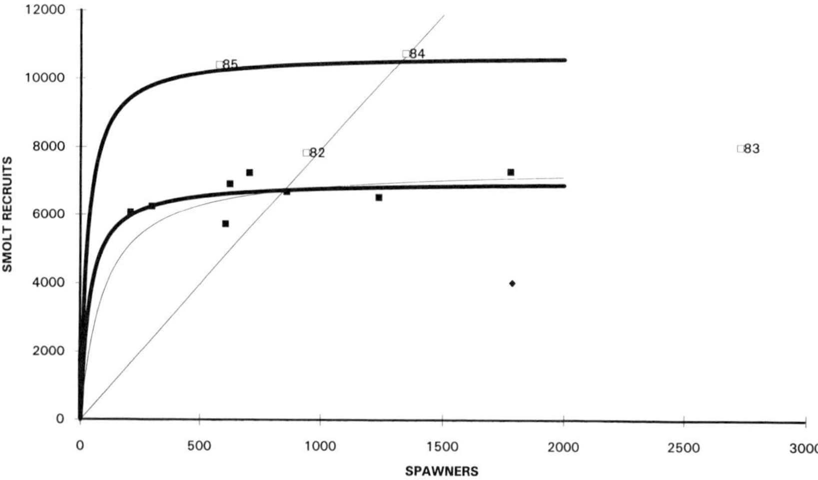



**Fig. 22.5**   The relationship between Keogh River steelhead spawners and recruits based on smolt size and survival (see text), for wild spawners and hatchery spawners. The curved line is the Beverton-Holt curve fit to the data. The straight line is the line of replacement stock.

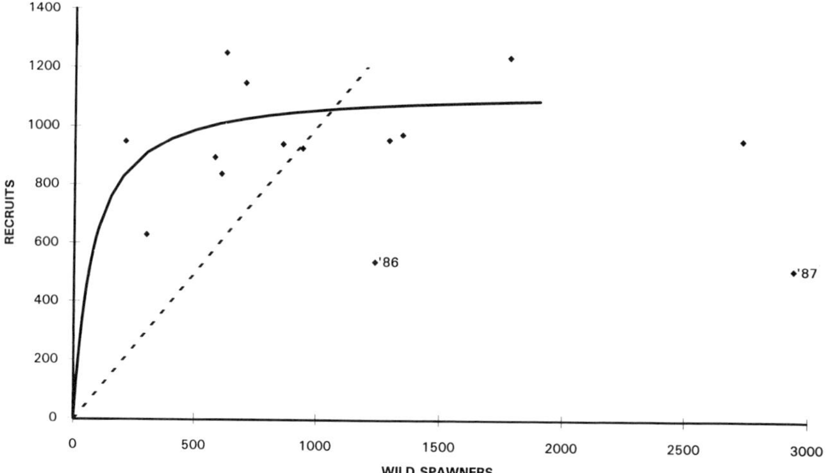



**Fig. 22.6**   The relationship between Keogh River wild steelhead spawners and smolt recruits. The curved line is the Beverton-Holt curve fit to the data. The straight line is the line of replacement stock. Open squares represent smolts which were wholly (1984) or partially (1982, 1983, 1985) affected by whole-river fertilization experiments.

6500 smolts were estimated from years unaffected by whole-river fertilization; maximum yield was obtained from smolts which had experienced fertilization of the Keogh River throughout their rearing phase in fresh water (Fig. 22.5), but the yield from only one brood year was available. The Beverton-Holt curve was forced through that data point (1984), assuming it represented expected recruitment under fertilized conditions. The whole-river fertilization experiment would have had to be conducted

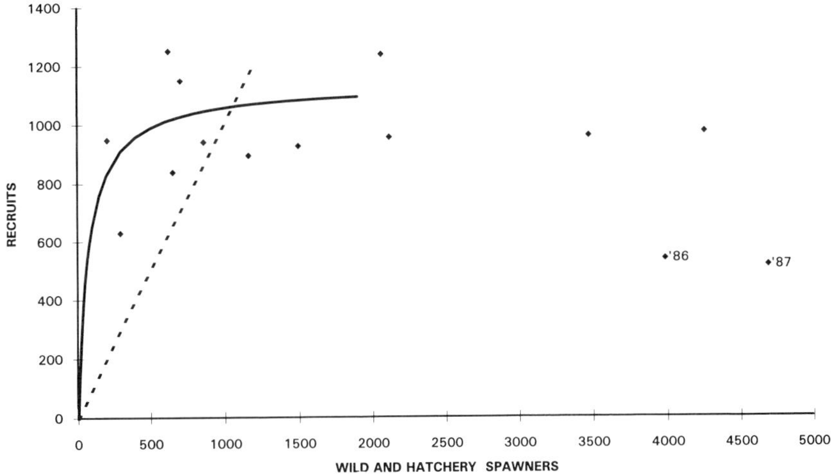



**Fig. 22.7**   The relationship between Keogh River steelhead spawners and recruits based on smolt size and survival (see text), for wild spawners only. The curved line is the Beverton-Holt curve fit to the data. The straight line is the line of replacement stock.

for several additional years to obtain more definitive estimates of asymptotic yield under enhanced nutrient conditions. Slaney and Ward (1993) discussed changes to fry, parr, smolt and adult numbers as a result of experimental fertilization of the nutrient deficient stream. The results indicated that recruitment and the carrying capacity of streams is a function of the productivity of the system at the lower trophic levels. Populations which resemble the Keogh under nutrient-enhanced conditions, i.e. with smolt ages averaging 2 yr *v.* 3 yr (Slaney & Ward 1993), could sustain harvest rates that are approximately 15% higher than that estimated for the Keogh River. Using geometric mean smolt-to-adult survival rates (12.6%), 500 spawners might yield, on average, 6542 smolts or 824 adult recruits naturally, but 10 155 smolts and 1280 adults if fertilized, assuming the smolts were of the same size and changes to smolt age structure did not alter return rates. Systems which yield older smolts (e.g. Skeena River; Narver 1969) would sustain lower harvest rates. The recruitment curves predicted for steelhead of the Keogh River are in close agreement with similar curves for Atlantic salmon of different smolt ages (Symons 1979).

Survival rates from smolt to adult were based on smolt size to derive recruit estimates of the smolt yield. The shape of the relationship between spawning stock and recruitment stock was similar with (Fig. 22.7) or without (Fig. 22.6) the inclusion of hatchery fish. Since the data were described by a Beverton-Holt curve, and hatchery fish were added to the spawning population size when the stock was at the asymptotic level, there was no effect to the curve shape. In both cases, the recruits of brood years 1986 and 1987 were below the replacement line (Figs 22.6, 22.7). The years 1986 and 1987 had drought conditions during critical periods (mid-summer), a condition which persisted on northern Vancouver Island for most of the 1980s. Recruitment data from these two years were excluded from parameter estimates, but were used to indicate adjustments required to harvest rates to account for environmental variability. Alternatively, these recruits were affected by some other variable, possibly including the density of hatchery spawners.

Two of twelve brood years may have been severely affected by drought. Wild fish stock recruitment at the Keogh River suggested that maximum recruitment was 1070 fish, and that $R_s$ and $P_s$ (recruits and spawners at maximum sustained yield, respectively) were 849 and 221 fish, respectively, for an MSY of 628. This suggests that in the absence of climatic perturbation, the allowable harvest rate, $\mu_s$ was 73.6%. However, to conservatively allow for known environmental effects, the $\mu_s$ must be set lower, at approx 57%, calculated as:

$$\mu_a = (R_{min} - P_s) \, / \, R_{min} \tag{8}$$

where $\mu_a$ is the adjusted allowable harvest, and $R_{min}$ is lowest recruitment. It remains to be tested that where exploitation is adjusted downwards to compensate for minimum recruitment due to climate, the expected benefits are attained, i.e. recruitment would increase with decreased harvest despite the disruptive event. As Elliott (1989) noted, the frequency of severe climatic events must be considered in recruitment studies. These results indicate that in at least two of the thirteen broods examined, recruitment was below that expected, either based on climate conditions or because the Beverton-Holt model was incorrect for these data.

Variation in age-at-return may justify a further decrease in the allowable harvest. Ward and Slaney (1988) and Ward (1989) indicated that 30 to 75% of female adults spent 3 yr in the ocean before returning to spawn. As a result of this variable age structure, expected recruitment does not equal run size, as may be the case in autumn-spawning salmon species of the west coast of Canada, where cohorts of equivalent age structure return simultaneously. Thus, some steelhead runs were very low, at 280 (1979) or 200 (1980) maiden-run fish, partly related to this age variation. In such years, exploitation at 60% would result in escapement far below $P_s$, and particularly when such low returns are followed sequentially. The remaining portions of cohorts that returned in the years before or after may compensate somewhat for this possible over-exploitation. Nevertheless, the spawning stock in some years (2 of 13 in this study) could be too low to provide any surplus. Thus, an exploitation rate that accounts for age variation may be justified.

Variation around the mean adult population size may be substantial based on variation in age alone. The mean run size from 1976 to 1986 was 1030 fish (Table 22.1), and geometric mean, 776. A return year could have as low as 460 to as many as 1086 fish due to age variation alone. A minimum allowable harvest rate to compensate for age variation would be 48%. This rate is similar to that estimated for steelhead stocks in general by Bjornn (1977, 1987). However, to allow for error, harvest rates may be adjusted downwards to provide a safety factor. Thus, a harvest rate no greater than 40% is suggested.

Age variation and its consequence to highly variable adult returns in a given year is not addressed by classical stock-recruit analysis. If Keogh stocks were harvested at, say, 40% from 1976 to 1993, the maiden runs (Ward & Slaney 1988) would be over-harvested in four of the 18 years. Inclusion of repeat spawners would increase the number of spawners by an average of 10% (Ward & Slaney 1988), but would still not provide adequate escapement. These results suggest, due to the current inability to predict age-at-return with confidence, that harvest must be dramatically reduced when forecasts for returns are equal to or less than $R_s$.

## 22.4   Discussion

Steelhead trout are iteroparous salmonids whose recruitment is dependent on the abundance of several age groups (Welch 1987). Variability in the abundance of the age groups is large (Ward & Slaney 1988), and factors which determine abundance were related both to density during critical periods and climatic events. Regulation of fish population stability has been shown to be linked to both density-dependent and climatic factors (Crecco *et al.* 1986; Reisenbichler 1986; Kope & Botsford 1988). In steelhead populations, the ability to detect stochastic climatic effects is hampered by the complexity of the life history. Compounding this is the likelihood that different climatic events affect different age classes, as demonstrated with ecologically similar Atlantic salmon (Frenette *et al.* 1984).

It is difficult to provide analytical power to these interpretations even if repeated observations were available from other streams. Walters and Collie (1988) argue for monitoring activities coupled with deliberate experimental management to more fully comprehend age-class fluctuations. As a method of understanding steelhead age-class variation, it would be costly, requiring many years of study for these fish which in a single run can potentially contain twelve age groups (from six different broods), varying from age 2.1 to 5.3 (Ward & Slaney 1988), each capable of repeat spawnings. Deliberate manipulation and observation in other streams may prove useful, while continuing studies in at least one stream. There is also need to carefully consider manipulation of spawner density if it can be coupled with more intensive evaluation of effects through each life stage, with external controls monitored simultaneously.

Bjornn (1977) suggested that the harvest rate of steelhead should not exceed 50% in mixed stock fisheries targeting on Clearwater River wild and hatchery stocks in Idaho, dependent on the success of a transportation programme. The results of this study confer with a low harvest rate on steelhead stocks, assuming applicability of these results to other stocks (Ward & Wightman 1989), which also contain broad age variation (Hooton *et al.* 1987).

Kope and Botsford (1988) noted that for populations with broad age structure, abundance indices like catch and spawning stock would not perform as well in providing a reasonable record of recruitment as would more detailed life stage sampling. This is a strong argument for continuation of studies that look more closely at population enumeration over many generations, such as that begun at the Keogh River, but with increased effort through each life stage.

Different conclusions may have been reached by Ward and Slaney (1993a) given the additional smolt yields by brood year (1983 to 1989). An apparent increased productivity and capacity was suggested from the smolt yield from brood years which were wholly (1984) or partly (1982, 1983, 1985) affected by the addition of nutrients to the whole river. Furthermore, there is a suggestion, albeit weak, when comparing the data with and without hatchery spawners, that there may be a negative or downward trend in recruitment with very high numbers of spawners. Thus, not only do hatchery smolts survive at lower rates than wild fish (Ward & Slaney 1990; Ward *et al.* 1994), but also their return in high numbers may have adverse effects on wild spawners, possibly due to crowding of the spawning areas. Alternatively, some abhorrent climatic event affected the recruitment from 1986 and 1987 spawners. It is noteworthy that in one other brood year, 1984, the contribution of hatchery spawners was also very high, yet recruitment was only slightly below the asymptote predicted from the Beverton-Holt curve (Fig. 22.7).

The leverage of the 1979 and 1980 brood year points was very high and thus strongly affects the parameters of a recruitment analysis. Clearly, more data are required in that area of the curve. High variability might be expected in this area of the curve and thus, uncertainty in parameter estimates. Hilborn and Walters (1992) discussed a number of difficulties of interpretation of stock-recruitment data, all of which probably apply to this analysis, including a lack of data at lower densities. Current status of steelhead returns is poor (Fig. 22.1), thus additional information on recruitment from low spawner density may be furthcoming if the project continues.

This preliminary analysis must be interpreted with caution. More data from the recruitment of broods of steelhead spawners is required, preferably based on age structure. The method described in this chapter, which relied upon the relation between smolt number and size, should yield the same Beverton-Holt parameters as one based on age of recruits. However, there were several sources of error in the limited data. Nine of the thirteen broods examined were unaffected by stream fertilization experiments (Johnston *et al.* 1990). Four of the broods were partially or wholly influenced by stream fertilization, which altered the productivity of the stream and culminated in lower age of smolts (Slaney & Ward 1993; Johnston *et al.* 1990; Slaney *et al.* 1986). The influence of the 1982/83 El Niño event was supposedly removed by deleting the 1982 smolts from the size-survival relationship. However, other smolt groups may have also been influenced by that or other environmental events. If so, it was not detectable from these results. Due to enhanced production in the river and possibly improved survival conditions for some smolt groups in the ocean, the allowable harvest estimate may have been inflated. Until a more complete data set containing 20 or more broods, and representative of the range in rearing conditions, is available, a conservative approach is urged.

Despite the preliminary nature of these results and uncertainty in parameter estimates, methods to improve stock assessment are suggested from the analysis. In assessment of a population of fish, the stock productivity, which is a function of age-at-maturity and fecundity, and the capacity of the habitat to support fish must be considered. An examination of the recruits produced from several broods is required in at least some index systems, representative of biogeoclimatic zones, or indicative of the population dynamics of the species. More detailed assessment at each life stage of steelhead has revealed where density dependence is paramount, the fry to parr stage (Ward & Slaney 1993a). Thus, autumn fry or spring $1+$ parr may provide a useful index of expected recruits, whereas emergent fry reflect spawner abundance. More detailed stock assessment, and future models, should consider fish size and territory size (e.g. Grant & Kramer 1990) in relation to stream productivity and capacity (nutrients and temperature), as well as other facets of the environment (e.g. Keeley & Grant 1995), and the interaction of all of these factors with fish density.

## Acknowledgements

The collection of information at the Keogh River has required many years of concerted team effort. The team is deeply thanked. I am particularly grateful to P.A. Slaney, D.W. Narver, A.F. Tautz, and the encouragement of Loon Ashley and Mike Iss. N.T. Johnston and K.I. Ashley are thanked for comments on a draft manuscript; C.J. Walters also suggested several improvements.

## References

Bjornn T.C. (1977) Mixed stocks of wild and hatchery steelhead trout. A management problem in Idaho. In: T.J. Hassler & R.R. Van Kirk [eds] *Proceedings of the genetic implications of steelhead management symposium*, Humboldt State Univ.: Arcata, California, pp 31–41.

Bjornn T.C. (1987) A life history model for predicting the abundance and yield of anadromous salmonids. *Idaho Coop. Fish and Wildl. Res. Unit Tech. Rep.* **87–2**, 42 pp.

Chadwick E.M.P. & Randall R.G. (1986) A stock-recruitment relationship for Atlantic salmon in the Mirimichi River, New Brunswick. *N. Am. J. Fish. Manage.* **6**, 200–203.

Crecco V., Savoy T. & Whitworth W. (1986) Effects of density-dependent and climatic factors on American shad, *Alosa sapidissma*, recruitment: a predictive approach. *Can. J. Fish. Aquat. Sci.* **43**, 457–463.

Elliott J.M. (1989). The critical period concept for juvenile survival and its relevance for population regulation in young sea trout, *Salmo trutta. J. Fish Biol.* **35(Suppl.A)**, 91–98.

Evans L.K. (1979) Incidental catches of steelhead trout in the commercial salmon fisheries of Barkley Sound, Johnstone Strait and the Dean and Bella Coola rivers area. *Prov. B.C. Fish. Tech. Circ.* No. 39, 30 pp.

Frenette M.M., Caron M., Julien P. &. Gibson R.J. (1984) Interaction entre le débit et les populations de tacons (*Salmo salar*) de la Rivière Matamec, Quebec. *Can. J. Fish. Aquat. Sci.* **41**, 954–963.

Grant J.W.A. & Kramer D.L. (1990) Territory size as a predictor of the upper limit to population density of juvenile salmonids in streams. *Can. J. Fish. Aquat. Sci.* **47**, 1724–1737.

Hilborn R. & Walters C.J. (1992) *Quantitative fisheries stock assessment: choice, dynamics and uncertainty.* London: Chapman and Hall, 570 pp.

Hooton R.S., Ward B.R., Lewynsky V.A., Lirette M.G. & Facchin A.R. (1987) Age and growth of steelhead in Vancouver Island populations. *Prov. B.C. Fish. Tech. Circ. No. 77*, 39 pp.

Irvine J.R. & Ward B.R. (1989) Patterns of timing and size of wild coho salmon smolts (*Oncorhynchus kisutch*) migrating from the Keogh River watershed on northern Vancouver Island. *Can. J. Fish. Aquat. Sci.* **46**, 1086–1094.

Johnston N.T., Irvine J.R. & Perrin C.J. (1986) A comparative evaluation of fence count, mark and recapture and Bendix Sonar estimates of salmon escapements in the Keogh River, a variable flow coastal B.C. river. *Can. Tech. Rep. Fish. Aquat. Sci.* No. **1453**, 44 pp.

Johnston N.T., Perrin C.J., Slaney P.A. & Ward B.R. (1990) Increased juvenile salmonid growth by whole river fertilization. *Can. J. Fish. Aquat. Sci.* **47**, 862–872.

Keeley E.R. & Grant J.W.A. (1995) Allometric and environmental correlates of territory size in juvenile Atlantic salmon (*Salmo salar*). *Can. J. Fish. Aquat. Sci.* **52**, 186–96.

Kope R.G. & Botsford K.W. (1988) Detection of environmental influences on recruitment data. *Can. J. Fish. Aquat. Sci.* **45**, 1448–1458.

Narver D.W. (1969) Age and size of steelhead trout in the Babine River, British Columbia. *J. Fish. Res. Board Can.* **26**, 2754–2760.

Perrin C. J., Bothwell M.L. & Slaney P.A. (1987) Experimental enrichment of a coastal stream in British Columbia: effects of organic and inorganic additions on autotrophic periphyton production. *Can. J. Fish. Aquat. Sci.* **44**, 1247–1256.

Reisenbichler R.R. (1986) Use of spawner-recruit relations to evaluate the effect of degraded environment and increased fishing on the abundance of fall-run chinook salmon, *Oncorhynchus tshawytscha*, in several California streams. PhD. thesis, University of Washington, Seattle, WA. 175 pp.

Ricker W.E. (1975) Computation and interpretation of biological statistics of fish populations. *Bull. Fish. Res. Board Can.* **191**, 382 pp.

Slaney P.A., Perrin C.J. & Ward B.R. (1986) Nutrient concentration as a limitation to steelhead smolt production in the Keogh River. *Proc. Annu. Conf. West. Assoc. Fish. Wildl. Agencies* **66**, 146–155.

Slaney P. A. &. Ward B.R. (1993) Experimental fertilization of nutrient deficient streams in British Columbia, pp. 128–141. In: G. Shooner et S. Asselin [éd.]. *Le développment du Saumon atlantique au Québec: connaître les règles du jeu pour réussir.* Colloque international de la Fédération Québécoise pour le saumon Atlantique. Québec, décembre 1992. Collection *Salmo salar* No 1.

Slaney P.A., Ward B.R. & Berg L. (1990) A preliminary assessment of the effect of external abrasion on the smolt-to-adult survival of net-pen cultured steelhead trout. *Prov. B.C. Fish. Tech. Circ.* No. 88, 11 pp.

Symons P.E.K. (1979) Estimated escapement of Atlantic salmon (*Salmo salar*) for maximum smolt production in rivers of different productivity. *J. Fish. Res. Board Can.* **36**, 132–140.

Walters C.J. & Collie J.S. (1988) Is research on environmental factors useful to fisheries management? *Can. J. Fish. Aquat. Sci.* **45**, 1848–1854.

Ward B.R. (1988) Trends and forecasts from winter-run steelhead at the Keogh River. *Prov. B.C. Fish. Proj. Rep.* No. RD6, 8 pp.

Ward B.R. (1989) Trends and forecasts from winter-run steelhead at the Keogh River: results from 1988. *Prov. B.C. Fish. Proj. Rep.* No. RD18, 6 pp.

Ward B.R. &. Slaney P.A. (1979) Evaluation of in-stream enhancement structures for the production of juvenile steelhead trout and coho salmon in the Keogh River: progress 1977 and 1978. *Prov. B.C. Fish. Tech. Circ.* No. 45, 47 pp.

Ward B.R. & Slaney P.R. (1988) Life history and smolt-to-adult survival of Keogh River steelhead trout (*Salmo gairdneri*) and the relationship to smolt size. *Can. J. Fish. Aquat. Sci.* **45**, 1110–1145.

Ward B.R. & Slaney P.R. (1990) Returns of pen-reared steelhead from riverine, estuarine, and marine releases. *Trans. Am. Fish. Soc.* **119**, 492–499.

Ward B.R. & Slaney P.R. (1993a) Egg-to-smolt survival and fry-to-smolt density dependence of Keogh River steelhead trout. In: R.J. Gibson & R.E. Cutting (eds) Production of juvenile Atlantic salmon, *Salmo salar*, in natural waters. *Can. Spec. Publ. Fish. Aquat. Sci.* **118**. pp. 209–217.

Ward B.R. & Slaney P.R. (1993b) Habitat manipulations for the rearing of fish in British Columbia, In: G. Shooner et S. Asselin [eds]. *Le développment du Saumon atlantique au Québec: connaître les règles du jeu pour réussir,* pp 142–148. Colloque international de la Fédération Québécoise pour le saumon Atlantique. Québec, décembre 1992. Collection *Salmo salar* No 1.

Ward B.R. & Wightman J.C. (1989) Monitoring steelhead trout at the Keogh River as an index of stock status and smolt-to-adult survival: correlations with other data sources. *Prov. B.C. Fish. Manage. Rep.* No. 95: 21 pp.

Ward B.R., Slaney P.A., Facchin A.R. & Land R.W. (1989) Size-biased survival in steelhead trout (*Oncorhynchus mykiss*): back-calculated lengths from adults' scales compared to migrating smolts at the Keogh River, British Columbia. *Can. J. Fish. Aquat. Sci.* **46**, 1853–1858.

Ward B.R., Burrows J.A. & Quamme D.L. (1990) Adult steelhead population size and salmonid migrants of the Keogh River during the spring, 1990. *Prov. B.C. Fish. Tech. Circ.* No. 89: 29 pp.

Ward B.R., Slaney P.A. & Tautz A.F. (1994) Comparison of survival and age at return from the sea of wild and hatchery steelhead smolts of the Keogh River. *Can. J. Fish. Aquat. Sci. Spec.* Publ. Presented at Internat. Conf. Wild and Enhanced Salmonids, June 17-20, 1991, Nanaimo, B.C.

Welch D.W. (1987) Influence of three life history strategies on the information content of stock-recruitment data. *Can. J. Fish. Aquat. Sci.* **44**, 1375–1379.

Withler I.L. (1966) Variability in life history characteristics of steelhead trout (*Salmo gairdneri*) along the Pacific coast of North America. *J. Fish. Res. Board Can.* **23**, 365–393.

# Chapter 23
# Mark-recapture experiment in a sea trout (*Salmo trutta* L.) ranching stream: catch effort and sample size for a confident population estimate through the season

H.R. KRISTIANSEN *Institute of Biology, University of Odense, Campusvej 55, DK-5230 Odense M, Denmark. Address for correspondence: Laboratory of Environmental Engineering, University of Aalborg, Sohngaardsholmsvaj 57, 9000 Aalborg C, Denmark*

**Abstract**   The sea trout (*Salmo trutta* L.) stock in the Danish lowland stream Kolding Å and its tributaries was investigated in 1989–90; 641 fish were individually marked. For each month the cumulative recapture rate, $R/(C-R)$, from the surveys was plotted against the effort. From August to early October the slopes were variable due to the combined effect of low fish density, accumulation of marked fish and limited inter-river migrations. From late October, the slopes approximated to zero after a catch effort of about 1.5 km. This was attributed to an increased fish density, uniform distributions and frequent inter-river migrations during the spawning period.

Seasonal plots of $R/(C-R)$ against cumulative catch indicated a minimum catch of about 50 fish for a confident estimate of N. The catch effort and sample size for a confident population estimate was expressed as $d(R/(C-R))/df = 0$ for $f > 1.5$ km, $C > 50$ fish and $R/(C-R) > 0.25$. From the rearranged formula $N = M*((C-R)/R + 1)$, the population size $N$ ($\pm 95$ % C.L.) was estimated as $3928 \pm 897$.

KEYWORDS: Catch effort, mark-recapture, population estimates, *Salmo trutta*

## 23.1   Introduction

Stock assessment in small trout streams is usually carried out by the removal method (Zippin 1958), although the mark-recapture method may give more reliable results (Bohlin & Sundström 1977; Heggberget & Hesthagen 1979). However, before using mark-recapture methods the assumptions should be critically evaluated (Cormack 1968; Ricker 1975; Begon 1979).

This chapter examines the use of a graphical mark-recapture method to assess sea trout (*Salmo trutta* L.) stocks in a Danish ranching system.

324

### 23.2     Materials & methods

Kolding Å is a typical lowland stream situated in eastern Jutland, Denmark (Fig. 23.1). The total catchment area, including the tributaries is 270 km². The daily discharge varies between 500 and 1000 l s$^{-1}$ with a peak reaching 3000 l s$^{-1}$ in late December. Some physical data from the catchment are presented in Table 23.1.

**Fig. 23.1**  Map showing a part of the Kolding Å River system. The study sites are shown with solid circles. Obstructions for upstream migration are shown by large solid circles.

**Table 23.1**  Distance from river mouth, length and width of study sites.

| Study site | Site 1 | Site 2 | Åkaer Å | Vr Nebel Å |
|---|---|---|---|---|
| Distance (km) | 6.2 | 11.3 | 12.5 | 12.5 |
| Length (m) | 1250 | 1250 | 1000 | 1300 |
| Width (m) | 5.4 | 6.1 | 5.0 | 4.5 |

Hatchery smolts have been released at the river mouth each spring since 1975 by the local Sports Angling Association. There was a steady rise in the stocking rate from 1000 two-year-old smolts to about 20 000, mainly one-year-old smolts, between 1987 and 1990.

Electrofishing was carried out at two sites in the main stream (Kolding Å) and on two sites in the lower part of two tributaries (Åkær Å and Vr. Nebel Å) at least once each month between 7 July 1989 and 14 January 1990. Most of the surveys were carried out on two consecutive days. *Ad hoc* electrofishing was also carried out at sites where sea trout occur (Fig. 23.1). Their distribution was restricted by five obstructions (Fig. 23.1).

The final mark-recapture surveys in 2.4 km of the tributaries Åkær Å and Vr Nebel Å were carried out in the period 12–14 January. The final recapture sample over 3.3 km of the main stream, including site 1 and 2, was collected on 18–19 January 1990.

Sea trout were captured in the main stream by upstream wading with a 3.0 kW a.c.

generator placed in a boat and rectified pulsed d.c. output to two electrodes. In the lower part of the tributaries, Åkær Å and Vr Nebel Å, a smaller generator (1.2 kW) and only one electrode was used.

The experimental streams were divided in 100 m sections by marking poles. The sea trout caught in each section were collected in a holding box with stream water, which was cooled and aerated during the summer. All fish were treated and released 50–100 m downstream of the section before sampling was continued.

Before marking, fish were anaesthetized for about 1 min in chlorobutanole ($0.5 \, \mathrm{g} \, \mathrm{l}^{-1}$). Size (total length to nearest cm), weight (g) and sex were recorded for each fish.

All sea trout caught during the marking procedures were individually marked. Eighty sea trout caught in July and August 1989 were marked with a Panjet inoculator using Alcian blue as the dye. The spots were positioned in different combinations on the belly. From September until the marking ceased Canadian double polyethylene monofilament individually coded green external tags (Rasmussen 1982) positioned in the musculature below the dorsal fin were used. Recaptures were treated like the unmarked fish except for the marking and were released with the original coded tag.

The mark-recapture equation (Ricker 1975) was rearranged as:

$$N = M*C/R = M*((C\text{-}R)/R + 1) \tag{1}$$

where $C\text{-}R/R$ is the proportion of unmarked to marked individuals.

For each survey the cumulative recapture rate $R/(C\text{-}R)$ for the sampled 300 m sections was plotted against length of the sampled stretch together with the cumulative catch. The minimum catch effort for a reliable estimate of $N$ was obtained as the point on the x-axis, where the slope of the $R/(C\text{-}R)$ plot was zero, allowing for small fluctuations of $\pm 0.05$. $R/(C\text{-}R)$ was also plotted against $C$ to get the minimum catch which was necessary for a confident population estimate.

The $M$ value used in the population estimate was corrected for known natural and fishing mortality. The $R/(C\text{-}R)$ plots were corrected for double recaptures before inclusion in the recapture equation.

### 23.3    Results

#### 23.3.1    *The sea trout run*

Fish densities were generally lower during the summer and early autumn. The proportion of marked to unmarked fish reached its maximum in August–September. In late October and early November the catches at both sites increased to 40–45 fish $\mathrm{km}^{-1}$, coincident with a slight increase ($100$–$200 \, \mathrm{l} \, \mathrm{s}^{-1}$) in discharge. Very few sea trout were caught in the tributaries between July and October. From November the numbers increased up to a maximum catch of 80 fish $\mathrm{km}^{-1}$ in January (Fig. 23.2).

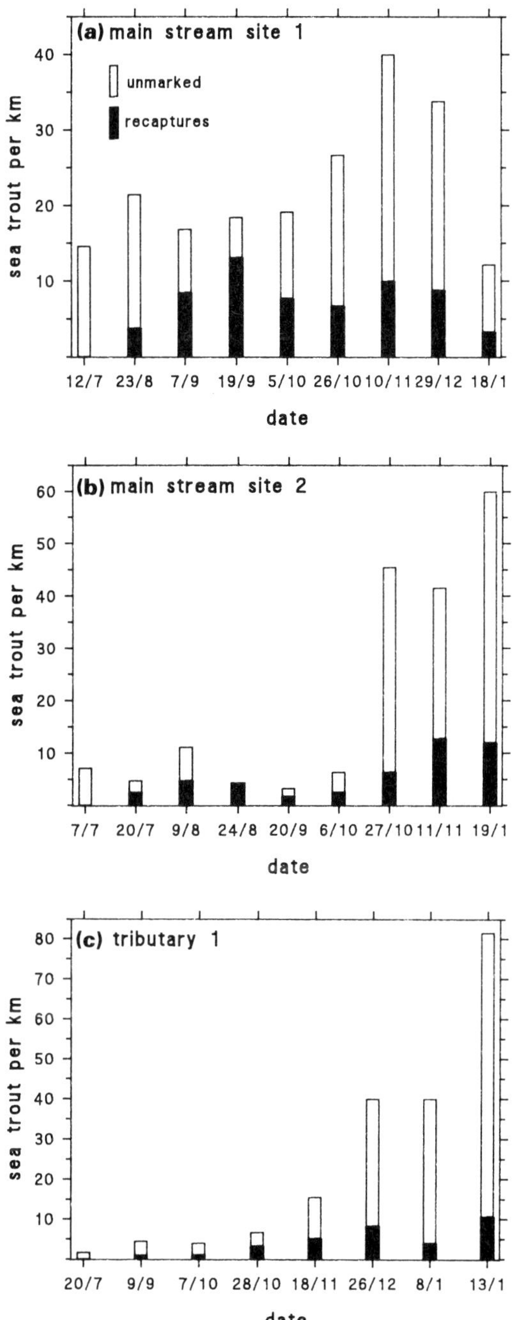


**Fig. 23.2**  Sea trout densities (fish km$^{-1}$) through the season at two study sites in the main stream and in two of the tributaries. (a) Main stream site 1. (b) Main stream site 2. (c) Tributary Åkær Å.

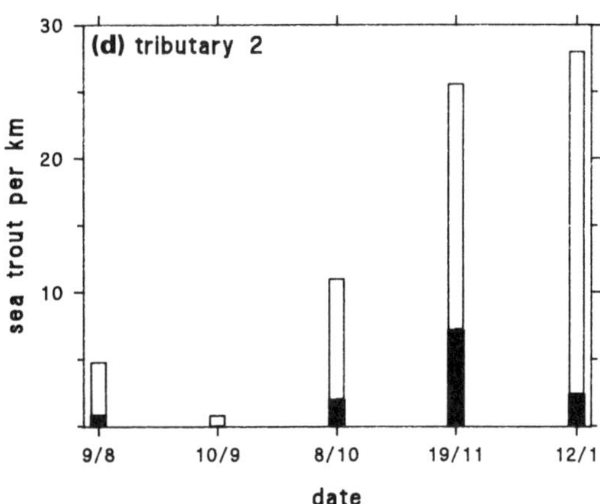


**Fig. 23.2**   (d) Tributary Vr Nebel Å. Note the different scale on the y-axes.

### 23.3.2    *Catch effort and sample size*

In August the cumulative proportion of marked to unmarked individuals ($R/(C\text{-}R)$) was unstable with a catch of 39 individuals on a sampled 2.7 km stretch (Fig. 23.3a). The same trend was repeated in September, except that $R/(C\text{-}R)$ reached values above 1 (Fig. 23.3b). Although sampling was carried out on a 4.2 km stretch in early October, and 48 sea trout were recorded, the $R/(C\text{-}R)$ curve was still unstable. Note that the first three points above 1 are not shown on the diagram (Fig. 23.3c).

In late October the slope of the $R/(C\text{-}R)$ curve became zero after 1.5 km was sampled, and 45 sea trout were recorded (Fig. 23.3d). In early November the $R/(C\text{-}R)$ curve again stabilized with a slope of zero after a sampling effort of 1.5 km with 52 recorded sea trout (Fig. 23.3e).

At the final recapture sample carried out in an upstream direction in January 1990 the slope of the $R/(C\text{-}R)$ curve approximated to zero after a sampling effort of 2.4 km with a catch of 37 fish. Reversing the cumulation of the data and plotting it in a downstream direction gave a stable $R/(C\text{-}R)$ curve after a sampling effort of 1.2 km with a catch of 74 fish (not illustrated).

Plots of the $R/(C\text{-}R)$ proportion against the cumulative catch showed no relationship for August, September and early October. By contrast the slope of the $R/(C\text{-}R)$ curves approximated to zero, allowing small fluctuations of 0.05 maximum, when the cumulative catch exceeded about 50 fish from late October onwards (Fig. 23.4).

### 23.3.3    *Population estimate*

The proportion of $R/(C\text{-}R)$, the minimum effort (*F-min*) and the minimum catch (*C-min*) for a reliable estimate of $N$ (Table 23.2) were determined from the plots of $R/(C\text{-}R)$ against effort. The estimated population in the period August to early

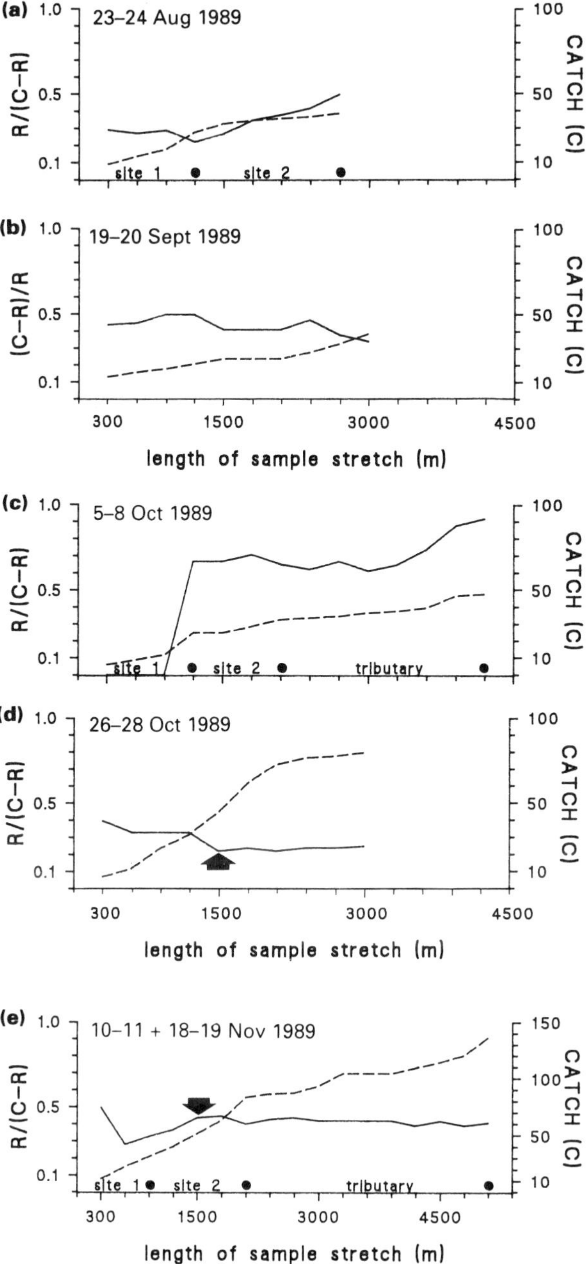



**Fig. 23.3** Cumulative proportion of marked/unmarked sea trout $R/(C\text{-}R)$ and the cumulative catch ($C$) against length of sampled stretch. (a) 23–24 Aug. (b) 19–20 Sept. (c) 5–8 Oct. (d) 26–28 Oct. (e) 10–11 Nov and 18–19 Nov. Arrows show the effort, where $\mathrm{d}(R/(C\text{-}R))/\mathrm{d}f$ remains zero $\pm 0.05$. The positions of the sites are indicated on the x-axis. Note that $R/(C\text{-}R)$ is plotted in September (b), and that the first three points are out of range (2.0, 1.25, 1.4) in early October (c).

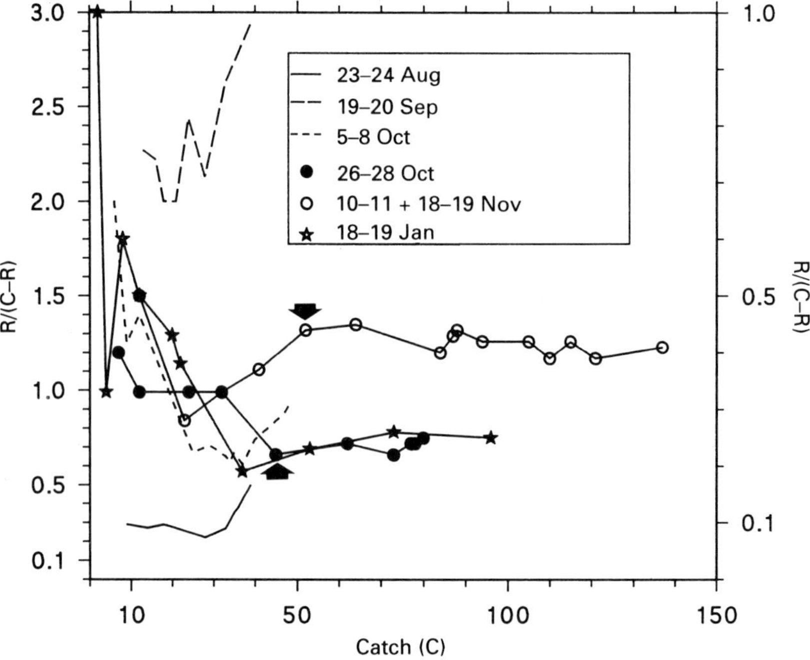



**Fig. 23.4** Cumulative proportion of marked/unmarked sea trout $R/(C\text{-}R)$ against the cumulative catch for six sampling surveys through the season: 23–24 Aug, 19–20 Sept and 5–8 Oct are read on the left y-axis; 26–28 Oct, 10–11 Nov, 18–19 Nov and 18–19 Jan are read on the right y-axis. Arrows show the catch, where $\mathrm{d}(R/(C\text{-}R))/\mathrm{d}f$ remains zero $\pm 0.05$.

**Table 23.2** Data used in the population estimate and the minimum catch ($C\text{-}min$) and effort ($E\text{-}min$) for a reliable estimate of $R/(C\text{-}R)$. *Reversed cumulation of data from main stream in downstream direction. # Data from the tributaries Åkær Å and Vr Nebel Å. Calculation of total stock is explained in the text.

| Sample dates | M | C | $R/(C\text{-}R)$ | $C\text{-}min$ | $E\text{-}min$ | N | 95% CL |
|---|---|---|---|---|---|---|---|
| 23–24 Aug | 47 | 39 | 0.5 | ? | ? | 141 | 55 |
| 19–20 Sep | 83 | 39 | 2.94 | ? | ? | 327 | 169 |
| 5–8 Oct | 104 | 48 | 0.92 | ? | ? | 217 | 58 |
| 26–28 Oct | 133 | 80 | 0.25 | 45 | 1500 | 665 | 279 |
| 10–19 Nov | 209 | 137 | 0.41 | 52 | 1500 | 719 | 173 |
| 18–19 Jan | 641 | 96 | 0.25 | 37 | 2400 | 3205 | 1289 |
| 18–19 Jan* | 641 | 96 | 0.25 | 74 | 1200 | 3205 | 1289 |
| 12–14 Jan# | 526 | 131 | 0.14 | 90 | 1100 | 4307 | 1987 |
| Total stock | 641 | 96 | 0.20 | | | 3928 | 897 |

October was below 10% of the total run. In late October and early November the population doubled to about 21% of the total run. The total stock of mature ascending sea trout in the Kolding Å and its tributaries was estimated at $3205 \pm 1289$ fish in January. Correcting for the higher observed proportion of unmarked fish in the tributaries (see Table 23.2) in 18–19 January raised the estimate to $3928 \pm 897$.

### 23.4    Discussion

#### 23.4.1    *Catch effort and sample size*

Mark-recapture methods have been used to estimate smolt output (Pratten & Shearer 1983; Kennedy *et al.* 1991), the stock of adult ascending sea trout (Jensen 1968) and stock size of brown trout (Swales 1986). However, there is a lack of information concerning application of the method and its constraints. The theoretical bias of $N$ related to random sampling (Robson & Regier 1964) was neglible in the present study as the proportion $M*C/N$ exceeded 4 in all calculations. According to Heggberget & Hesthagen (1979) $N$ was possibly underestimated by about 10 %, when 25% of the population was marked.

Vulnerability-size relationships of sampling gear (Ricker 1969) and unequal catchability among individuals (Cormack 1972; Bohlin & Sundström 1977; Begon 1979) were not considered to influence the population estimate. The effect of learning and refractory behaviour from electrofishing (Cross & Stott 1975) should have been negligible because of the time lag between marking and recapture.

Assuming that the marked population size is known the plots of $R/(C\text{-}R)$ against the catch effort are valuable tools to evaluate the reliability of the data used in the estimation equation. When the slope of the $R/(C\text{-}R)$ curve approaches zero the effort is sufficient for an unbiased estimate of $N$, according to:

$$\mathrm{d}(R/(C\text{-}R))/\mathrm{d}f = 0$$

where $f$ is the effort. The summer and early autumn surveys gave variable slopes, although between 2.7 and 4.2 km were sampled (Figs 23.3 a, b, c). This was possibly due to the combined effect of a low sea trout density and limited inter-river migrations, with most of the fish holding stations at selected areas. This was supported by Figs 23.2 a and b, which showed the marked fish accumulated in the samples in August and September.

By contrast the slopes of the $R/(C\text{-}R)$ plots were zero ($\pm 0.05$) after about 1.5 km of the main stream was sampled in late autumn (Figs 23.3d, e). This change coincided with a marked increase in the sea trout densities, most evident at site 2. The nearly linear catch curves with slopes approaching 1 were due to a uniform distribution of the sea trout, which indirectly indicated that inter-riverine migrations had increased. These data suggest that the high densities, uniform distributions and frequent inter-river migrations, which occur immediately prior to the spawning period commencing in early November, make this the most suitable time for a reliable estimation of the population. The importance of high density is corroborated by the reversed cumulation of the data, as the effort could be reduced by 50% if only areas with aggregations are sampled.

### 23.4.2    *Population estimate*

One of the aims was to develop a reliable method for sea trout stock assessment. Several environmental and biological factors have an impact on the annual smolt output and the subsequent survival of the smolts and post-smolts in the sea. Thus the spawning population varies from year to year.

A problem in sea trout stock assessment is short-term migrations directly from the sea to the spawning tributaries, when the spawning commences. This gave rise to a dilution of the marked population in the tributaries at the time of the final recapture sample, which was not seen in the main stream. The proportion of $R/(C\text{-}R)$ in the tributaries was much lower than in the main stream (Table 23.2); thus the estimation of the total stock was possibly negatively biased. This bias was adjusted by using the average of $R/(C\text{-}R)$ for the main stream and the tributaries, respectively, in the final population estimation.

In the present study, emigration from the river system during the surveys was not considered. If the emigration rates were similar for marked and unmarked fish this should not affect the $R/(C\text{-}R)$ ratios, and in turn the estimated $N$. The effect of rapid immigration and emigration on the sea trout stock assessment should be studied more carefully in the future.

The data suggest that the equation $d(R/(C\text{–}R))/df$ was satisfied for $f > 1.5$ km, for $C > 50$ fish with $R/(C\text{-}R)$ ratios between 0.25 and 0.41. It would be worthwhile developing a more general stock assessment model to include more data and the conditions discussed.

### Acknowledgements

I gratefully acknowledge help in the field given by members of the local angling association and many of my personal friends . The study was supported by the County of Vejle (Jan Nielsen), the University of Odense (Erik Kristensen) and the Inland Freshwater Fisheries Laboratory (Peter Geerz Hansen). Participation in the stock assessment symposium was kindly supported by Professor J.C. Rankin, University of Odense. Finally, I thank Ewen McLean, Aalborg University, for assisting in the translation.

### References

Begon M. (1979) *Investigating animal abundance. Capture-recapture for biologists.* Baltimore. University Park Press, 97 pp.

Bohlin T. & Sundström B. (1977) Influence of unequal catchability on population estimates using the Lincoln index and the removal method applied to electro-fishing. *Oikos* **28**, 123–129.

Cormack R.M. (1968) The statistics of capture-recapture methods. *Reviews in Oceanography and Marine Biology* **6**, 455–506.

Cormack R.M. (1972) The logic of capture-recapture estimates. *Biometrics* **28**, 337–343.

Cross D.G. & Stott B. (1975) The effect of electric fishing on the subsequent capture of fish. *Journal of Fish Biology* **7**, 349–357.

Heggberget T.G. & Hesthagen T. (1979) Population estimates of young Atlantic salmon, *Salmo salar* L., and brown trout, *Salmo trutta* L., by electrofishing in two small streams in north Norway. *Report of the Institute of Freshwater Research, Drottningholm* **58**, 27–33.

Jensen K.W. (1968) Sea trout (*Salmo trutta*, L.) of the River Istra, Western Norway. *Report of the Institute of Freshwater Research, Drottningholm* **48**, 187–213.

Jonsson N. (1991) Influence of water flow, water temperature and light on fish migration in rivers. *Nordic Journal of Freshwater Research* **66**, 20–35.

Kennedy G.J.A., Strange C.D. & Johnston P.M. (1991) Evaluation of Carlin tagging as a mark-recapture technique for estimating total river runs of salmon smolts, *Salmo salar* L. *Aquaculture and Fisheries Management* **22**, 363–368.

Pratten D.J. & Shearer W.M. (1983) Sea trout of the North Esk. *Fisheries Management* **14**, 49–65.

Rasmussen G. (1982) A pilot experiment with different types of external tags. International Council for the Exploration of the sea. Anadromous and Catadromous Fish Committee. C.M. 1982/M:8.

Ricker W.E. (1969) Effects of size-selective mortality and sampling bias on estimates of growth, mortality, production and yield. *Journal of the Fisheries Research Board of Canada* **26**, 479–541.

Ricker W.E. (1975) Computation and interpretation of biological statistics of fish populations. *Bulletin Fisheries Research Board of Canada*. **191**. 382 pp.

Robson D.S. & Regier H.A. (1964) Sample size in Petersen mark-recapture experiments. *Transactions of the American Fisheries Society* **93**, 215–26.

Swales S. (1986) Population dynamics, production and angling catch of brown trout, *Salmo trutta*, in a mature upland reservoir in mid-Wales. *Environmental Biology of Fishes* **16**, 279–293.

Zippin C. (1958) The removal method of population estimation. *Journal of Wildlife Management* **22**, 82–90.

# V CASE STUDIES – LAKES AND RESERVOIRS

# Chapter 24
# Stock assessment considerations in large lakes and reservoirs

E.H. ALLISON *School of Development Studies, University of East Anglia, Norwich, NR4 7TJ, UK*

## 24.1    Introduction

The assessment of a fish stock – its biomass and population dynamics – is the activity which occupies the majority of fisheries scientists' time. Stock assessment is (or should be), merely a component of an integrated management plan (Cowx 1991 & Chapter 39), so that the fishery scientist may participate in only a limited part of the process. It is important, however, that all involved in stock assessment should have a clear appreciation of the other components of the management process.

## 24.2    Management objectives

Management objectives should be clearly defined before initiating a stock assessment programme, as the type of information required and the way it is presented are determined by the goal of the study. This is largely the duty of fishery managers, although experts from the fields of politics, socio-economics, environment and conservation and from the fishing industry will frequently be called upon to advise on setting appropriate and realistic objectives.

Management objectives can be grouped into three main categories, which are interrelated to some extent.

### 24.2.1    *Meeting basic human needs*

Fish is the major source of animal protein in many developing countries. The management objective in this situation is to maximize sustainable yield of fish, and the stock assessment scientist will be required to design the research programme to answer questions on how this might be achieved. Such questions may include how to maximize the sustainable yield from the existing exploited stocks through regulation of fishing effort, or the likely effects of such interventionist management strategies as fish stocking programmes, species introductions or production enhancement through nutrient enrichment.

In many countries, basic food-security is assured to those in productive employment, and the management goal may be to maximize involvement in fishing and related activities. Stock assessment programmes in this situation will include assessment of the relative effects of artisanal and industrial fisheries on stock size and species composition, and will have an emphasis on suggesting management targets which permit maximum involvement in fishing, perhaps at the expense of efficiency or even of total catch.

### 24.2.2    *Income generation*

Inland fisheries are an important source of revenue (including foreign currency) to many countries, through sale and export of fish. Licence revenues are also important in both recreational and commercial fisheries. Where income generation from fish sales are of primary importance, stock assessment programmes can be designed to maximize the flow of information from the commercial enterprises that benefit from the fishery.

In developed countries, inland fisheries are seldom a major component of the economy, but may be supported as part of efforts to maintain a diverse economic base, support traditional ways of life, or boost related activities such as processed food industries, boat and fishing gear manufacture and sale, and a variety of leisure industries associated with recreational fishing.

### 24.2.3    *Environmental quality*

A basic level of water quality is required to support any fishery, but where the management goal includes, for example, maintenance of biodiversity, or water clarity for recreational use, assessment studies must structure their focus to include information on the nutrient dynamics and biomass, and productivity of phyto- and zooplankton.

### 24.3    **Stock assessment methods, problems and recommendations**

Stock assessment in temperate and tropical regions tends to be based on similar methodology. However, collecting the basic data required for assessments generally appears more difficult in the tropics, and imposes constraints on the methods of assessment that can successfully be applied. The most commonly used stock assessment tools in temperate commercial fisheries are yield per recruit ($Y/R$) and virtual population analysis (VPA) (e.g. Löffler, Chapter 29) which require data on age-structure, growth and mortality, as well as data on catches and/or fishing mortality. In the tropics, growth is not always sufficiently seasonal for marks on fish hard-parts to be used to age them. However, age-structure and population dynamics parameters can be derived indirectly through length-frequency analysis (Pauly & Morgan 1987), or directly through identification of daily growth rings (see Beamish & McFarlane 1987 for review). While length-frequency analysis is straightforward, it is only applicable when relatively distinct cohorts can be identified. In fish with protracted spawning seasons this is seldom the case and daily ring counts, time-consuming and difficult in longer-lived fish, must be carried out on each species important to the fishery. This is feasible for those fisheries based on large stocks of one or a few species (e.g. the L. Tanganyika and L. Kariba fisheries for *Stolothrissa tanganicae* and *Limnothrissa miodon*, Marshall, Chapter 26) but not practicable for fisheries exploiting large numbers of species which cannot, in any case, be routinely separated to species level in fishery statistical surveys (e.g. the bottom-trawl fisheries in Southern Arm of Malawi; Banda, Tomasson & Tweddle, Chapter 5).

For diverse multi-species fisheries, surplus-production models are often applied to the entire stock, or to the most important species only, although the theoretical MSY can only be identified with any certainty once the stock has been overexploited (Hilborn & Walters 1992). Surplus production models are empirical models based on the implicit acceptance of a stock-recruitment relationship. It may be possible, making the same implicit assumption, to manage some of these fisheries on the basis of size limits which allow escapement of part of the spawning stock. Setting gear mesh sizes to allow the more important fish in the community to spawn at least once before becoming liable to capture may be the simplest management measure to adopt in this situation; only data on the length at first maturity are required, and the change in size-distribution of the population can potentially be used as an index of exploitation rate (Pope *et al.* 1988).

Management based on size limits, with no effort restrictions, will lead to elimination of the larger species (Pauly 1979). From the point of view of maximizing yield, this may not be detrimental, as allowing the stock to become dominated by small, fast-growing species will enhance fish productivity (Turner, Chapter 37). It must also be considered, however, that low-diversity stocks of small, usually pelagic, fish tend to be unstable – as studies of the fisheries for small marine pelagics have indicated (see Cushing 1982 for review). Also, fisheries for small pelagic species usually support industrial fisheries at the expense of artisanal ones, which appear to be associated with diverse fish stocks.

In diverse multi-species stocks, aggregation of species into trophic groups may be desirable. The dynamics of such 'functional groups' may be more predictable using theoretical models than the dynamics of individual species (Hay 1994), and they are therefore more likely to be successfully managed than individual species.

Although the type of assessment programme should be designed according to management requirements, in reality approaches to stock assessment seem to be determined largely by resources and expertise. Despite the diversity of assessment approaches which exist, an idealized assessment programme is likely to comprise the following components, whatever the management requirements.

### 24.3.1    *Frame survey*

Data on the number and type of boats and fishing gear are collected. This is often the only source of fishing effort data, particularly in artisanal fisheries in the tropics, where no licensing scheme exists, through which fishermen and fleet information can be obtained.

### 24.3.2    *Catch-effort data*

Ideally, both catch and effort data are collected from all fishermen, but often sub-sampling of fish landing areas, and extrapolation using frame-survey data, is carried out (Orach-Meza 1991).

### 24.3.3    *Fish population parameters*

Obtained from surveys (e.g. population size) and/or commercial catch sampling (age or size structure of exploited stock). Additional parameters (growth, natural and fishing mortality, length-weight relationships, size/age – maturity relationships and diet and food consumption rates) can be obtained either from experimental surveys or from samples of commercial catches.

### 24.3.4    *Environmental variables*

Lake morphometry, thermodynamics, nutrient status, primary and secondary production are useful in understanding factors regulating production within a lake or reservoir over time, or in comparative study with other lakes and reservoirs, where predictive relationships such as the morphoedaphic index are used for first-order estimates of potential fish yields (e.g. Cryer, Chapter 28; Moreau & De Silva 1991; Crul 1992 for reviews).

### 24.3.5    *Socio-economic factors*

Information on *per capita* fish consumption, fishing costs and profits, and lake-catchment land use are all useful for sector planning purposes, and the objectives of managed fisheries should always relate to a potential social or economic benefit.

### 24.4    **Data collection programmes**

The need for regular review of assessment programmes is imperative. Lake and reservoir fisheries may change rapidly, through changes in the stocks, the quality of the water, the population in the catchment area, or the introduction of new fishing technologies. Management objectives must be regularly evaluated and re-defined, and assessment programmes altered to account for changes, if required. In L. Malawi, for example, assessment programmes designed in the 1970s focused on large, commercially important species such as the cyprinid *Labeo mesops*. This species is now absent from commercial catches (S. Alimoso pers. comm.), which are presently dominated by small cichlids, many of which are not recorded by genera or functional group, and have not been studied (Tweddle & Magasa 1989). The fisheries data collection programme must now reallocate resources to emphasise the study of these more important fisheries.

The requirement for long-term studies on the fisheries, fish populations and the environment is also essential. Factors controlling fish populations can only be established by such long-term parallel studies, an excellent example of which is the study of brown trout (*Salmo trutta* L.) in the Lake District (Elliott 1994). Even if such studies do not lead to further understanding of factors regulating fish populations, they are valuable in quantifying variability and uncertainty (Buijse & Dekker, Chapter 19). It is now widely recognized that stock assessment studies should present some estimate of precision associated with their advice to management agencies (see Hilborn *et al.* 1993 for review). Long-term studies allow precision and accuracy, and therefore predictive power, to evolve as the assessment programme develops.

All assessment methods are potentially subject to bias, through violation of assumptions inherent in their application, and to uncertainty through natural variability or measurement error. The use of more than one method of assessment can be very useful in cross-validating sampling and analytical methods. The use of both fishery dependent and fishery independent methods in inland fish stock assessment studies has been stressed previously (Cowx 1991) and both are vital to understanding how a fishery performs. For example, Cryer (Chapter 28) compared yield estimates from simple empirical models, such as the morphoedaphic index (Ryder 1965) with angler catch statistics for L. Taupo, New Zealand, and Allison (Chapter 17) compared acoustically estimated fish biomasses in L. Malawi with estimates derived from the biomass and size-distribution of the plankton, using Platt & Denman's (1978) and Borgmann's (1982, 1987) size-spectrum theory.

Calibrating the size selectivity and efficiency of sampling gear are essential parts of any assessment study. Much experimental work has been done on sampling gear selectivity and efficiency, of for example, gillnets (Hamley 1980) and plankton nets (Nichols & Thompson 1991). This body of experimental work has not always been transferred to stock assessment programmes, so that only indices of abundance which are site and gear specific have been derived in many cases, when with little extra work, absolute abundances could be obtained. This would be very valuable for other workers in comparative studies, and can allow standardization of survey data in time-series where multiple types of gear are used.


## 24.5    Dissemination of stock assessment results

Scientists engaged in stock assessment studies should make every effort to ensure that their work is published in a form available to others. While not every study is of sufficient general interest to be published in primary scientific journals, all well-documented studies carried out competently are likely to be acceptable for publication in national and regional journals, regional and international workshop proceedings, or as experience papers in international conference proceedings. These outlets are preferable to publication of internal departmental reports, where results remain unavailable to others. Both FAO and ICLARM are supportive of publications in tropical fisheries, and may provide assistance in publication and distribution of reports.

Presentations of the results of assessments to managers must be concise, simple but sufficiently comprehensive to allow them to make a decision based on the evidence, including an estimate of the risk attached to each decision, derived from the uncertainty in the data and assessment models.

It is important to involve fishing communities in stock assessment and management programmes if regulations are to have any chance of being successfully implemented. The dialogue between managers and fishermen is particularly well developed in recreational or sport fisheries, but has been less successful in commercial fisheries. To educate fishermen on the need for management, or to involve them in assessment programmes, a variety of innovative methods have been used. These include the use of song, dance and dramatic reconstruction in small-scale aquaculture and fisheries and extension work in Africa (e.g. a video presentation project by ICLARM/GTZ on pond aquaculture for smallholder farmers in Malawi). The more conventional approaches to fisheries extension work must continue to receive emphasis, as the process of stock assessment is of little value unless it leads to better management.

With good communication between scientists, managers and fishermen, enforcement of regulations has more chance of success. The likelihood of successful restrictive management of fisheries is even greater if the restrictions are adopted and enforced by fishing communities themselves. This 'community based management', successful for some time in Asia (Ali, Chapter 38), is now being applied in Africa (Mandima, Chapter 27). The essential prerequisite is to restrict access to resources to particular communities. These closed-access fisheries can be managed successfully with minimum enforcement costs, and self-supporting stocking schemes can be implemented (Lorenzen 1995). Private and restricted-access fisheries are common in developed countries, where a group of recreational fishermen may have sole rights to particular water bodies.

Stock assessment in inland waters requires a wide diversity of information, necessitating a multi-disciplinary approach. In large fisheries, a team approach can be adopted, with the integration of expertise on limnology, fish biology, stock assessment and fishery statistics, and socio-economics, management and enforcement. Many freshwater fisheries are small, in remote areas, and not economically important enough to justify such intensive study. In these situations, one individual may have to fulfil all the above functions. As no one can be an expert in all fields, use should be made of widely available expertise through consultation and communication.

Finally, it must be asked: does improved stock assessment result in better management? Are decisions made on the basis of stock assessment advice? Experience suggests that the answer is, at best, 'not always'. For example, a common experience in fisheries is that when catch rates fall, smaller meshes are used to maintain catch rates contrary to advice. In other cases scientific advice is ignored for political, economic or social reasons, often resulting in degradation of fish stocks. This breakdown in communication between fisheries scientists, fishermen and managers recently resulted in the refusal of European Community marine fishery scientists to specify quotas for pressure-stocks. Cynics claim that when the fishery is operating profitably, stock assessment advice is ignored, and when the fishery collapses, assessment scientists are blamed for not foreseeing it. Perhaps stock assessment presentations to managers should include a detailed analysis of the effects of the advice being ignored.

Is our focus on improving stock assessment misplaced? If stock assessment procedures improve to the extent that they have useful predictive powers, managers and fishermen will have increased confidence in the predictions made by scientists and they are more likely to listen.

## References


Beamish R.J. & McFarlane G.A. (1987) Current trends in age-determination methodology. In R.C. Summerfelt & G.E. Hall, *The age and growth of fish*. Iowa State University Press, Ames, Iowa, USA, pp15–42.

Borgmann U. (1982) Particle-size conversion efficiency and total animal production in pelagic ecosystems. *Canadian Journal of Fisheries and Aquatic Sciences* **39**, 668–674.

Borgmann U. (1987) Models of the slope of, and the biomass flow up, the biomass size spectrum. *Canadian Journal of Fisheries and Aquatic Sciences* **44 (Suppl. 2)**, 136–140.

Cowx I.G. (1991) Catch effort sampling strategies: conclusions and recommendations for management. In: I.G. Cowx (ed.), *Catch Effort Sampling strategies: their application in freshwater fisheries management*. Oxford: Fishing News Books, pp 404–413.

Crul R.C.M. (1992) Models for estimating potential fish yields of African inland waters. *FAO/CIFA Occasional Paper* **17**, 21 pp.

Cushing D.H. (1982) *Climate and fisheries*. London: Academic Press, 387 pp.

Elliott J.M. (1994) *Quantitative ecology and the brown trout*. Oxford University Press, 387 pp.

Hamley J.M. (1980) Sampling with gill nets. In: T. Backiel & R.L. Welcomme (eds) Guidelines for sampling fish in inland waters. *FAO/EIFAC Technical Paper* **33**: 37–53.

Hay M.E. (1994) Species as 'noise' in community ecology: do seaweeds block our view of the kelp forest? *Trends Ecol. Evol.* **9**, 414–416.

Hilborn R, Pikitch E.K. & Francis R.C.C. (1993) Current trends in including risk and uncertainty in stock assessment and harvest decisions. *Canadian Journal of Fisheries and Aquatic Sciences* **50**, 874–880.

Hilborn R. & Walters C.J. (1992) *Quantitative fisheries stock assessment: Choice, dynamics and uncertainty*. London: Chapman & Hall, 570 pp.

Lorenzen K. (1995) Tropical culture-based fisheries: population dynamics modelling and adaptive management. *Fisheries Management and Ecology* **2**, 61–74.

Moreau J. & De Silva S.S. (1991) Predictive fish yield models for lakes and reservoirs of the Philippines, Sri Lanka and Thailand. *FAO Fisheries Technical Paper* **319**, 42 pp.

Nichols J.H. & Thompson A.B. (1991) Mesh selection of copepodite and nauplius stages of four calanoid copepod species. *Journal of Plankton Research* **13**, 661–671.

Orach-Meza F.L. (1991) Statistical sampling methods for improving the catch assessment of lake fisheries. In: I.G. Cowx (ed.) *Catch Effort Sampling strategies: their application in freshwater fisheries management*. Oxford: Fishing News Books, pp 323–334.

Pauly D. (1979) Theory and management of tropical multi-species stocks: a review, with emphasis on the southeast Asian demersal fisheries. ICLARM Studies and Reviews 1, Manilla, Philippines, 35 pp.

Pauly D. & Morgan G.R. (1987) Length-based methods in fisheries research. ICLARM, Manilla, Philippines & KISR, Safat, Kuwait, 468 pp.

Platt T. & Denman K. (1978) The structure of pelagic marine ecosystems. *Rapp. p-V. Reun. Cons. int. Explor. Mer* **173**, 60–65.

Pope J.G., Stokes T.K., Murawski S.A. & Idoine S.I. (1988) A comparison of fish size-composition in the North Sea and on Georges Bank. In: W. Wolff, C.J. Soeder & F.R. Drepper (eds), *Ecodynamics: Contributions to theoretical ecology*. Berlin: Springer-Verlag, pp 146–152.

Ryder A. (1965) A method for estimating the potential fish production of north-temperate lakes. *Transactions of the American Fisheries Society* **94**, 214–218.

Tweddle D. & Magasa J.H. (1989) Assessment of multi-species cichlid fisheries of the Southeast Arm of Lake Malawi, Africa. *J. Cons. int. Explor. Mer* **45**, 209–222.

# Chapter 25
# Stock assessment in Sri Lankan reservoirs: a review

U.S. AMARASINGHE *Department of Zoology, University of Kelaniya, Kelaniya, Sri Lanka.*

**Abstract**   Three approaches, *viz.* empirical models, surplus production models and length-based analytical models, have been used for stock assessment in the Sri Lankan reservoir fisheries. Official data on fish production and fishing effort are not reliable in most reservoirs of the country. As such, empirical models relating biological productivity levels and reservoir morphometry to fish yield, and the surplus production models used in a handful of reservoirs where reliable data on fish production are available, can be used to set limits for the size of the fisheries. The length-based stock assessment procedures are considered applicable in the reservoir fisheries where the exotic cichlid, *Oreochromis mossambicus* (Peters), is the dominant species, accounting for over 70% of the landings.

KEYWORDS: Cichlidae, man-made lakes, maximum sustainable yield, population dynamics, reservoirs, stock assessment

## 25.1   Introduction

Tropical reservoirs are productive biological systems which support profitable fisheries in most regions (Fernando & Holcik 1991). In Asia, however, reservoir fisheries are not managed scientifically (De Silva 1987a). On the other hand, Sri Lankan reservoir fisheries are among the best documented in the South East Asian region (De Silva 1988). This chapter reviews research on stock assessment in Sri Lankan reservoirs.

## 25.2   Reservoir fisheries of Sri Lanka

In Sri Lanka, the total surface area of ancient reservoirs which have been rehabilitated or restored and new reservoirs is about 175 000 ha. This gives a value of about 2.7 ha for every km$^2$ of the country. Most of these reservoirs are very small ( < 250 ha) village tanks but about 50 reservoirs are larger than 500 ha (De Silva 1987b, 1988). Fisheries in Sri Lanka which are mainly dependent on the major perennial reservoirs ( > 500 ha), are a relatively recent development after the introduction of the exotic cichlid species *Oreochromis mossambicus* (Peters) in 1952. The trends in the growth of the reservoir fisheries of Sri Lanka have been documented by various workers (e.g. Fernando &

Indrasena 1969; De Silva 1988; Amarasinghe 1992). Inland fisheries currently account for about 20% of the country's total fish production (De Silva 1988).

The reservoir fisheries are essentially artisanal. The craft are non-mechanised fibre-glass outrigger canoes using gillnets to fish. The stretched mesh sizes of gillnets range from 7.5 to 12 cm and the dimensions and filament characteristics of gillnets are almost identical in all the reservoirs. Generally two fishers working on each craft set their gillnets overnight in reservoirs. In some reservoirs, fishers beat the water with wooden poles or weighted ropes to drive fish towards gillnets (Amarasinghe & Pitcher 1986; Amarasinghe & De Silva 1992a). Beach seines and cast nets are also used sporadically in some reservoirs.

Fish yields in individual reservoirs vary considerably. According to Moreau & De Silva (1991), annual fish yields in 19 reservoirs range from 40 to 650 kg ha$^{-1}$ with an overall mean value of 258 kg ha$^{-1}$. Some fisheries regulations have been imposed by the Ministry of Fisheries. Beach seining is illegal in major reservoirs and mechanization of craft is not allowed. Also minimum permissible mesh size of gillnets is 7.5 cm. Although these regulations are not scientifically determined, they appear useful, to a certain extent, to control fishing mortality.

### 25.3    Empirical approaches for stock assessment in reservoirs

In empirical approaches for stock assessment, first approximations of the potential or state of exploitation of fishery resources are determined on the basis of information that is less complete than would normally be required for standard assessment models. They are particularly useful for situations where reliable data for stock assessment using standard methods are not available as is the case for most Sri Lankan reservoirs.

Wijeyaratne & Amarasinghe (1987) showed that maximum sustainable yield (MSY) in several reservoirs, computed from the Schaefer model using long-term catch and effort data, was correlated with the morphoedaphic index (MEI = conductivity ($\mu$S cm$^{-1}$) / mean depth (m)), and is described by:

$$\ln MSY = 0.9005 \ln MEI + 1.922$$

Subsequently, however, De Silva (1988) indicated that *MEI* could not be used for yield prediction in Sri Lankan reservoirs due to the relative shallowness, high draw-down and flushing rates of these water bodies, which do not permit them to act as basins releasing nutrients.

Moreau & De Silva (1991) used data from 19 Sri Lankan reservoirs to develop empirical models for predicting fish yields using morphometric, physico-chemical, biological and fisheries data. Some morphometric and physico-chemical character-istics and features of the fisheries of Sri Lankan reservoirs are given in Table 25.1.

The total annual fish production in Sri Lankan reservoirs is significantly correlated with the reservoir area, catchment area and fishing effort (Moreau & De Silva 1991) as follows:

**Table 25.1** Some morphometric, physicochemical and biological characteristics and features of the fisheries of Sri Lankan reservoirs. Source: Moreau & De Silva (1991).

| Reservoir | Area ha | Catchment area $km^2$ | Mean depth m | Alkalinity $mgl^{-1}$ | Chl-a $mg\ l^{-1}$ | Effort boats | Fishing intensity boats $km^{-2}$ | Catch t $yr^{-1}$ | Yield kg $ha^{-1}$ $yr^{-1}$ | Yield $boat^{-1}$ $yr^{-1}$ |
|---|---|---|---|---|---|---|---|---|---|---|
| 1 Badagiriya | 482 | 346 | 4.3 | — | 39.7 | 24 | 5.0 | 230 | 476 | 9.6 |
| 2 Huruluwewa | 225 | 197 | 6.2 | 2.24 | — | 19 | 0.9 | 84 | 40 | 4.4 |
| 3 Kalawewa | 2583 | 827 | 4.8 | 2.10 | — | 83 | 3.2 | 897 | 347 | 10.8 |
| 4 Kaudulla | 2537 | 82 | 9.2 | 2.47 | 58.5 | 72 | 2.8 | 536 | 211 | 7.4 |
| 5 Lunugamwehera | 3023 | 904 | 12.0 | — | — | 20 | 0.6 | 400 | 132 | 20.0 |
| 6 Maduru Oya | 2548 | 453 | — | — | — | 80 | 3.4 | 692 | 272 | 8.6 |
| 7 Mahakanadarawa | 1457 | 323 | 5.8 | 2.54 | — | 25 | 1.7 | 122 | 84 | 4.9 |
| 8 Mahawilachchiya | 927 | 364 | 4.2 | 2.72 | — | 38 | 2.6 | 251 | 258 | 6.6 |
| 9 Minneriya | 2550 | 240 | 11.7 | 2.74 | 12.5 | 74 | 2.9 | 478 | 188 | 6.5 |
| 10 Nachchaduwa | 1785 | 604 | 3.1 | 1.74 | — | 66 | 3.7 | 874 | 497 | 13.2 |
| 11 Padaviya | 2357 | 272 | 12.2 | — | — | 47 | 2.0 | 308 | 131 | 6.5 |
| 12 Parakrama Samudra | 2662 | 1382 | 5.3 | 2.42 | 23.8 | 102 | 3.8 | 844 | 317 | 8.4 |
| 13 Pimburettewa | 830 | 112 | 6.0 | 3.52 | 58.5 | 93 | 11.2 | 539 | 650 | 5.8 |
| 14 Rajangana | 1600 | 1592 | 6.4 | 3.16 | — | 105 | 6.6 | 638 | 399 | 6.1 |
| 15 Ridiyagama | 888 | 31 | 3.8 | 2.61 | 19.7 | 26 | 0.3 | 171 | 193 | 6.6 |
| 16 Senanayaka Samudra | 7825 | 983 | 12.8 | — | — | 54 | 0.7 | 1165 | 149 | 21.6 |
| 17 Udawalawe | 3362 | 1162 | 15.3 | 1.40 | 12.0 | 41 | 1.7 | 496 | 208 | 12.1 |
| 18 Weerawila | 570 | 82 | 4.8 | — | — | 8 | 1.4 | 46 | 81 | 5.8 |
| 19 Yodawewa | 488 | 46 | 2.6 | — | 23.4 | 12 | 2.4 | 77 | 157 | 6.4 |

$$Y = -67.3 + 0.111\ A + 5.898\ f \qquad\qquad (r^2 = 0.857;\ P < 0.05)$$

$$Y = -67.6 + 0.101\ A + 5.539\ f + 0.064\ CA \quad (r^2 = 0.864;\ P < 0.05)$$

where $Y$ is the annual fish production (t), $A$ the reservoir area (ha), $CA$ the catchment area (km$^2$) and $f$ the fishing effort (number of fishermen). According to these relationships, over 85% of changes in fish production in Sri Lankan reservoirs can be explained by changes in morphometric features and fishing effort. The addition of the catchment area, however, did not increase the predictive power of the model. Furthermore, Moreau & De Silva (1991) derived regression relationships between the annual fish yields (kg ha$^{-1}$) and morphometric, biological and fishery data of Sri Lankan reservoirs (Table 25.2). They indicated that morphometric parameters are useful for yield prediction in Sri Lankan reservoirs.

**Table 25.2**  Relationships between annual fish yields (kg ha$^{-1}$) and morphometric, biological and fishery data of Sri Lankan reservoirs. $CA$ = catchment area (km$^2$); $A$ = reservoir area (ha); $Z$ = mean depth (m); $FI$ = fishing intensity (fishermen km$^{-2}$); $Chl$ = Chlorophyll-a content (mg l$^{-1}$). Source: Moreau & De Silva (1991).

| Relationship | n | $r^2$ |
|---|---|---|
| Yield = 117.5 + 4.242 $(CA/A)$ | 17 | 0.433 |
| Yield = 71.2 + 1093 $Z$ | 19 | 0.27 |
| Yield = 88.9 + 54.6 $FI$ | 19 | 0.749 |
| Yield = 28.2 + 10.6 $Chl$ | 8 | 0.902 |

It should be noted that statistically significant relationships between fish yield and another variable are useful for predictive purposes only where there is a plausible biological basis for such a relationship. Silva & Davies (1987) have shown that $CA/A$ ratio, which is termed 'Index of Allochthonous Input', has a positive influence on primary productivity of reservoirs. Also the independent variables used in the above-mentioned relationships have been shown to be biologically reasonable for predicting fish yields in lakes and reservoirs (e.g. Ryder 1965, 1982; Marshall 1984).

### 25.4     Surplus production models

In Sri Lankan reservoirs, each fisher typically uses about 300 m of netting of 1.5 to 2 m depth. They rely on fibre-glass out-rigger canoes as craft. These craft generally carry two men and a limited amount of gear which simplifies the computation of effort.

Amarasinghe & Pitcher (1986) pointed out that the best measure of effort is probably the number of nets used by each fisher. They further stated that number of operating boats or number of fishers also provides a reasonable estimate of effort because in standard survey techniques, net counts tend to be subject to a certain amount of false reporting.

Amarasinghe & De Silva (1992a) employed a multiple regression technique similar to the method used by Marten (1979) for the inshore fishery of Lake Victoria (East Africa), to evaluate the relative efficiencies of gill netting and beach seining in reservoirs. The curvilinear multiple regression relationship between fish yield and effort was:

$$Y = \sum_i a_i f_i + \sum_i b_i f_i^2 + \sum_i \sum_j c_{ij} f_i f_j$$

where $Y$ is the yield (kg ha$^{-1}$ month$^{-1}$) and $f_i$ is the fishing intensity per month for gear $i$. The regression coefficients of the linear terms ($a_i$) reflect the contribution of each fishing method to the total yield. When the linear term is positive, the coefficient of the negative square term ($b_i$) indicates how quickly increased fishing effort leads to diminished returns due to reduced stocks. The cross-product term ($c_{ij}$) indicates the interaction between fishing gears. Using this method, Amarasinghe & De Silva (1992a) indicated that beach seining was detrimental to the fish stocks in reservoirs.

Reliable data on catch and effort are available in several reservoirs for a considerable period (over 8 years in most reservoirs). De Silva *et al.* (1991) showed a positive linear relationship between fish yields (kg ha$^{-1}$, yr$^{-1}$) and fishing intensities (boats km$^{-2}$) in 19 Sri Lankan reservoirs (Fig. 25.1). However, it appears that a single data point corresponding to Pimburettewa reservoir (designated as 13) has a great

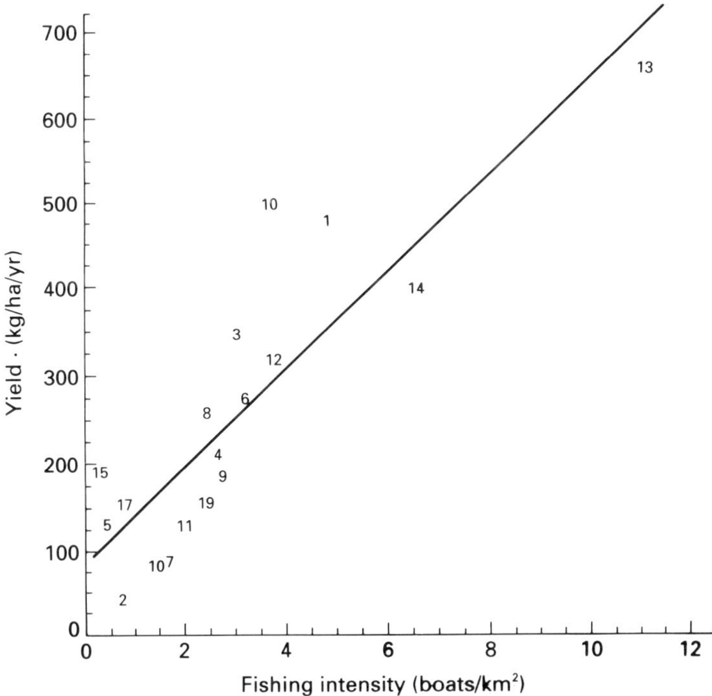


**Fig. 25.1**  Relationship of the yield to fishing intensity for different reservoirs in Sri Lanka. The numbers 1–19 refer to different reservoirs as in Table 25.1 (redrawn from De Silva *et al.* 1991).

influence on the relationship. According to Amarasinghe (1987), annual fish yield in Pimburettewa reservoir in 1985 was 441 kg ha$^{-1}$, and fishing intensity was 7.8 boats km$^{-2}$. If this is the case, a second-order curve would be more appropriate to relate yield to fishing intensity. In both cases, fish stocks in most individual reservoirs appear to be exploited at a sub-optimal level.

De Silva *et al.* (1991) also showed that when all reservoirs are considered as a single unit, mean fish yields on an annual basis and fishing intensity conformed to the Schaefer model (Fig. 25.2). The *MSY* and optimal fishing intensity were deduced as 256 kg ha$^{-1}$ yr$^{-1}$ and 3.2 boats km$^{-2}$, respectively (De Silva *et al.* 1991). Although fish stocks in most individual reservoirs appear to be sub-optimally exploited (Fig. 25.1), the overall reservoir fishery tended towards overexploitation in 1984 (Fig. 25.2).

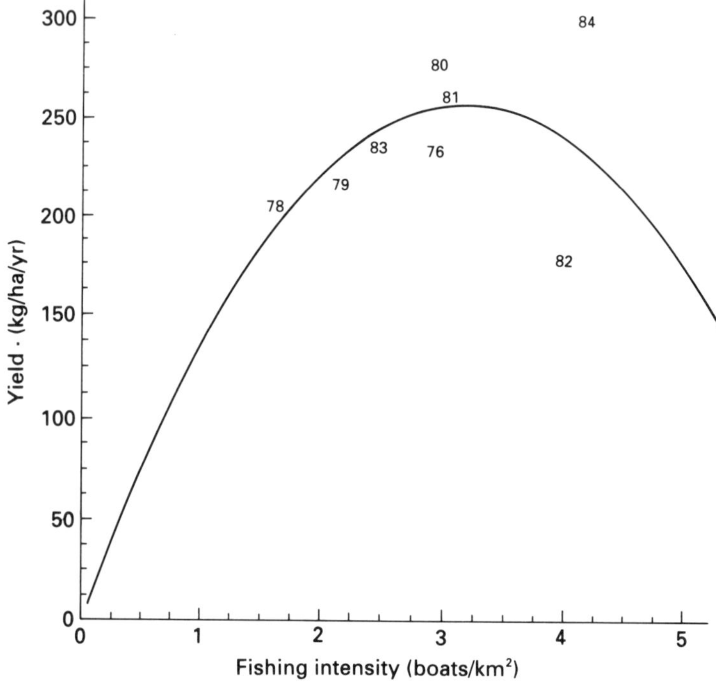


**Fig. 25.2**   The relationship of mean annual yield to fishing intensity in the Sri Lankan reservoir fishery. The numbers 76–84 correspond to the calender years (redrawn from De Silva *et al.* 1991).

Lack of reliable fish production statistics is a major limitation for stock assessment in Sri Lankan reservoirs using surplus production models. Amarasinghe (1992) indicated that official statistics in most reservoirs were overestimates compared to the average annual fish yields based upon catch and effort surveys for the majority of crafts actually fishing. Since the number of crafts operating in Sri Lankan reservoirs is allocated according to these overestimated fisheries statistics, it could lead to overexploitation through an increase in effort above the optimal level (Amarasinghe 1992).

### 25.5     Dynamic pool models

Recently developed length-based stock assessment procedures (Pauly & Morgan 1987; Sparre & Venema 1992) offer choice of the most appropriate method for assessment of tropical fish stocks using dynamic pool models. In Sri Lankan reservoirs, as in any gillnet fishery, before using length frequency data for parameter estimation, it is imperative to investigate whether gillnet selection influences estimates. Such an attempt was made by Amarasinghe & De Silva (1992b) who have compared overall length frequency distributions of *Oreochromis mossambicus* and *Oreochromis niloticus* in gillnet and beach seine catches in two Sri Lankan reservoirs (Fig. 25.3). Furthermore, using length-structured virtual population analysis Gayanilo *et al.* (1989), showed that fishing mortality was constant throughout the exploited phase of cichlid fish stocks in the two reservoirs which suggests that the probability of capture in the exploited phase of the stock was close to unity (Amarasinghe & De Silva 1992b).

Estimations of growth parameters of cichlid populations in Sri Lankan reservoirs (Table 25.3) were carried out using ELEFAN (Gayanilo *et al.* 1989). However,

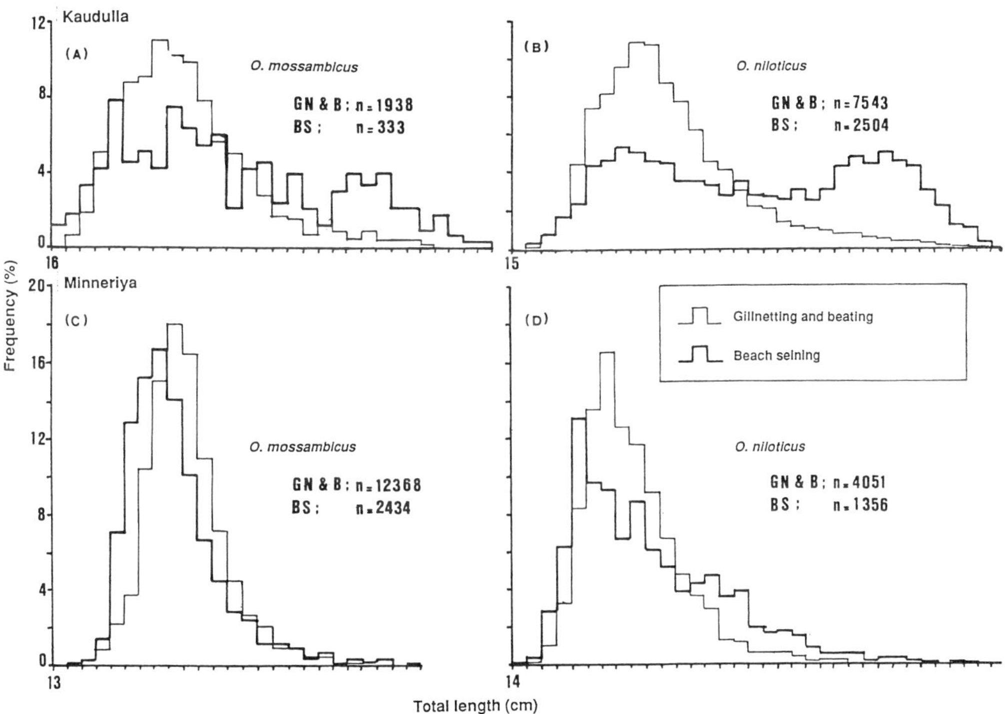



**Fig. 25.3** Overall length frequency distributions of two cichlid species in gillnet and beach seine catches in two Sri Lankan reservoirs. A – *O. mossambicus* in Kaudulla; B – *O. niloticus* in Kaudulla; C – *O. mossambicus* in Minneriya; D – *O. niloticus* in Minneriya. Note that in Kaudulla, where limited areas were fished by beach seines due to presence of tree stumps, large size classes were dominated by males due to 'behavioural selection' (from Amarasinghe & De Silva 1992b).

**Table 25.3**  Growth and mortality parameters of cichlids in Sri Lankan reservoirs $Z1$ – Total mortality from length-coverted catch curve method in ELEFAN; $Z2$ – Total mortality from Beverton & Holt (1956) method; $Z3$ – Total mortality from Ssentongo & Larkin (1973) method. (A – Amarasinghe 1987; B – Amarasinghe *et al.* 1989; C – Amarasinghe & De Silva 1992b; D – De Silva & Senaratne 1988).

|  | Pimburettewa | Parakrama Samudra | Kaudulla | Minneriyaa | 11 reservoirs |
|---|---|---|---|---|---|
| Authority | A | B | C | C | D |
| Asymptotic total length (cm) |  |  |  |  |  |
| *O. mossambicus* | 39.3 | 34.9–40.6 | 43.7 | 45.0 | 26.7–32.7 |
| *O. niloticus* | — | — | 54.5 | 54.5 | — |
| Growth constant ($\mathrm{yr}^{-1}$) |  |  |  |  |  |
| *O. mossambicus* | 0.34 | 0.22–0.30 | 0.52 | 0.45 | 0.32–0.70 |
| *O. niloticus* | — | — | 0.34 | 0.43 | — |
| Total mortality |  |  |  |  |  |
| *O. mossambicus* $Z1$ | 2.53 | 1.07–2.42 | 1.40 | 3.03 | — |
| *O. mossambicus* $Z2$ | 2.28 | — | 1.71 | 3.19 | — |
| *O. mossambicus* $Z3$ | 2.45 | — | — | — | — |
| *O. niloticus* $Z1$ | — | — | 1.82 | 3.49 | — |
| *O. niloticus* $Z2$ | — | — | 1.96 | 3.58 | — |
| Natural mortality |  |  |  |  |  |
| *O. mossambicus* | 0.82 | 0.61–0.78 | 1.01 | 0.96 | — |
| *O. niloticus* | — | — | 0.75 | 0.88 | — |

Amarasinghe & De Silva (1992b) indicated that asymptotic lengths ($L_\infty$) of *Oreochromis mossambicus* estimated by De Silva & Senaratne (1988) for a series of Sri Lankan reservoirs which ranged from 26.7 to 32.7 cm in total length were influenced by small sample sizes obtained from gillnet catches. As indicated by Basson *et al.* (1988), ELEFAN and similar methods for estimating growth parameters from length-frequency data are most effective when one of the parameters of the growth equation, usually $L_\infty$, is known or approximates to a narrow range. This requirement was fulfilled by Amarasinghe & De Silva (1992b) using a root value of $L_\infty$ from the Wetherall method (Wetherall 1986; Pauly 1986).

Figure 25.4 illustrates length frequency distributions for *Oreochromis mossambicus* in Minneriya, a Sri Lankan reservoir, fitted with a growth curve by means of ELE-FAN. Growth performance index ($\varnothing = \log K + 2 \log L_\infty$; Moreau *et al.* 1986) values for *Oreochromis mossambicus* (2.37–2.8) and *Oreochromis niloticus* (2.81–2.91) in Sri Lankan reservoirs (Amarasinghe 1987; Amarasinghe *et al.* 1989; Amarasinghe & De Silva 1992b), calculated using standard lengths (approximately 0.8 total length) are within the ranges of values estimated by Moreau *et al.* (1986) for populations in African lakes (2.05–2.8 for *Oreochromis mossambicus*, and 2.41–3.11 for *Oreochromis niloticus*).

Total mortality rates of *Oreochromis mossambicus* and *Oreochromis niloticus* in

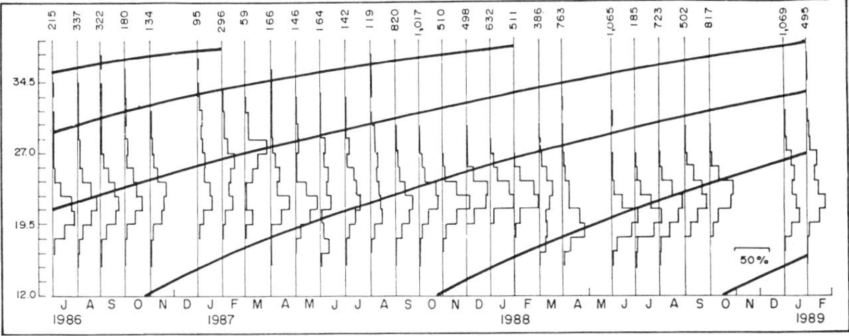

**Fig. 25.4**  Length frequency (%) distributions of *O. mossambicus* in Minneriya reservoir fitted with a growth curve by means of ELEFAN (from Amarasinghe & De Silva 1992b).

some reservoirs estimated using various length based methods are also shown in Table 25.3. As time series data on total mortality and fishing effort data were not available, natural mortality ($M$) rates (Table 25.3) were estimated using the empirical relationship between $M$, growth parameters and water temperature (Pauly 1980).

Yield-per-recruit analyses for cichlid species in several Sri Lankan reservoirs were performed by Amarasinghe (1987), Amarasinghe *et al.* (1989), and Amarasinghe & De Silva (1992b). The relative yield-per-recruit ($Y'/R$) model (Beverton & Holt 1966) is ideally suited for use in data-sparse, tropical situations. Pauly & Soriano (1986) suggested an expression of the $Y'/R$ model to compensate for the effect of a wide selection range especially because the assumption of knife-edge selection is not valid in many fisheries. This technique is exemplified by the $Y'/R$ analyses for two cichlid species in two Sri Lankan reservoirs (Fig. 25.5).

Gillnet selection studies (Amarasinghe 1988a; Amarasinghe & De Silva, in press) indicate that the desirable mesh size corresponding to optimal size of first capture (about 24 cm) of cichlids in Sri Lankan reservoirs is 10 cm.

### 25.6    Conclusion

Lack of reliable data on the fisheries of Sri Lankan reservoirs, scattered over the dry zone of the country (De Silva 1988) is a major limitation for assessment of fish stocks. As Gulland (1983) pointed out, comparative approaches, as attempted by Wijeyaratne & Amarasinghe (1987), Moreau & De Silva (1991) and De Silva *et al.* (1991), which are reviewed above, are particularly valuable for reservoir fisheries for which a detailed assessment of each reservoir individually would be impracticable. Also, since the Sri Lankan reservoir fisheries are not managed scientifically, such comparative approaches which do not require great precision, are adequate to evaluate the status of the fisheries.

It has been suggested that active involvement of fishers in the submission of accurate statistics, either individually or through the societies to which they are

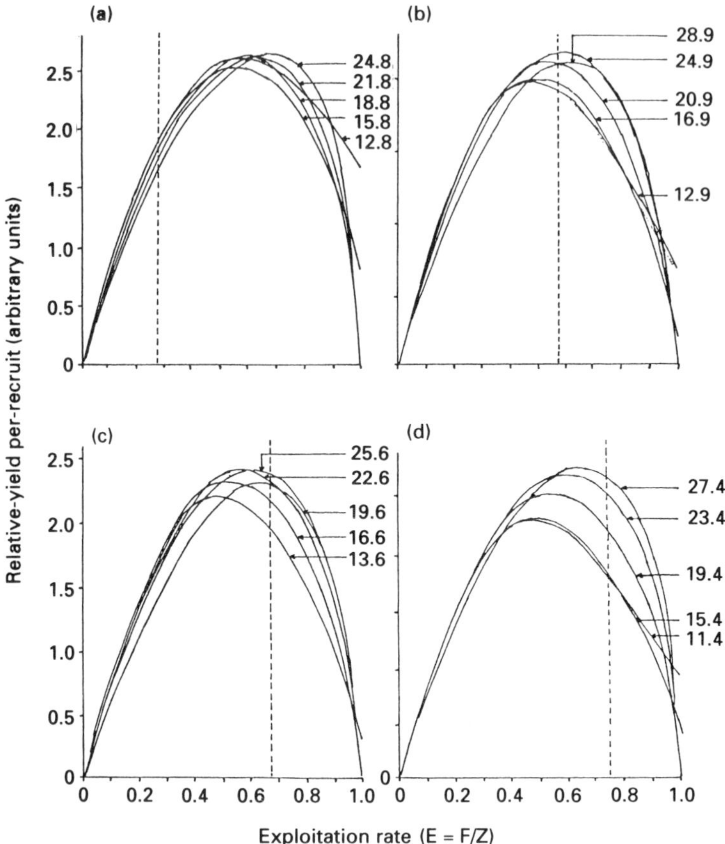

**Fig. 25.5** Relative yield-per-recruit values as functions of exploitation rates. A – *O. mossambicus* in Kaudulla; B – *O. niloticus* in Kaudulla; C – *O. mossambicus* in Minneriya; D – *O. niloticus* in Minneriya (from Amarasinghe & De Silva 1992b).

affiliated, is useful to overcome the problem of collection of fishery statistics in Sri Lankan reservoirs (De Silva 1988; Amarasinghe 1992).

As indicated above, analytical models are applicable for the assessment of fish stocks in individual reservoirs, and are useful to investigate long-term effects of changing exploitation rates and sizes of first capture on the fishery. However, Amarasinghe & De Silva (1992b) mentioned that growth of cichlids has perhaps been influenced by introgressive hybridization between *Oreochromis mossambicus* and *Oreochromis niloticus*. Investigation of the effect of hybridization on the cichlid fisheries in reservoirs is therefore useful for effective management. As suggested by Amarasinghe (1988b), it is necessary that management strategies be implemented through fisheries co-operative societies so that parallel studies on socio-economics are useful for the purpose.

## References

Amarasinghe U.S. (1987) Status of the fishery of Pimburettewa wewa, a man-made lake in Sri Lanka. *Aquaculture and Fisheries Management* **18**, 375–385.

Amarasinghe U.S. (1988a) Empirical determination of a desirable mesh size for the gillnet fishery of *Oreochromis mossambicus* (Peters) in a man-made lake in Sri Lanka. *Asian Fisheries Science* **2**, 59–69.

Amarasinghe U.S. (1988b) The role of fishermen in implementing management strategies in reservoirs of Sri Lanka. In: S.S. De Silva (ed.) *Reservoir Fishery Management and Development in Asia.* Ottawa: International Development Research Centre, pp 158–163.

Amarasinghe U.S. (1992) Recent trends in the inland fishery of Sri Lanka. In: E.A. Baluyut (ed.) Workshop on Tilapia in Capture and Culture-Enhanced Fisheries in the Indo-Pacific Fishery Commission Countries, Bogor, Indonesia, 27-29 June 1991. *FAO Fisheries Report* **458**, Supplement. Rome: Food and Agriculture Organization, pp 84–105.

Amarasinghe U.S. & De Silva S.S. (1992a) Empirical approaches for evaluating the efficiencies of different fishing methods in tropical, shallow reservoirs: a Sri Lankan case study. In: S.S. De Silva (ed.) *Reservoir Fisheries of Asia.* Ottawa: International Development Research Centre, pp 217–227.

Amarasinghe U.S. & De Silva S.S. (1992b) Population dynamics of *Oreochromis mossambicus* and *O. niloticus* (Cichlidae) in two reservoirs of Sri Lanka. *Asian Fisheries Science* **5**, 37–61.

Amarasinghe U.S. & De Silva S.S. (in press) Selectivity pattern in the multi-mesh gillnet fishery for cichlid species in a Sri Lankan reservoir. In: *Proceedings of the Third Asian Fisheries Forum.* Manila: Asian Fisheries Society.

Amarasinghe U.S. & Pitcher T.J. (1986) Assessment of fishing effort in Parakrama Samudra, an ancient man-made lake in Sri Lanka. *Fisheries Research* **4**, 271–282.

Amarasinghe U.S., De Silva S.S. & Moreau J. (1989) Spatial changes in growth and mortality and effect on the fishery of *Oreochromis mossambicus* (Peters) (Pisces, Cichlidae) in a man-made lake in Sri Lanka. *Asian Fisheries Science* **3**, 57–68.

Basson M., Rossenberg A.A. & Beddington J.R. (1988) The accuracy and reliability of two new methods for estimating growth parameters from length frequency data. *Journal du Conseil International Exploration de la Mer* **44**, 277–285.

Beverton R.J.H. & Holt S.J. (1956) A review of methods for estimating mortality rates in fish populations with special reference to source of bias in catch sampling. *Rapport et Proces-Verbaux des Reunions, Conseil International pour l'Exploration de la Mer* **140**, 67–83.

Beverton R.J.H. & Holt S.J. (1966) Manual of methods for fish stock assessment. Part 2. Tables of yield functions. *FAO Fisheries Technical Paper* **38** Revision I. Rome: Food and Agriculture Organization. 67 pp.

De Silva S.S. (1987a) Reservoir and lake fisheries. *Archiv für Hydrobiologie* **28**, 269–271.

De Silva S.S. (1987b) Impact of exotics on the inland fishery resources of Sri Lanka. *Archiv für Hydrobiologie* **28**, 273–293.

De Silva S.S (1988) Reservoirs of Sri Lanka and their fisheries. *FAO Fisheries Technical Paper* **298**. Rome: Food and Agriculture Organization, 128 pp.

De Silva S.S. & Senaratne K.A.D.W. (1988) *Oreochromis mossambicus* is not universally a nuisance species: the Sri Lankan experience. In: R.S.V. Pullin, T. Bhukaswan, T. Tonguthai & J.L. Maclean (eds) *The Second International Symposium on Tilapia in Aquaculture.* ICLARM Conference Proceedings 18. Manila: International Centre for Living Aquatic Resources Management, pp 445–450.

De Silva S.S., Moreau J., Amarasinghe U.S., Chukajorn T. & Gurrero R.D. (1991) A comparative assessment of the fisheries in lacustrine inland waters in three Asian countries based on catch and effort data. *Fisheries Research* **11**, 177–189.

Fernando C.H. & Holcik J. (1991) Fish in Reservoirs. *Internationale Revue der gesamten Hydrobiologie und Hydrographie* **76**, 149–167.

Fernando C.H. & Indrasena H.H.A. (1969) The freshwater fisheries of Ceylon. *Bulletin of the Fisheries Research Station, Ceylon* **20**, 101–134.

Gayanilo F.C. Jr., Soriano M. & Pauly D. (1989) A draft guide to the Compleat ELEFFAN. ICLARM Software 2, 70 pp. Manila: International Centre for Living Aquatic Resources Management.

Gulland J.A. (1983) *Fish Stock Assessment.* Chichester: Wiley, 223 pp.

Marshall B.E. (1984) Towards predicting ecology and fish yields in African reservoirs from preimpoundment physico-chemical data. *CIFA Technical Paper 12*, Rome: Food and Agriculture Organization, 36 pp.

Marten G.G. (1979) Impact of fishing on the inshore fishery of Lake Victoria (East Africa). *Journal of the Fisheries Research Board Canada* **36**, 891–900.

Moreau J. & De Silva S.S. (1991) Predictive fish yield models for lakes and reservoirs of the Philippines, Sri Lanka and Thailand. *FAO Fisheries Technical Paper* **319**. Rome: Food and Agriculture Organization.

Moreau J., Bambino C. & Pauly D. (1986) Indices of overall growth performance of 100 tilapia (Cichlidae) populations. In: J.L. Maclean, L.B. Dizon and L.V. Hosillos (eds) *The First Asian Fisheries Forum*. Manila: Asian Fisheries Society, pp 201–206.

Pauly D. (1980) On the relationship between natural mortality, growth parameters and mean environmental temperature in 175 fish stocks. *Journal du Conseil. Conseil International Exploration de la Mer* **39**, 175–192.

Pauly D. (1986) On improving operation and use of the ELEFAN programs. Part II. Improving the estimation of $L\infty$. *Fishbyte* **4**, 11–13.

Pauly D. & Morgan G.R. (eds) (1987) *Length-based methods in fisheries research*. ICLARM Conference Proceedings 13, Manila: International Centre for Living Aquatic Resources Management, 418 pp.

Pauly D. & Soriano M.L. (1986) Some practical extensions to Beverton and Holt's relative yield-per-recruit model. In: J.L. Maclean, L.B. Dizon and L.V. Hosillos (eds) *The First Asian Fisheries Forum*. Manila: Asian Fisheries Society, pp 491–495.

Ryder R.A. (1965) A method for estimating the potential fish production in north-temperate lakes. *Transactions of the American Fisheries Society* **94**, 214–218.

Ryder R.A. (1982) The morphoedaphic index – use, abuse and fundamental concepts. *Transactions of the American Fisheries Society* **111**, 154–164.

Silva E.I.L. & Davies R.W. (1987) The seasonality of monsoonal primary productivity in Sri Lanka. *Hydrobiologia* **150**, 165–175.

Sparre P. & Venema S.C. (1992) Introduction to tropical fish stock assessment Part 1 – Manual. *FAO Fisheries Technical Paper* **306/1** Revision 1. Rome: Food and Agriculture Organization, 376 pp.

Ssentongo G.W. & Larkin P.A. (1973) Some simple methods of estimating mortality rates of exploited fish populations. *Journal of the Fisheries Research Board Canada* **30**, 695–698.

Wetherall J.A. (1986) A new method for estimating growth and mortality parameters from length-frequency data. *Fishbyte* **4**, 12–14.

Wijeyaratne M.J.S. & Amarasinghe U.S. (1987) Estimations of maximum sustainable fish yields and stocking densities of fish fingerlings in freshwater lakes and reservoirs. *Archiv für Hydrobiologie* **28**, 305–308.

# Chapter 26
# Progress and problems in the stock assessment of Lake Tanganyika sardines

B.E. MARSHALL *Dept of Biological Sciences, University of Zimbabwe, P.O. Box MP167, Mount Pleasant, Harare, Zimbabwe*

**Abstract** *Limnothrissa miodon* and *Stolothrissa tanganicae* are endemic to Lake Tanganyika, although the former is now found in Lakes Kivu, Kariba and Cahora Bassa. Their basic demographic characteristics are known but, because of wide variations in the estimates of their growth parameters, possibly not very precisely. Mortality rates and $P/B$ ratios are high so they can support productive fisheries but yields tend to be below the predicted values. In Lake Kariba only adult fish are caught, compared with Lake Tanganyika where intensive fishing of juveniles takes place. This is a major risk to these clupeid fisheries and more effort should be devoted to the stock assessment of the juvenile portion of the stock.

KEYWORDS: biomass, growth, Lake Cahora Bassa, Lake Kariba, Lake Kivu, Lake Tanganyika, *Limnothrissa*, mortality, *Stolothrissa*, yield

## 26.1    Introduction

The pelagic fish community of Lake Tanganyika is dominated by two endemic clupeid species. *Stolothrissa tanganicae* Regan is smaller and a more specialized planktivore than *Limnothrissa miodon* (Boulenger) which feeds on a variety of prey items, including *Stolothrissa* juveniles. *Limnothrissa* was introduced into Lake Kivu in 1960 (Collart 1960) and the man-made Lake Kariba in 1967–68 (Bell-Cross & Bell-Cross 1971). It escaped from Lake Kariba to invade Lake Cahora Bassa, another man-made lake on the Zambezi River (Bernacsek & Lopes 1984). In 1992 it was introduced into another reservoir, Lake Itezhi-Tezhi in Zambia (Mubamba 1993) but it may not have survived there. These fish support pelagic fisheries that are at least ten times more productive than those based on the native fish species (Marshall 1995). Because appropriate management decisions cannot be made without an understanding of the fish populations themselves there have been a number of investigations into their population dynamics and stock assessment in recent years. The current state of this work and the main problems that still need to be addressed are reviewed.

## 26.2    The present state of stock assessment

The main problem in the stock assessment of these clupeids is determination of their age and growth rates. They are small, short-lived species with more or less continuous reproduction, and their age-classes and different cohorts are indistinct. Conventional length-based methods can only be used with difficulty but recent developments in the analysis of length-frequency distributions, such as ELEFAN, have made it easier to obtain information from large samples of these fish. Consequently, a number of recent papers and reports (e.g. Mannini 1990; Moreau *et al.* 1991; Anon. 1992; Mambona & Fryd 1993; Mulimbwa & Mannini 1993) have used these methods to give data on growth, mortality and other aspects of stock assessment.

Almost all of this work is based on length-frequency data and few attempts have been made to utilize the hard parts of the body. Useful information is arising from recent work on the daily rings on their otoliths (Chifamba 1992; Mtsambiwa 1993). Most of this work is based on samples from commercial catches, consequently relatively little is known about their pre-recruitment biology.

The fundamental parameters of stock assessment – growth, mortality and biomass – are now available for the stocks in three of the four lakes (Table 26.1). Although much of the literature stresses the importance of stock assessment to management, very little has been done to apply the data to the fisheries. Perhaps this is partly because the biology of these fish differs radically from that of larger species and so the common approaches based on growth and mortality may not be helpful (Marshall 1992, 1993a).

**Table 26.1** The availability of stock assessment data for the various populations of the Tanganyika sardines. $QD$ = qualitative description of growth, $L_\infty$ and $K$ = parameters of the von Bertalanffy growth equation, $Z$ and $M$ = instantaneous rates of total and natural mortality, $B$ = biomass. The numbers identify the sources, as follows: (1) Matthes (1965–66); (2) Roest (1978), Chapman & van Well (1978), Moreau *et al.* (1991), Mambona & Fryd (1993), Mulimbwa & Mannini (1993); (3) Coulter (1981); (4) Spliethoff *et al.* (1983), Mannini (1990); (5) Lamboeuf *et al.* (1990); (6) Cochrane (1984), Marshall (1987), Chifamba (1992); (7) Marshall (1988a); (8) Gliwicz (1984).

| Population | $QD$ | $L\infty$ | $K$ | $Z$ | $M$ | $B$ |
|---|---|---|---|---|---|---|
| *Stolothrissa* (Tanganyika) | 1 | 2 | 2 | 2 | 2 | 3 |
| *Limnothrissa* (Tanganyika) | 1 | 2 | 2 | 2 | 2 | 3 |
| *Limnothrissa* (Kivu) | | 4 | 4 | 4 | | 5 |
| *Limnothrissa* (Kariba) | | 6 | 6 | 6 | 6 | 7 |
| *Limnothrissa* (Cahora Bassa) | 8 | | | | | |

## 26.3    Problems of growth and mortality

Growth curves for sardines can now be constructed relatively easily but our understanding of their growth and the factors that influence it is still incomplete. Growth rates are extremely variable, for example *Stolothrissa*, where 68% and 34% differences in size between the largest and smallest fish at 6 and 12 months old, respectively, were found (Fig. 26.1). Since these data come from a very large lake and were collected over a number of years these growth curves may reflect environmental factors. On the other hand, similar differences occur in specimens collected at about the same time and from the same area of Lake Kariba (Table 26.2) which suggests that such variability is inherent in all stocks.

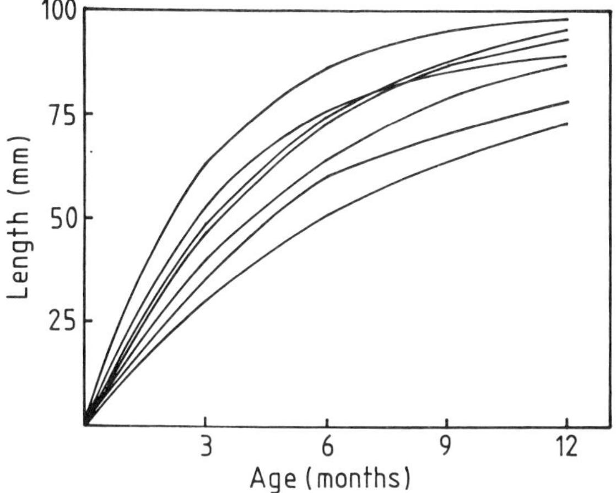


**Fig. 26.1** Growth curves of *Stolothrissa tanganicae* in Lake Tanganyika. Based on samples from Zambia (1962) and Burundi (1972–82) and drawn from data in Moreau *et al.* (1991).

**Table 26.2** The maximum and minimum lengths (mm) of *Limnothrissa miodon* in Lake Kariba in relation to their age. Based on data from otolith readings in Chifamba (1992) and Mtsambiwa (1993).

| Age (days) | Minimum | Maximum | % difference |
| --- | --- | --- | --- |
| $20 \pm 2$ | 7 | 10 | 43 |
| $40 \pm 2$ | 10 | 14 | 40 |
| $100 \pm 5$ | 25 | 53 | 112 |
| $200 \pm 10$ | 48 | 72 | 50 |
| $300 \pm 10$ | 66 | 81 | 23 |

This makes it difficult to interpret differences between stocks. For example, *Limnothrissa* consistently appear to grow about 20% larger in the Burundian waters of Lake Tanganyika than they do on the Zaïrean side (Mambona & Fryd 1993). This may reflect the existence of two separate stocks, despite the samples being taken less than 20 km apart, but the differences are more likely to have arisen because fish from Burundi were measured to the nearest 2 mm and those from Zaïre to the nearest 5 mm. Minor measurement errors assume much greater significance in small fish and may contribute to the apparent differences in their growth.

The stunted populations of *Limnothrissa* in Lakes Kariba and Cahora Bassa present more problems in relation to growth. The cause of their stunting is uncertain and attempts to explain it (e.g. Marshall 1988b, 1993b) are not entirely satisfactory. Its effect, from a stock assessment point of view, is a debate over the quantification of growth, especially with regard to the parameters of the von Bertalanffy equation. The earliest growth curves (Cochrane 1984; Marshall 1987) gave estimates of $L_\infty$ = 74 and 81 mm while later ones (Anon. 1992; Chifamba 1992) gave $L_\infty$ = 168 and 136 mm (Fig. 26.2a). Although these estimates are much closer to those from Lakes Kivu and Tanganyika (Figs 26.2b, 26.2c), they describe the Lake Kariba population inadequately. Similar variations exist in the estimation of $k$.

This suggests that the von Bertalanffy equation may not accurately describe growth in Lake Kariba *Limnothrissa* (or in Lake Cahora Bassa). Nonetheless, stock assessment is usually based on the assumption that it does accurately describe growth in those lakes, and this may give misleading results. Studies on the growth of these fish

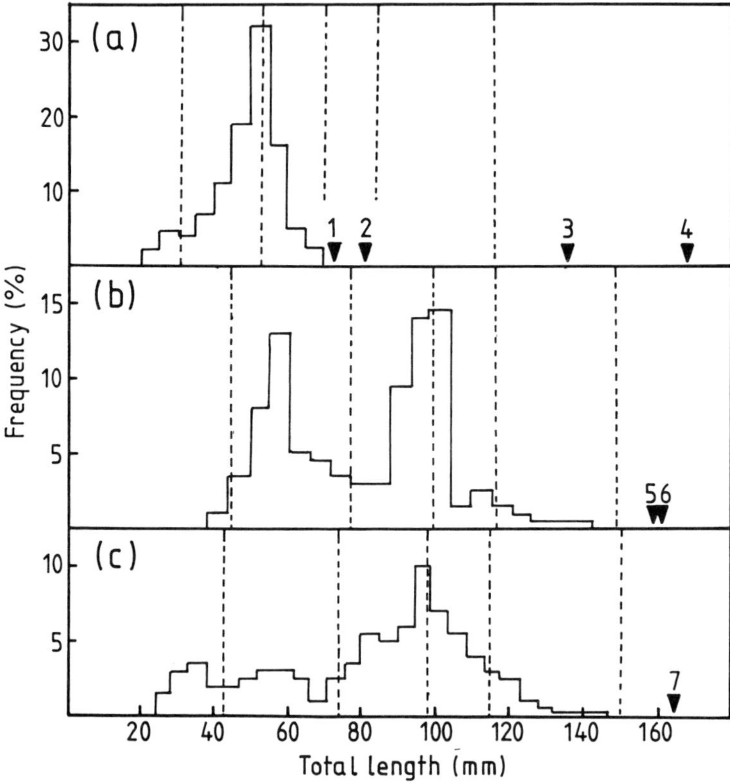

**Fig. 26.2**  Length-frequency distributions of *Limnothrissa miodon* in (a) Lake Kariba (Marshall 1993a), (b) Lake Kivu (Mannini 1990) and (c) Lake Tanganyika (Mulimbwa & Mannini 1993). Data for (b) and (c) were converted to total length from standard or fork length. The inverted triangles represent estimates of $L_\infty$ and are from: (1) Cochrane (1984), (2) Marshall (1987), (3) Chifamba (1992), (4) Anon. (1992), (5) Mannini (1990), (6) Spliethoff *et al.* (1983), and (7) Mulimbwa & Mannini (1993). The broken vertical lines denote the mean lengths of fish at 3, 6, 9, 12 and 24 months of age (the values for Kariba are based on Chifamba's growth curve).

continue and there is a risk that much effort will be wasted in the search for a 'correct' estimate of $L_\infty$ and $k$.

The uncertainties and variability of growth estimations will affect those of mortality. Very few individuals live for more than one year in Lake Kariba or two years in Lakes Kivu and Tanganyika (Fig. 26.2), and their mortality rates are very high. Estimates of mortality rates vary widely and they appear to be independent of the asymptotic length, at least over the 10 cm range found in these two species (Fig. 26.3). There seems, furthermore, to be no obvious relationship between fishing effort and mortality rates and the latter may be more sensitive to environmental changes. This can be illustrated using data from Lake Kariba, where the size, mortality and biomass of the sardines were apparently influenced by the flow of the Zambezi River (Table 26.3).

It is probably unlikely that mortality rates can be estimated with much more precision than they are at present, Nevertheless, controversy about mortality rates still exists. For example, Marshall's (1987) estimates of mortality of *Limnothrissa* in Lake Kariba were considered to be too high and were reduced to an annual rate of $Z = 5$–$6$ (Anon. 1992) but the original values of $Z$ were very similar to those of other

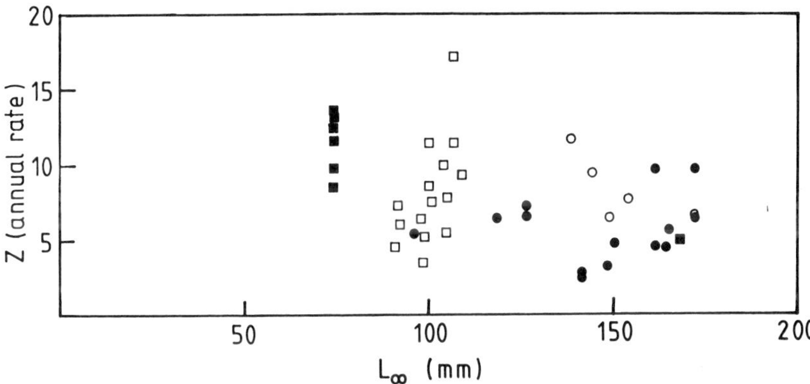

**Fig. 26.3**  Published rates of total mortality ($Z$) in relation to asymptotic length ($L_\infty$) in *Limnothrissa* in Lakes Tanganyika (●), Kivu (○) and Kariba (■) and for *Stolothrissa* (□). Data from Marshall (1987), Mannini (1990), Moreau *et al.* (1991), Mambona & Fryd (1993) and Mulimbwa & Mannini (1993).

**Table 26.3**  The relationship between the discharge of the Zambezi River into Lake Kariba and the mean length, mortality rate and biomass of *Limnothrissa miodon* in the lake, 1981–83. Data from Marshall (1988b). $a$ – this value gives the river flow in the hydrological years, which begin in October, so the data are for 1980–81, 1981–82 and 1982–83.

| | River discharge (km³)[a] | Length (mm) | Mortality (yr⁻¹) | Biomass (kg ha⁻¹) |
|---|---|---|---|---|
| 1981 | 41.78 | 55.2 | 11.64 | 90.9 |
| 1982 | 24.95 | 50.8 | 12.24 | 48.1 |
| 1983 | 25.72 | 49.9 | 13.80 | 38.7 |

populations, and there seems very little justification for attempting to achieve greater precision in this area. This reflects, perhaps, the desire to have a 'true' estimate of mortality even though this is difficult to determine in these small, short-lived fish.

### 26.4     Potential yield

One of the aims of stock assessment is to provide objective data that can be used in the management of a fishery. This is normally achieved by determining the effects of fishing on the population dynamics of the fish and using the data to justify management practices like effort regulation, fishing quotas or closed seasons. Since the basic parameters of the Tanganyika sardines are difficult to estimate, highly variable and possibly unrelated to fishing effort this approach is probably limited in its applicability.

An alternative approach is to estimate potential yield from the biomass and production/biomass ($P/B$) ratio of the stock. The $P/B$ ratios are very high, consequently the Lake Kariba catch can exceed the biomass (Table 26.4), although they have not been determined precisely. The P/B ratio is related to the instantaneous growth coefficient ($G$) or, under some circumstances, to the instantaneous rate of total mortality ($Z$) (Allen 1971), so the estimates of the $P/B$ ratios of these sardines will vary as much as those of growth and mortality. The ratio is unlikely, however, to be less than 5.0 which means that the potential yield will be equivalent to twice the biomass, assuming that 40% of their annual production can be utilized by fishermen (Table 26.4).

**Table 26.4**   The estimated biomass, current yield and potential yield (all in kg ha$^{-1}$) of pelagic fish in Lakes Tanganyika, Kivu and Kariba. Based on data in Marshall (1988a), Moreau & Nyakageni (1988), Lindqvist & Mikkola (1989), Mughanda & Mutamba (1993) and Lupikisha (1993).
$a$ – a mean value for the entire pelagic community. Data from the other lakes refer to *Limnothrissa* only.
$b$ – estimate is twice that of the biomass (see text). No reliable data are available for Lake Cahora Bassa since fishing has only just begun on that lake.

|                     | Tanganyika[a] | Kivu | Kariba |
| ------------------- | ------------- | ---- | ------ |
| Biomass             | 100           | 18   | 59     |
| Current yield       | 26            | 14   | 74     |
| Potential yield[b]  | 200           | 36   | 118    |

### 26.5     Relationships between catch and effort

Events in Lakes Kariba and Tanganyika give apparently contradictory answers to the relationship between catch and effort. Fishing boats on Lake Kariba are relatively large, constructed of steel, have diesel engines and lighting plants and powered lift nets. The fishing effort has increased steadily since 1974 and there are now some 450 vessels, or about 0.09 km$^{-2}$, operating on the lake (Lupikisha 1993). Apart from a decline in the early years, which was probably due to environmental changes in the lake (Marshall 1992), the catch per unit effort in the fishery has remained relatively unchanged (Fig. 26.4). This reflects an increase in efficiency over the years but the fish stock appears to be able to withstand intensive fishing. There is no evidence that the fishery has reached its limit (Marshall 1992) and it could, perhaps quite easily, attain its predicted potential.

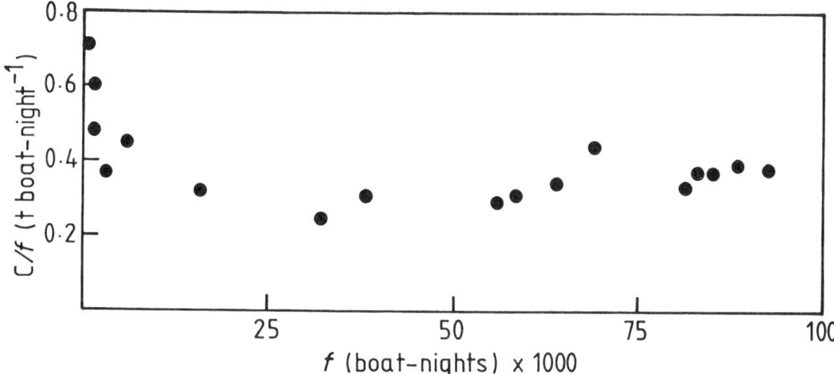

**Fig. 26.4**   The relationship between fishing effort (*f*) and catch per unit effort (*C/f*) in the Lake Kariba sardine fishery 1974–90. Based on data in Lupikisha (1993).

In Lake Tanganyika, the catch per unit effort of the industrial fisheries in Zambia and Burundi decreased steadily from 1972 (Fig. 26.5). Industrial fisheries declined rapidly in the Tanzanian sector during the same period even though the increase in effort was relatively small (Fig. 26.6). If the estimates of pelagic biomass in Lake Tanganyika are accurate then the lake is far from being overfished and it is not clear why its fish stocks should have responded so differently from those in Lake Kariba.

It is possible that these data indicate a rapid decrease in the operational efficiency

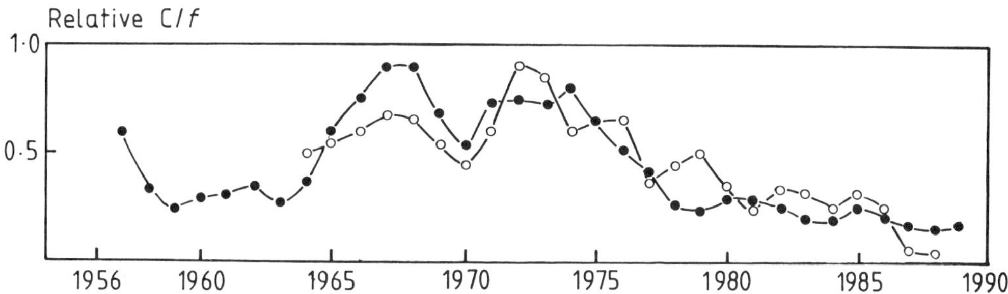

**Fig. 26.5**   The relative catch per unit effort (*C/f*) in the industrial fisheries in the Burundi (●) and Zambian (○) sectors of Lake Tanganyika. Redrawn from data in Roest (1992) and shown as two-year smoothed means.

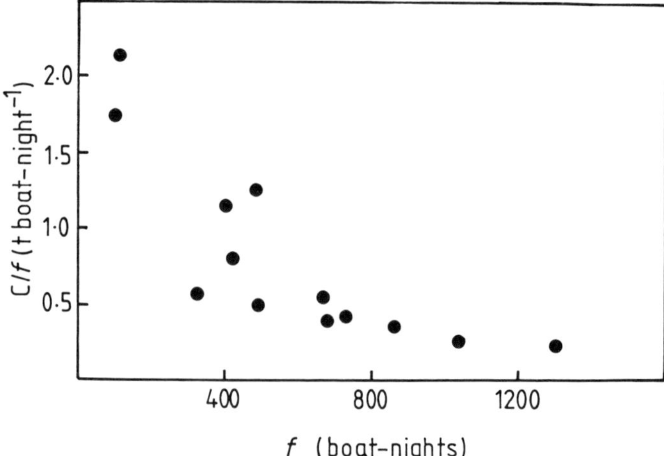


**Fig. 26.6**   The relationship between catch per unit effort (*C*/*f*) and fishing effort (*f*) in the industrial fishery of the Tanzanian sector of Lake Tanganyika 1974–87. Based on data in Bayona (1993).

of the larger fishing vessels. They may also reflect the relative complexity and high levels of predation in Lake Tanganyika's pelagic community. The abundance of *Stolothrissa* appears to be inversely related to that of *Lates stappersii* (Boulenger), perhaps its most important predator (Roest 1988). *Limnothrissa* also preys upon *Stolothrissa* but its impact and importance is not fully understood. Fishing may therefore be an additional source of mortality in Lake Tanganyika, rather than a substitutional one as it is in Lake Kariba.

The response of the stocks in the two lakes may also be determined by the way the fisheries operate. There were an estimated 58 industrial purse seiners, 1851 artisanal lift-nets and 10 000 coastal canoes fishing on Lake Tanganyika in the late 1980s (Lindqvist & Mikkola 1989), which amounts to 0.36 vessels km⁻², i.e. four times the density on Lake Kariba. Although the coastal canoes were the most numerous there appears to be no data on their catches but they take mostly juvenile fish (Roest 1978) since both sardine species breed inshore. Furthermore, there appears to have been a considerable increase in the number of fishermen who use small-meshed beach seines, sometimes made of mosquito net, to catch sardine fry (Phiri 1993). The extent of this practice has not been documented but it is evidently widespread in some areas.

The major fishing effort is therefore expended on the juvenile sardines in the littoral zone. Since this zone is a relatively small portion of the lake these fishing practices may have a major impact and could now be affecting the recruitment of sardines to the open waters. By contrast, the Lake Kariba fishery catches only adults of which very few survive to one year because *Limnothrissa* is essentially an annual species in this lake. Thus the fishery is taking individuals that will die in any case (Marshall 1993a, 1993b).

### 26.6     Where do we go from here?

Stock assessment is of little value if it cannot answer the questions asked by fishery managers. Whilst the stock assessment of the Tanganyika sardines is relatively advanced its application to the fisheries has been rarely attempted. Catch and effort data are important components of stock assessment but they have not been very helpful, partly because of their inadequacy but also because the sardine populations do not behave like those of other species. This was demonstrated in the Lake Kariba fishery where it was shown that the catch and effort data cannot be used to predict optimum levels of effort or maximum sustainable yield (Marshall 1992).

The demographic features of the commercially-exploited stocks are sufficiently well understood for practical purposes and probably cannot be greatly improved because of their high degree of variability. Nevertheless, a continuing output of reports with these data is to be expected because ELEFAN and other computer programmes make it easy to produce them and more work will be done with other methods such as the use of daily rings in otoliths. Although this is a laudable development it carries the risk of stagnation since researchers will become locked into an endless attempt to refine their estimates of growth and mortality.

Information from the Lake Kariba fishery suggests that this would be a wasted effort since the adult fish are destined to die within a short time. It is the fate of the juveniles that matters as the major threat to clupeid fisheries is reproductive failure leading to a decrease in recruitment. Reproductive failure can be caused by environmental factors which include changes to the food supply, increased predation and deteriorating conditions in breeding areas. The increasing level of sediment pollution in Lake Tanganyika (Cohen *et al.* 1993) is an example of such a potential threat. Intensive fishing of larvae and juveniles may also lead to recruitment failure, which may already be occurring in Lake Tanganyika.

Most of the work on the Tanganyika sardines to date has been done on commercially-exploited adults. Investigations into their larval and pre-recruitment biology have only just begun but this work is likely to be of much more significance. Further progress will only be made once the importance of this stage of the sardine's life-history is appreciated.

### References

Allen K.R. (1971) Relation between production and biomass. *Journal of the Fisheries Research Board of Canada* **28**, 1573–1581.

Anon. (1992) Working group on assessment of kapenta (*Limnothrissa miodon*) in Lake Kariba (Zambia and Zimbabwe). SADCC Zambia-Zimbabwe Fisheries Project, Kariba, 4–17 March 1992. Kariba, Zimbabwe: Lake Kariba Fisheries Research Institute, 75 pp.

Bayona J.D.R. (1993) Variation in abundance and distribution of *Limnothrissa miodon* in the Tanzanian sector of Lake Tanganyika: the need for continued stock assessment. *FAO/CIFA Occasional Paper* **19**, 121–140.

Bell-Cross G. & Bell-Cross B. (1971) Introduction of *Limnothrissa miodon* and *Limnocaridina tanganicae* from Lake Tanganyika into Lake Kariba. *Fisheries Research Bulletin of Zambia* **5**, 207–214.

Bernacsek G.M. & Lopes S. (1984) Cahora Bassa (Mozambique). *FAO/CIFA Technical Paper* **10**, 21–42.

Chapman D.W. & van Well P. (1978) Growth and mortality of *Stolothrissa tanganicae*. *Transactions of the American Fisheries Society* **107**, 26–35.

Chifamba P.C. (1992) Daily rings on otoliths as a method for ageing the sardine, *Limnothrissa miodon*, in Lake Kariba. *Transactions of the Zimbabwe Scientific Association* **66**, 15–17.

Cochrane K.L. (1984) The influence of food availability, breeding seasons and growth rate on commercial catches of *Limnothrissa miodon* (Boulenger) in Lake Kariba. *Journal of Fish Biology* **24**, 623–635.

Cohen A.S., Bills R., Cocquyt C.Z. & Caljon A.K. (1993) The impact of sediment pollution on biodiversity in Lake Tanganyika. *Conservation Biology* **7**, 667–677.

Collart A. (1960) L'Introduction de *Stolothrissa tanganicae* (Ndagala) au lac Kivu. *Bulletin Agricole du Congo Belge* **51**, 975–985.

Coulter G.W. (1981) Biomass, production, and potential yield of the Lake Tanganyika pelagic fish community. *Transactions of the American Fisheries Society* **110**, 325–335.

Gliwicz Z.M. (1984) Limnological study of Cahora Bassa reservoir with special regard to sardine fishery expansion. FAO/GCP/MOZ/006/SWE Field Document 8. Rome: FAO, 71 pp.

Lamboeuf M., Mulindabigwi A. & Mutamba A. (1990) Estimation de l'abondance du stock d'Isambaza (*Limnothrissa miodon*), resultats de la prospection acoustique de Mars 1990. Projet de Développement de la Peche au lac Kivu, Aide-Memoire 26. Gisenyi, Rwanda: FAO, 8 pp.

Lindqvist O.V. & Mikkola H. (1989) Lake Tanganyika: review of limnology, stock assessment, biology of fishes and fisheries. Report to the Regional Project for the Management of Fisheries on Lake Tanganyika, FI:GCP/RAF/229/FIN. Rome: FAO, 51 pp.

Lupikisha J.M.C. (1993) Catch trends of *Limnothrissa miodon* in Lake Kariba. *FAO/CIFA Occasional Paper* **19**, 51–67.

Mambona W.B. & Fryd M. (1993) Population parameters of *Stolothrissa tanganicae* and *Limnothrissa miodon* in the northern part of Lake Tanganyika. *FAO/CIFA Occasional Paper* **19**, 157–167.

Mannini P. (1990) Paramètres de la population de *Limnothrissa miodon* du lac Kivu (1980–1989). Projet de Développement de la Peche au lac Kivu, Document de Travail 32. Gisenyi, Rwanda: FAO, 41 pp.

Marshall B.E. (1987) Growth and mortality of the introduced Lake Tanganyika clupeid, *Limnothrissa miodon*, in Lake Kariba. *Journal of Fish Biology* **31**, 603–615.

Marshall B.E. (1988a) A preliminary assessment of the biomass of the pelagic sardine *Limnothrissa miodon* in Lake Kariba. *Journal of Fish Biology* **32**, 515–524.

Marshall B.E. (1988b) Why are the sardines in Kariba so small? *Zimbabwe Science News* **22**, 31–35.

Marshall B.E. (1992) The relationship between catch and effort and its effect on predicted yields of sardine fisheries in Lake Kariba. *NAGA, the ICLARM Quarterly* **15(4)**, 36–39.

Marshall B.E. (1993a) *Limnothrissa* in man-made lakes: do we understand the implications of their small size? *FAO/CIFA Occasional Paper* **19**, 1–17.

Marshall B.E. (1993b) The biology of the African clupeid *Limnothrissa miodon* with reference to its small size in man-made lakes. *Reviews in Fish Biology and Fisheries* **3**, 17–38.

Marshall B.E. (1995) Why is *Limnothrissa miodon* such a successful introduced species and is there anywhere else we should put it? In: T.J. Pitcher & P.J.B. Hart (eds) *Impact of Species Change in African Lakes*. London: Chapman & Hall.

Matthes H. (1965-66) Preliminary investigations into the biology of Lake Tanganyika Clupeidae. *Fisheries Research Bulletin of Zambia* **4**, 39–45.

Moreau J. & Nyakageni B. (1988) Les relations trophiques dans la zone pélagique du lac Tanganyika (secteur Burundi). Essai d'evaluation. *Revue d'Hydrobiologie Tropicale* **23**, 157–164.

Moreau J., Munyandorero J. & Nyakageni B. (1991) Evaluation des paramètres dèmographiques chez *Stolothrissa tanganicae* et *Limnothrissa miodon* du lac Tanganyika. *Verhandlungen Internationale Vereinigung für Theoretische und Angewandte Limnologie* **24**, 2552–2558.

Mtsambiwa M.Z. (1993) Fitting a von Bertalanffy growth model to length at age data for larval *Limnothrissa miodon* from Lake Kariba. *FAO/CIFA Occasional Paper* **19**, 68–74.

Mubamba R. (1993) Introduction of Lake Tanganyika sardines into Itezhi-Tezhi Reservoir, Zambia. *FAO/CIFA Occasional Paper* **19**, 44–50.

Mughanda M. & Mutamba A. (1993) Thirteen years of exploitation of *Limnothrissa miodon* Blgr. in Lake Kivu (Rwanda and Zaïre). *FAO/CIFA Occasional Paper* **19**, 213–229.

Mulimbwa N. & Mannini P. (1993) Demographic characteristics of *Stolothrissa tanganicae*, *Limnothrissa miodon* and *Lates stappersii* in the northwestern (Zaïrean) waters of Lake Tanganyika. *FAO/CIFA*

*Occasional Paper* **19**, 176–195.

Phiri H. (1993) The effect of increased fishing pressure on the abundance of *Limnothrissa miodon* in southern Lake Tanganyika. *FAO/CIFA Occasional Paper* **19**, 196–203.

Roest F.C. (1978) *Stolothrissa tanganicae*: population dynamics, biomass evolution and life history in the Burundi waters of Lake Tanganyika. *FAO/CIFA Technical Paper* **5**, 42–62.

Roest F.C. (1988) Predator-prey relations in northern Lake Tanganyika and fluctuations in the pelagic fish stocks. *FAO/CIFA Occasional Paper* **15**, 104–129.

Roest F.C. (1992) The pelagic fisheries resources of Lake Tanganyika. *Mitteilungen Internationale Vereinigung für Theoretische und Angewandte Limnologie* **23**, 11–15.

Spliethoff P.C., de Iongh H.H. & Frank V. (1983) Success of the introduction of the freshwater clupeid *Limnothrissa miodon* (Boulenger) in Lake Kivu. *Fisheries Management* **14**, 17–31.

# Chapter 27
# Limitations of catch and effort data collection methods used for inland fisheries in the SADC region, with particular reference to Zimbabwe

J.J. MANDIMA *University of Zimbabwe, Lake Kariba Research Station, PO Box 48 Kariba, Zimbabwe*

**Abstract**   Methods used for collecting catch and effort data for inland fisheries in the Southern Africa Development Community (SADC) region are reviewed to highlight the existing limitations. A catch per unit effort (CPUE) index is used to indicate fish abundance. It is calculated from data collected during monthly catch assessment surveys (CAS), annual total frame surveys (TFS) and supplementary frame surveys (SFS) for the artisanal sector, and monthly catch returns for the commercial sector. Gillnets are the standard gear used in all surveys, with a 100 m length making up a unit of effort and CPUE being expressed as kg 100 m net$^{-1}$ d$^{-1}$.

The review shows that CAS are limited to few fishing areas, thus biasing the coverage of any one water body; the length and timing of enumeration period each month may not be optimal and this causes bias in total catch estimates where within month variations exist. Furthermore, the dependency on self completed catch records from the commercial sector results in misreporting of gillnet yardage. Additionally, use of law enforcement personnel to conduct the surveys creates a conflict of interests causing distortion of information by the fishermen.

KEYWORDS: Catch effort, fisheries, Zimbabwe

## 27.1    Introduction

In the Southern Africa Development Community (SADC) region (Fig. 27.1), inland fisheries consist of lake, reservoir and riverine fisheries. Of these, most major commercial fishing is confined to the large natural and man-made lakes (e.g. Tanganyika, Malawi, Kariba and Cahora Bassa) where the fish resources are most abundant. Small reservoirs, commonly called small water bodies (SWBs) in the SADC region, have increased in number over the years, with Zimbabwe alone now having in excess of 10 000 SWBs. This is a move aimed at conserving water to counter what have now become recurrent drought seasons. Fisheries exploitation from these SWBs by artisanal fishermen, and commercial fishermen operating as co-operatives, is on the

**Fig. 27.1**   Distribution of major water bodies in the SADC region.

increase, such that in Zimbabwe most rural and farm communities rely on these water bodies for their fish requirements. Riverine fisheries are restricted due to the limited number of major rivers (e.g. the Zambezi, Kafue, Shire) in the whole region, but the few seasonal, medium-sized rivers have fisheries which are exploited by artisanal fisherfolk on a local level. Their fish production is, however, not monitored regularly.

The demand for fish has generally increased in Zimbabwe (Balarin 1984) and all other SADC countries. This has been accompanied by an increase in the exploitation of fisheries resource using diverse fishing gears. This has caused governments to institute strong fisheries management policies aimed at monitoring stock abundances. Catch and effort data are collected through monthly returns of catch records by the commercial sector, and through total frame surveys (TFS) and catch assessment surveys (CAS) for the artisanal sector. These data are used to calculate catch per unit effort (CPUE) which provides an index of stock abundance and forms the basis for fisheries legislation and management. In Zimbabwe, these data are supplemented by test fishings in major lakes and reservoirs using gillnets of various mesh sizes and, on rare occasions, other gears like seine nets and long lines.

From a fisheries management point of view, gillnets are used as the reference gear for most inshore fishing, but in practice many other traditional gears are used by artisanal fishermen (Chimbuya 1993). These gears have different degrees of selectivity and catching efficiency. Some are destructive when used during upstream fish migrations and spawning periods. When fishermen use gillnets, the mesh sizes almost always depart significantly from the officially recommended limits, with a general tendency to use small meshes when catches decline. The diversity of fishing gears used by artisanal fishermen, the high incidence of poaching, and the uncertainty in defining the unit of effort, all render the data, on which most fisheries management in the SADC region is based, questionable.

This chapter reviews the variety of catch and effort data collection methods used in the SADC region, with particular reference to cases in Zimbabwe. The gear diversity at the artisanal fishermen level and the different characteristics of fishing groups (freelance fishermen, co-operatives, rod and line fishermen) are examined to highlight how such factors ultimately create inaccuracies in catch and effort assessments.

## 27.2    Materials and methods

Data and information for this study were collected from various large lakes and medium-sized reservoirs throughout Zimbabwe during short-term study visits to government research stations and sites fished by both commercial and artisanal fishermen. On-site discussions and gear demonstrations were conducted with fishermen who were generally willing to release information about their fishing patterns on request. The catch and effort data were obtained from monthly catch records submitted by commercial fishermen and test fishing records from government research stations. Information from other SADC countries was accessed from a literature review.

## 27.3    Results

### 27.3.1    *Fishing gears and fishing methods*

**Gillnets**    These are the most commonly used fishing gear in SADC inland fisheries. In Zimbabwe gillnets are made of multi-filament nylon twine. Mesh sizes used for test fishing were 38, 51, 64, 76, 89, 102, 114 and 127 mm stretched mesh (van der Mheen, unpublished). For commercial fishing, the officially recommended minimum mesh size is 76 mm, but in practice fishermen use gillnets of mesh sizes as low as 51 mm. This results from fishermen being unable to purchase ready made gillnets from commercial suppliers so they buy nylon material and make nets to their own specifications. When purchased commercially, each gillnet unit is 90 m long, and reduced to 45 m when mounted at a 50% hanging ratio. The number of gillnets per fisherman is limited by government and this varies according to the size of water body, number of active fishermen, and the abundance of fish resources determined by test fishing or other stock assessment methods. On the Zimbabwean side of Lake Kariba (Fig. 27.2), each fisherman is allowed a maximum of 5 gillnets. Recommendations have been made to relate the number of nets and the mesh size allowable per small communal dam to the size of the water body (van der Mheen, unpublished) (Table 27.1).

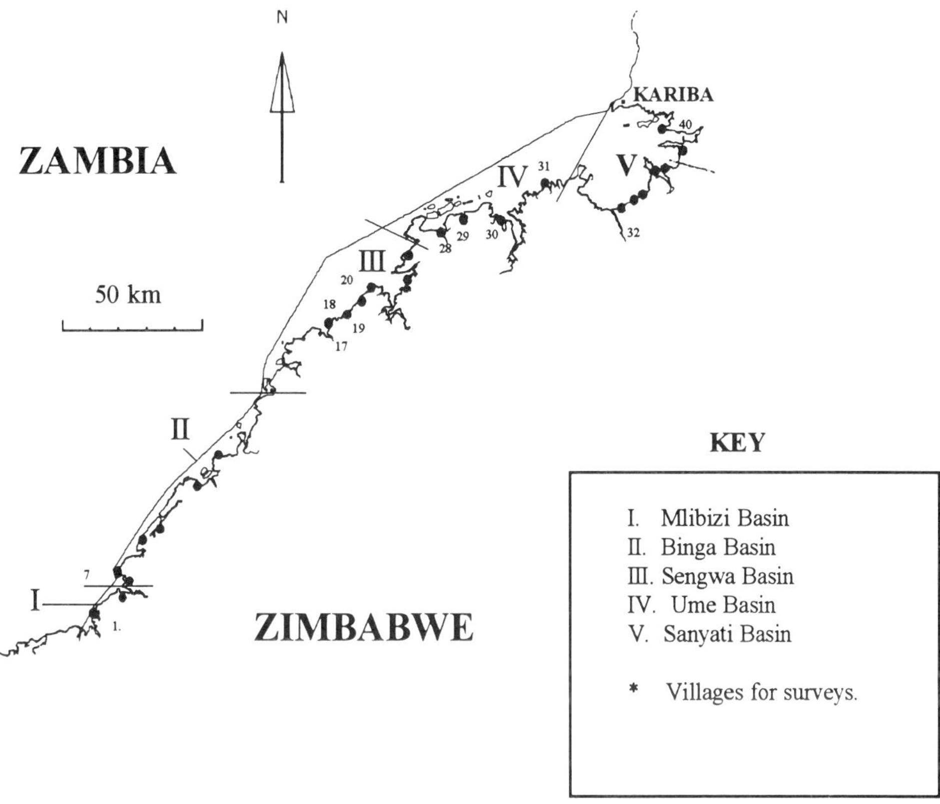


**Fig. 27.2**   Zimbabwean side of Lake Kariba showing location of inshore fishing villages.

The gillnets are normally set in the late afternoon and retrieved the following morning. Gillnets can be top and/or bottom set, depending on the habit of the species sought. The process of setting the gillnets requires the use of a small boat and, with most fishermen using non-motorized boats, there is limited coverage of fishable areas. In some water bodies like Lake Kariba, nets are prone to damage by crocodiles. When this happens fishermen make their own repairs but when carried out repeatedly, the mesh size, hanging ratio and length changes gradually until it no longer resembles the original specifications.

**Seine nets**   These are used as either beach seines for the inshore fishery or as pelagic seines for mid-water fish species like the small clupeid, *Limnothrissa miodon* (Boul.) in Lakes Kariba and Cahora Bassa (Fig. 27.1). Use of seine nets for both the inshore and

**Table 27.1**  Guideline for numbers of gillnets on communal SWBs in Zimbabwe.

| Size of SWB (ha) | Mesh size (mm) | No. of nets | Mesh size (mm) | No. of nets |
| --- | --- | --- | --- | --- |
| 0–10 | — | 0 | — | 0 |
| 10–25 | 76 | 2 | 114 | 2 |
| 25–50 | 76 | 4 | 114 | 4 |
| 50–100 | 89 | 6 | 127 | 6 |

pelagic fisheries has, however, been limited because they are expensive and also more difficult to operate without a reliable fast boat. When used, for example in Lake Malawi for the inshore fishery, seines have the advantage of being less selective and able to take all the fish above a certain minimum size.

**Long lines**   In Zimbabwe, this gear is used for test fishing to catch predator species like the vundu *Heterobranchus longifilis* (Cuv.) in Lake Kariba. Scattered use of long lines by artisanal fishermen takes place throughout the country, but their contribution to overall fish production is minor.

**Other gears**   Other traditional gears and methods commonly used in various reservoir fisheries in Zimbabwe are the throw net, scoop net, rod and line, basket traps, tin traps, spears, poisons and explosives. Details of how they are used can be found in Chimbuya (1993).

### 27.3.2   *Status of fisherman*

Two fishery sub-sectors exist in the SADC region: (i) the artisanal sub-sector and (ii) the commercial sub-sector. In the former, various levels of operation exist according to the time spent fishing, thus fisherfolk are classified as freelance, part time/occasional or full time. Most of these fish are for subsistence, but seasonal surplus fish are caught for sale. In Zambia and Zimbabwe, fishing villages have been established along the shores of Lake Kariba. These settlements are permanent on the Zimbabwean side, while significant movement of fishermen between villages is a common feature in Zambia (Fig. 27.2).

In other inland SWBs, fishing is carried out by members of communities close to the water body. Fishing is mainly by rod and line which is not licensed. This kind of fishing is quite extensive and the total fish production from this activity has been difficult to estimate. However, during the agricultural lay-off (May–October) each year, considerable fishing goes on and personal communication with fishermen revealed that substantial catches are made during this period; all such production is never recorded. In some communal dams, local people have formed fishing co-operatives which are licensed to fish using a specified number of gillnets, depending on the size of the water body.

The commercial sub-sector consists of corporate companies that have established fishing villages that they supply with fishing gear and craft, pay basic wages to the full time fishermen they engage, and take care of marketing. Individual companies and co-operatives also operate commercially on a full-time basis. In all these commercial operations, the fishing gear, craft and catch are easy to monitor.

Also prevalent is recreational fishing for the tigerfish, *Hydrocynus vittatus* (Cuv.) in Lake Kariba and trout, *Salmo trutta* (L.) in the Eastern Highlands of Zimbabwe. The tilapias are also exploited by sports fishermen in most inland reservoirs, for example in Sebakwe, Gwenoro and Amapongokwe dams in the Midlands province of Zimbabwe and Lake Mutirikwi in Masvingo (Table 27.2). While most sport fishing is seasonal, the quantity of fish caught has been increasing annually, rendering it essential to consider this mechanism of exploitation in fisheries management.

Overall, the status of fishermen in the SADC region is very diverse, a characteristic which makes it difficult to assess the exploitation levels of the fisheries resources in most inland waters.

**Table 27.2**  Sebakwe anglers' survey for September and October 1993.

|  | September | | October | |
| --- | --- | --- | --- | --- |
|  | Number | Weight (g) | Number | Weight (g) |
| *Tilapia rendalli* | 22 | 1295 | 40 | 6025 |
| *Oreochromis machrochir* | — | — | 64 | 9725 |
| *Oreochromis mossambicus* | — | — | 16 | 3825 |
| *Labeo altivelis* | — | — | 1 | 250 |
| *Alestes imberi* | — | — | 10 | 575 |
| *Pharyngochromis darlingi* | 35 | 600 | 221 | 875 |
| *Serranochromis robustus* | 10 | 325 | 11 | 1875 |
| *Clarias gariepinus* | 38 | 3875 | 22 | 5125 |
| Total number of anglers | 16 | | 42 | |
| Total number of fish | 105 | | 385 | |
| Total weight of fish (kg) | 6.095 | | 28.275 | |
| Total hours fished | 63 | | 90 | |
| Number of fish per angler | 6.56 | | 9.17 | |
| Fish weight (g) per angler | 380.94 | | 673.2 | |
| Number of fish per hour | 1.67 | | 4.28 | |

### 27.3.3  *Catch and effort data collection methods*

In Zimbabwe, Zambia and Malawi, fisheries catch and effort data are collected through use of three survey methods: i) annual total frame survey (TFS); ii) monthly catch assessment survey (CAS); and iii) periodic supplementary frame survey (SFS). All these surveys are conducted for the artisanal sub-sector where fishing gear, craft and fishermen's status are so diverse that there is very limited scope for maintaining reliable catch and effort records that are valid for long-term use.

*Annual total frame surveys* (TFS)

This is a comprehensive exercise conducted to provide basic information that has a bearing on the fishing pressure, hence effort, directed to a particular fishery. In Zambia and Zimbabwe, this is carried out for the inshore fishery of Lake Kariba. Information is gathered on the number and location of fishing villages, number of fishermen and their respective assistants per village, boats and gears by type per fisherman, mesh sizes of nets, fishing methods and status of fishermen, i.e. whether they are freelance, part-time, full-time or co-operative members (Sanyanga *et al.* 1991; Thorsteinsson 1992). TFS are also a common approach in Malawi (Alimoso 1991). Typical results from a TFS for Zambia and Zimbabwe are given in Table 27.3.

**Table 27.3**    Results for 1990 total frame survey in Zambia and Zimbabwe (Thorsteinsson 1992).

| Variable | Zambia | Zimbabwe |
|---|---|---|
| Number of villages | 256 | 39 |
| Number of fishermen | 2167 | 746 |
| Number of boats | 1835 | 409 |
| Number of nets | 21568 | 1487 |

The dynamic nature of most artisanal fisheries in the SADC region make it very difficult for information collected once per year to be useful for long-term management. To minimize this, shorter surveys are conducted, for example in Zambia, to validate findings of the TFS.

*Supplementary frame surveys (SFS)*

These are half-yearly or quarterly short surveys aimed at validating findings of the TFS. In Zambia, SFS show a lot of movement of fishermen between villages, an increase in the number of fishermen and an increase in the number of nets in use. This means that the reliability of effort estimates obtained from frame surveys is questionable. To minimize this error, SFS have been recommended by the Zambian Department of Fisheries (DoF), but from time to time these intentions are hindered by lack of financial and manpower resources.

*Monthly catch assessment surveys (CAS)*

This is an activity common on large lakes in the SADC region, and, for Zambia and Zimbabwe, CAS are carried out for the artisanal freelance fishermen exploiting the inshore fishery of Lake Kariba, while Lakes Malawi and Tanganyika are the targets for CAS in Malawi and Tanzania respectively. On the Zimbabwean side of Lake Kariba, enumeration teams are sent out by the Department of National Parks and Wildlife Management (DNPWM) to fishing villages for 10 days once every month.

Ten villages out of the 39 fishing villages are selected for this exercise, and for logistical reasons these are all located in lake basins 4 and 5 which are closer to Kariba town (Fig. 27.2). During these surveys, records of the number and weight of each fish species and the number of nets per fisherman are made. With these records the CPUE, which is defined as kg fish 100 m net$^{-1}$ d$^{-1}$, is calculated.

Catch assessment surveys conducted in Zambia and Malawi have a wider coverage of the target water body. The Zambian approach is based on the Bazigos systems 1971–1973 (Bazigos 1974). In both countries, the shoreline is first divided into major and minor strata on the basis of fishing intensity, fishing methods and the physical characteristics of the area (Alimoso 1991; Thorsteinsson 1992). In Zambia, the number of fishing boats per village is used as an important criterion in the stratification process.

In the various strata and substrata, three and four villages in Zambia and Malawi respectively are selected randomly for the monthly surveys in which the catch and effort data are recorded and CPUE determined. In Zambia, CPUE is defined as kg fish boat$^{-1}$ d$^{-1}$.

*Commercial sector*

For this sector, monthly catch records reflecting daily catches are the principal source of catch data (Alimoso 1991). These are submitted on the understanding that for this sector, there is minimal variation of effort, hence results are reliable. In Zimbabwe, commercial fishermen have designated fishing areas which make it easy to follow their fishing routines.

### 27.4    Discussion

Catch and effort data collection methods in current use in the SADC region are time-consuming and financially taxing, so to minimize these costs, a compromise has been made whereby scientific accuracy is traded off against lower expenses. Data from TFS and CAS are usually so limited that surplus production models only are suitable for analysing these data. The dynamic pool models which require detailed growth, recruitment and mortality information for each species cannot be used (Alimoso 1991). From a fisheries modelling point of view, this situation proves that fisheries data collection methods now in use have many limitations.

In Zimbabwe's Lake Kariba, where CAS is limited to basins 4 and 5 for logistic reasons, the entire lake coverage is compromised. Conducting the catch surveys on the same days each month can potentially bias estimates of total catch where regular within-month variations of catches exist, as stated by the respondent fisherfolk. Furthermore, 10 days may not be an optimal period of time in which to carry out enumeration. Also, visiting the same fishing villages on each CAS tends to completely exclude all the other fishing camps where different micro-environments, and hence species, exist, thus necessitating the institution of a different management policy. Kell

(1991) and Malvestuto (1991) emphasized that surveys monitoring fish populations should take into account variations due to fish behaviour, and the physical and temporal characteristics of the habitat.

Total effort is understated by commercial fishermen, and there is a big risk of mis-identification of fish species on records submitted monthly. This is clearly depicted in commercial co-operatives where the reported daily effort is always similar to that allowed on their fishing permit and constant throughout the season as if no nets are damaged by crocodiles and hippopotami or stolen from time to time (Thorsteinsson 1992). This reflects Gerdeaux's (1991) observation that data collected by enumerators are of a better quality than the declarations recorded in the fishermen's notebooks. Malvestuto (1991), however, highlights various problems associated with recall data events for recreational fisheries.

The situation in Zambia, where sample villages are randomly selected, gives all fishing villages equal chance for assessment but it lacks continuity, thus preventing a time series catch assessment which is the ultimate goal. The definition of CPUE as kg boat$^{-1}$ d$^{-1}$, as in Zambia, is unsuitable since the number of nets per boat can be very variable such that there is risk of understating or overstating fishing effort. Such a unit of effort could only work where fishermen are reliable and maintain the licensed number of gillnets, a situation which is rare in most artisanal fisheries worldwide. Instead, the definition of CPUE as kg 100 m$^{-1}$ d$^{-1}$, as used by Zimbabwe and Malawi, gives a fair estimation of effort given that all gillnets have standard length and depth dimensions, but it needs refining since it assumes that effort is the same for gillnets of different meshes.

Use of the same manpower for law enforcement and scientific data collection brings into question the validity of such data since fishermen will tend to relate to them more as law enforcement personnel than as enumerators. They then give them access only to catch and effort information and data that correspond to their permit, when in reality, traditional gears are used and closed areas are fished. This means that fisheries managers end up with gross catch underestimates, which is misleading in policy making. Such cases are substantiated by the great discrepancy between catch and effort records collected by independent enumerators and government fisheries personnel (Chimbuya pers. comm. 1993).

On the whole, it is clear that a number of loopholes exist in the current methods used to collect catch and effort data. At best, these methods give an underestimate of the real effort and catch. Other inland water resources like rivers are not monitored, yet there is considerable fisheries production from these resources. There is thus a need for fisheries scientists and managers in the region and worldwide, to devise more adaptive management strategies which would allow for more reliable data collection methods.

In the light of the several limitations of the catch and effort data collection methods in use in the SADC region, it is proposed that fishermen's co-operatives or organi-zations are involved in management, as has been the case in some European and North American fisheries. Jentoft (1989) suggested the practice of fisheries co-management whereby government agencies and fishermen's co-operative organizations share the responsibility for management functions. He notes how the Norwegian fishery has benefited from this approach. Co-management seeks to utilize the informal management systems traditionally practised by fishing communities and it emphasizes the need for government to recognize the existing traditional informal regulations and the transfer of regulatory power to fishermen's organizations. Such a state of affairs would facilitate the collection of catch and effort data regularly and cost-effectively. One would also expect more reliable data to come out since the participants in the industry will be supplying such data with the insight of its being beneficial to them directly.

## Acknowledgements

Many thanks are due to the Department of National Parks and Wildlife Management personnel in Zimbabwe based at Lakes Kariba, Chivero, Sebakwe and Mutirikwi who were useful resource persons during this study. Thanks also go to Mr E Dhlomo who prepared the figures. I unreservedly wish to thank the University of Zimbabwe for funding this study through Research Board Vote Number 2.9999.10.3643.

## References

Alimoso S.B. (1991) Catch effort data and their use in the management of fisheries in Malawi. In: I.G. Cowx *Catch Effort Sampling Strategies: their Application in Freshwater Fisheries Management*. Oxford: Fishing News Books, pp. 393–403.

Balarin J.D. (1984) National reviews for aquaculture development in Africa: 1. Zimbabwe, *FAO Fish. Circ.* (770.1), 69 pp.

Bazigos G.P. (1974) The design of fisheries statistical surveys; inland waters. *FAO Fish. Tech. Pap.* **133**, 122 pp.

Chimbuya S. (1993) Traditional fishing gear in Zimbabwe. *ALCOM News* Issue No. 12, October 1993, pp 2–7.

Gerdeaux D (1991) Study of the daily catch statistics for the professional and recreational fisheries on lakes Geneva and Annecy. In: I.G. Cowx *Catch Effort Sampling Strategies: their Application in Freshwater Fisheries Management*. Oxford: Fishing News Books, pp 118–126.

Jentoft S. (1989) Fisheries co management: Delegating government responsibility to fishermen's organisations. *Marine Policy* April.

Kell L. (1991) A comparison of methods used for coarse fish population estimation. In: I.G. Cowx *Catch Effort Sampling Strategies: their Application in Freshwater Fisheries Management*. Oxford: Fishing News Books, pp 184–200.

Malvestuto S.P. (1991) The customisation of recreational fishery surveys for management purposes in the United States. In: I.G. Cowx *Catch Effort Sampling Strategies: their Application in Freshwater Fisheries Management*. Oxford: Fishing News Books, pp 201–213.

Sanyanga R.A., Lupikisha J. & Thorsteinsson V. (1991) The proposed catch and effort data recording system for LKFRI (Zimbabwe) and DoF (Zambia) for the inshore fishery of Lake Kariba. Zambia/Zimbabwe *SADC Fisheries Project Report* No. 7, September, 6 pp.

Thorsteinsson V. (1992) The Catch and Effort Data Recording Systems, Frame surveys and Databases for fisheries data on Kariba in Zambia and Zimbabwe. Zambia/Zimbabwe *SADC Fisheries Project Report* No. 13.

# Chapter 28
# Trout yield in Lake Taupo, New Zealand: anglers' eldorado or just another fishery?

M. CRYER *Science and Research Directorate, Department of Conservation, Private Bag, Turangi, New Zealand*

*Present address: MAF Fisheries, PO Box 3437, Auckland, New Zealand*

**Abstract**   Lake Taupo is a large oligotrophic lake on New Zealand's central volcanic plateau. It supports one of the world's great fisheries for rainbow trout, *Oncorhynchus mykiss*. Angler harvest for 1983 was estimated by mail survey to be 877 t, whereas acoustic surveys in 1988 and 1989 suggested that Lake Taupo could probably support only about one fifth of this catch. Ten empirical estimators of yield from freshwater lakes were examined to assess the likely yield for a lake of Taupo's size and characteristics. The applicability of some models to Lake Taupo was questionable, and the extremes of the predicted range were 70 and 700 t. This gives some idea of the uncertainty in predicting yield from crude information, but half of the models suggested a yield of between 200 and 250 t. Acoustic and creel surveys suggested a similar yield, whereas historical mail surveys all lead to much larger estimates. The reasons for the divergence are discussed.

KEYWORDS: Lake Taupo, MSY, morphoedaphic index, *Oncorhynchus mykiss*, recreational fishery

## 28.1   Introduction

Lake Taupo supports one of the world's great recreational fisheries for rainbow trout, *Oncorhynchus mykiss* (Walbaum), with smaller catches of brown trout, *Salmo trutta* L. The value of the fishery as an asset of international repute is beyond question.

The lake is large (616 km$^2$), deep (mean 100 m; max. 185 m), and oligotrophic. Its geological history is very short, the current basin being the result of a spectacular caldera eruption 2000 years ago on New Zealand's central volcanic plateau. Licence sales and, by implication, fishing effort showed an upward trend from the turn of the century but stabilized in recent years (Fig. 28.1).

Indigenous species such as koaro, *Galaxias brevipinnis* Günther, common bully, *Gobiomorphus cotidianus* McDowall, and koura (freshwater crayfish), *Paranephrops planifrons* White, were unable to support intense predation from trout introduced in the late nineteenth century (Parsons 1979; Burstall 1983), and New Zealand smelt, *Retropinna retropinna* Richardson, were successfully introduced to the lake as a

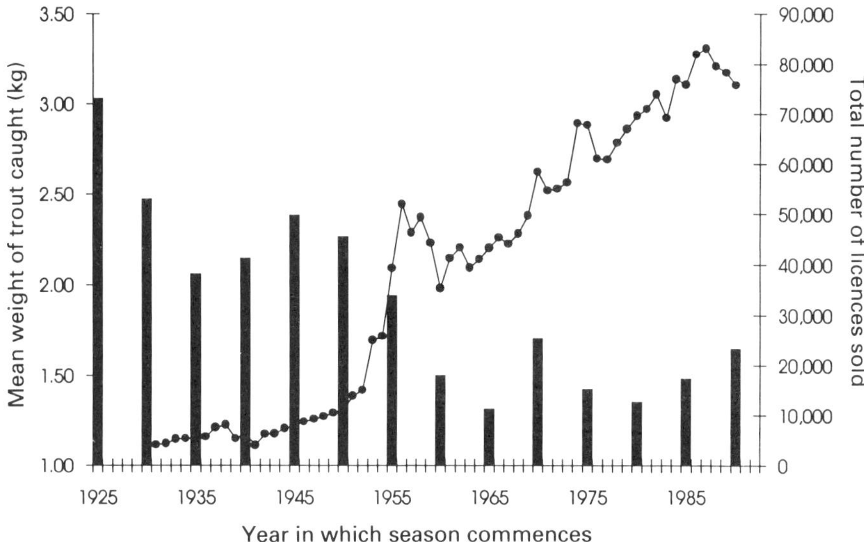



**Fig. 28.1**   Total licence sales (all categories combined) for the Taupo trout fishery since 1930 (closed symbols) and the mean weight of fish caught every fifth year since 1925 (histograms).

forage species in the 1930s. The current fishery relies very heavily on trout predation on smelt in an essentially pelagic environment. It is, however, still a wild fishery and is entirely unsupported by hatchery releases.

Using a mail (event recall) survey aimed primarily at assessing the economic turnover associated with the fishery, Shaw *et al.* (1985) estimated that 45 113 anglers took 626 000 trout for an aggregate weight of 877 t in 1982–83. This was one third more than the 651 t estimated for 1966–67, and double the estimate of 440 t in 1951–52 (Burstall 1983; Davies *et al.* 1988). Moreover, mean catch rates of 0.35 (1966–67) and 0.25 fish per hour (1982–83) suggested a decline in angler success rates, and the mean weight of fish taken in 1982-83 was only 1.4 kg, compared with 2.1 kg recorded in 1951–52.

This combination of apparently increasing effort and catch combined with decreasing catch rates led to a period of intense study on the lake in the 1980s. Trout distribution and production in the lake proper was studied using acoustic surveying, gill nets and diver counts (Cryer 1991). Forage species were sampled by purse seine, 'drop' net, and acoustic indexing (Cryer 1991). Spawning, recruitment and early life history were studied by trapping, diver counts, redd counts, and electrofishing in tributary streams (Stephens 1989; Tully 1989; Pitkethley 1990; Rosenau 1992). Stephens (1989) also examined the implications of hydroelectric flow regimes on the fishery using incremental flow methodology. Angler catch per unit effort (CPUE) was monitored annually (Cryer & MacLean 1991) and an intensive on-site intercept survey was planned for execution in 1990 (DOC 1991). The overall aim of these studies was to determine the current and potential productivity of the fishery.

However, historical estimates of yield for the Taupo fishery from mail surveys did not seem to be compatible with data from acoustic surveys and studies of food chain dynamics. Mail surveys suggested an annual yield of almost 900 t whereas acoustic surveying suggested an average recruited biomass (in the lake proper) of only 120 t. This chapter summarizes a 'desk study' of published information which was undertaken to appraise the plausibility of these disparate estimates of productivity. Was this really the Anglers' Eldorado as suggested by mail surveys, or was its fishery yield more commonplace?

### 28.2     Methods

The literature was searched for cases which attempted to predict or correlate yield, catch, or productivity with features of lakes which might be easier to measure than fisheries yield. This does not purport to be a comprehensive search or review of such methods, only a summary of those studies that were found to be useful for estimating likely yield for Taupo, given the information available on the system.

### 28.3     Results

Ryder (1965) found that his morphoedaphic index (MEI; the ratio of dissolved solids to mean depth) was empirically related to fish yield. The transfer of his predictive equation to SI units was, however, flawed, the correct version being:

$$Y = 1.382 \; MEI^{0.4461} \tag{1}$$

where $Y$ is the annual fisheries yield (kg ha$^{-1}$) and $MEI$ is the ratio of total dissolved solids (mg l$^{-1}$) to mean depth (m). Timperley and Vigor-Brown (1986) recorded a total dissolved solids of 88 mg l$^{-1}$, and this together with Taupo's mean depth of $\sim 100$ m and surface area of 616 km$^2$, leads to a predicted yield of 1.305 kg ha$^{-1}$ yr$^{-1}$, or about 80 000 kg for the whole lake. Strictly, the MEI was developed for use with Canadian lakes, and its applicability to Taupo, where the growing season is long and there is a peak of primary productivity in winter, is open to question.

Oglesby (1977) gave an equation relating gross primary production to fish yield and, as this was developed using lakes of varying trophic status and latitude, it is probably a better predictive model for Taupo.

$$Y = 0.00038 \; X^{2.21} \tag{2}$$

where $Y$ is the predicted fish yield in mg dry wt m$^{-2}$ yr$^{-1}$, and $X$ is gross primary production in g C m$^{-2}$ yr$^{-1}$. Gross primary production has been estimated at 172 to 190 g C m$^{-2}$ yr$^{-1}$ for Lake Taupo (A. Viner, Department of Scientific and Industrial Research, Taupo, personal communication), and this leads to a predicted fish yield of 33 to 41 mg dry wt m$^{-2}$ yr$^{-1}$, or 200 000 and 250 000 kg wet weight respectively, assuming a wet weight biomass ten times greater than dry weight.

Schlesinger and Regier (1982) found that mean annual air temperature (*TEMP*) explained a greater proportion of global variability in fisheries yield than *MEI*. They derived equations to predict maximum sustainable yield (*MSY*) in intensively fished waters from temperature alone (3), from temperature and *MEI* (4), and finally from temperature and a modification of *MEI* wherein mean depth was constrained to a maximum of 25 m (5).

$$\log_{10} MSY = 0.061\ TEMP + 0.043 \qquad\qquad (r^2 = 0.744) \qquad \textbf{(3)}$$

$$\log_{10} MSY = 0.051\ TEMP + 0.280\ \log_{10} MEI + 0.236 \qquad (r^2 = 0.810) \qquad \textbf{(4)}$$

$$\log_{10} MSY = 0.044\ TEMP + 0.482\ \log_{10} MEI_{25} + 0.021 \qquad (r^2 = 0.830) \qquad \textbf{(5)}$$

Predictions of annual MSY from these equations were 5.13, 6.08, and 11.26 kg ha$^{-1}$ respectively, or 316, 375, and 694 t. Alternative equations for estimating the yield from less intensively fished lakes incorporated a dummy variable for intensity of fishing into a larger data set. Effort (*E*) takes a value of 2.0 for intensively fished lakes, and a value of 1.0 for those under light, moderate, or unknown fishing pressure.

$$\log_{10} Y = 0.051\ TEMP + 0.358\ E + 0.161\ \log_{10} MEI - 0.383 \qquad (r^2 = 0.708) \quad \textbf{(6)}$$

$$\log_{10} Y = 0.050\ TEMP + 0.349\ E + 0.146\ \log_{10} MEI_{25} - 0.367 \qquad (r^2 = 0.698) \quad \textbf{(7)}$$

Values of $r^2$ for these latter equations are lower than those for intensively fished lakes alone, but this is probably caused by the larger sample size (123 lakes compared with 43). Predictions from Equations (6) and (7), assuming a moderate or unknown fishing pressure are 3.38 and 4.09 kg ha$^{-1}$ yr$^{-1}$; 208 and 252 t.

Youngs and Heimbuch (1982) examined the MEI concept from first principles. They found that the relation between MEI and fishery yield was spurious, stemming from two components: large lakes tend to produce more fish than small ones; and large lakes tend to be deeper than small ones (see also Kronmal 1993 on the problems of 'ratio standards'). However, Youngs and Heimbuch developed predictive equations based on lake area (8), lake area and total dissolved solids (9), and finally area (*A*), dissolved solids (*TDS*), and mean depth (*Z*) (10). They found that these were more powerful models than those using the composite variable *MEI*.

$$\ln Y = 7.01 + 0.83\ \ln A \qquad\qquad (r^2 = 0.939) \qquad \textbf{(8)}$$

$$\ln Y = 5.24 + 0.84\ \ln A + 0.36\ \ln TDS \qquad\qquad (r^2 = 0.949) \qquad \textbf{(9)}$$

$$\ln Y = 5.56 + 1.02\ \ln A + 0.34\ \ln TDS - 0.54\ \ln Z \qquad\qquad (r^2 = 0.972) \qquad \textbf{(10)}$$

Predictions of fisheries yield in Taupo from this series of equations are 229, 208, and 70 t annually. These equations do not predict yield per unit area directly, but for comparison with Equations (1) to (8) above, the relevant values are 3.72, 3.38, and 1.13 kg ha$^{-1}$ yr$^{-1}$. Predictions for all ten models are summarized in Table 28.1.

**Table 28.1**  Empirical models used to predict fishery yield in lakes of varying size, depth, primary productivity, and trophic state. Predictions for the yield of trout from Lake Taupo (kg yr$^{-1}$) are given based on its size of 616 km$^2$, mean depth of $\sim$100 m, primary production of 172 to 190 g C m$^{-2}$ yr$^{-1}$, and *TDS* of 88 mg l$^{-1}$. *Denotes models where mean depth is constrained to 25 m. Recent direct estimates are included for comparison.

| Model | Variables included | | | | | | |
|---|---|---|---|---|---|---|---|
| | Size | TDS | Depth | Temp | Phyto | Effort | Prediction for Taupo |
| (1) | Y | Y | Y | — | — | — | 80 000 |
| (2) | Y | — | — | — | Y | — | 225 000 |
| (3) | Y | — | — | Y | — | — | 316 000 |
| (4) | Y | Y | Y | Y | — | — | 375 000 |
| (5) | Y | Y | Y* | Y | — | — | 694 000 |
| (6) | Y | Y | Y | Y | — | Y | 208 000 |
| (7) | Y | Y | Y* | Y | — | Y | 252 000 |
| (8) | Y | — | — | — | — | — | 229 000 |
| (9) | Y | Y | — | — | — | — | 208 000 |
| (10) | Y | Y | Y | — | — | — | 70 000 |
| Mail 'event recall' survey: | | | | | | | 877 000 |
| Creel and overflight 'site intercept' survey: | | | | | | | 175 000 |
| MSY based on acoustic survey and simple models: | | | | | | | 157 000 |

## 28.4    Discussion

Lakes of similar surface area to Lake Taupo, for which reliable yield data are available (Youngs & Heimbuch 1982), have annual fish yields of between 140 and 400 t. However, all are relatively shallow (mean about 10 m) and rich in dissolved solids (120 mg l$^{-1}$ compared with Taupo's 88 mg l$^{-1}$), attributes which would tend to favour higher productivity than in Taupo. Conversely, lakes which show a similar yield (877 t) to that of Shaw *et al.* (1985) range in size from double to nine times the surface area of Taupo, the smallest of these being very shallow (3 m) and quite rich (210 mg l$^{-1}$ dissolved solids). Finally, lakes with similar trophic status to Taupo vary in size, but their mean fishery production is 3.4 kg ha$^{-1}$ yr$^{-1}$ which suggests an annual yield of about 210 t from Taupo.

Most of the model predictions in Table 28.1 relate to *MSY* or some similar measure of yield for intensively fished lakes, although models (6) and (7) are generalized to include unknown or moderate fishing pressure. Predicted values for Taupo range from 70 to 700 t per season, which gives some idea of the uncertainty involved in estimating fish yield from simple data. However, five of the ten models, based on three different data types predict an annual harvest of between 200 and 250 t. The average of the ten predictions is 265 t.

A combined creel survey and aerial overflight programme carried out by the New Zealand Department of Conservation in 1990–91 (DOC 1991) revealed that the actual catch of trout from the fishery in that season was approximately 113 000 trout for an aggregate weight of about 175 t. Licence sales have been relatively stable over the eight years between the estimate for 1982–83 (Shaw *et al.* 1985) and the overflight survey in 1990–91, and catch rates vary only moderately among years. This 'stability' in the fishery suggests that the two estimates of catch, one by event recall and the other by on-site intercept, should be similar.

An estimate of potential yield for the fishery was calculated from biomass estimates derived from acoustic surveys (Cryer 1991, Chapter 15). The methods for this estimation are simple, but have been derived by the New Zealand Ministry of Agriculture and Fisheries for assisting in the management of marine fisheries for which the availability of information is highly variable (e.g. Annala 1989).

$$MCY = M\ B_{min} \tag{11}$$

where $MCY$ is the maximum catch that can be taken from the fishery under a constant harvest strategy, $M$ is an estimate of the instantaneous rate of natural mortality, and $B_{min}$ is the minimum observed historic biomass. $M$ was estimated at 0.77, based on a maximum age in rainbow trout of about 6 years, and the minimum (annual mean) recruited biomass was 123 t (including some fish not available to acoustic surveying, Cryer 1991). Simulations conducted by Mace (1988) showed that $MCY$ was usually in the range 0.6 to 0.7 $MSY$, so an elementary estimate of $MSY$ can be derived from:

$$MSY = M\ B_{min}\ /\ 0.6 \tag{12}$$

For Lake Taupo trout, $MSY$ was estimated to be 157 t. This estimate is uncertain, but lies within the range predicted by empirical models (albeit towards the lower end), and is close to the estimate of yield made by the on-site intercept survey in 1990–91. It is, however, only about one fifth of the estimate of 877 t derived by the event recall mail survey in 1982–83.

Hiett and Ghosh (1977) and Hiett and Worrall (1977) demonstrated that event recall methods, such as that used by Shaw *et al.* (1985), became increasingly inaccurate with increasing recall time. After 15 days, anglers overestimated their most recent catch by 64%, and their effort by 58%. After 60 days, only 5% of anglers could accurately recall their last trip. The recall period for the 1982–83 surveys (Shaw *et al.* 1985) (anglers were asked about the whole of their last season's fishing) varied from 6 months for anglers interviewed early in the season to almost 18 months for those interviewed later. Recall periods of this duration without correction can be expected to inflate estimates of catch and effort by a factor of between 3 and 5 (Thompson & Hubert 1990). Correcting for a bias of this magnitude would make historical estimates of yield for Lake Taupo comparable with empirical estimators, and with recent direct measurements of recruited biomass and fishery yield.

In summary, the catch of 175 t from the Lake Taupo trout fishery in 1990–91 is close to the estimated maximum sustainable yield for the fishery and is unexceptional for a lake of Taupo's size, latitude, and trophic status. However, many who fish the lake still regard it as the anglers' Eldorado, and none who have taken its sparkling 3 kg rainbows on a dry fly will ever regard it as just another fishery.


## Acknowledgements

Thanks are due to a multitude of generous people who assisted me in preparation, planning, in the field, and in laboratory during this study. My deep gratitude goes also to the anglers of Taupo. Their encouragement, ideas, and financial assistance (through the Taupo Ward CNIWCC, Taupo Flyfishers and Hamilton Anglers) will not be forgotten.


## References

Annala J.H. (1989) Report from the fishery assessment plenary May 1989. Stock assessments and yield estimates. *New Zealand Ministry of Agriculture and Fisheries Plenary Report 1989*. Wellington.

Burstall P. (1983) Trout fishery – History and management. In: D.J. Forsyth & C. Howard-Williams (eds) *Lake Taupo: Ecology of a New Zealand Lake*. Department of Scientific and Industrial Research Information Series No. 158. Wellington: New Zealand Department of Scientific and Industrial Research.

Cryer M. (1991) Lake Taupo trout production: a four year study of the rainbow trout fishery of Lake Taupo, New Zealand. *New Zealand Department of Conservation Science and Research Series* No. 26, 191 pp.

Cryer M. & MacLean G.D. (1991) Catch for effort in a New Zealand recreational trout fishery: a model and implications for survey design. In: I.G. Cowx (ed.) *Catch Effort Sampling Techniques and their Application in Freshwater Fisheries Management*. Oxford: Fishing News Books, pp 61–71.

Davies P.E., Thompson W. & Sloane R.D. (1988) Tasmania's trout fisheries – current status and changes since 1945. *Tasmanian Inland Fisheries Commission Newsletter* **17(2)**.

DOC (1991). How many fish are caught at Taupo? Target Taupo. Special edition 1, Turangi: New Zealand Department of Conservation.

Hiett R.L. & Worrall J.W. (1977) Marine recreational fishermen's ability to estimate catch and recall catch and effort over time. *Human Sciences Research Inc. Research Report No. HSR-RR-77/13*. Virginia: McLean, 23 pp.

Hiett R.L. & Ghosh D.N. (1977) A recommended approach to the collection of marine recreational fin-fishing and shellfishing data on the Pacific coast. *Human Sciences Research Inc. Research Report No. HSR-RR-77/18*. Virginia: McLean, 57 pp.

Kronmal R.A. (1993) Spurious correlation and the fallacy of the ratio standard revisited. *Journal of the Royal Statistical Society* **A 156**, 379–392.

Mace P.M. (1988) The relevance of MSY and other biological reference points to stock assessments in New Zealand. *New Zealand Ministry of Agriculture and Fisheries Fisheries Assessment Research Document* No. 88/6. Wellington.

Oglesby R.T. (1977) Fish Yield as a Monitoring Parameter and its Prediction for Lakes. In: J.S. Alabaster (ed.) *Biological Monitoring of Inland Fisheries*. Barking: Applied Science Publishers.

Parsons J. (1979) *A Taupo Season*. Wellington: Collins.

Pitkethley R.J. (1990) Population Dynamics of Juvenile Trout in Two Tributary Streams of the Tongariro River. *Unpublished MSc Thesis*, University of Waikato, Hamilton, New Zealand, 162 pp.

Rosenau M.L. (1992) An Assessment of the Importance of Tributary Stream Rearing Habitat for Recruitment of Rainbow Trout into Lake Taupo. *Unpublished D.Phil Thesis*, University of Waikato, Hamilton, New Zealand.

Ryder R.A. (1965) A method for estimating the potential fish production of north-temperate lakes. *Transactions of the American Fisheries Society* **94**, 214–218.

Schlesinger D.A. & Regier H.A. (1982) Climatic and morphoedaphic indices of fish yield in natural lakes. *Transactions of the American Fisheries Society* **111**, 141–150.

Shaw D.J., Fletcher M. & Gibbs E.J. (1985) Taupo: a Treasury of Trout. Taupo: Taupo Times Ltd.

Stephens R.T.T. (1989) Flow Management in the Tongariro River. Wellington: *New Zealand Department of Conservation Science and Research Series* No. 16, 115 pp.

Thompson T. & Hubert W.A. (1990) Influence of survey method on estimates of statewide fishing activity. *North American Journal of Fisheries Management* **10**, 111–113.

Timperley M.H. & Vigor-Brown R.J. (1986) Water chemistry of lakes in the Taupo Volcanic Zone, New Zealand. *New Zealand Journal of Marine and Freshwater Research* **20**, 173–183.

Tully D. (1989) The Spawning Migration and Reproductive Biology of Rainbow Trout (*Oncorhynchus mykiss*) in Three Tributary Streams of Lake Taupo. *Unpublished MSc Thesis*, University of Waikato, Hamilton, New Zealand, 126 pp.

Youngs D.W. & Heimbuch D.G. (1982) Another consideration of the morphoedaphic index. *Transactions of the American Fisheries Society* **111**, 151–153.

# Chapter 29
# Assessment and prediction of whitefish stock (*Coregonus lavaretus*) of Lake Constance (Bodensee)

HERBERT LÖFFLER *Institute for Lake Research, Environmental Protection Agency Baden-Württemberg, Untere Seestraße 81, D 88085 Langenargen, Federal Republic of Germany*

## 29.1    Introduction

Whitefish (*Coregonus lavaretus*) form a major part of the annual yield in Lake Constance. A review of the history, biology, ecology and management of the fishery for the past 100 years is given by Löffler (1990) and Wagner *et al.* (1993).

During the last 10 years whitefish normally contributed more than 50% to the nominal yield, but the percentage was higher before the cultural eutrophication of the lake. The whitefish yield, especially of the main (offshore spawning) form 'Blaufelchen', showed marked annual fluctuations (Fig. 29.1). This variation in yield appears to have increased progressively each decade for the period 1921–1990 (Table 29.1).

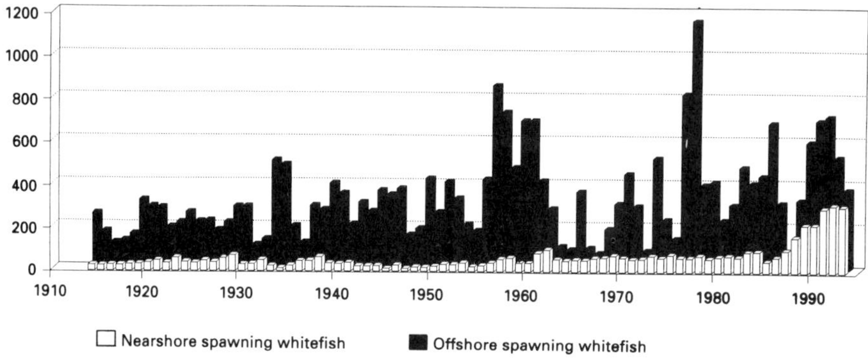


**Fig. 29.1**  Whitefish yield of Lake Constance (Bodensee) since 1914.

The uncertainty with respect to annual yield is partially explained by early attempts to assess the size of the stock and predict the yield. The first attempts in the 1930s considered parameters such as spawning yield and success, age structure of the stock, and cycles in the annual yield and year-class strength. As most of the predictions focused on the actual season and the following season they proved reliable (Table 29.2).

386

**Table 29.1**  Range of whitefish yield in Lake Constance between 1921 and 1990.

| Decade | Minima (kg) | Maxima (kg) | Ratio |
|---|---|---|---|
| 1921–1930 | 175 155 | 286 528 | 1 : 1.6 |
| 1931–1940 | 105 999 | 499 112 | 1 : 4.7 |
| 1941–1950 | 183 422 | 417 818 | 1 : 2.3 |
| 1951–1960 | 172 356 | 847 862 | 1 : 4.9 |
| 1961–1970 | 68 604 | 439 175 | 1 : 6.4 |
| 1971–1980 | 81 085 | 1 152 191 | 1 : 14.2 |
| 1981–1990 | 64 963 | 688 789 | 1 : 10.6 |

**Table 29.2**  Prediction of whitefish yield of Lake Constance.

| Author | Method applied | Predicted yield | Result |
|---|---|---|---|
| Elster (1935) | Year-class strength | 1935: strong decline | 1934: 480 t |
|  | Spawning results | 1937: very good | 1935: 195 t |
|  |  |  | 1937: 290 t |
| Elster (1936) |  | 1936: decline | 1936: 117 t |
| Nümann (1937) | Yield of 1936 | 1937: good | 1937: 290 t |
|  |  | 1938: (good) | 1938: 272 t |
|  |  | 1939: poor | 1939: 394 t |
|  | 4-year cycle | 1940: bad | 1940: 348 t |
| Nümann (1938) | Year-class distribution | 1938 > > 1937 | 272 t/290 t |
| Nümann (1947) | Year-class distribution | 1947: poor | 1947: 154 t |
| Nümann (1948) |  | 1948: below average | 1948: 183 t |
| Nümann (1949) |  | 1949: good | 1949: 418 t |

## 29.2    Recent stock assessment procedures

### 29.2.1    *Virtual population analysis*

Obligatory catch data and regular survey fishing programs have been carried out. It was thus possible to calculate year class strength by virtual population analysis (VPA) (Fig. 29.2) and to monitor the share of the different year classes to the annual catch (Fig. 29.3). It was also possible to evaluate the effect various biotic and abiotic parameters had on the yield (Table 29.3)

Although the predictions from one year to the next satisfied the models the predictions of YCS were completely inaccurate in half of the 16 cases. Even where the trend in predicted YCS closely followed the observed values (Eckmann *et al.* 1988) they were of little use as a prognostic tool because they were several orders of magnitude different (Table 29.3).

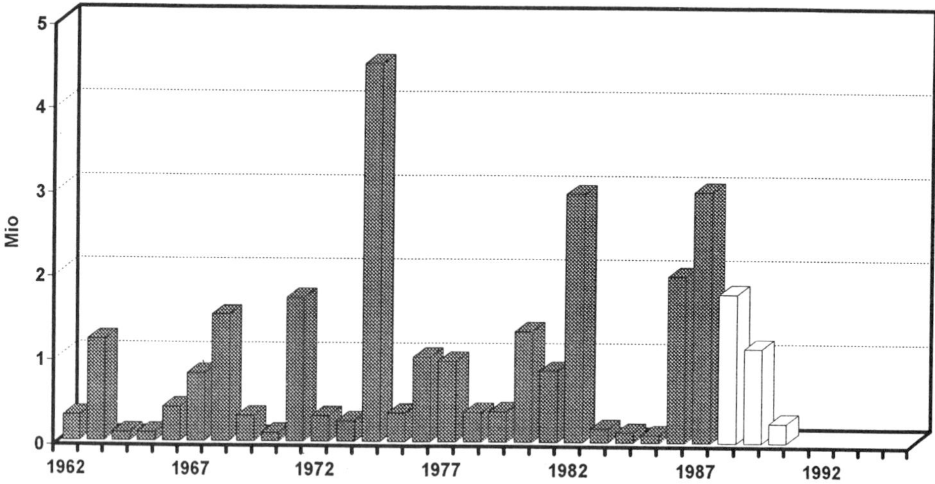

**Fig. 29.2**  Year-class strengths of whitefish of Lake Constance (Bodensee).

### 29.2.2   *Hydroacoustics*

In 1983 an EIFAC (European Inland Fisheries Advisory Commission) experiment on pelagic fish stock assessment examined hydroacoustic methods and compared them with results from the VPA. Lake Constance was divided into 21 hydroacoustic transects. The total whitefish population (AG > I) (shore area and surface layer excluded) in October 1983 was estimated at $2.5 \times 10^6$ fish (95% confidence limits of $1.7 \times 10^6$ and $3.7 \times 10^6$) (Dahm *et al.* 1985). The final virtual-population analysis, however, gave a stock of $7.2 \times 10^6$ whitefish (Hartmann *et al.* 1987). A regular hydroacoustic survey programme is now in place on Lake Constance, and supported by trawl surveys of young fish.

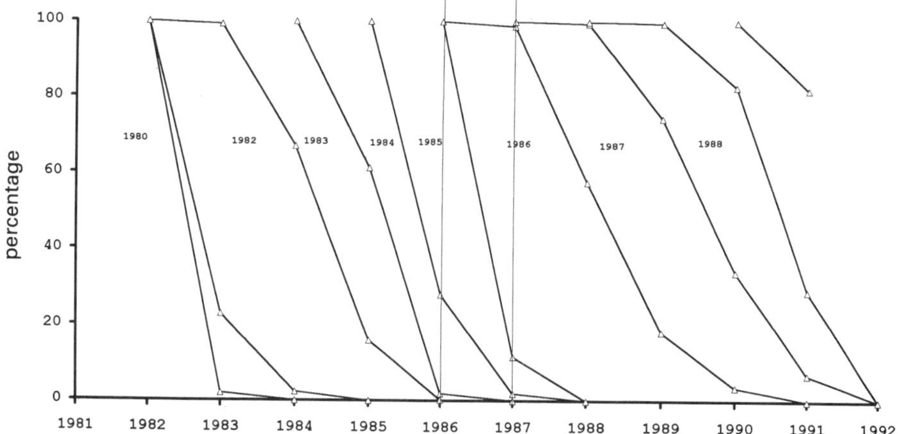

**Fig. 29.3**  Year-class composition of the annual catches.

**Table 29.3**  Predictions of the year-class-strengths of whitefish in Lake Constance.

| Author | Parameter | Prediction | Result |
|---|---|---|---|
| Hartmann (1980) | Surface temperature April/May | YCS 1977: $\sim 0.4 \times 10^6$ | $0.95 \times 10^6$ |
|  |  | YCS 1978: $\sim 0.4 \times 10^6$ | $0.35 \times 10^6$ |
|  |  | YCS 1979: $\sim 0.4 \times 10^6$ | $0.76 \times 10^6$ |
| Hartmann (1982) | Spring temperature | YCS 1977: $0.5 \times 1^6$ | $0.95 \times 10^6$ |
|  |  | YCS 1978: $0.4 \times 10^6$ | $0.35 \times 10^6$ |
|  |  | YCS 1979: $0.4 \times 10^6$ | $0.76 \times 10^6$ |
|  |  | YCS 1980: $0.4 \times 10^6$ | $1.31 \times 10^6$ |
|  |  | YCS 1981: $0.7 \times 10^6$ | $0.84 \times 10^6$ |
| Hartmann (1983) | Temperature/wind April–May | YCS 1981: $1.7 \times 10^6$ | $0.84 \times 10^6$ |
|  |  | YCS 1982: $0.7 \times 10^6$ | $2.95 \times 10^6$ |
|  |  | YCS 1983: $0.3 \times 10^6$ | $0.16 \times 10^6$ |
| Eckmann *et al.* (1988) | Meteorological conditions April | YCS 1983: $0.34 \times 10^6$ | $0.16 \times 10^6$ |
|  |  | YCS 1984: $1.81 \times 10^6$ | $0.12 \times 10^6$ |
|  |  | YCS 1985: $1.30 \times 10^6$ | $0.10 \times 10^6$ |
| Hartmann (1988) | Meteorological conditions April/May | YCS 1985: $0.4 \times 10^6$ | $0.09 \times 10^6$ |
|  |  | YCS 1986: $0.4 \times 10^6$ | $1.97 \times 10^6$ |

### 29.2.3  *Egg dredging*

In an attempt to determine fertilization and egg mortality regular dredge hauls were performed. In recent years natural mortality of whitefish eggs was as high as 95% (Hartmann & Quoβ 1989), because of unfavourable bottom oxygen conditions (Deufel & Quoβ 1992) (Fig. 29.4). Despite considerable variation between the different hauls and high inter-annual variation, this method appears to be a valuable tool for stock assessment (Hartmann & Quoβ 1989). However, it has not been possible to identify the causes of the variation as no correlation between physical and

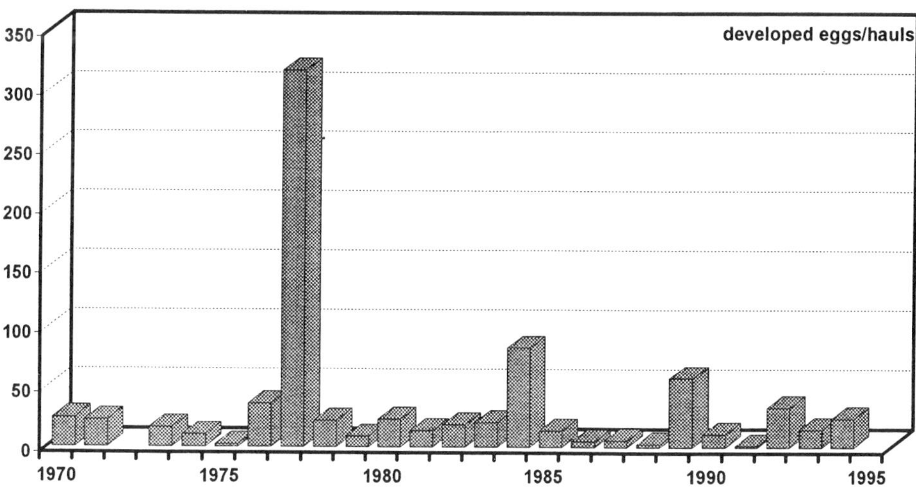


**Fig. 29.4**  Percentage of developed whitefish eggs from dredge hauls.

chemical water parameters, such as temperature, bottom minimum oxygen concentration, methane release and the number of spawners in December or spawning yield was found to explain the mechanisms of egg survival and resulting year class success.

### 29.2.4    *Recruitment*

Since 1986 attempts have been made to monitor the abundance of juvenile whitefish by regular sampling with a floating multi-mesh gillnet (10–56 mm knot-to-knot mesh size, 80 × 6 m). The samples were taken monthly during the fishing season (February–December). The number of 0+ whitefish caught (Table 29.4) confirmed the dominance of the 1986 and 1987 year classes.

**Table 29.4**   Number of whitefish $\leqslant$ 15 cm per 10 hauls in a floating multimesh gillnet.

| Oct 1986–Jun 1987 | Oct 1987–Jun 1988 | Oct 1988–Jun 1989 | Oct 1989–Jun 1990 | Oct 1990–Jun 1991 | Oct 1991–Jun 1992 | Oct 1992–Jun 1993 |
|---|---|---|---|---|---|---|
| 130.0 | 58.9 | 19.4 | 85.6 | 11.0 | 13.3 | 47.5 |
| Rank: 1 | Rank: 3 | Rank: 5 | Rank: 2 | Rank: 7 | Rank: 6 | Rank: 4 |
| YC 1986: | YC 1987: | YC 1988 | YC 1989: | YC 1990: | YC 1991: | YC 1992: |
| 1.97 | 2.98 | $\geqslant$1.76 | $\geqslant$1.14 | $\geqslant$0.27 | $\geqslant$0.37 | Not yet assessed |

### 29.3    **Conclusions**

Despite efforts to understand the mechanisms of recruitment, there is still a high degree of uncertainty relating to the factors affecting year-class strength. For example, between 1983 and 1987 nine different parameters were suggested as directly or indirectly influencing year class strength of whitefish (Hartmann 1988).

Mathematical models for fish stock assessment are limited because they are based on hypotheses and do not explain causal effects (Hartmann 1989a,b). For example, uncertainties resulting from inaccuracy in the catch statistics lead to an error between nominal and actual catch or from unexplainable larval mortality (Hartmann 1989b).

It appears that conventional fishing gears to sample fish populations are a poor mechanism for predicting stock size but sampling yearlings is a reasonable index of year class strength (Dahm *et al.* 1992; Curtis *et al.* 1993).

The survival of naturally spawned eggs is influenced by a combination of unknown factors which make the interpretation of the observations difficult. Additionally, the contribution of natural and artificially reared fish in the lake remains unknown, thus the influence of stocking cannot be assessed (Eckmann *et al.* 1988; Hartmann & Quoß 1989; Hartmann 1990; Todd 1986).

From the results of the EIFAC experiments on stock assessment in lakes, echo sounding appears to be a reliable technique (Dahm *et al.* 1992). However, these surveys need to be supported by independent population estimates and information from trawl and gillnet catches.

## References


Curtis G.L., Bronte C.R. & Selgeby J.H. (1993) Forecasting contributions of lake whitefish year-classes to a Lake Superior commercial fishery from estimates of yearling abundance. *North American J. Fish. Management* **13**, 349–352.

Dahm E., Hartmann J., Lindem T. & Löffler H. (1985) EIFAC experiments on pelagic fish stock assessment by acoustic methods in Lake Constance. *EIFAC Occas. Pap.* **15**, 14 pp.

Dahm E., Hartmann J., Jurvelius, J., Löffler H. & Völzke V. (1992) Review of the European Inland Fisheries Advisory Commission (EIFAC) experiments on stock assessment in lakes. *Journal of Applied Ichthyology* **8**, 1–9.

Deufel J. & Quoß H. (1992) Über die Größen und Ursachen der Vernichtungsziffern bei Blaufelchen des Bodensees vor und während der Eutrophierung. *Fischer & Teichwirt* **43**, 305–306.

Eckmann R., Gaedke U. & Wetzlar H.J. (1988) Effects of climatic and density-dependent factors on year-class strength of *Coregonus lavaretus* in Lake Constance. *Canadian Journal of Fisheries and Aquatic Sciences* **45**, 1088–1093.

Elster H. J. (1935) Weitere Beiträrge zur Fischereibiologie der Blaufelchen. *Allg. Fisch. Ztg.* **60**, 115–119; 132–136.

Elster H.J. (1936) Bemerkungen über die Blaufelchenfischerei im Bodensee 1935. *Allg. Fisch. Ztg.* **61**, 98–102.

Hartmann J. (1980) Zur Jahrgangs- und Bestandsstärke von Felchen und Barsch des Bodensees. *Fischwirt* **30**, 3–4.

Hartmann J. (1982) Fangprognosen für den Bodensee. *Fischwirt* **32**, 26–26.

Hartmann J. (1983) Fangaussichten bei Felchen (Jahrgänge 1981–83) und Barsch (Jgg. 1980–82) des Bodensees. *Fischwirt* **33**, 69–70.

Hartmann J. (1988) Ist die Rekrutierung (Jahrgangsstärke) beim Bodenseefelchen (*Coregonus lavaretus*) schon verstanden? *Österr. Fisch.* **41**, 135–142.

Hartmann J. (1989a) Fallstricke bei Jahrgangsstärke-Modellen – Beispiel Bodensee. *Österr. Fisch.* **42**, 207–212.

Hartmann J. (1989b) Is recruitment of the whitefish (*Coregonus lavaretus*) of Lake Constance already understood? *Canadian Journal of Fisheries and Aquatic Sciences* **46**, 725–726.

Hartmann J. (1990) Vergrößert der Larvenbesatz den Felchenertrag des Bodensees wesentlich? *Österr. Fisch.* **43**, 48–53.

Hartmann J., Dahm E., Dawson J., Doering P., Jürgensen S., Lindem T., Löffler H., Raemhild G. & Voelzke V. (1987) EIFAC experiments on pelagic fish stock assessment by acoustic methods in Lake Tegel. *EIFAC Occas. Pap.* **17**, 26 pp.

Hartmann J. & Quoß H. (1989) Gedredgte Eier, Laicherbestand, Brutanstaltquote, Jahrgangserfolg im Bodensee. *Österr. Fisch.* **42**, 84–87.

Löffler H. (1990) Fisheries management of Lake Constance: an example of international cooperation. In: W.L.T. van Densen, B. Steinmetz & R.H. Hughes (eds.) *Management of freshwater fisheries* Wageningen: Pudoc, pp 38–52.

Nümann W. (1937) Die Fangaussichten in der Blaufelchenfischerei des Bodensees nach den Ergebnissen des Jahres 1936. *Fisch. Ztg.* **40**, 217–219.

Nümann W. (1938) Blaufelchenuntersuchungen 1937 und die Prognose für 1938. *Allg. Fisch. Ztg.* **63**, 161–169.

Nümann W. (1947) Ergebnisse in der Blaufelchenfischerei am Bodensee nach dem Kriege. *Allg. Fisch. Ztg.* **72**, 185–192.

Nümann W. (1948) Die Blaufelchenfischerei am Bodensee während des Jahres 1947. *Allg. Fisch. Ztg.* **73**, 109–113.

Nümann W. (1949) Die Blaufelchenfischerei am Bodensee 1948 und Fangprognose für 1949. *Allg. Fisch. Ztg.* **74**, 173–175.

Todd, T.N. (1986) Stocking and natural recruitment of larval coregonines in the Bodensee. *Arch. Hydrobiol. Beih.* **22**, 337–342

Wagner B., Löffler H., Kindle T., Klein M. & Staub E. (eds)(1993) Bodenseefischerei: Geschichte – Biologie und Ökologie – Bewirtschaftung; zum 100 jährigen Jubiläum der Internationalen Bevollmächtigten-Konferenz für die Bodenseefischerei. Sigmaringen, Thorbecke.

# Chapter 30
# The use of depletion fisheries data in fish stock assessment in Lake Wolderwijd, The Netherlands

J.J.G.M. BACKX *Witteveen + Bos Consulting Engineers, P.O. Box 233, 7400 AE Deventer, The Netherlands.*

**Abstract**   In a biomanipulation exercise to restore water quality in Lake Wolderwijd, about 735 t of fish were removed during winter fishings between November 1990 and August 1993. The original population size was estimated in 1989 based on trawl survey data and on empirical prediction models. This estimate was subsequently revised based on trawl surveys carried out in July and September 1991 and 1992 to monitor recruitment (Y-O-Y) and the development of the stock of older fish. In 1993 the stock assessment was carried out in September only. The fish stock size each January, and the remaining population at the start of the growing season, were reconstructed from the results of the stock assessment and data collected during winter removal fisheries. Linear regression models were used for stock assessment. However, as in other experiments, the initial biomass of ruffe (*Gymnocephalus cernua*) was underestimated by 20–36 % and the biomass of bream (*Abramis brama*) $\geqslant 25$ cm by 3–19%. Catch/effort mapping during depletion fisheries provided a unique opportunity to collect information about fish distribution. It was found that no amount of intensive and expensive sampling could replace this information on seasonal species-specific distribution and gear selectivity. This information was indispensable for fish stock assessment.

KEYWORDS: Depletion estimates, stock assessment, species-specific distribution, trawl survey

## 30.1   Introduction

To improve water quality in Lake Wolderwijd (2700 ha), The Netherlands, a reduction in the total biomass of planktivorous and benthivorous fish to 20 kg ha$^{-1}$ and 25 kg ha$^{-1}$, respectively, was considered necessary (Hosper & Jagtman 1990). In 1989 a feasibility study on mass removal of fish was started for Lake Wolderwijd – Nuldernauw (Grimm & Backx 1994). This was followed by an assessment of the stocks from 1989 onwards to estimate the target fish stock reduction.

The most common and generally most acceptable, techniques for estimating population size are: (1) direct counts and surveys; (2) reconstruction from historical catch data (VPA, catch-at-age analysis); (3) mark-recapture and change in ratio methods (e.g. Peterson, Schnabel estimates): and depletion estimates by analysis of catch per unit of effort (CPUE) (Hilborn & Walters 1992). Tagging is not recommended unless it is possible to mark at least 25% of the population. The mark-recapture method is also not acceptable in a large lake such as Wolderwijd because of the large amount of time and effort required to conduct such a study for all species. In 1989 sufficient historical data and catch effort data were not available for assessing the stock size in Wolderwijd; consequently the fish biomass was determined by stratified sampling according to depth using a small, fine meshed trawl. Based on these estimates, targets were set for reducing the fish biomass and a fish stock reduction programme started in November 1990 (Grimm & Backx 1994).

By June 1991 the mass removal of fish was considered a success in terms of the goals achieved (Backx & Grimm 1994): within 7 months the initial fish biomass (203 kg ha$^{-1}$) was reduced by 77.5% (425 t). In 1992 and 1993 additional depletion fisheries were carried out and 101 and 196 t respectively were caught. The catch effort data collected during the three successive fishing campaigns on Lake Wolderwijd provided an opportunity to estimate fish abundance in the lake by analysis of the CPUE. In this chapter estimates of fish abundance based on the stratified trawl surveys are compared with estimates based on catch effort analysis. The value of depletion fisheries data is discussed.


### 30.2    Description of Lake Wolderwijd

The total area (2700 ha) of the lake is divided into Lake Wolderwijd (2000 ha), Lake Nuldernauw (660 ha) and five harbours (40 ha). The lake is isolated from the neighbouring lakes by a lock in the north and a lock and a sluice in the south. Two small islands are located in the roughly circular Lake Wolderwijd (about $4 \times 3.5$ km). Lake Nuldernauw is approximately 9 km long and has a width of 0.4–1.0 km. The total shore length is approximately 36 km. The navigation channel has an average width of 200 m. The average depth of these features is given in Figure 30.1.


### 30.3    Fish stock management 1989–1993

#### 30.3.1    *Stock assessment*

In 1989 the total fish biomass in Lake Wolderwijd was estimated by stratified sampling according to depth using a small, fine-meshed trawl. The efficiency of this gear was estimated in earlier experiments (Grimm & Backx 1994). The trawl survey data in February 1989 were strongly biased towards bream because of the seasonal species-specific distribution. Even after additional monitoring in May 1989, March 1990, May 1990 and July 1990 only the biomass of bream could be reliably calculated from the results of the trawl surveys (Backx & Grimm 1994). The initial population was reconstructed using trawl survey data from 1989 and 1990 and the empirical relation

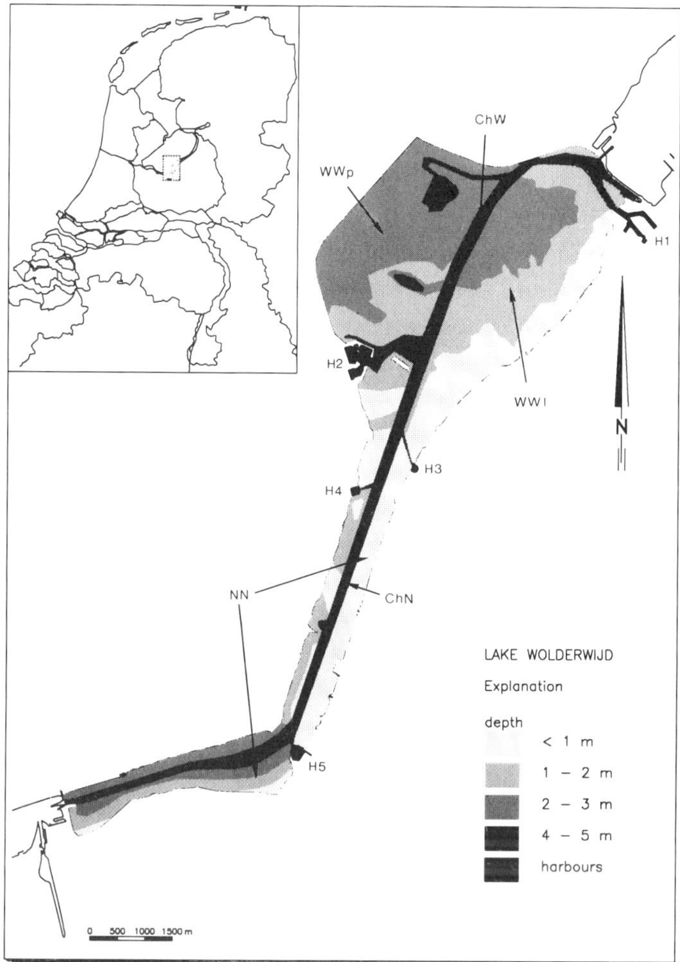



**Fig. 30.1**  Lake Wolderwijd-Nuldernauw, showing depth and surface area (ha) of different sectors. ChW = channel Wolderwijd (150 ha); Wwp = polder-side Wolderwijd (850 ha); Wwl = land-side Wolderwijd (1000 ha); ChN = channel Nuldernauw (150 ha); NN = rest Nuldernauw (510 ha); H1-H5 = harbours (40 ha).

between the total phosphorus concentration and the fish biomass (Hanson & Leggett 1982), which has been shown to be valid for Dutch lakes (Grimm & Backx 1990). The biomass and composition of the population in 1991 was approximated from experience on similar lakes (Grimm & Backx 1994).

To assess recruitment of young of the year (Y-O-Y) and to monitor the development of the stock of older fish, assessment took place in July and September in 1991 and 1992. In 1993 assessment took place only in September. The fish stock in January and the remainder of the population at the start of the growing season, were reconstructed annually from assessments in July and September in the previous year and data collected during winter removal fisheries (Table 30.1).

**Table 30.1**  Results of the stratified trawl surveys in September (S) in Lake Wolderwijd between 1989 and 1993 and the reconstructed size (kg ha$^{-1}$) in January (J). The 'remainder' population (R) is the January stock size reduced with the catches during the winter depletion fisheries.

| | 1989/90 | 1991 | | | 1992 | | | 1993 | | |
|---|---|---|---|---|---|---|---|---|---|---|
| | S+ empirical | J | R | S | J | R | S | J | R | S |
| *Planktivorous* | 87 | | | | | | | | | |
| Bream <15 cm | | 21.2 | 4.8 | 9.7 | 6.0 | 4.6 | 7.0 | 7.0 | 2.5 | 5.7 |
| Roach <15 cm | | 23.5 | 5.3 | 7.2 | 9.9 | 7.1 | 16.9 | 24.3 | 17.6 | 9.4 |
| Smelt | | 1.5 | 0.3 | 1.4 | 2.1 | 1.3 | 1.6 | 1.0 | 0.6 | 0.9 |
| 3-spined Stickleback | | 0.0 | 0.0 | 0.4 | 0.6 | 0.5 | 0.0 | 0.0 | 0.0 | 0.1 |
| *Potential planktivorous* | | | | | | | | | | |
| Ruffe | 1 | 32.8 | 7.4 | 15.2 | 33.6 | 15.0 | 26.0 | 26.0 | 12.9 | 17.0 |
| Bream 15–24 cm | 8 | 8.6 | 1.9 | 4.5 | 21.7 | 15.0 | 2.5 | 3.0 | 1.2 | 4.4 |
| *Benthivorous* | | | | | | | | | | |
| Bream ⩾25 cm | 90 | 95.6 | 21.5 | 22.4 | 25.6 | 17.8 | 19.0 | 68.2 | 29.9 | 26.3 |
| Roach ⩾15 cm | 5 | 15.6 | 3.5 | 3.1 | 4.5 | 1.5 | 4.3 | 21.4 | 16.5 | 6.8 |
| Carp | | 0.0 | 0.0 | 0.0 | 0.1 | 0.0 | 0.0 | 0.1 | 0.0 | 0.1 |
| *Piscivorous* | | | | | | | | | | |
| Perch | 2 | 3.2 | 0.7 | 4.8 | 4.6 | 3.7 | 26.8 | 10.7 | 8.0 | 7.7 |
| Pike perch | 1 | 1.0 | 0.2 | 1.4 | 2.0 | 1.8 | 0.4 | 2.0 | 1.9 | 0.1 |
| Total* | 194 | 203.0 | 45.6 | 70.0 | 110.7 | 68.3 | 104.5 | 163.7 | 91.1 | 84.8 |

* Eel not included.

### 30.3.2    *Population size and fish stock reduction*

In January 1991 a first attempt was made to reconstruct the initial population. Catch effort data and efficiency of the gears were used and the total standing stock was estimated as 142.8–149.4 kg ha$^{-1}$ (Grimm & Backx 1994). This appeared too low because 157.4 kg ha$^{-1}$ was removed at the end of the fishing. The only conclusion was that the ruffe (*Gymnocephalus cernua* (L.)) population was strongly underestimated during the 1989 and 1990 stock assessment The initial stock size in Lake Wolderwijd was first estimated properly in January 1991 (Table 30.1), based on the empirical relation and the catches composition during the depletion fisheries in 1991 (Backx & Grimm 1994) at 203 kg ha$^{-1}$.

From November 1990 to June 1991 the initial fish stock was reduced to *ca* 45.6 kg ha$^{-1}$ by removing 425.1 t fish (157.4 kg ha$^{-1}$), of which 23% was planktivorous (mainly roach *Rutilus rutilus* (L.) and bream *Abramis brama* (L.) <15 cm), 20% potential planktivorous species (bream 15–24 cm and ruffe) and 55% benthivorous bream >25 cm and roach >15 cm (Backx & Grimm 1994). From stock assessments in July and September 1991, it was concluded that ruffe was the first species which responded to the reduction in the fish stock. In September the total biomass Y-O-Y fish was 48 kg ha$^{-1}$, of which almost 30 kg ha$^{-1}$ was ruffe. Due to growth, the stock of bream and roach >15 cm increased *ca* 7 kg ha$^{-1}$ during the summer (Table 30.1).

In January 1992 the fish stock was about 110 kg ha$^{-1}$ (Table 30.1). Dominant species were ruffe (30%) and bream >25 cm (23%). Ruffe and bream (15–24 cm) were both underestimated during the September 1991 trawl survey by approximately 15 kg ha$^{-1}$. Winter fishing in 1992 only aimed to reduce the stock of planktivorous fish. Fishing effort was limited and the fishing was ended in mid-April. In total 114.5 t fish (42.2 kg ha$^{-1}$) were caught and removed in 1992, of which 72% was potential planktivorous (mainly roach and bream <15 cm, bream 15–24 cm and ruffe) and 26% benthivorous bream >25 cm and roach >15 cm. The remaining population was estimated to be 68 kg ha$^{-1}$ of which 43.5 kg ha$^{-1}$ was potential planktivorous fish and 19.3 benthivorous fish (Table 30.1).

At the end of the growing season, in September 1992, about 105 kg ha$^{-1}$ was present in Lake Wolderwijd (Table 30.1). During the summer 43 kg ha$^{-1}$ of Y-O-Y was recruited, dominated by perch (58%), ruffe (23%) and roach (9%). The stock of roach increased in 1992 by *ca* 7 kg ha$^{-1}$, despite the removal of 6 kg ha$^{-1}$, due to strong recruitment in 1992 and growth (Table 30.1).

Based upon the results of the stock assessment in September 1992 it was decided to reduce only the planktivorous fish stock in winter 1993. As in winter 1992, fishing effort was limited. However, bream catches by commercial fishermen were high in January and February and after additional monitoring in February 1993 it was concluded that the stock of bream >25 cm was ±50 kg ha$^{-1}$ and the roach stock was about 20 kg ha$^{-1}$, higher than recorded during the trawl surveys in September 1992. Immigration was presumed to be the cause of the increase (Backx 1993) and extra effort was performed to reduce the bream biomass. In January 1993 the total fish stock was 164 kg ha$^{-1}$. Bream >25 cm (42%), roach (28%) and ruffe (16%) were the dominant species (Table 30.1).

In 1993, 196.3 t of fish (72.7 kg ha$^{-1}$) were caught, of which 36% was potential planktivores and 60% benthivores. The remaining population was about 91 kg ha$^{-1}$ of which 39 kg ha$^{-1}$ was potential planktivores and 46 kg ha$^{-1}$ benthivores (Table 30.1). Results of the stock assessment in September 1993 showed a change in the species composition. Perch and roach increased and a strong year-class of pikeperch appeared with a length of about 13 cm (fork length) at the beginning of September. Total population size was estimated to be 85 kg ha$^{-1}$ (Table 30.1). After depletion, stock size was presumed to be 91 kg ha$^{-1}$. It is possible that the immigration of roach was overestimated during the surveys in spring 1993. Unfortunately no intensive depletion fisheries were carried out for 1994 to verify this.

## 30.4     Depletion estimates

### 30.4.1     *Methods*

During three successive years, 1991, 1992 and 1993, the fish stock of Lake Wolderwijd was reduced. In total 735 t of fish were caught. Details on allocation of gears, effort and catch composition during the depletion fisheries in 1991 are given in Backx &

Grimm (1994). Analysis of the CPUE provided an alternative to calculate population size. The linear regression model described by Leslie and Davis (1939) was chosen because measures of effective catch were more accurate than measures of fishing effort on which, for example, the DeLury method is based (Ricker 1975). Hilborn & Walters (1992) also stated that the Leslie estimation was generally preferable.

Calculations were made for ruffe and bream $\geqslant 25$ cm. These were the main species caught by one single gear and at fixed location during the winter depletion fisheries. Ruffe were mainly caught with small fykes (60% of the total catch in three years) and by trawling (24%) in the channel (Chw and Chn, Fig. 30.1). Small fykes were the only gear fished intensively in all the years (Table 30.2). Intensive trawling was only performed in 1991. Depletion estimates for ruffe were based on the linear regression of weekly CPUE (kg fyke nights$^{-1}$ for small fykes and kg ha$^{-1}$ for trawl catches) and the cumulative total catch of ruffe in all the gears.

**Table 30.2**  Results of the fishing with small fykes on ruffe during the depletion fisheries in Lake Wolderwijd from winter 1990 until 1993 and the percentage of the total ruffe population removed with small fykes.

| Year | Period | Effort (net nights) | Catch (kg) | Average CPUE (kg net night$^{-1}$) | % removed |
|------|--------|------|------|------|------|
| 1991 | Mar–Jun | 45798 | 25707 | 0.56 | 29 |
| 1992 | Feb–Apr | 23570 | 36147 | 1.53 | 40 |
| 1993 | Feb–Mar | 44480 | 30458 | 0.68 | 43 |

Bream $\geqslant 25$ cm were mainly caught with large (770 m and 1200 m long) seines in the open water (WWp, WWl and NN, Fig. 30.1). These seines were fished in 1991 and 1993 only. Effort was much higher in 1991 (Table 30.4). Depletion estimates were made by plotting weekly catch per area fished (kg ha$^{-1}$) against cumulative catch of bream $\geqslant 25$ cm in all the gears.

## 30.5    Results

### 30.5.1    *Ruffe*

In 1991 average CPUE of ruffe with small fykes was low (0.56 kg fyke night$^{-1}$, Table 30.2) mainly because the ruffe population was being fished by trawling. Trawl catches declined rapidly and small fykes were first introduced in March when 20 700 kg of ruffe was already caught by trawling. CPUE was highest in 1992 (1.53 kg fyke night$^{-1}$) due to strong recruitment of 0 + ruffe in summer 1991 and the relatively short period of fishing (Table 30.2). In the years 1991 to 1993 respectively, 29%, 40% and 43% of the ruffe population was removed (Table 30.2) with small fykes. In total 68.48 t, 50.26 t and 35.38 t of ruffe were caught in the successive years (Table 30.3).

**Table 30.3**  Population estimates (kg ha$^{-1}$) of ruffe based on stratified trawl surveys and on the analysis of CPUE during removal fisheries with trawls and small fykes in Lake Wolderwijd from winter 1990 until 1993. The % of the total ruffe population caught during depletion fishing and the % deviation between the reconstructed population size and the depletion estimates.

| Year | Trawl survey September (t–1) (kg ha$^{-1}$) | Catch depletion fisheries (kg ha$^{-1}$) | | | Initial stock size January (kg ha$^{-1}$) | | | Deviation % |
| --- | --- | --- | --- | --- | --- | --- | --- | --- |
| | | | | | linear regression | | reconstructed | |
| | | total | fykes | % | trawl | fykes | | |
| 1991 | ? ($\pm$1) | 25.4 | 9.5 | 77 | 18.7 | 26.3 | 32.8 | 20 |
| 1992 | 15.2 | 18.6 | 13.4 | 55 | — | 21.5 | 33.6 | 36 |
| 1993 | 26.0 | 13.1 | 11.3 | 50 | — | 18.3 | 26.0 | 30 |

Population estimates based on trawl surveys and the results of the linear regression analysis were compared with the reconstructed population size in January (Table 30.3). In 1991 the catches with the trawl declined from 275 kg ha$^{-1}$ to 25 kg ha$^{-1}$ within 8 weeks (Figure 2a). Between February and June, trawl catches stayed low ($<25$ kg ha$^{-1}$). In February small fykes were introduced and the catch immediately declined to $<1$ kg fyke night$^{-1}$ in April, due to migration of ruffe out of the channel. At the beginning of May, before spawning, CPUE increased in both gears (Fig. 30.2a). In 1992 CPUE declined from 4 kg fyke night$^{-1}$ in February to 0.6 kg fyke night$^{-1}$ in the middle of April (Fig. 30.2b) when the fishing with small fykes was ended (Table 30.2). In 1993 catches were low (Fig. 30.2c). The relative low biomass in 1993 and perhaps gear avoidance were the main reasons.

The population estimates for ruffe based on the decline in CPUE (Table 30.3) underestimated the initial stock size. The estimate based on the trawl catches (Table 30.3) was lower (18.7 kg ha$^{-1}$) than the total ruffe biomass removed in 1991 (25.4 kg ha$^{-1}$). The estimates based on the small fyke catches deviated from the reconstructed population by 20% in 1991, 36% in 1992 and 30% in 1993 (Table 30.3). In 1991 and 1992 depletion estimates provided a better estimate than the trawl surveys. Deviation of the trawl survey from the reconstructed population size was greatest in 1991 and least in 1993 (Table 30.3). In 1993 the trawl survey was similar to the reconstructed population size. Trawl surveys were adjusted throughout the years to the species-specific distribution recorded during the successive depletion fisheries.

### 30.5.2   *Bream* $\geqslant$ *25 cm*

In 1991 70% of the population of bream $\geqslant$ 25 cm was caught with large seines (Table 30.4). In 1992 less effort was performed (1575 ha fished) than in 1991 (4628 ha fished). CPUE was higher in 1992 (58.2 kg ha$^{-1}$) compared to 1991 (38.5). In total 200.1 t (74.1 kg ha$^{-1}$) was caught in 1991 and 103.4 t (38.3 kg ha$^{-1}$) in 1993 (Table 30.5).

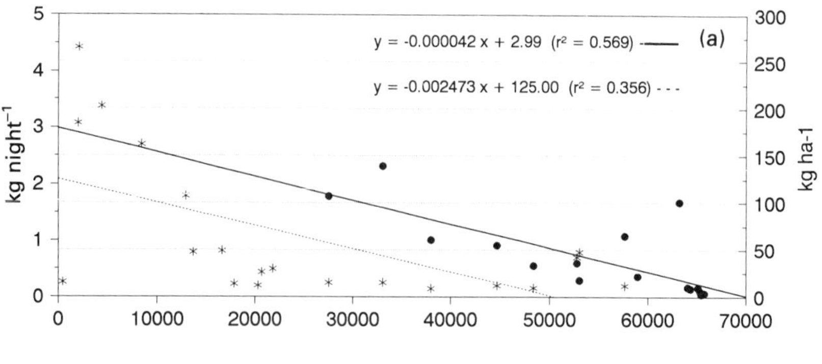


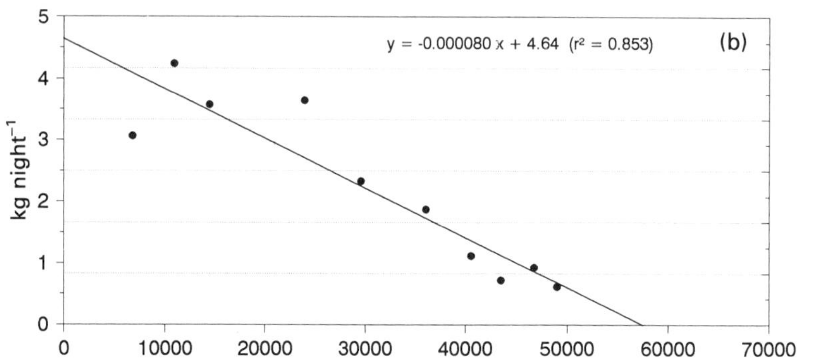


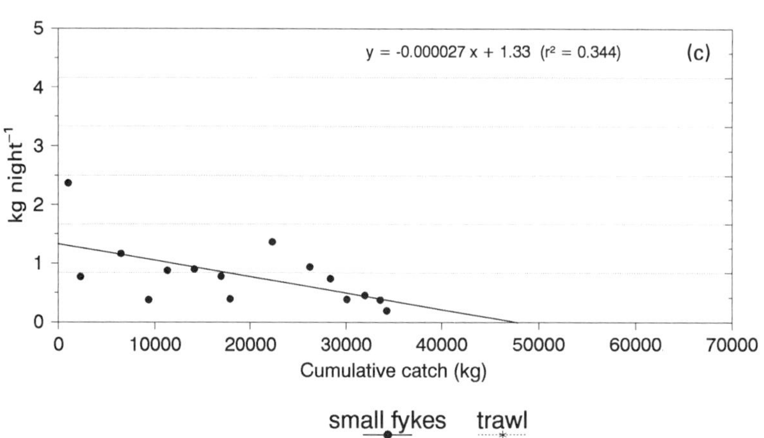


**Fig. 30.2**　Decline in the CPUE of ruffe during the fishing with trawls (dashed line, kg ha⁻¹) and fyke nets (solid line, kg night⁻¹) in Lake Wolderwijd in the years 1991 (a), 1992 (b) and 1993 (c).

**Table 30.4**  Results of the fishing with large seines on bream $\geqslant 25$ cm during the depletion fisheries in Lake Wolderwijd from winter 1990 until 1993 and the % of the total bream $\geqslant 25$ cm population removed with large seines.

| Year | Period | Effort (ha fished) | Catch (kg) | Average CPUE (kg ha$^{-1}$) | % removed |
|---|---|---|---|---|---|
| 1990/91 | Nov–May | 4628 | 164709 | 38.5 | 70 |
| 1993 | Jan–Aug | 1575 | 90861 | 58.2 | 49 |

**Table 30.5**  Population estimates (kg ha$^{-1}$) of bream $\geqslant 25$ cm based on stratified trawl surveys and on the analysis of CPUE during removal fisheries with seines in Lake Wolderwijd in winter 1990/1991 and in 1993. The % of the total bream $\geqslant 25$ cm population caught during depletion fishing and the % deviation between the reconstructed population size and the depletion estimates.

| Year | Trawl survey September (t–1) (kg ha$^{-1}$) | Catch depletion fisheries (kg ha$^1$) | | | Initial stock size January (kg ha$^{-1}$) | | Deviation % |
|---|---|---|---|---|---|---|---|
| | | total | seines | % | linear regression seines | reconstructed | |
| 1991 | 90.0 | 74.1 | 61.0 | 78 | 92.7 | 95.6 | –/–3 |
| 1992 | 22.4 | 7.8 | — | 30 | — | 25.6 | |
| 1993 | 19.0 | 38.3 | 33.7 | 56 | 55.1 | 68.2 | –/–19 |

The best estimate of the initial population size based on the catch effort analysis (Fig. 30.3 a & b) was in 1991 (3% deviation). In 1993 the initial population was underestimated by 19%. Trawl survey results deviated 6% in 1991 and 72% in 1993 (Table 30.5). As mentioned earlier, the stock of bream $\geqslant 25$ cm increased in 1993 because of immigration through the locks. The analysis of CPUE confirmed this. During the 1993 depletion more bream $\geqslant 25$ cm was caught (38.3 kg ha$^{-1}$) than monitored during the trawl survey in September 1992 (19.0 kg ha$^{-1}$). The depletion estimate of the initial population size was 55.1 kg ha$^{-1}$ (Table 30.5).

## 30.6   Discussion

As commonly found, initial biomass was underestimated when linear regression analysis was used. Petersen & Cederholm (1984) and Mahon (1980) found underestimations of approximately 13% and 28%, and underestimates of 30% to 50% are not unlikely (Hilborn & Walters 1992). For this reason, population sizes derived by the regression method are normally considered minimum values. As Zippin (1958) and Mahon (1980) showed, the error in estimating population size is highly influenced by the proportion of the total population taken. Helminen *et al.* (1993) stated that

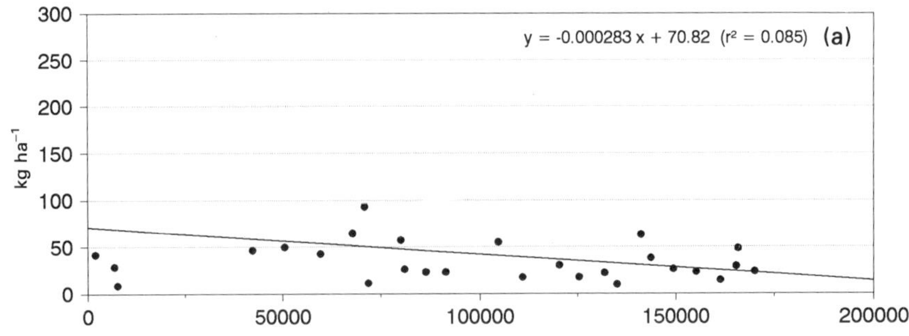


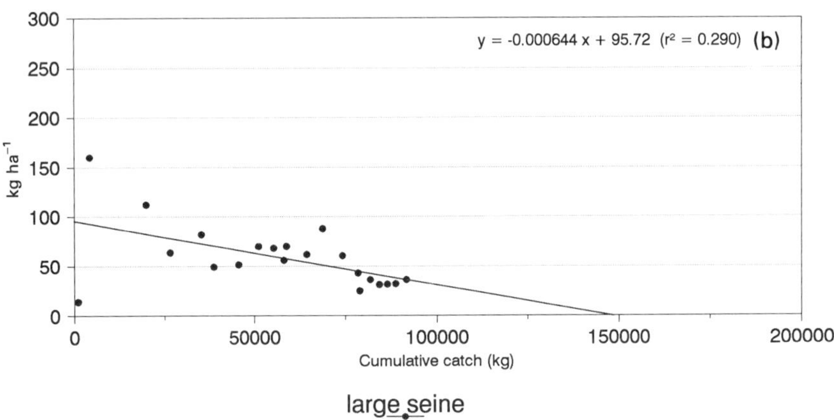


**Fig. 30.3**  Decline in the CPUE of bream $\geqslant 25$ cm during the fishing with large seines in Lake Wolderwijd in winter 1990/1991 (a) and in 1993 (b).

efficiency must be high ($> 50$ %) before this method is applicable. The current study confirms the relation between proportion taken and reliability. The estimates for bream $\geqslant 25$ cm differ less from the reconstructed population (3–19%) than the ruffe estimates (20–36%), probably because 49–70% of the population bream $\geqslant 25$ cm and only 29–40% of the ruffe population were removed during the depletion.

The accuracy of any estimator of population parameters depends mainly on how successfully the assumptions underlying the model are met in practice. Cowx (1983) and Ricker (1975) summarized the conditions upon which population estimations by the survey removal method are based. The two most important are:

(1)  the population must be isolated, i.e. the influence of migration, losses caused by natural mortality and recruitment must be insignificant during the sampling period;

(2)  the probability of capture remains constant throughout the entire period and is the same for all individuals in the population.

The Wolderwijd experiment was carried out annually over a relatively short period of time in winter. Effects of recruitment and natural mortality were minimized. However, the 1993 depletion analysis confirmed immigration during spring. Although assumption (1) was not met, the analysis of CPUE would have indicated the increased stock size. Variable catchability (2), segregation and behavioural stress, were probably the greatest sources of error in applying methods of estimation based on change in CPUE. The curvature (flattening) of the regression based on the ruffe trawl catches, indicated the presence of large numbers of fish with low catchability. In this study the flattening was associated with the introduction of an alternative gear. After introducing small fykes more ruffe were caught than were estimated by the CPUE analysis of the trawl catches.

Optimal depletion fisheries will never meet the above mentioned assumptions, thus fishing campaigns must be as efficient as possible. The operational planning, the choice of gears and planned effort, are dictated by species-specific distribution and must be as efficient as possible. During fishing in Lake Wolderwijd the 'look-and-catch' strategy proved worthwhile. Once located, concentrations of target species were fished. In this way data from depletion fisheries can provide valuable information on temporal and spatial species-specific distribution. This information is indispensable for fish stock assessment. It is concluded that depletion experiments are probably one of the most valuable but under-utilized techniques available to fisheries scientists (Hilborn & Walters 1992). Large scale overall abundance indexing during fishing provides a unique opportunity to collect information about fish distribution and behaviour through the season.

## References

Backx J.J.G.M. & Grimm M.P. (1994) Mass-removal of fish from lake Wolderwijd (2700 ha), The Netherlands. Part II: Implementation phase. In: I.G. Cowx (ed.) *Rehabilitation of Freshwater Fisheries.* Oxford: Fishing News Books, pp 401–414.

Backx J.J.G.M. (1993) Waarnemingen aan de visstand in het WolderwijdNuldernauw in het voorjaar van 1993 en een evaluatie van het spuibeheer bij Nijkerk. *Witteveen + Bos report* No. HD13.17, 20 pp (in Dutch).

Cowx I.G. (1983) Review of the methods for estimating fish population size from survey removal data. *Fisheries Management* **14**, 67–82.

Grimm M.P. & Backx J.J.G.M. (1994) Mass-removal of fish from lake Wolderwijd (2700 ha), The Netherlands. Part I: planning and strategy of a largescale biomanipulation project. In: I.G. Cowx (ed.) *Rehabilitation of Freshwater Fisheries.* Oxford: Fishing News Books, pp 390–400.

Hanson J.M. & Leggett W.C. (1982) Empirical prediction of fish biomass and yield. *Canadian Journal of Fisheries and Aquatic Sciences* **39**, 257–263.

Helminen H., Ennola K., Hirvonen A. & Sarvala J. (1993) Fish stock assessment in lakes based on mass removal. *Journal of Fish Biology* **42**, 255–263.

Hilborn R. & Walters C.J. (1992) *Quantitative fisheries stock assessment. Choice, dynamics & uncertainty.* New York, London: Chapman and Hall, 570 pp.

Hosper S.H. & Jagtman E. (1990) Biomanipulation additional to nutrient control for restoration of shallow lakes in The Netherlands. In: R.D. Gulati, E.H.H.R.Lammens, M.L.Meijer & E. van Donk (eds) *Biomanipulation – Tool for Water Management. Hydrobiologia* **200/201**, 523–534.

Leslie P.H. & Davis D.H.S. (1939) An attempt to determine the absolute number of rats in a given area. *Journal of Animal Ecology* **8**, 94–113.

Mahon R. (1980) Accuracy of catch-effort methods for estimating fish density and biomass in streams. *Environmental Biology of Fishes* **5**, 343–360.

Petersen N.P. & Cederholm C.J. (1984). A comparison of the removal and mark-recapture methods of population estimation for juvenile coho salmon in a small stream. *North American Journal of Fisheries Management* **4**, 99–102.

Ricker W.E. (1975) Computation and interpretation of biological statistics of fish populations. *Bulletin of the Fisheries Research Board of Canada* **191**, 382 pp.

Zippin C. (1958) The removal method of population estimation. *Journal of Wildlife Management* **22**, 82–90.

# VI  CASE STUDIES – RIVERS

# Chapter 31
# Stock assessment considerations for riverine fisheries

D.C. JACKSON *Department of Wildlife and Fisheries, Mississippi State University, Box 9690, Mississippi State, Mississippi 39762, USA.*

## 31.1    Introduction

Riverine fisheries, the systems that support them, and management problems are tremendously diverse in their characteristics, and, as such, system-specific and problem-specific orientations regarding the assessment of fish stocks prevail. It is critical, however, that assessments on these systems bear in mind that stocks do not exist in isolation but are integrated into the overall system dynamics, incorporating environmental capacity, fish biology and productivity, and human interactions via harvest and other activities.

## 31.2    Stock assessment problems

There are certain common denominators linking riverine fish stock assessment universally. Paramount among these are the uncertainties of open systems driven by outside influences on the aquatic environment. These influences come from lateral as well as upstream sources and, in the case of anadromous fisheries, also from downstream. This creates considerable system-related variance which also fluctuates on a temporal scale. River fish stock assessment must therefore always consider previous events as well as dealing with the present if they are to be effective in dealing with probable events on distant horizons. Short-term studies pertaining to specific fisheries tend to contribute little towards the ultimate charge of generating positive, responsible relationships between our various societies and the living aquatic resources of riverine ecosystems.

River fisheries stock assessment tends to be based on scientific methodologies which, at their best, can provide critical but only approximate measures of reality. Sampling and hypotheses testing are critical to these endeavours, and, because science requires that experiments be repeatable, standardization of techniques is an imperative. Without this, comparisons are difficult and the evolution of our understanding will not be capable of addressing critical fisheries-related issues. It is therefore recommended that a handbook of standard methods for the assessment of riverine fish stocks be developed.

Tremendous investments have been made towards understanding the dynamics of riverine fisheries resources. It is uncertain, however, if this support will continue without accountability on the part of river fisheries professionals in response to social, agency and institutional demands. The question therefore arises as to the availability and utilization of results from past studies. There is too much grey literature. Research contracts must expect publication in the primary literature. The process of science is not complete unless there is peer review and communication of results.

The bio-political arena is fully justified in requesting professional guidance in the realm of river fisheries management. Time frames are situationally dependent and controlled by political and biological phenomena that dictate stock assessment resolution. It is recommended that standard methods incorporate a hierarchical approach to assist river fisheries professionals in establishing priorities under pressure. Existing data should be utilized to generate testable hypotheses at an agreed level of resolution. However, these existing data can only rarely provide definitive answers to contemporary problems.

Complexity does not necessarily equate to professional competency. The capacity to communicate beyond the scientific perspective is urgently required. Concise, direct, clear communication of results in a language that can be understood across disciplines is recommended. Likewise, effort should be expended to enhance the elegance of our professional activities through simplification. For example, many fisheries biologists tend to conceive of and work on problems on scales that are too small for the questions at hand. Greater stability as well as appropriate inference is often obtainable by working at larger scales, especially in stock assessment work. Only through such efforts will river fisheries biologists be able to make an impact on development processes affecting riverine systems.

## 31.3    Recommendations

(1)  Develop a handbook of standard methods for riverine fish stock assessments that incorporates a hierarchical decision process.
(2)  Emphasize (if not require) primary journal publication on a component of contract accountability if professional recognition/backing by fisheries societies/ peer professionals is to be expected or utilized for support purposes.
(3)  Utilize existing data for hypotheses development; do not expect it to provide definitive answers but don't ignore it.
(4)  Maintain long-term programmes.
(5)  Develop the discipline to communicate concisely, directly and clearly beyond the ranks of our professional guild.

# Chapter 32
# Growth and mortality rates of an endangered species, *Synodontis victoriae*, in Sondu/Miriu river, Kenya

J. MULI and W. OJWANG *Kenya Marine & Fisheries Research Institute, PO Box 1881, Kisumu, Kenya*

**Abstract**   Growth and mortality parameters of *Synodontis victoriae* were estimated from length-frequency data collected from River Sondu/Miriu of Lake Victoria. The von Bertalanffy growth parameters were estimated as $L_\infty = 16$ cm and $K = 0.906$ yr$^{-1}$. The growth performance index $\phi'$ was 2.365. Total mortality was estimated as $Z = 3.983$ yr$^{-1}$. The natural mortality was estimated using Pauly's empirical formula as $M = 1.4$ yr$^{-1}$ for an average annual temperature of 21°C. The exploitation rate $E = 0.6$ indicated the fishery was overexploited.

KEYWORDS: endangered species, growth, mortality rates, exploitation rate

## 32.1   Introduction

The fisheries of the afferent rivers of the Nyanza Gulf, Kenya, were once important (Whitehead, 1959), but have been declining since 1940 (Van Someren 1959; Cadwallr 1965; Benda 1979; Marten 1979; Kibaara 1981; Ogutu-Ohwayo 1990). The annual catch of 2500 t (Whitehead 1959) from the Nyanza rivers is no longer possible. For example, the yield of River Sondu/Miriu has dropped from 668 t in 1959 to the current 81.5–108 t (Ochumba & Manyala 1992). The species *Shilbe mystus* and *Synodontis victoriae* (Boulenger) have replaced the once abundant *Barbus altianalis* (Boulenger) and *Labeo victorianus* (Boulenger). Decline in fish stocks in the river can be attributed to four sources: overfishing by destructive fishing methods (Van Someren 1959); papyrus encroachment and habitat degradation (Balirwa & Bugenyi 1980); introduction of Nile perch (*Lates niloticus*) and Nile tilapia (*Oreochromis niloticus*) (Ogutu-Ohwayo 1990); and poor management (Ochumba & Manyala 1992). Pollution was also listed as a possible cause of the decline of riverine fishes by Kibaara (1981).

This chapter examines the demographic aspects of *Synodontis victoriae*, one of the major siluroid catfish species present in Lake Victoria and its afferent rivers (Rinne & Wanjala 1983) which is considered to be endangered (Kibaara 1981). This information will be used in the sustainable management of this fishery.

### 32.2    Study area

The Sondu/Miriu River basin is located between latitude 0°17'S and 0°22'S, and longitude 34°04'E and 34°49'E (Fig. 32.1). It forms the fourth largest Kenyan river basin that drains into Lake Victoria, covering an area of 3470 km$^2$. The main tributaries of the river are the Kapsonoi and Yunith. It has an average population density of 268 people km$^2$. The study area on the lower Sondu/Miriu River was a 15 km stretch between the river-mouth and Odino falls. It was arbitrarily divided into 14 sections, each with a sampling station covering the main river channel. The width of the river varies between 16 and 19 m and the depth between 1.0 and 3.4 m, although occasionally deep depressions up to 10 m occur. A detailed description of the area is available in Ochumba & Manyala (1992).

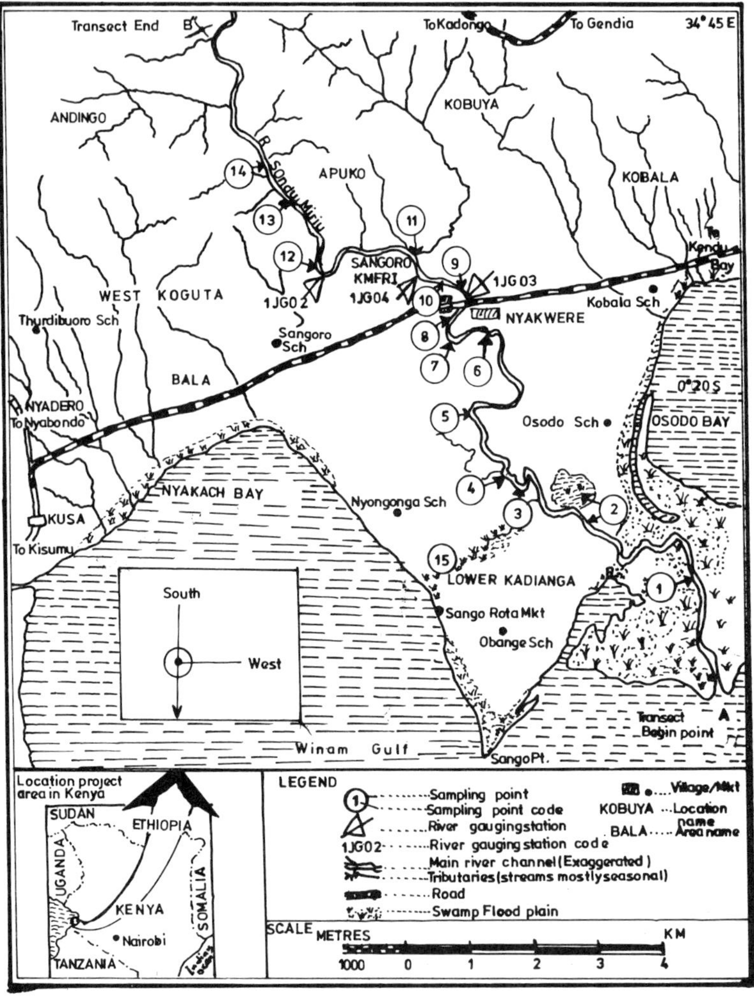


**Fig. 32.1**   The lower Sondu Miriu River catchment (with sampling points shown).

The annual rainfall distribution in the area varies between 470 mm and 1410 mm (Ogalo 1981) with a double peak pattern, in which 30% of the total rainfall occurs in the long rainy season (March–May) and 15% in the short rainy season (August–October).

### 32.3    Material and methods

Length-frequency data were collected during weekly fishings along the river Sondu/Miriu from February to December, 1991. A beach seine of 1170 m × 2.8 m and 28 mm stretched mesh size, and multifilament gillnets of 45.7 m × 21.3 m and stretched mesh sizes of 63.5, 50.8, 47.6 and 38.1 mm were used. A standard haul was considered to sweep an area of approximately 9000 $m^2$ after drifting the beach seine and gillnets for about 15 min. Identification of the fish species landed was based on Greenwood (1966). The standard length (*SL*) and total length (*TL*) were measured to the nearest mm below. Weight of all fish was taken to the nearest g. Surface water temperature in the river Sondu/Miriu was measured weekly for all the stations using a mercury thermometer.

Samples were pooled by month and for the length frequency analysis a size class of one cm was used. Von Bertalanffy growth parameters were estimated using these pooled data.

The analysis assumed von Bertalanffy growth according to:

$$L_t = L_\infty (1 - \exp(-K(t - t_o)))$$

Where $L_t$ is length at time $t$, $L_\infty$ is the asymptotic length an individual would reach if it lived indefinitely, and $K$ is the rate at which $L_\infty$ is approached. The parameter $t_o$ is a location parameter necessary to relate size to absolute age, and cannot be estimated using length frequency analysis. Powell's (1979) method, discussed by Wetherall *et al.* (1987), was used to obtain $L_\infty$ and the $Z/K$ ratio, where $Z$ is the instantaneous rate of total mortality. The analysis was carried out using length-frequency stock assessment (LFSA) package of microcomputer programmes.

$t_o$ was calculated using the empirical formula given by Pauly (1980a)

$$\log(-t_o) = 0.3922 - 0.2752 \log L_\infty - 1.038 \log K$$

The growth constant K was estimated using Ebert's (1973) method after Vakily *et al.* 1986.

$$K = \ln(L_\infty{}^D - L_2{}^D / L_\infty{}^D - L_1{}^D)(1 / (t_1 - t_2)^D)$$

Where $L_1$ and $L_2$ are the mean lengths at times $t_1$ and $t_2$ when length frequency samples were taken and $D$ is the surface parameter. The growth performance index was computed according to Moreau *et al.* (1986).

$$\phi' = \log K + 2 \log L_\infty$$

The total mortality rate ($Z$) was estimated by multiplying the $K$ obtained by the value of $Z/K$. Natural mortality ($M$) was calculated using Pauly's empirical formula (1980b).

$$\ln M = 0.0152 - 0.279\ln L_\infty + 0.6543\ln K + 0.463\ln T$$

Where $T$ is the mean annual temperature (°C) of surface water.

### 32.4     Results

Sixteen fish species (Table 32.1) were caught out of the expected 28 reported by Ochumba & Manyala (1992). This suggests a decline in species diversity. The most abundant species were *S. mystus*, *S. victoriae*, *S. afrofischeri* and *Lates niloticus*.

The length frequency distribution of *S. victoriae* (Fig. 32.2) shows a length range of 4.5–13.5 cm with sample size of 947 specimens. The length-weight relationship derived for *S. victoriae* was

$$W = 0.041 \; SL^{2.613} \qquad\qquad (r = 0.853; \; n = 40)$$

**Table 32.1**  Monthly catch rates by number of fish sampled in the Sondu/Miriu River from February to December 1991 (blanks indicate no fish were caught).

| Species | Feb | Mar | Apr | May | Jun | Jul | Aug | Sep | Oct | Nov | Dec | Total |
|---|---|---|---|---|---|---|---|---|---|---|---|---|
| *Lates niloticus* | 5 | | | | | | | | | | 1 | 46 |
| *Clarias gariepinus* | 2 | | | | | | | | | | | 3 |
| *Schilbe mystus* | 106 | 114 | 660 | 537 | 29 | 462 | 191 | 130 | 64 | | 3 | 2359 |
| *Bagrus docmac* | | 1 | | | | 1 | | | | | | 2 |
| *Oreochromis niloticus* | | | | | | 1 | 5 | | | | 1 | 7 |
| *Oreochromis variabilis* | | | | 1 | | | | | | | | 1 |
| *Oreochromis leucostictus* | | | 6 | | | | | | | | | 6 |
| *Tilapia zilli* | | | | | 6 | 1 | 1 | | | | | 8 |
| *Barbus neglectus* | 7 | | | | | 3 | | | | | | 11 |
| *Barbus altianalis* | | 3 | 2 | 3 | 3 | 4 | | 2 | 1 | 1 | 1 | 17 |
| *Barbus jacksonii* | | | | | | | | | | | | |
| *Barbus Kerstenii* | | | | | | | | | | | | |
| *Synodontis victoriae* | 480 | 127 | 150 | 108 | 23 | 13 | 8 | 17 | | 6 | 15 | 947 |
| *Synodontis afrofischeri* | 2 | 13 | 38 | 3 | | 4 | | | 2 | | 14 | 76 |
| *Alestes sadleri* | | 2 | | | | | | | | | | 2 |
| *Alestes jacksonii* | | | | | | | | | | | | |
| *Xenoclarias* sp. | | | | | | | | | | | | |
| *Mastacembalus frenatus* | | | | | | | | | | | | |
| *Ctenopoma muirei* | | 5 | | | | | | 1 | | | | 6 |
| *Haplochromis* spp. | | | | | | | | | | | | |
| *Aplocheilichthys eduardis* | | | | | | | | | | | | |
| *Micropterus salmoides* | | | | | | | | | | | | |
| *Protopterus aethiopicus* | | | | | | | | | | | | |
| *Mormyrus Kannume* | | 1 | 3 | | | | | | | | | 4 |
| *Petrecephalus cutostoma* | | | | | | | | | | | | |
| *Gnathonemus longiberbis* | | | | | | | | | | | | |
| *Marcusenius grahami* | | | | | | | | | | | | |
| *Labeo victorianus* | 8 | 16 | 8 | 7 | 3 | | | 1 | | | | 43 |

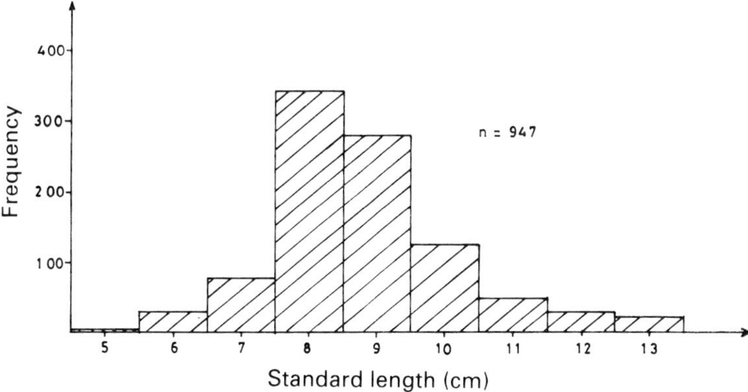



**Fig. 32.2**

The exponent (*b*) was significantly different from 3, indicating allometric growth, and the asymptotic weight was estimated as $W_\infty = 57.4$ g.

Estimation of $L_\infty$ and $Z/K$, using the Powell's (1979) method (Fig. 32.3) gave values of $L_\infty = 16$ cm *SL* and $Z/K = 4.396$, using the length range 8.5–13.5 cm to obtain the regression relationship.

The growth constant *K*, estimated using Ebert's (1973) method, was 0.906 yr$^{-1}$. An $L_\infty$ of 16 cm *SL* corresponds to 19.3 cm *TL*, according to the relation $TL = 0.72 + 1.163SL$. (n = 200). For *S. victoriae* in river Sondu/Miriu, the average length of first maturity (*Lm*) was 8.1 cm (personnal observation). As $L_\infty = 16$ cm, the ratio $Lm/L_\infty$ is low (= 0.5) inferring that the growth rate of *S. victoriae* in this river was high.

$t_o$ was obtained by Pauly's (1980a) empirical equation as –0.2. This value is low and not significantly different from zero. For convenience $t_o$ was set equal to zero (Sparre *et al.* 1989).

From the values of $Z/K$ and *K*, total mortality rate was calculated as $Z = 3.983$ yr$^{-1}$. Natural mortality *M* was calculated using Pauly's (1980b) empirical formula and a mean annual water temperature of the river Sondu/Miriu of 21°C. The

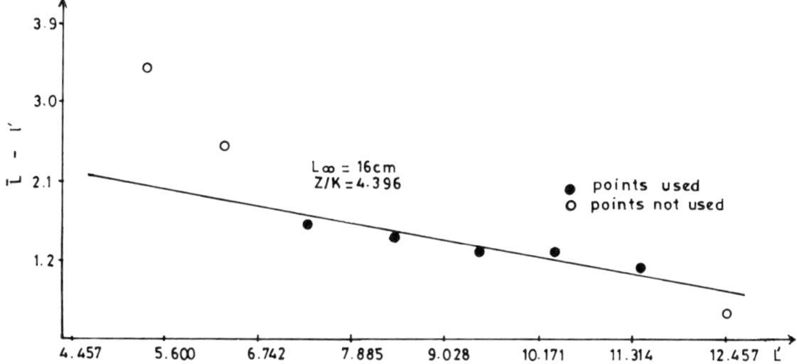



**Fig. 32.3**

result was multiplied by 0.8 to correct for schooling and any other statistical noise (Sparre *et al.* 1989). Thus natural mortality becomes $M = 1.4\,\mathrm{yr}^{-1}$, fishing mortality $F = 2.583\,\mathrm{yr}^{-1}$ and the exploitation rate $E$ was 0.649 ($E = F/Z$). The growth performance was computed as $\phi` = 2.365$.

## 32.5    Discussion

Comparing the growth parameters $L_\infty = 16$ cm and $K = 0.906\,\mathrm{yr}^{-1}$ with those obtained from a study in 1985/86 (i.e. $L_\infty = 21$ cm and $K = 0.6\,\mathrm{yr}^{-1}$; Ochumba & Manyala 1989) there was a reduction in average size and an increase in the growth rate (Fig. 32.4). These changes can be attributed mainly to overfishing and to a lesser extent to diet type and food utilization. The overfishing process has been well documented, e.g. Garrod (1961), Cadwallr (1965), Ssentongo & Welcomme (1985) and Getabu (1987, 1992). It usually results in the reduction of the average size of the fish in the stock and a faster growth rate (Garcia 1986). The average size is therefore attained much faster causing the growth constant ($K$) to increase. The contribution of feeding to the growth of fish has been well documented. Pauly (1982) showed that feeding contributed to growth, which in turn determined the yield. The food resources and their digestibility have been found to determine the different growth patterns of *O. niloticus* in natural waters (Pullin 1988). This could also apply to *S. victoriae*.

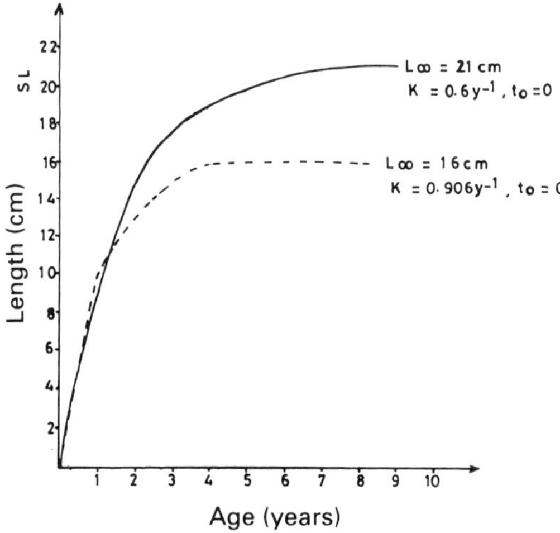


**Fig. 32.4**

According to Beverton and Holt (1959), $M/K$ ratio is expected to be in the range 1.5 to 2.5 and the value of 1.545 (0.906/1.4) for this study suggests the estimate of $M$ is reasonable.

The exploitation rate $E = 0.6$ (optimum being 0.5) suggests the fishery is being overexploited. There is a need to reduce the fishing effort to below the maximum sustainable yield (MSY) level. Ochumba & Manyala (1992) proposed this could be regulated by:

(1)  licensing of nets in addition to registration of boats;
(2)  minimum sizes of landed fish should be controlled by co-operatives, clans, partly elders and other social organizations; reducing the use of small mesh gillnets and beach seines during the spawning periods; protection of established nursery grounds;
(3)  rehabilitation of native traps for catching spent fish on their downstream migration; and
(4)  stock enhancement.

In the Sondu/Miriu river, and other afferent rivers of Lake Victoria there is a need for similar studies on endangered species, especially *Labeo victorianus* and *Barbus altianalis* (Kibaara 1981).


## Acknowledgement

I wish to thank Dr E. Okemwa (Director, KMFRI) for permission to publish this work. Thanks too to Mr Mwatha for permission to use AIRS computer, and Messrs J.O. Manyala and D. Atula who assisted in the data analysis. Messrs S.O. Osewe and R. Okech assisted in the word processing. Messrs Odanga and Agoro drew the figures. Special thanks go to Messrs J. Achuku, R. Munuve and the fishermen at Sangoro Laboratory who provided the fish specimens.


## References

Balirwa J.S. & Bugenyi F.W.B. (1980) Notes on the fisheries of the River Nzoia, Kenya. *Biological Conservation* **18**, 53–58.

Benda L.C. (1979) Analysis of data from 1968–1976 from nine fish landings in the Kenya waters of Lake Victoria. *Journal of Fish Biology* **15**, 385–387.

Beverton R.J.H. & Holt S.J. (1959) A review of the lifespans and mortality rates of fish in nature, and their relation to growth and other physiological characteristics. In: G.E.W. Wolstensholme & M. O'connor (eds). Ciba foundation. *The colloquium on ageing. The lifespan of animals.* Vol 5, Churchill, London, pp 142–177.

Cadwallr D.A. (1965) The decline in *Labeo victorianus* Blgr. (Pisces. Cyprinidae) fishery of Lake Victoria and associated deterioration in some indigenous fishing methods in the Nzoia River, Kenya. *East African Agricultural and Forest Journal* **30**, 249–256.

Ebert T.A. (1973) Estimating growth and mortality rates from size data. *Oecologia* **11**, 282–298.

Garcia S. (1986) Seasonal trawling bans can be very successful in heavily overfished areas. The Cyprus effect. *Fishbyte* **4**, 7–12.

Garrod D.J. (1961) The rational exploitation of the *Tilapia esculentus* stock of the North Buvuma island area, Lake Victoria. *East African Forest Journal* **27**, 64–76.

Getabu A. (1987) Aspect of the Lake Victoria Fisheries with emphasis on *Oreochromis niloticus* and *Alestes sadleri* for the Nyanza Gulf. *FAO Fish Rep.* **389**, 416–431.

Getabu A. (1992) Growth parameters and total mortality in *Oreochromis niloticus* (Linnaeus) from Nyanza Gulf, Lake Victoria. *Hydrobiologia* **232**, 91–97.

Greenwood P.H. (1966) *The fishes of Uganda*, 2nd edn. The Uganda Society, Kampala.

Kibaara D. (1981) Endangered fish species of Kenya's inland waters with emphasis on *Labeo spp.* In: Proceeding of the workshop of the Kenya Marine and Fisheries Research Institute on aquatic resources of Kenya, 1981, 13–19 July, Mombasa, pp 157–164.

Marten G.G. (1979) Impact of fishing on the inshore fishery of Lake Victoria (East Africa). *Journal of the Fisheries Research Board of Canada* **36**, 891–900.

Moreau J.C., Bambino C. & Pauly D. (1986) A comparison of four population (Fam; Cichlidae). ICLARM contribution No. **292**, 14 pp.

Ochumba P.B.O. & Manyala J. (1989) Growth and related statistics of six riverine fish species in the Sondu/Miriu river of Lake Victoria, Kenya. In: A handbook to FAO/DANIDA and FAO/TAFIRI regional seminar on the fish stocks of Lake Victoria. January/February 1989. Mwanza. Tanzania.

Ochumba P.B.O. and Manyala J.O. (1992) Distribution of fishes along the Sondu-Miriu River of Lake Victoria, Kenya with special reference to upstream migration, biology and yield. *Aquaculture and Fisheries Management* **23**, 709–719.

Ogalo L. (1981) Trends of rainfall in East Africa. *Kenya Journal of Science and Technology* (A) **2**, 83–90.

Ogutu-Ohwayo R. (1990) The decline of the native fishes of Lake Victoria and Kyoga (East Africa) and the impact of introduced species, especially the Nile perch, *Lates niloticus* and the Nile tilapia, *Oreochromis niloticus*. *Environmental Biology of Fishes* **32**, 85–94.

Pauly D. (1980a) A new methodology for rapidly acquiring basic information of tropical fish stocks: growth mortality and stock recruitment relationships. In: S.B. Saila and P.M. Redel (eds). *Stock assessment for tropical small-scale fisheries*. ICMRD, Univ. Rhode Island, Kingston.

Pauly D. (1980b) A selection of simple methods for the assessment of tropical fish stocks. *FAO. Fish. Circ.* **729**, 54 pp.

Pauly D. (1982) Studying single species dynamics in a tropical multispecies context. In: D. Pauly & C.I. Murphy (eds). *Theory and management of tropical fisheries ICLARM Conference proceedings* **9**, 33–70.

Powell B.G. (1979) Estimation of mortality and growth parameters from the length frequency of a catch. *Rapp. p-v Reun. CIEM* **175**, 167–169.

Pullin R.S.V. (1988) Tilapia Genetic Resources for Aquaculture. In: *Proceeding of Workshop on Tilapia Genetic Resources for Aquaculture* 23–24 March 1987, Bangkok, Thailand ICLARM conference proceedings **16**, 108 pp.

Rinne J.N. & Wanjala B. (1983) Mortality, fecundity and breeding seasons of major catfishes (suborder: Siluroidea) in Lake Victoria. *East African Journal of Fish Biology* **23**, 357–363.

Sparre P., Ursin E. & Venema S.C. (1989) Introduction to tropical Fish stock assessment: Part 1. *FAO Fisheries Technical Paper* **No. 306/1**, 20–156.

Sparre P. (1987) Computer programmes for fish stock assessment length-based fish stock assessment (LFSA) for Apple II computer. *FAO Fish. Tech. Pap.* **101, suppl. 2**, 218 pp.

Ssentongo G. & Welcomme R.L. (1985) Past history and current trends in the fisheries of Lake Victoria. *FAO Fish. Rep.* **335**, 123–138.

Vakily J.M., Palomares M.L. & Pauly D. (1986) Computer programmes for fish stock assessment. Application for the HP 41 CV calculator. *FAO Fisheries Technical Paper* **101, Supp. 1**, 255 pp.

Van Someren V.D. (1959) A study of a small basket trap river fishery in Kenya. *East Africa Journal for Agriculture* **24**, 257–267.

Wetherall J.A., Polovin J.J. & Ralston S. (1987) Estimating growth and mortality in steady-state fish stocks from length-frequency data. In: D. Pauly & G. Morgan (eds.). Length based methods in fisheries research ICLARM Manila, pp. 53–74.

Whitehead P.J.P. (1959) The river fisheries of Kenya, Nyanza Province. *East Africa Agriculture Journal* **24**, 274–278.

# Chapter 33
# Stock assessment and fisheries population dynamics in Spanish mountain streams

D. GARCÍA DE JALÓN, M. MAYO and F. HERVELLA *Laboratorio de Hidrobiología, ETSI Montes, Universidad Politécnica de Madrid, 28040-Madrid, Spain*

**Abstract**   A review of the characteristics of the mountain stream fisheries of the Iberian Peninsula is presented. Biological and abiotic factors that control the fish population dynamics are summarized. Data from inventories carried out in different Spanish mountain streams over the last eight years are used to assess stock paramenters. A population simulation model (SIMTRUT), based upon intrapopulation characteristics, is also presented. This model helps us to understand the importance of angling intensity and minimum catchable size as management mechanisms.

KEYWORDS: trout, cyprinids, Spanish streams, population model

## 33.1   Introduction

The composition and dynamics of Spanish fisheries populations have been studied in different mountain streams: for example, García de Jalón *et al.* 1987; García de Jalón *et al.* 1990; Lobón-Cerviá & Penczak 1984; Lobón-Cervia *et al.* 1986.

Most Spanish mountain streams are inhabited by fish communities composed mainly of salmonids and rheophilous cyprinids. Among the former, brown trout (*Salmo trutta* L.) is always dominant and often the only species present. Generally, salmonids live alone in headwater streams and high altitude reaches, while in lower reaches they are found with other species. Only in southern Spanish mountain streams are salmonids absent, but these streams are not considered in this chapter.

The other fish species that live with trout in middle mountain and piedemont stream-reaches are mainly rheophilous and lithophilous cyprinids, with long and hydrodynamic-shaped bodies. Among these cyprinids, three types are found (García de Jalón 1992):

(1)   Small-sized species whose diet is mainly composed of benthic invertebrates, e.g. dace (*Leuciscus carolitertii, L. cephalus, L. pyrenaicus*), minnow (*Phoxinus phoxinus*), Spanish roach (*Rutilus arcasii*) and gudgeon (*Gobio gobio*);

(2)   Medium-sized species with a diet basically phytophagous, e.g. nase (*Chondrostoma polylepis, C. toxostoma*);

(3)  Large-sized species with omnivorous diets, represented by barbels (*Barbus bocagei, B. graellsii, B. haasi, B. meridionalis*).


Type (1) species maintain permanent populations in mountain streams; types (2) and (3) usually live in the lower reaches but use the upper zones seasonally, ascending during spring spawning migrations and many of them remaining during summer low-flow conditions. There are also other small-sized fishes, e.g. loaches (*Barbatulus barbatula, Cobitis calderoni, C. marrocana*) which live on the stream bottom and feed on benthic invertebrates. Their populations are always relatively small compared with the other types of fish.


### 33.2    Population dynamics of mountain stream fish communities

Spanish mountain streams are characterized by torrential flow regimes and strong fluctuations in water temperature. Both factors determine the composition, structure and function of the stream communities. Therefore, each fish population has developed a specific pattern of occupying the fluvial habitat according to its natural adaptations. This pattern has two dimensions: a spatial dimension, shown by the distribution along the longitudinal river continuum, and a temporary dimension, represented by upstream and downstream migrations.

Stream salmonid populations have shorter longevity, earlier maturity age, higher annual production/biomass ratio ($P/B$) and generally a greater gonadosomatic index, than most of the stream cyprinid populations. A gradient of population characteristics is also found among the different types of cyprinids: longevity and sexual maturity age increase from cyprinids of the first type (a) to those of the second (b), and from the second type to the third; conversely fecundity and $P/B$ ratio decrease in the same order (García de Jalón 1992).

By comparing the life history strategies of salmonids and type (1) cyprinids to those of type (2) and (3) fish it can be observed that they:

- expend greater energy on their reproduction (greater number of eggs per female, and/or greater egg size, and reproduction at younger ages);
- present a simpler age-structure (as they have fewer year-classes), and;
- have a higher global biomass-growth (as they have a greater annual $P/B$ ratio).


These life strategy characteristics correspond to those defined by Pianka (1970) and Stearns (1976) as 'r' selection strategy, while type (2) and (3) cyprinids correspond to 'K' selection strategy.

Thus, taking into account the distribution of species along the gradient (Fig. 33.1), it may be concluded that a correlation exists between strategies used by different fish populations and their longitudinal distribution along the fluvial continuum. In headwater streams trout live alone using 'r' strategies. Downstream the trout live with small type (1) cyprinids which also use 'r' strategies. Further downstream and in the

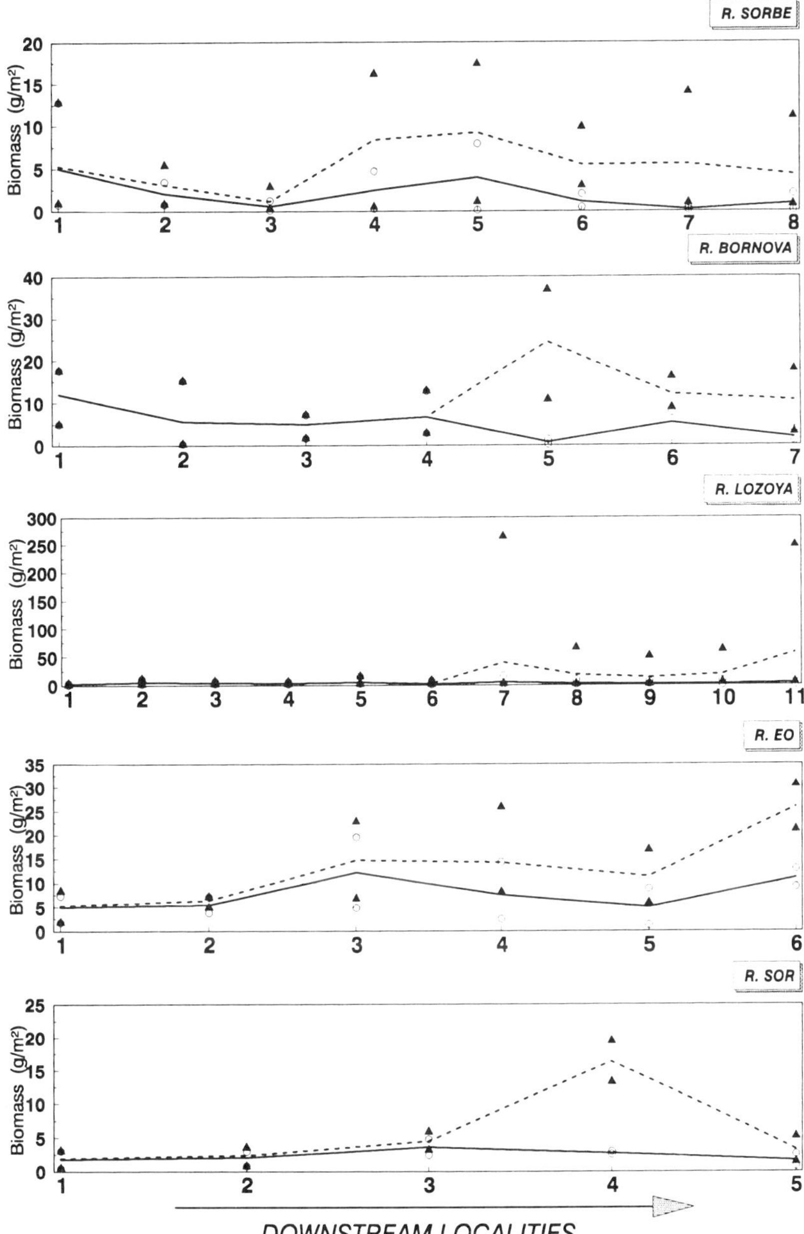


**Fig. 33.1**  Mean annual biomass in different Spanish mountain streams from headwaters to downstream. Continuous line represents trout biomass and dashed line the total community. Triangles and circles represent extreme values for trout and total stock.

piedemont reaches, type (2) and (3) cyprinids are present in increasing proportions, and use 'K' strategies.

This indicates that fish communities in the upper reaches are regulated mainly by abiotic factors (density independent), while in the lower reaches they are regulated by biotic factors (density dependent). This pattern of fluvial fish population regulation is similar to that shown on Polish rivers (Zalewski & Naiman 1985), and is explained by the predictability of the abiotic controlling factors increasing from the head streams to lower reaches.

### 33.2.1    *Seasonal fluctuations*

Data from fish stock evaluations in Spanish streams usually show a strong variability about samples from the same locality at different seasons. As an example, mean annual trout and total fishery biomass along different mountain streams are shown in Fig. 33.1, together with the extreme biomass values recorded at different seasons of the year. Seasonal coefficients of variation of trout biomass in each river are presented in Table 33.1.

**Table 33.1**    Seasonal and inter-annual coefficient of variation (%) of trout biomass in different Spanish streams.

| Seasonal fluctuation | | Inter-annual fluctuation | |
| --- | --- | --- | --- |
| R. Lor | 56 | R. Gállego | 47 |
| R. Sor | 42 | R. Jucar | 30 |
| R. Eo | 53 | R. Escabas | 24 |
| R. Lozoya (upper) | 51 | | |
| R. Lozoya (lower) | 106 | | |
| R. Sorbe | 80 | | |
| R. Bornova | 71 | | |

The rivers Sorbe, Bornova and Lozoya are high altitude (850–1660 m) streams from central Spanish mountains, while the rivers Sor and Eo correspond to coastal mountain streams (30–400 m altitude) from the north west. All show the species distribution pattern previously discussed, with only trout in headwater streams and increasingly mixed stocks downstream, which are finally dominated by cyprinids.

Seasonal fisheries biomass fluctuations also increase downstream, especially due to the fluctuation of cyprinid stock dynamics. The River Sor is the only one that presents a moderate seasonal fluctuation. Among the rest, extreme fluctuations are found in the River Lozoya, especially at stations 7 and 11, changing from 2 $gm^{-2}$ in February to 250 $gm^{-2}$ in June (Table 33.3).

A characteristic of Spanish mountain stream fisheries is their continuous change, probably due to temperature and flow fluctuation. Spring upstream migrations of cyprinids (especially types (2) and (3)) invade trout reaches (Martínez & García de

Jalón 1988). When these migrations represent dense populations, they disrupt the territorial behaviour of trout, and often cause upstream displacement of trout populations (Casado *et al.* 1989). This seasonal upstream occupation has been intensified by the construction of reservoirs that maintain greater cyprinid populations than can be sustained on natural streams (Barceló & García de Jalón 1986; García de Jalón *et al.* 1992).

Summer conditions in mountain streams, with low flows and high water temperatures, represent an adverse habitat for salmonids and favour cyprinids that remain in these upper reaches until the first autumn floods.

At the end of autumn, some trout, especially those from the lower reaches, migrate to upper reaches of the main stream or to small tributaries for spawning. Many fry and juveniles born in these small upland streams migrate downstream, because their habitat requirements exceed those provided by small streams as they grow, especially during summer low flow conditions. Therefore, there is a continuous movement of the trout stocks from the headwaters and tributaries to the principal stream channel and back.

The sharing of the same habitat by trout and different types of cyprinids, be it permanent or temporary, suggests competition. This issue has been studied in several Spanish streams looking at common food-resources (macroinvertebrates) through comparison of diet (Table 33.2). The results suggest there is little dietary overlap, even though there are many common items in the diet of trout and cyprinids, as each population selects some food items exclusively.

**Table 33.2**   Degree of dietary overlap (Schoener Index) between brown trout (*S. trutta*) and four cyprinid species in Spanish mountain streams. A value of > 0.6 suggests significant overlap of diet (Zaret & Rand 1971).

|  | River Osia | River Veral | River Sorbe |
| --- | --- | --- | --- |
| *Barbus bocagei* | 0.62 | 0.45 | — |
| *Barbus haasi* | — | 0.49 | — |
| *Phoxinus phoxinus* | — | 0.53 | — |
| *Leuciscus pyrenaicus* | — | — | 0.41 |

Physical factors, especially water temperature, explain this coexistence more effectively. Cold winter temperatures in Spanish mountain streams (2–12°C) allow salmonids to maintain an active metabolism, while cyprinids tend to hibernate on the bottom. By contrast, hot summer water temperatures (more than 20°C) may adversely affect the salmonids but are tolerated by most cyprinids. However, not all interactions between salmonids and cyprinids are adverse. Large trout are ichthyophagus and those that live at the lower mountain reaches intensively predate on early life stages of cyprinids.

**33.2.2**    *Annual fluctuations*

Data about population fluctuations between years are scarce, but evidence (Fig. 33.2) suggests inter-annual changes are less important than the seasonal shifts (Table 33.1). The populations from the chalk streams of central eastern Spain, e.g. River Jucar and River Escabas, are relatively stable with a mean coefficient of variation < 30% and < 24%, respectively. Conversely, the River Gállego, a Pyrenean stream, has a greater inter-annual fluctuation (47%), a value similar to the seasonal fluctuation (Table 33.1). These annual fluctuations are mainly caused by the annual variability of flow regimes which are markedly torrential. However, especially in the case of trout populations, there is a clear effect caused by stocking and angling pressure.

**33.3    Stock assessment**

In Spanish streams (Table 33.3), total fishery biomass increases downstream reaching a maximum generally in the lower reaches. This maximum biomass exceeds 50 g m$^{-2}$ in the River Lozoya and the River Gallo, while in most streams it is around 20–30 gm$^{-2}$. Mean biomass values are 5–6 gm$^{-2}$ in unproductive streams (Sorbe, Tajo and Sor), but are usually between 14 and 18 g m$^{-2}$. The River Gallo is an exceptionally rich stream with mean biomass values higher than 30 gm$^{-2}$.

Most mountain stream trout populations increase their mean biomass from the headstream to the middle-reaches, and decrease towards their lower mountain reaches (Table 33.3). Mean annual trout biomass in the most unproductive streams ranges between 2 and 3 gm$^{-2}$, but in most streams it is around 5–7 gm$^{-2}$. Again, the River Gallo is exceptionally rich with a mean biomass of 24 gm$^{-2}$.

Fish production shows how energy, expressed in terms of live weight, flows through and is used by stream fisheries (Table 33.4). Fish production follows the same pattern as biomass, showing similar maximum and minimum values at the same localities. However, trout production in the River Lozoya and in the River Eo are highest in their lowest reaches, probably because they are influenced by populations living immediately below in the Pinilla Reservoir and Cantabric Sea, respectively.

These production figures are generally much smaller than those obtained during other studies on Spanish streams. For example, River Ucero production reached 60 g m$^{-2}$yr$^{-1}$ (Lobón–Cerviá *et al.* 1986) and in the River Jarama production was estimated at 10.4–58.3 g m$^{-2}$yr$^{-1}$ (Lobón-Cerviá & Penczak 1984). These last values are similar to those of the most productive stations in the River Lozoya and River Gallo.

**33.4    Stock management**

One of the main objectives in fisheries management is to regulate angling pressure to maintain population levels with an age structure according to anglers' demands. This angling pressure is controlled by fishing intensity (angling-h stream km$^{-1}$ yr$^{-1}$) and by

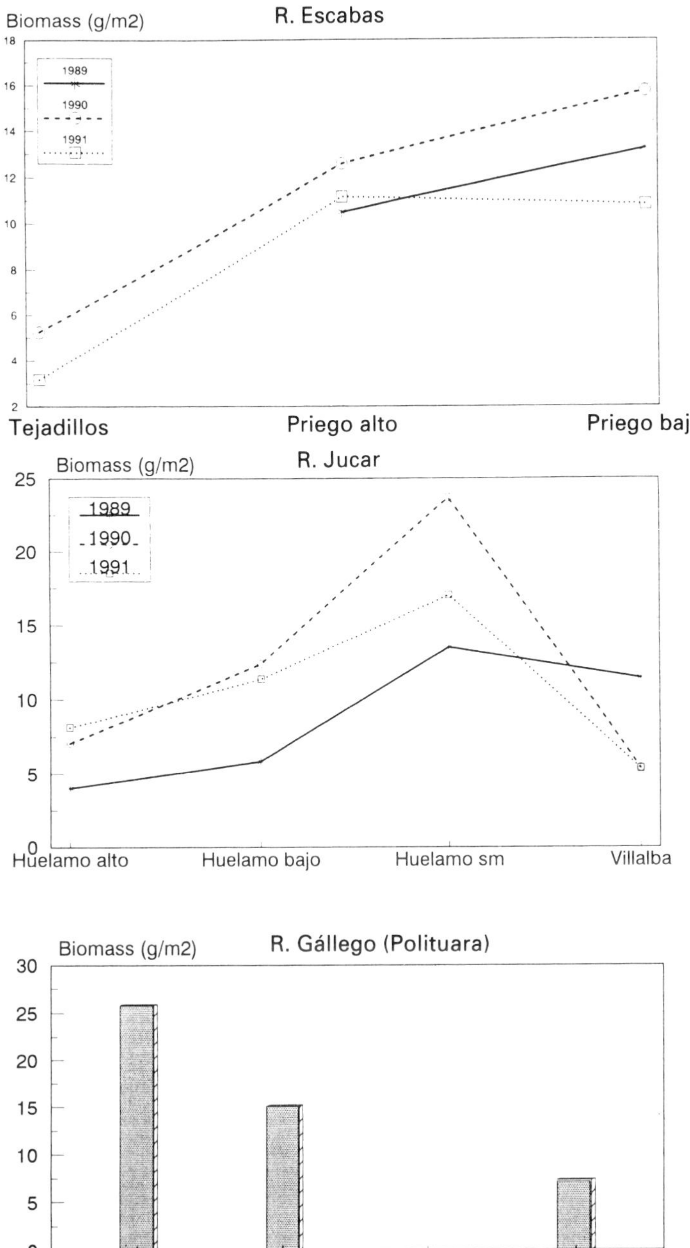


**Fig. 33.2** Annual fluctuations in stock biomass in different Spanish mountain streams.

**Table 33.3** Mean trout and total fish annual biomass (g m$^{-2}$) in Spanish mountain streams at different localities along their longitudinal gradient.

| River | | 1 | 2 | 3 | 4 | 5 | 6 | 7 | 8 | 9 | 10 |
|---|---|---|---|---|---|---|---|---|---|---|---|
| Sorbe | trout | 6.1 | 4.55 | 2.56 | 2.70 | 6.02 | 4.28 | 0.17 | 0.28 | | |
| | total | 6.3 | 5.86 | 3.45 | 8.26 | 10.85 | 6.67 | 3.78 | 2.60 | | |
| Bornova | trout | 18.90 | 5.77 | 7.60 | 10.90 | 0.50 | 5.63 | 1.91 | | | |
| | total | 18.90 | 5.77 | 7.60 | 10.92 | 16.72 | 13.77 | 8.53 | | | |
| Lozoya | trout | 2.81 | 5.47 | 4.08 | 3.31 | 5.56 | 2.17 | 4.96 | 2.52 | 2.23 | 3.9 |
| | total | 2.81 | 5.47 | 4.08 | 3.31 | 5.56 | 2.70 | 39.9 | 17.97 | 19.2 | 56.35 |
| Tajo | trout | 0.86 | 10.0 | 5.77 | 1.60 | 2.37 | 0.78 | 1.52 | 0.47 | | |
| | total | 0.86 | 10.0 | 5.80 | 1.70 | 5.34 | 1.51 | 6.48 | 12.15 | | |
| Gallo | trout | 23.37 | 33.07 | 28.13 | 11.16 | | | | | | |
| | total | 23.37 | 34.93 | 52.46 | 13.94 | | | | | | |
| Eo | trout | 4.99 | 5.47 | 12.21 | 7.46 | 4.94 | 11.02 | | | | |
| | total | 5.25 | 56.39 | 14.74 | 14.24 | 11.3 | 25.76 | | | | |
| Sor | trout | 1.82 | 2.14 | 3.61 | 2.79 | 1.68 | | | | | |
| | total | 1.97 | 2.47 | 4.55 | 16.4 | 3.3 | | | | | |
| Lor | trout | 4.38 | 8.79 | 7.21 | | | | | | | |
| | total | 4.91 | 9.7 | 32.61 | | | | | | | |
| Arnoia | trout | 6.07 | 6.25 | 3.54 | | | | | | | |
| | total | 7.08 | 20.7 | 15.26 | | | | | | | |
| Aragón- | trout | 3.51 | 17.64 | 9.58 | 3.09 | 1.85 | 1.73 | | | | |
| Subordan | total | 3.51 | 17.64 | 18.02 | 18.72 | 26.28 | 23.58 | | | | |
| Veral | trout | 3.35 | 3.89 | 8.83 | 11.63 | 4.22 | 2.3 | | | | |
| | total | 3.35 | 3.89 | 8.83 | 12.29 | 4.96 | 19.79 | | | | |

limiting minimum catchable size. Spanish fishing authorities are mainly interested in salmonid populations, and thus a simulation model of trout population dynamics was developed to help managers make decisions.

### 33.4.1   *The SIMTRUT model*

SIMTRUT is a deterministic model that simulates a trout population evolution through time, controlled by two management parameters: minimum size and angling intensity (García de Jalón 1992). SIMTRUT has been built according to the Theory of System Dynamics (Forrester 1968), and implemented in an IBM compatible computer through a software program AMDS developed by González & Fernández (1986).

In the model the trout population is characterized by parameters such as natural mortality, fecundity, morphic coefficient, and by those that define growth (Von Bertalanffy) and recruitment curves. The basic structure of SIMTRUT is composed of variables:

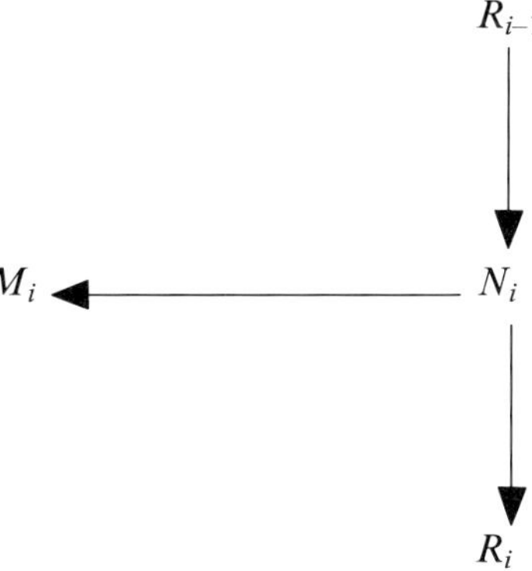



**Table 33.4** Mean trout and total fish annual production (g m$^{-2}$) in Spanish mountain streams at different localities along their longitudinal gradient.

| River | | 1 | 2 | 3 | 4 | 5 | 6 | 7 | 8 | 9 | 10 |
|---|---|---|---|---|---|---|---|---|---|---|---|
| Sorbe | trout | 5.6 | 5.45 | 3.18 | 2.28 | 1.98 | 3.81 | 0.21 | 0.35 | | |
| | total | 6.0 | 7.32 | 4.03 | 6.34 | 6.91 | 6.17 | 4.40 | 2.90 | | |
| Bornova | trout | 14.23 | 3.76 | 8.29 | 10.03 | 0.46 | 5.17 | 2.30 | | | |
| | total | 14.23 | 3.76 | 8.29 | 10.04 | 12.39 | 10.01 | 6.5 | | | |
| Lozoya | trout | 2.66 | 6.16 | 4.06 | 3.26 | 5.46 | 2.19 | 4.05 | 1.85 | 3.01 | 9.94 |
| | total | 2.66 | 6.16 | 4.06 | 3.26 | 5.46 | 2.36 | 29.39 | 28.18 | 30.45 | 55.6 |
| Tajo | trout | 1.13 | 7.01 | 5.32 | 2.95 | 4.07 | 1.12 | 2.64 | 1.35 | | |
| | total | 1.13 | 7.01 | 5.34 | 3.03 | 6.84 | 1.86 | 9.56 | 19.13 | | |
| Gallo | trout | 18.47 | 27.74 | 27.35 | 13.39 | | | | | | |
| | total | 18.47 | 31.47 | 50.33 | 16.54 | | | | | | |
| Eo | trout | 3.55 | 5.59 | 5.70 | 3.21 | 6.81 | 12.16 | | | | |
| | total | 3.61 | 6.19 | 12.81 | 9.99 | 12.23 | 23.66 | | | | |
| Sor | trout | 1.34 | 1.58 | 2.0 | 4.57 | 3.08 | | | | | |
| | total | 1.38 | 1.68 | 2.24 | 4.81 | 3.38 | | | | | |
| Lor | trout | 5.09 | 5.16 | 1.45 | | | | | | | |
| | total | 5.18 | 5.30 | 5.37 | | | | | | | |
| Arnoia | trout | 9.58 | 12.64 | 5.89 | | | | | | | |
| | total | 10.10 | 15.40 | 10.98 | | | | | | | |
| Aragón- | trout | 3.22 | 10.27 | 6.35 | 4.10 | 1.73 | 1.49 | | | | |
| Subordan | total | 3.22 | 10.27 | 10.34 | 12.00 | 11.96 | 13.11 | | | | |
| Veral | trout | 3.32 | 3.24 | 7.54 | 13.84 | 6.40 | 1.33 | | | | |
| | total | 3.32 | 3.24 | 7.54 | 15.14 | 7.19 | 9.63 | | | | |

where $N_i$ (state variable) represents the number of individuals aged from $i-1$ to $i$ years old; $R_{i-1}$ (flow variable) represents the number of individuals recruited from $i-1$ to $i$ year class; $R_i$ (flow variable) represents the number of individuals recruited from $i$ to $i+1$ year class; and $M_i$ (flow variable) represents the number of individuals of the i year class that are lost by natural death.

The year classes that are fished are also subjected to other flow variables ($F_i$) that represent the number of individuals fished from class $i$. Recruitment variables ($R_i$) act instantaneously only once a year, while fishing ($F_i$) acts only during the angling season, but natural mortality acts continuously. The model supposes that mortality rate and fishing pressure rate are constant for all ages.

The equation that defines the evolution of the state variables ($N_i$) through time (evaluated by their derivative with respect to time) depends on entrances by recruitment and on the loss by mortality, fishing and recruitment to the next age class:

$$N_i' = R_{i-1} - M_i - F_i - R_i$$

Finally, the model closes the life cycle with annual recruitment of alevins, evaluated by the Ricker (1954) equation. A simplified Forrester diagram of the SIMTRUT structure is presented in Figure 33.3.

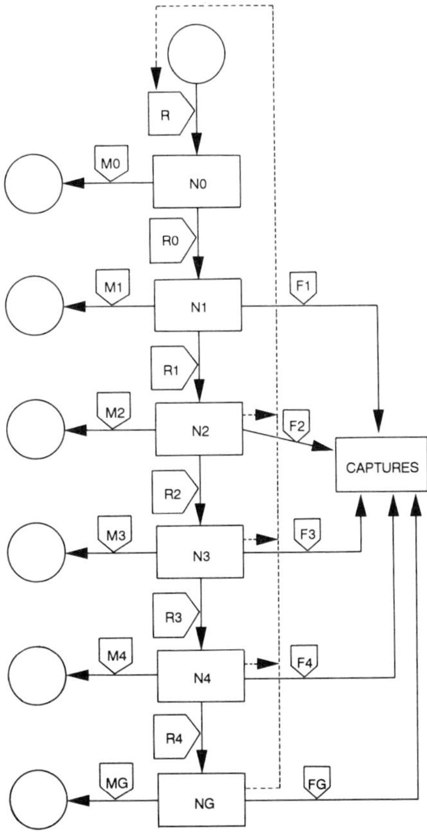


**Fig. 33.3**  Forrester diagram of the dynamic simulation model SIMTRUT. Rectangles represent state variables; ($N_i$) is the number of individuals from age class $i$; pentagons represent flow variables; $M_i$ number of individuals from $i$ class lost by natural mortality; and $F_i$ the number of individuals from class $i$ fished.

**33.4.2**   *Assessment to stock management*

SIMTRUT is a technical instrument that helps fisheries managers make decisions. It is used to improve the understanding of population functions and dynamics, rather than to predict exact evolution of the population. The model is especially useful for analysing the effects of managing the environmental conditions.

To be used for this purposes, SIMTRUT needs input data represented by the parameters that characterize the trout population to be analyzed. To simulate the dynamics of the population, SIMTRUT also needs the initial conditions of the state variables (densities of different age classes), therefore previous quantitative fisheries inventories are required.

Once the stream population has been characterized the model is run under all possible management conditions, in relation to fishing intensity and minimum catchable size. This is done by starting from a set of values and changing them before different SIMTRUT runs. These management parameters are changed by adding (or substracting) finite increments through their reasonable range. For example, minimum catchable size ranges from 16 to 40 cm and the model is run for 16, 18, 20, . . ., 40 cm for each value of fishing intensity.

In each simulation the model is run for a fixed period of time. A long period (30 yr) is selected and used to show the effect once the population has stabilized.

Fisheries managers are interested in satisfying the angling demand by optimizing the production (crop) of fisheries resources, which are always limited and must be preserved. This angling demand is expressed in two dimensions: (a) to catch as many fish as possible, and (b) to catch fish that are as big as possible. Depending on the social type of the angler, priority will be given to one or other dimension. Thus, stock production may be optimized in terms of density (a) or in terms of biomass (b), depending on the objectives of management.

The management conditions that optimize either the number of catches or their mean size are generally different. SIMTRUT has been used to assess this optimization and to determine under which management conditions they are obtained. Data recorded in each simulation are the density of fish angled during the simulation period and its biomass.

The results of a set of simulations corresponding to a population are presented in 3-D graphs (generally a topographic representation is used), where either the density or the biomass captured are plotted against the two management parameters, e.g. the River Jucar (Huelamo) trout stocks (Fig. 33.4). For each minimum catchable size there is a fishing intensity value that maximizes catches. Combining all these maximum points, two maximum curves are defined: one for catches express in density and the other in biomass. Plotting these two curves on the same graph (Fig. 33.5), three sectors on the plane defined by the management conditions are found: (a) sector defined by large minimum catchable size and low fishing intensity conditions; (b) sector defined by small minimum sizes and high fishing intensity; and (c) sector located between both curves.

Captured biomass (g/m²)

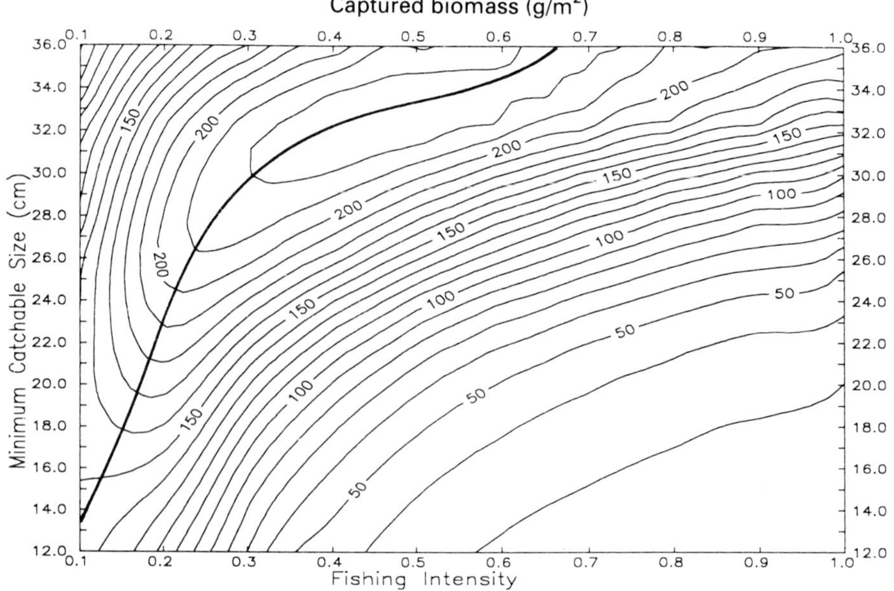

Number of captures

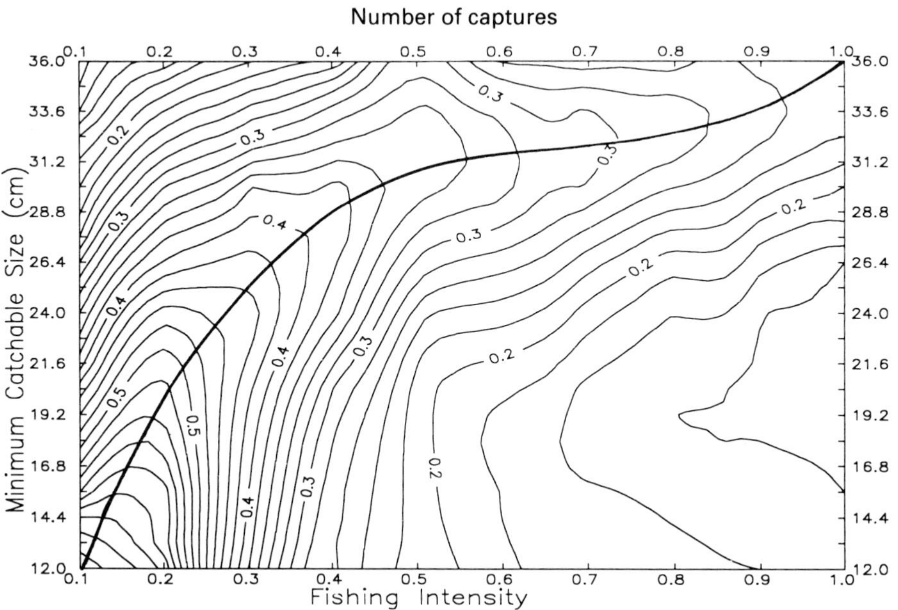

**Fig. 33.4** 30-yr simulations by SIMTRUT with a trout population from River Jucar (Huelamo). Each model has been run under different management conditions (minimum size and fishing intensity). The numbers on the isolines represent accumulated biomass (g m⁻²) and density (fish m⁻²) fished. Maximum capture curves, in terms of biomass and density, are represented.

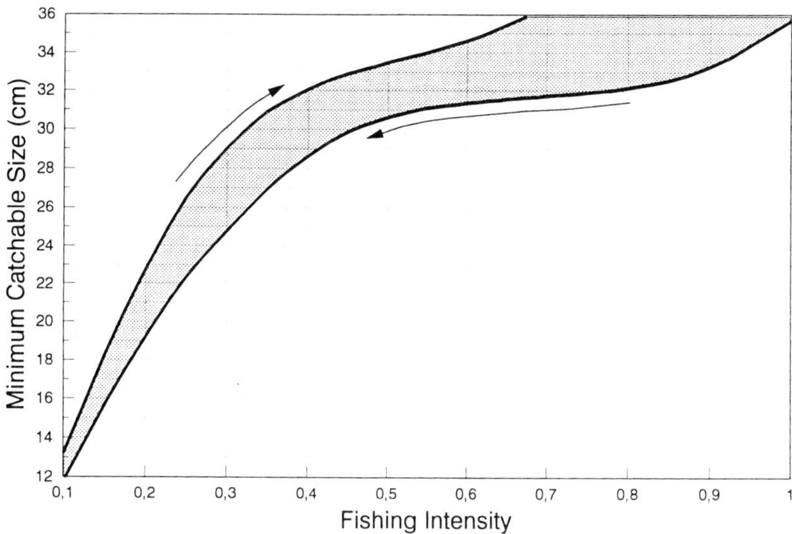


**Fig. 33.5** Stock assessment diagram for River Jucar's trout population. The shaded sector represents the recommended optimum managing conditions. The arrows show increasing values for captured biomass (left-top) and numbers (right-below).

It is in this last sector where fisheries managers must find the optimum conditions for the resource exploitation. Thus, once the real fisheries management are known, recomendations to change them can be included in that sector. Within this sector, managers may select the most suitable condition according to anglers' demands and conservation issues. Large size captures are optimized at 36 cm minimum size and 65% of fishing intensity (see Fig. 33.4), while maximum number of captures are obtained at less than 20 cm and 10% respectively.

## References

Barcelo E. & García de Jalón D. (1986) Edad y crecimiento de la boga de río (*Chondrostoma polylepis* Steind., 1865) en la Cuenca del Duero. *Limnetica* **2**, 235–240.

Casado C., García de Jalón D., Montes C., Barcelo E. & Menes F. (1989). The effect of an irrigation and hydroelectric reservoir on its downstream communities. *Regulated Rivers. Research and Management* **4**, 275–284.

Forrester J.W. (1968) *Principles of Systems*. New York: Wright-Allen Press.

García de Jalón D. (1992) Dinámica de las poblaciones piscícolas en los río de montaña ibéricos. *Ecología* **6**, 281–296.

García de Jalón D., Tolosana, E. & Alcalde F.C. (1987) Estimación de algunos parámetros poblacionales de la trucha común (*Salmo trutta fario* L.) en los ríos pirenaicos. *Boletín Estación Central de Ecología* **29**, 51–58.

García de Jalón D., Mayo M., Hervele F., Barceló F. & Villeta F. (1990) *Pesca fluvial en Galicia. Bases limnológicas para su gestión*. Santiago: Xunta de Galicia, 145 pp.

García de Jalón D., González Tánago M. & Casado C. (1992) Ecology of regulated streams in Spain: An overview. *Limnetica* **8**, 161–166.

González, J.C. & Fernández G. (1986). AMDS: An Expert Aid for System Dynamics Modelling. In:

J.Aracil, J.A. Machuca & M. Karsky, (eds.). *Proceedings International Conference System Dynamics Society*, Sevilla, pp 1019–1028.

Lobón-Cerviá J. & Penczak T. (1984) Fish production in the Jarama River, Central Spain. *Holarctic Ecology* **7**, 128–137.

Lobón-Cerviá J., De Sostoa, A. & Montañés C. (1986) Fish production and its relation with the community structure in an aquifer-fed stream of Old Castile (Spain). *Pol. Arch. Hydrobiol.* **34**, 333–343.

Martínez J. & García de Jalón D. (1988) Estudio de las poblaciones trucheras del río Tormes. *Ecología* **2**, 303–313.

Pianka E.R. (1970) On 'r' and 'K' selection. *American Naturalist* **104**, 592–597.

Ricker W.E. (1954) Stock and recruitment. *J. Fish. Res. Bd Can.* **11**, 559–623.

Stearns S.C. (1976) Life history tactics: a review of the ideas. *Quarterly Reviews in Biology* **51**, 3–47.

Zalewski M. & Naiman R. (1985) The regulation of riverine fish communities by a continuum of abiotic-biotic factors. In: J.S. Alabaster (ed.) *Habitat modification and freshwater fisheries*. London: Butterworths, pp 3–9.

Zaret T.M. & Rand A.S. (1971) Competition in tropical stream fishes: Support for the competitive exclusion principle. *Ecology* **52**, 336–342.

# Chapter 34
# Stock assessment of European eel, *Anguilla anguilla* L.

B. KNIGHTS, E. WHITE and I.A. NAISMITH *Applied Ecology Research Group, University of Westminster, London W1M 8JS, UK.*

**Abstract**   *Anguilla anguilla* L. is an important component of natural freshwater ecosystems as well as being exploited commercially in each of the glass eel/elver, yellow eel and silver eel life stages. Elver recruitment and freshwater stocks throughout Europe have diminished markedly over the last 10–15 years. Similar declines have been observed for the American eel in Canada. Assessment of stocks is difficult because of the multi-stage life cycle and problems inherent in accurately and reliably sampling each stage. Fishery-independent methods (trapping of immigrant elvers/juveniles and emigrant mature silver eels, fyke netting and electric fishing) were compared with fishery-dependent methods (cohort analyses, catch return and commercial sales data). Mark-recapture techniques were also discussed. Limitations of sampling and methodology make accurate and reliable stock assessment difficult in most fresh waters but key studies are discussed and recommendations made for future approaches. Possible causes of declining recruitment of *A. anguilla* to Europe are discussed. Changes in oceanic factors, such as currents and water temperatures, are strongly implicated. Monitoring recruitment and stocks is emphasized in the context of the possible importance of eels as indicators of long-term climate change.

KEYWORDS: *Anguilla*, management, recruitment, stock assessment

## 34.1   Introduction

The European eel, *Anguilla anguilla* L., is noted for its catadromous life cycle, breeding in the Sargasso Sea but spending much of its life feeding and growing in fresh water or estuarine coastal waters (Fig. 34.1). Other species also show long migrations, e.g. the Japanese eel *A. japonica* which breeds in the E. China Sea. High fecundity is necessary in eels to compensate for variable and, potentially, very high natural mortality at all stages of the life cycle.

Mature ('silver') eels from all European waters probably breed indiscriminately in the Sargasso Sea to form a single genetic stock. However, some hybridization with American eels (*A. rostrata*) can occur (Avise *et al.* 1990). Therefore a full stock assessment ideally requires information on the population dynamics and recruitment

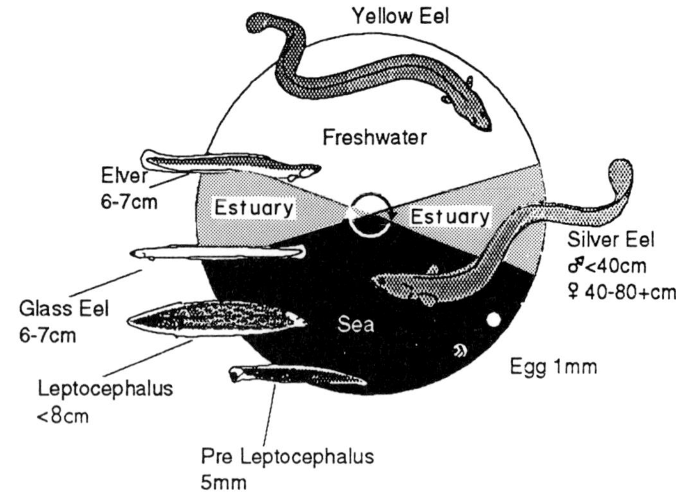


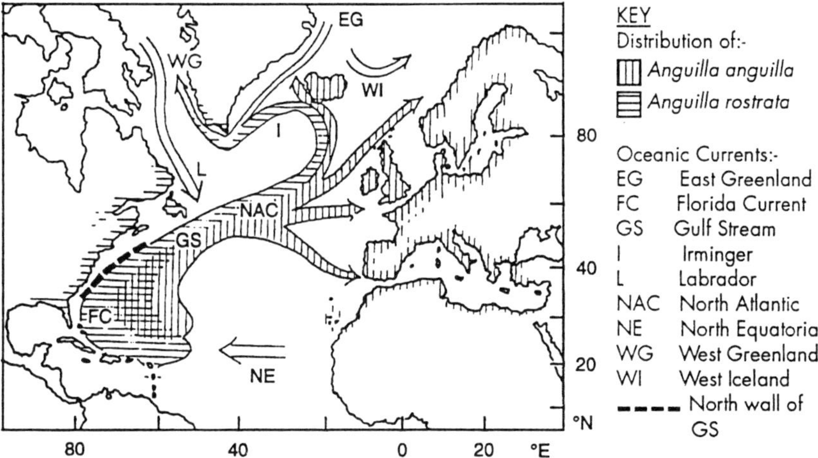


**Fig. 34.1**  Life cycle, oceanic currents, migration pathways and distribution of the European eel (and the American eel).

for each life stage for all European waters, plus the oceanic stages. However, as this review will highlight, assessments are made difficult by the widespread distribution of life stages, migratory mobility and fundamental biology of eels.

Stock assessments are needed because eels form an important component of natural freshwater ecosystems and fish communities. Stocks also support important commercial fisheries throughout Europe. Each continental life stage is exploited to some extent. Glass eels/elvers are caught to supply aquaculture and restocking needs and specialist food markets (especially in Spain). Immature yellow and mature silver eels supply other specialist added-value food markets, e.g. for smoking and jellying. Commercial fishery values are therefore high, estimated at £12.5 m as fresh fish, £40 m after processing (C. Moriarty, personal communication).

The need for accurate information is particularly pressing because there have been widespread falls in recruitment and stocks during the last 15 years (Moriarty 1990a, 1993; Knights & White 1993; White 1994; White & Knights 1994). The European Inland Fisheries Commission (EIFAC) (FAO) Working Party on Eel has voiced its concerns to EIFAC (FAO) and the International Council for Exploration of the Sea, ICES (EIFAC 1993). Fundamental to the recommendations in the 1993 report is the need to measure recruitment and stocks as accurately as possible. It is essential that this is carried out in a co-ordinated way and on a continent-wide basis.

Collection of reliable long-term data sets is also essential because of the potential importance of the European eel as an indicator of trends in climate change. This is because the success of the trans-oceanic migration of leptocephalus larvae and their recruitment to continental waters must be strongly influenced by changes in currents and water temperatures in the N. Atlantic. The eel is also essentially a warm-water species (Tesch 1977; Deelder 1984), hence growth and survival in fresh water and eventual recruitment to the breeding stock are also strongly dependent on climate.

This chapter critically reviews stock sampling and assessment methods currently used and their inherent problems. Particular reference will be made to recent work on the Rivers Severn and Avon in England and other intensive/extensive detailed studies. Recommendations for future studies are made.

## 34.2    Fishery-independent methods

### 34.2.1    *Studies of breeding and leptocephalus stages*

Problems inherent in sampling fish and planktonic larvae at relatively low densities in the Sargasso Sea and North Atlantic have made it impossible to gain any detailed and meaningful quantitative data on breeding and fecundity. Similarly, information currently available on the trans-oceanic migration of European and American leptocephali is scanty and often equivocal (e.g. see Fig. 34.1 and review by McCleave 1993). Some correlations were found between relative index values for immigrants sampled in the Bay of Biscay and the commercial glass eel catches in some French estuaries between 1977 and 1988 (Moriarty 1990a).

### 34.2.2    *Netting of immigrant eels*

Having migrated across the Atlantic via ocean currents, leptocephali metamorphose on the continental shelf into transparent glass eels (Fig. 34.1). These enter estuaries, often in large runs, moving upriver in the spring. They mainly utilize selective tidal stream transport, hiding in the substratum at other times (McCleave & Kleckner 1982). Glass eels have limited energy reserves and commence feeding before metamorphosing, developing pigmentation and becoming elvers. These will be designated as 0 + eels in this review to indicate that they are in their first year of non-oceanic life. Some elvers will remain in coastal or estuarine waters, feeding and growing until maturing ('silvering') and returning across the Atlantic Ocean to breed. Others will migrate upstream, directly or after one or more years, as pigmented juveniles (immature migratory 'yellow' eels). Migrations deeper into catchments can continue in successive years until eels reach sizes and ages as large as 40–45 cm and 10 + years, respectively (Moriarty 1990b). It has been suggested that increasing density and competition may promote migration, low densities having the reverse effect (Knights 1987; White 1994). A lack of elver and juvenile recruitment is reflected by low stock densities deeper in the catchments and an increasing preponderance of older and female eels (Aprahamian 1988; Naismith & Knights 1993). Such females, when they eventually mature and migrate back to the Sargasso (usually when > 45–50 cm), form an important component of the breeding stock.

Glass eels can be sampled by netting in estuaries and tidal waters from boats, but this is only effective when they congregate in shoals in mid-water as the tide is rising. Good comparative sets of data also exist for samples hand-netted at sluices at Den Oever, Netherlands (Dekker 1986: Moriarty 1993). Normally, however, catches are very variable and effort and costs are high for specific survey purposes (White 1994; White & Knights 1994). Therefore suitably cost-effective data can only usually be gained from commercial fishery catches, as discussed later in the section on fishery-dependent methods. Particularly detailed studies, involving mark-recapture, have been carried out on a Taiwanese net fishery for elvers of Japanese eel, *A. japonica* (Section 34.3.3).

### 34.2.3    *Trapping of immigrant eels*

Various forms of baited and un-baited barrel, wire cone and box traps have been used for catching eels (Tesch 1977; Harrison 1988). These have been found to be very inefficient for quantitative sampling of large yellow eels and very small juveniles (Naismith & Knights 1990a).

Other forms of traps can be combined with passes on weirs, dams, sluices and other barriers in estuaries and rivers to capture migrant eels (White & Knights 1994; Knights & White 1995). The anguilliform body pattern is well adapted for swimming (especially in boundary layers) and for climbing over obstacles via rough surfaces and/or by entwining in vegetation or fibrous materials such as straw 'ropes' or synthetic geotextile mats. Immigrant eels also show strong rheotactic behaviour during migration. Thus, they can be easily attracted by suitable flows of water to the base of a channel or pipe which can be provided with a 'climbing' medium to help them ascend quite considerable heights and distances. Such systems have been used throughout Europe and in other continents to form passes over barriers, with permanent traps added in some instances to capture eels for stocking purposes (Rigaud *et al.* 1988; White 1994; White & Knights 1994). Laboratory and field studies have shown that simple, but robust, experimental pass-traps can be made from plastic guttering channels, provided with commonly available garden netting of 20 mm square mesh (Fig. 34.2; Naismith & Knights 1988; White 1994; White & Knights 1994). This climbing material was found in the latter study to be less size-selective than geotextile matting which has been recommended by Dahl (1991) for eel passes in Denmark.

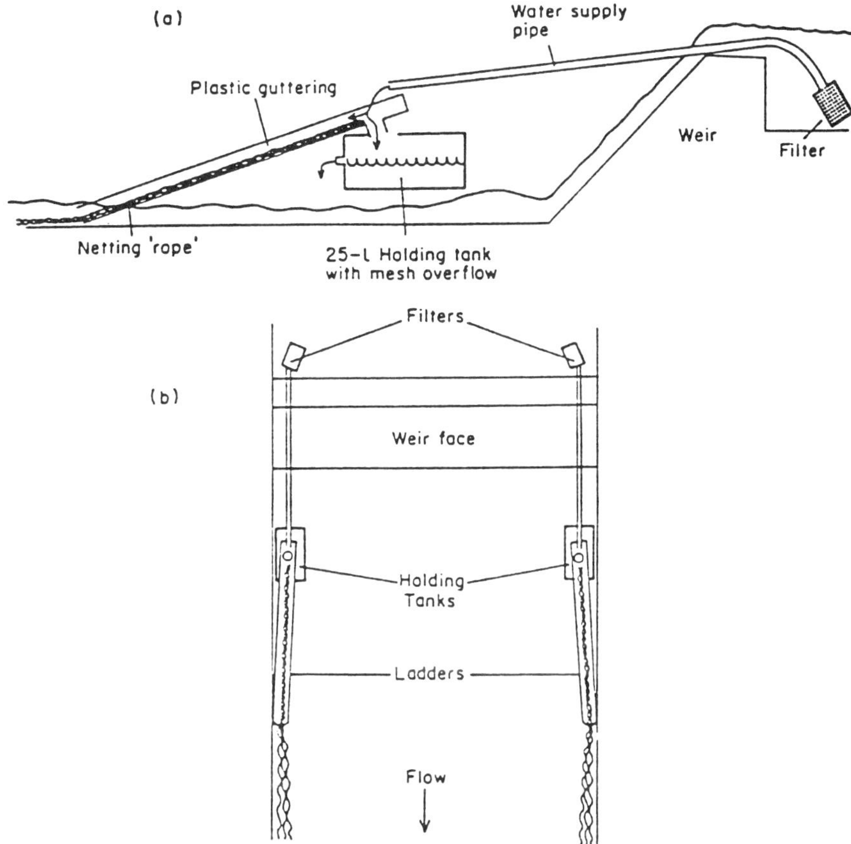


**Fig. 34.2**  Design of simple experimental traps to catch migrating elvers and juvenile eels (a) in section, (b) in plan (see text for further detail).

It is impossible, however, to determine whether traps actually catch every migrant. In a three-year study in the Rivers Severn and Avon in England, 6418 out of > 346 000 eels trapped were marked and released (White 1994; White & Knights 1994). Only about 2% were recaptured but of these recaptures, only five had by-passed traps on barriers between the points of release and recapture. It therefore appears that such traps are effective in sampling migrants but that migratory tendencies are very variable. Similar recapture rates have been found in studies in the Thames, England (Naismith & Knights 1988) and Meuse, Belgium (Baras, Philippart & Salmon, Chapter 7). It appears therefore that such traps can only give a general index of abundance of migrants and recruitment at each trapping site. Studies based on such trap sampling show decreases in numbers of migrants but an increase in average size and age of eel with distance upstream and the number of barriers to be surmounted (Moriarty 1986, 1990b; Naismith & Knights 1988; White 1994; White & Knights 1994). Numbers of barriers and their surmountability had greater influences than distance alone according to the latter study. This indicates that provision of passes (and/or stocking) would be an effective long-term means of enhancing migration and stocks in river systems with many barriers. For example, in the Severn and Avon, passage for over 346 000 migrants was provided by the experimental pass-traps on 15 barriers during 1991–93. Passes are in any case essential if there is a major obstacle, such as a high dam or barrage (White & Knights 1994), particularly in estuaries or lower river stretches where the numbers and densities of migrants are highest.

Trapping of glass eels and elvers can be particularly effective in estuaries or rivers where traps can be placed so as to catch a very large proportion of runs. This has allowed accurate enumeration of migrations into the Imsa River in Norway (Vøllestad & Jonsson 1988). Such a system also forms the basis of the 'ranching' eel fishery based on Lough Neagh in Northern Ireland, elvers caught in the River Bann being used to stock and maintain the fishery (see Section 34.3.4).

### 34.2.4    *Sampling of yellow eels*

Ideally, assessments of immigrant recruitment need to be related to stock levels eventually achieved in upstream fresh waters. Sampling problems again make this very difficult. For example, seine netting can be used in fairly shallow waters with smooth bottoms but this was not found to be an efficient means of quantitative sampling (Naismith & Knights 1990a). Eels can probably escape easily under the net and smaller individuals can pass through meshes. Long-lining is too labour intensive for experimental studies, but catch data from commercial long-line fisheries have been used in some studies, e.g. of Lough Neagh, N. Ireland (Moriarty 1990b) and Thau lagoon, S. France (Bach *et al.* 1992).

Other methods available are electric fishing and fyke netting. These were compared, in combination with mark-recapture, by Naismith & Knights (1990a) in an artificially-stocked pond mesocosm study and in a river. Fyke netting was also studied in a river, in the Thames Estuary and in a 1.4 ha lake. In the densely-stocked rectangular study pond with a flat hard gravel bottom, electric fishing and fyke net sampling efficiencies were low. Problems probably arose because electrically-stunned eels did not tend to float, making them more difficult to see and net, especially in deeper and more turbid waters. In the wild, they commonly burrow into the substratum, or hide in crevices, and are less prone to capture. Fyke nets are passive-capture devices and catch sizes are generally dependent on stock density, fishing effort and eel activity levels. The latter are in turn related to water temperature and lunar rhythms (e.g. see Bach *et al.* 1992).

These problems were reflected in the mesocosm study; recapture rates were low and tended to vary between successive samplings. This made application of a range of standard catch-depletion estimations of population numbers unreliable. Recapture rates were also low in natural open waters ( < 2%), probably because of mobility of populations. By contrast, Moriarty (1986) concluded that recapture rates of 5.5–18.5% could be expected in non-migratory populations, especially in smaller and more restricted water bodies where population densities are relatively high. This would correlate with mark-recapture studies in the Thames Estuary which showed only a 0.2% recapture from 999 releases in the more open middle reaches, compared to 7.4% from 443 in the narrower upper reaches (Naismith & Knights 1990b).

Sampling methods also have to be chosen in relation to the basic characteristics of the types of waters being studied. Thus Naismith & Knights (1990a) noted that electric fishing was only effective in fairly shallow and transparent fresh waters. Fyke nets in such situations are prone to theft and vandalism. They can be used in all other types of water, except in swift currents where they are prone to clogging by debris and to being washed away.

Another sampling problem is that both electric fishing and fyke netting tend to be size-selective for larger eels (Berg 1989a; Naismith & Knights 1990a, b). Smaller eels are more difficult to see and net. Typical commercial fyke nets, such as those used in the Thames estuary, use 19 mm knotted or 12 mm knotless stretched mesh to catch eels > 33 cm (i.e. > 100 g). Berg (1989a) and Naismith & Knights (1990a) found that fykes often caught some eels that were below the mode of body lengths of the total catch. These represented variable sampling of eels which were small enough to normally escape through the meshes. Both studies therefore recommend that only data for eels above the modal body length of the catch should be used in any quantitative studies. In attempts to sample small eels, small mesh nets were tested by Naismith & Knights (1990a) but these were avoided by eels of all sizes.

The conclusion from the exhaustive studies of these authors was that electric fishing catch-depletion methods for stock assessment were unreliable. Mark-recapture studies with fyke nets can, however, yield reasonably accurate cumulative estimates of population size, but only in small closed waters one or, preferably, three to six sampling days after release and providing 15–20% of the population can be marked and 20–30% recaptured. The studies indicate that such capture-recapture rates can only be achieved in relatively small and enclosed waters. In more open waters, catch estimations are affected to a greater extent by immigration and emigration. The area sampled by nets is unknown and therefore data can only be expressed as catch per unit effort (CPUE) to give a relative measure of population density. Electric fishing data can be expressed as number $m^{-2}$ but again can only be used as a relative measure of population density. Despite these limitations, such methods do reveal important relative differences in stock densities and population structures in space and time (Aprahamian 1986; Moriarty 1990c; Moriarty *et al.* 1990; Naismith & Knights 1993).

**34.2.5**    *Trapping of emigrant eels*

For many centuries, silver eels migrating back to sea have been caught for eating using various forms of bar-grid or conical basket or netting traps, especially at water mills (Tesch 1977; Naismith 1992). Indeed, Naismith & Knights (1993) found records of such traps in the Thames catchment dating back to 1086. These records indicate that catches were very large, implying stock levels were historically very much higher than at present.

Silver eel migrations tend to occur during spates and many silver eels probably avoid capture. However, some systems are probably relatively efficient. For example, at several sites on the River Bann, a series of large conical nets can be lowered into the water from gantry systems that span the river (Moriarty 1990b, c). In the River Imsa, a Wolf-trap has been used (Vøllestad & Jonsson 1988). Because both immigrant elvers and emigrant silver eels can be relatively efficiently sampled, the Bann and Imsa have provided some of the most complete data sets for studies of eel population sizes and dynamics in river catchments.

Attempts have been made to track eels as they migrate back into the sea, but cost and technical difficulties have limited studies to relatively shallow waters such as the Baltic (e.g. Tesch 1993).

**34.2.6**    *Marking eels*

Mark-recapture is an important component of many fishery independent and dependent methods because it allows estimation of population size, growth rates and mortality. Implanted tag methods have drawbacks when used on eels. Jaw tags were retained for as long as 10 years (Berg 1986) but Floy tags were frequently lost, being pulled out when eels burrow in the substratum (Naismith 1992). Berg (1986) also found that eels were stressed by tagging, growth rates being depressed by about 50%.

Larger eels can be marked by Panjet injection of Alcian blue dye on the belly. Small eels are, however, easily damaged by the force of the jet. Freeze-branding and marking with silver nitrate pencil tend to cause open wounds because of the lack of large protective scales in eels (Naismith 1992).

Various vital dyes can be used to stain the skin of unpigmented glass eels, including ones visible under UV light, but such stains tend to be short-lived (Tzeng 1984a; White 1994; White & Knights 1994). Although the method is time-consuming, sub-epidermal injections of acrylic paint cause minimal stress and produce long-lasting marks (Naismith 1992; White 1994; White & Knights 1994). A further advantage of this method is that different colours, shapes and body positioning of marks can be produced to extend the range of information for batch or unique individual marking. Mark-recapture studies are, however, generally limited by poor recapture rates.

**34.3    Fishery-dependent methods**

**34.3.1    *Cohort analysis***

Cohort analysis involves separating and quantifying fish populations into year-classes on the basis of body length and/or direct ageing. This allows the construction of life tables and hence assessment of changes in relative numbers of fish of different ages. Estimates can then be made about fecundity, recruitment of year classes, the effects of natural (and, where relevant, fishing) mortality and the possible roles of density-dependent factors on mortality.

Application of standard fisheries cohort analysis to eel stocks is made difficult because of the sampling problems discussed. Growth rates, and hence length- and weight-for-age relationships, are also generally very variable (Deelder 1984). Even in the early stages of freshwater life it is often difficult to predict age from size. For example, White & Knights (1994) identified small migrants trapped up to 30.5 km from the tidal head in the River Severn as at least 1 + eels on the basis of size and pigmentation. On otolith ageing, some, however, proved to be 0 + elvers.

Eels do not possess easily-removed scales for annual ring counting and therefore have to be killed for otolith ageing. Burning-and-cracking methods have been recommended. However, rings are often not easy to distinguish accurately (Moriarty 1973; Aprahamian 1987; Naismith 1992; White 1994; White & Knights 1994). Scanning electron microscopical methods have been developed to count daily rings (Lecomte-Finiger 1992). These suggest that leptocephali can reach continental Europe in less than one year, as compared to earlier cohort analyses of leptocephali caught in the Atlantic which suggested transit times of up to three years (McCleave 1993).

A further complication is that age and size tend to vary between the stocks in different parts of catchments. Stocks in productive and warm brackish water estuaries and still waters tend to show higher densities, and to be dominated by faster growing eels that mature and emigrate as males. The sexes do not have any external distinguishing characteristics, except that males usually mature and emigrate at a body length < 40 cm. These stock characteristics tend to be reversed in colder and less productive head waters. Local migrations and longer-distance immigration and emigration within and between river stretches and estuarine zones also confuse the issue. Yellow eels marked in fresh water have been recaptured in a trawl some 20 + km away in coastal waters (C. Morrice, personal communication).

Despite these problems, some attempts have been made to use length-cohort analysis for eel stock assessment. For example, Naismith & Knights (1990b) developed a model which accounted for the episodic nature of the maturation and emigration processes of the sexes. This was used to assess stocks and the effects of exploitation in the Thames Estuary. Eels ( > 8500) were fyke netted in different parts of the estuary and divided into sex and length-frequency/age classes. These classifications were based mainly on measurements of body lengths, correlated with ageing and sexing of selected samples. As discussed earlier, fyke nets are size-selective, therefore only eels above the modal body length of the catches were considered. Instantaneous mortality coefficients ($Z$) were calculated for 3 cm length-classes for (a) immature recruits (30–32 cm), (b) maturing males (33–35 cm), (c) male emigrants (36–41 cm), (d) immature females (33–35 cm), (e) immature females (36–59 cm) and (f) female emigrants ( > 60 cm). It was thus possible to compare mortality coefficients ($Z$ values) between components of stocks in different parts of the estuary and for different periods during the fishing season. It was concluded that differences in Z between the unfished inner estuary stock and fished middle and outer estuary ones were too small and variable to indicate a very high fishing mortality, and hence exploitation rate. They also calculated a mean Z value of 0.37 per 3 cm length-class from the 7 cm elver stage onwards. This was similar to the range of mean values of 0.17–0.65 calculated across wider ranges of age and size in other European stocks (Rasmussen & Therkildsen 1979; Vøllestad & Jonsson 1988; Berg 1989b). These values can, however, vary considerably between year classes, depending on initial recruitment and subsequent density-dependent mortality effects (Vøllestad & Jonsson 1988).

Derivations of mortality coefficients from length-class frequency data also allow estimates to be made of the recruitment needed to produce fishable stocks in study waters. For example, Naismith & Knights (1990b) calculated from the mean Z value that approximately 80 glass eels would be needed to produce one fishable 41–43 cm (100 g) yellow eel (i.e. about 8 million to support the annual yield of the Thames Estuary fishery of about 10 t). This compares to figures in the literature of between 0.5 and 45 elvers (or 0.2 and 0.5 juveniles) per 100 g yield when stocking directly into fresh waters, the lower rates applying to more productive and warmer waters and low rates of dispersion from the stocking site (Knights & Higgs 1988). The figure of 80 elvers for the Thames estuary appears very high but density-dependent mortality is probably high. The Z calculations also cannot adequately account for losses due to immigration and emigration.

Moriarty (1990b, c) quotes approximately five elvers need to be stocked to produce one catchable eel in Lough Neagh. This contrasts with about 13 elvers per catchable eel in various Polish lakes (Moriarty *et al.* 1990). The difference between these two studies was probably mainly due to higher competition with cyprinids in the Polish Lakes.

Vøllestad & Jonsson (1988) calculated that a mean of 0.9 recruits was required for 100 g yield in the Imsa river (SW Norway) between 1975 and 1987. They were able to study both ascending elvers and descending silver eels caught in traps at the river mouth. Recruitment appeared to be very successful and mortality low, considering that elvers had to actively migrate upriver and spend an average of 8 years before silvering and emigrating. However, density in the river was very low (some 10–20 times lower than in other European waters) and density-dependent mortality was only important in years of good glass eel recruitment.

### 34.3.2  *Anglers' catch data*

Eels are not favoured prey for anglers, although they may be of some value in enhancing total catch weights in fishing matches in sites such as estuaries. Accurate catch data are rarely recorded and/or easily obtained, and are generally not extensive enough for quantitative stock assessment. They can, however, be of some limited use in relating catchability to water temperature and in revealing trends in relative catches over a number of years. For example, North & Hickley (1989) analysed catch data for the Severn between 1975 and 1986 and suggested a decrease in relative abundance of eels with distance upstream and a decline in catches between 1983 and 1986. These results would accord with the declines in Severn stocks discussed elsewhere in this review.

### 34.3.3  *Commercial catch data*

Eel fisheries tend to be fragmented and often secretive operations, without any centralized marketing systems where data on sales are collated. In some cases, data have been gained from fishermen by building up relationships and trust, even if these sometimes have to be expressed in relative catch terms rather than absolute figures for the whole stock. In some commercial eel fisheries, a regulatory agency may require fishermen to be licensed and/or to provide catch returns. For example, in England, fyke nets and pair-trawls used in the Thames estuary yellow eel fishery are licensed by the National Rivers Authority (NRA), Thames Region, and a condition of the licence is that annual catch returns have to be made. Similarly, elver nets used in the Severn Estuary are subject to a licensing and catch return system administered by the NRA, Severn-Trent Region. The catch return data are, however, often inaccurate because of licence evasion, poor record-keeping and/or mis-reporting by fishermen. For example, from 508 returns in 1992 and 545 returns in 1993 for the Severn elver fishery, yields were estimated at $4.3 \pm 11.2$ t and $5.2 \pm 13.3$ t respectively whereas values of 17.7 and 21 t were derived from export data (White 1994; White & Knights 1994). Direct questioning of large samples of fishermen also produced poor estimates. The catch return data appeared to be of such poor quality that it is questionable whether their collection was worth the financial and time costs involved.

Catch data appear more reliable in some other European fisheries, e.g. in those French ones based on the Loire and Vilaine estuaries (Guerault & Desaunay 1989). Generally the most reliable sets result from situations where researchers have built up relationships with fishermen or where data are collected in the course of research, e.g. as in the Imsa trap studies of Vollestad & Jonsson (1988). The best sets tend to be for glass eels/elver fisheries, which provide information about potential recruitment to catchments (e.g. see Table 34.1). They are also useful in monitoring relative changes between years (Section 34.4).

One of the most intensive studies involved a Japanese glass eel/elver fishery in Taiwan (Tzeng 1984a, b, 1985, 1989). A vital skin dye was used to mark 4225 glass

**Table 34.1**  Mean annual catches of 0+ eels from various sites, calculated for each five-year period between 1978 and 1992 (by weight per year or, for Den Oever, by index to zero mean (1993–1985) and for Bay of Biscay, larval catch per hour and for Moses-Saunders Dam, thousands of eels). Also shown is the index for the position of the north wall of the Gulf Stream (larger values signify a more northerly position). See text for further discussion and sources.

|  | 1978–82 | 1983–87 | 1988–92 |
| --- | --- | --- | --- |
| Viskan (Sweden) (t) | 401 | 265 | 47 |
| Vidaa (Denmark) (kg) | 298 | 252 | 31 |
| Yser (Belgium) (kg) | 327 | 18 | 62 |
| Bann (N. Ireland) (t) | 5 | 1.7 | 2.4 |
| Shannon (Ireland) (t) | 4 | 0.9 | 0.2 |
| Den Oever (Netherlands) (Index) | 0 | −0.54 | −0.58 |
| Severn (England) (t) | 34 | 17 | 16 |
| Loire (France) (t) | 437 | 167 | 67 |
| Minho (Portugal) (t) | 29 | 20 | 8 |
| Moses-Saunders Dam (Canada) (N × 1000) | 890 | 710 | 120 |
| Position of north wall of Gulf Stream (relative index) | −0.63 | 0.31 | 0.69 |

eels which were then released into coastal waters off the Shuang-Chi River in northern Taiwan. Recapture rates achieved by 70 commercial fishermen averaged 30.0% in coastal waters, 9.5% in the river mouth and 7.6% in the inner river, i.e. a total of over 47% (Tzeng 1984a). The exploitation rate ranged between 44.1 and 75.4% for seven waves of immigrants. These were exceptionally high exploitation rates compared to those in most of the large and open European estuarine and river fisheries. The majority of the Taiwanese eels were caught by hand-trawl as they congregated in discrete shoals in coastal waters and very few were caught more than 5 km upstream from the mouth of the estuary. The fishing effort was also probably exceptionally intense, with 70 men exploiting a population of glass eels estimated to only number about 65 400 individuals (less than 200 kg) (Tzeng 1984b). The commercial catch was 38 661 individuals (i.e. an exploitation rate of about 60%). These elver runs appear very low compared with those for rivers like the Severn and Loire which are measured in tonnes (Table 34.1).

### 34.3.4    *Commercial sales data*

It is often difficult to collect commercial sales data on eels for use in stock assessment because of the lack of centralized markets and the commercial secretiveness of major buyers and dealers. However, some sources are of use. For example, the majority of elvers caught in the Severn area in England are exported to Europe and other countries for restocking natural waters and for aquaculture. Export statistics are collected by H.M. Customs & Excise Tariffs & Statistics Office and these were obtained from agents in the study by White & Knights (1994). The export data category 'live eels' was examined. This does not distinguish elvers from yellow or silver eels. However, elver exports by month, port/airport of export and country of destination were generally distinguishable by their price per kilogram (up to 20–25 times higher than those for yellow and silver eels). Thus it was possible to estimate exploitation data for comparison with catch returns (discussed above) and for comparisons with catches since 1979 in other European fisheries (Table 34.1 and Fig. 34.3). It is possible similar sources of information might be available for other eel fisheries in Europe.

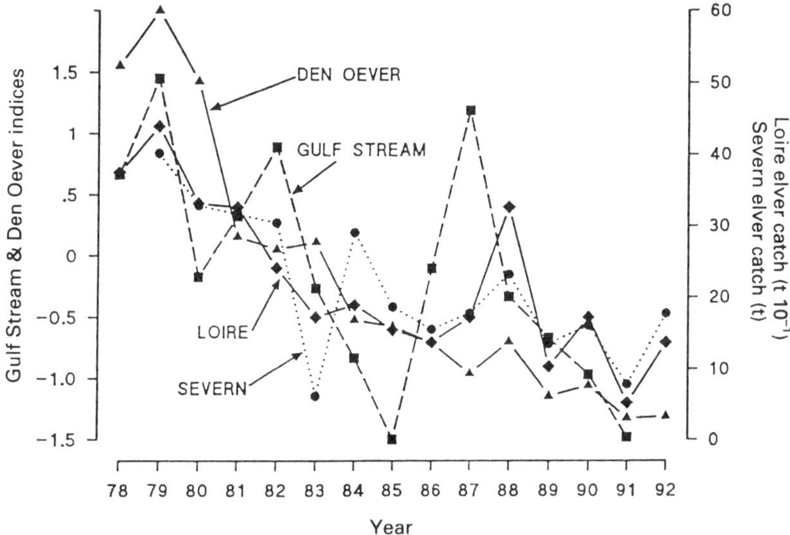



**Fig. 34.3** Severn, Loire and Den Oever annual elver catch estimates for 1978–1992 compared to the index for the position of the north wall of the Gulf Stream. The north wall index is plotted such that a downward trend relates to a more northerly position. See text for further discussion and sources.

Another example of a commercial fishery for which useful data can be derived is that based on the River Bann/Lough Neagh in Northern Ireland. This is an important example of a managed fishery, run as a co-operative (Moriarty 1990b, c). Glass eels and elvers are trapped at a weir near the river mouth and transported overland for stocking into the Lough. Baited hooks are the only form of fishing allowed, minimum size limits are imposed and all catches are brought to one central station before export to Europe. Migrant silver eels can also be trapped by large nets lowered from gantries into the Bann. Free access to full catch information is limited because of commercial sensitivity but some data can be gained from annual accounts and reports of the co-operative. Studies based on these data have been useful in assessing stock dynamics. For example, Moriarty (1990b, c) quoted a productivity of about 20 kg ha$^{-1}$ and a yield of approximately 49 g of eel per stocked recruit. The annual catch was in the region of 850 t, involving stocking of some 17–18 million glass eels (Moriarty 1990b).

### 34.4    Recent declines in recruitment of European eels

Comparisons of the most reliable data sets for different glass eel/elver fisheries reveal a marked decline in recruitment throughout Europe since the late-1970s, corresponding with worrying falls in riverine stocks (Knights & White 1993; Moriarty 1993; White 1994; White & Knights 1994). Trapping studies on the Moses-Saunders Dam on the St Lawrence River in Canada showed (Table 34.1) that the relative catches of the American eel, *A. rostrata*, have also fallen (Castonguay *et al.* 1994). Year-on-year changes in recruitment tend to be closely related to one another and also to the position of the north wall of the Gulf Stream (Taylor *et al.* 1992) (Fig. 34.3). This implies that changes in currents, temperatures and other oceanic events have been affecting the migration of leptocephalus larvae.

Falling recruitment could be due, amongst other factors such as pollution and migration barriers, to overfishing of elvers (EIFAC 1993). However, Tzeng (1989) found that catch data for *A. japonica* varied considerably over a number of years independently of very high and increasing levels of exploitation.

### 34.5    Conclusions

The assessment of inland stocks of eels is particularly difficult because of their catadromous life cycle and inherent biological characteristics. Having entered fresh water, some individuals show high mobility and continuing migration. Eels also tend to be nocturnal and can burrow in the substratum. Despite these inherent problems, it is important that eel stocks are assessed and monitored as fully and as accurately as possible. They are important components of aquatic ecosystems and fish communities and are of greater commercial importance than is generally recognized. The best information possible is needed to manage stocks and ensure sustainable yields are achieved.

Finally, the studies discussed above show conclusively that recruitment and stocks of European (and Canadian) eels have fallen since the late-1970s. Thus monitoring must continue and the causes of these falls must be determined. Because eels from all European waters probably form a single breeding stock, this must involve international co-operation. It appears likely that changes in oceanic systems have been having a major effect on the transatlantic migration of leptocephali. Such changes may be indicative of anthropogenic global warming or due to natural climatic events. Whatever the causes, it is important that continuing efforts be made to assess stocks of European eels (and other species worldwide) because these could yield important information about long-term trends in climate change and their possible causes.

### 34.6    Recommendations for future studies

The following recommendations can be made regarding stock assessment of the European eel.

(1) A co-ordinated European effort is needed to compile a report on the present and possible future status of eel stocks using available data, especially those held in unpublished sources (e.g. for general fisheries surveys).

(2) Integrated co-operative research programmes and monitoring of stocks and recruitment are needed. Despite the problems inherent in stock assessment of eels, general broad scale studies comparing relative densities and population dynamics are useful. More detailed studies should be focused on stocks for which there are good data already available, e.g. as in the River Bann/Lough Neagh system and the rivers Imsa, Thames and Severn/Avon.

(3) Information on stocks and recruitment needs to be correlated with the factors possibly causing the declines, e.g. changes in oceanic factors, overfishing, migration barriers, pollution and parasites/disease.


**Acknowledgements**

The authors would like to thank the National Rivers Authority (NRA) for permission to utilize data from the R&D contract on the Rivers Severn and Avon granted to the Applied Ecology Research Group at the University of Westminster. Thanks are due particularly to Alan Churchward of NRA Severn-Trent. The authors would also like to express their gratitude to NRA fisheries and other staff, the former Thames Water Authority, members of the EIFAC Working Party on Eel and other eel workers for their support.


**References**

Aprahamian M.W. (1986) Eel (*Anguilla anguilla* L.) production in the River Severn, England. *Polskie Archiwum Hydrobiologii* **33**, 373–389.

Aprahamian M.W. (1987) Use of the burning technique for age determination in eel (*Anguilla anguilla* L.) derived from the stocking of elvers. *Fisheries Research* **6**, 93–96.

Aprahamian M.W. (1988) Age structure of eel, *Anguilla anguilla* L., populations in the River Severn, England, and the River Dee, Wales. *Aquaculture and Fisheries Management* **19**, 365–376.

Avise J.C., Williams S.N., Arnold J., Koehn R.K., Williams G.C. & Thorsteinsson V. (1990) The evolutionary genetic status of Icelandic eels. *Evolution* **44**, 1254–1262.

Bach P., Legendre P., Amanieu M. & Lassere G. (1992) Strategy of eel (*Anguilla anguilla* L.) in the Thau lagoon. *Estuarine, Coastal and Shelf Science* **35**, 55–73.

Berg R. (1986) Field studies on eel *Anguilla anguilla* L. in Lake Constance: tagging effects causing retardation of growth. *Vie et Milieu* **36**, 285–286.

Berg R. (1989a) The assessment of size-class proportions and fisheries mortality of eels using various catching equipment. EIFAC Working Party on Eel, Porto, 1989. Rome: EIFAC, FAO, 15 pp.

Berg R. (1989b) The growth of eels: A critical assessment of data from open waters. EIFAC Working Party on Eel, Porto, 1989. Rome: EIFAC, FAO, 10 pp.

Castonguay M., Hodson P.V., Moriarty C., Drinkwater K.F. & Jessop J. (1994) Is there a role of ocean environment in American and European eel decline? *Fisheries Oceanography* **3**, 197–203.

Dahl J. (1991) Eel passes in Denmark; why and how. EIFAC Working Party on Eel, Dublin, 1991. Rome: EIFAC, FAO, 20 pp.

Deelder C.L. (1984) Synopsis of biological data on the eel. *FAO Fisheries Synopsis*, **No. 80**. Rome: FAO, 87 pp.

Dekker W. (1986) Regional variations in glass eel catches. An evaluation of multiple sample sites. *Vie et*

*Milieu* **36**, 251–254.

EIFAC (1993) Report of the Eighth Session of the Working Party on Eel, Olsztyn, Poland, 24–29 May 1993. Report to the International Council for Exploration of the Sea. Rome: EIFAC, FAO, 25 pp.

Guerault D. & Desaunay Y. (1989) Evolution de l'abondance de la civelle (*Anguilla anguilla*) dans les estuares de la Loire et de la Vilaine. EIFAC Working Party on Eel, Porto, 1989. Rome: EIFAC, FAO, 15 pp.

Harrison B. (1988) Commercial eel capture. Institute of Fisheries Management, Management Advisory Booklet AB006. IFM, West Bridgford NG2 7LF, England.

Knights B. (1987) Agonistic behaviour and growth in the European eel, *Anguilla anguilla* L., in relation to warm-water aquaculture. *Journal of Fish Biology* **31**, 263–276.

Knights B. & White E. (1993) Anguillaphiles meet in Poland – and ask 'Is the European eel an endangered species?' *Fish* **32**, 15–16.

Knights, B. & White, E. (1995) A review of eel passes. In: *Fish passes in barrage and weir design*. HR Wallingford, Wallingford OX10 8BA, England.

Knights B. & Higgs J. (1988) The potential for exploitation of eels in Lee Valley Regional Park Waters. Report to Lee Valley Park Regional Authority. Stanstead Abbots, Essex, England: LVRPA, 27 pp.

Lecomte-Finiger R. (1992) Growth history and age at recruitment of European glass eels (*Anguilla anguilla*) as revealed by otolith microstructure. *Marine Biology* **114**, 205–210.

McCleave J.D. (1993) Physical and behavioural controls on the oceanic distribution and migration of leptocephali. *Journal of Fish Biology* **40 (Supplement A)**, 243–273.

McCleave J.D. & Kleckner R.C. (1982) Selective tidal stream transport in the estuarine migration of glass eels of the American eel (*Anguilla rostrata*). *Journal du Conseil permanent International pour l'Exploration de la Mer* **40**, 267–271.

Moriarty C. (1973) A technique for examining eel otoliths. *Journal of Fish Biology* **5**, 183–184.

Moriarty C. (1986) Riverine migration of young eels (*Anguilla anguilla* L.). *Fisheries Research* **4**, 43–58.

Moriarty C. (1990a) European catches of elver 1928–1988. *International Revue gestalt Hydrobiologie* **75**, 701–706.

Moriarty C. (1990b) Eel stocks, capture and culture. In: K.T. O'Grady, A.J.B. Butterworth, P.B. Spillett and J.C.J. Domaniewski (eds) *Fisheries in the year 2000*. Proceedings of the 21st Anniversary Conference of the Institute of Fisheries Management Institute of Fisheries Management, West Bridgford NG2 7LF, England.

Moriarty C. (1990c) Eel management practice in three lake systems in Ireland. In: W.L.T. van Densen, B. Steinmetz & R.H. Hughes (eds) *Management of Freshwater Fisheries*. Wageningen, Netherlands: Pudoc, 12 pp.

Moriarty C. (1993) The decline in catches of European elvers. EIFAC Working Party on Eel, Olsztyn, 1993. Rome: EIFAC, FAO, 9 pp.

Moriarty C., Bninska M. & Leopold M. (1990) Eel, *Anguilla anguilla* L., stock and yield in Polish lakes. *Aquaculture and Fisheries Management* **21**, 347–356.

Naismith I.A. (1992) The assessment and exploitation of eel (*Anguilla anguilla* L.) stocks in the River Thames and its catchment. Unpublished PhD Thesis, University of Westminster, London, 246 pp.

Naismith I.A. & Knights B. (1988) Migrations of elvers and juvenile European eels, *Anguilla anguilla* L., in the River Thames. *Journal of Fish Biology* **33(Supplement A)**, 161–175.

Naismith I.A. & Knights B. (1990a). Studies of sampling methods and of techniques for estimating populations of eels, *Anguilla anguilla* L. *Aquaculture & Fisheries Management* **21**, 357–367.

Naismith I.A. & Knights B. (1990b) Modelling of unexploited and exploited populations of eels, *Anguilla anguilla* (L.), in the Thames Estuary. *Journal of Fish Biology* **37**, 975–986.

Naismith I.A. & Knights B. (1993) The distribution, density and growth of European eels, *Anguilla anguilla* L., in the River Thames catchment. *Journal of Fish Biology* **42**, 217–226.

North E. & Hickley P. (1989). An appraisal of anglers' catches in the River Severn, England. *Journal of Fish Biology*, **34**, 299–306.

Rasmussen G. & Therkildsen B. (1979) Food, growth and production of *Anguilla anguilla* L. in a small Danish stream. *Rapports et Proces-Verbaux des Reunions, Conseil International pour l'Exploration de la Mer* **174**, 115–123.

Rigaud C., Fontanelle G., Gascuel D. & Legault A. (1988) Le franchissement des ouvrages hydrauliques par les anguilles (*Anguilla anguilla*): Presentation des dispositifs installes en Europe. Les Publications du Departement d'Halieutique No. 9. 147 pp. ENSA de Rennes, France.

Taylor A.H., Colebrook J.M., Stephens J.A. & Baker N.G. (1992) Latitudinal displacements of the Gulf Stream and the abundance of plankton in the north-east Atlantic. *Journal of the Marine Biological Association UK* **72**, 919–921.

Tesch F.-W. (1977) *The eel: Biology and management of anguillid eels*. London: Chapman and Hall.

Tesch F.-W. (1993) Vertical movements of migrating silver eels in the sea. EIFAC Working Party on Eel, Olsztyn, 1993, Rome: EIFAC, FAO, 10 pp.

Tzeng W.-N. (1984a) Dispersal and upstream migration of marked anguillid eel, *Anguilla japonica*, elvers in the estuary of the Shuang River, Taiwan. *Bulletin of the Japanese Society of Fisheries and Oceanography* **45**, 10–20.

Tzeng W.-N. (1984b) An estimation of the exploitation rate of *Anguilla japonica* immigrating into the coastal waters off Shuang Chi River, Taiwan. *Bulletin of the International Zoological Academy Sinica* **23**, 173–180.

Tzeng W.-N. (1985) Immigration timing and activity rhythms of the eel, *Anguilla japonica*, in the estuaries of N. Taiwan, with emphasis on environmental influences. *Bulletin of the Japanese Society of Fisheries and Oceanography* **47–48**, 11–28.

Tzeng, W.-N. (1989) Resource and biology of the eel *Anguilla japonica* in the estuaries of Taiwan. EIFAC (FAO) Working Party on Eel Meeting, Oporto, 1989. Rome: EIFAC, FAO, 12 pp.

Vøllestad L.A. & Jonsson B. (1988). A 13-year study of the population dynamics and growth of the European eel *Anguilla anguilla* in a Norwegian river: evidence for density-dependent mortality and development of a model for predicting yield. *Journal of Animal Ecology* **57**, 983–997.

White E. & Knights B. (1994) Elver and eel stock assessment in the Severn and Avon. R&D Project Record 256/13/ST. Bristol, England: National Rivers Authority, 141 pp.

White E. (1994) A study of the exploitation, migration and management of elvers and juvenile eels (*Anguilla anguilla* L.) in the Rivers Severn and Avon. Unpublished PhD thesis. London: University of Westminster, 186 pp.

# Chapter 35
# Exploitation of *Alosa pontica* in the Danube Delta, Romania

I. NAVODARU *Fisheries Research Group, Danube Delta Institute, Babadog Str. 165, Tulcea, Romania*

## 35.1 Introduction

Danube shad (*Alosa pontica* Eichwald) are an important component of the commercial catches of the Danube in Romania. They feed in the western Black Sea near to the shore in their first year, before moving into deeper water off the coast as they get older. After three to four years the fish mature and migrate for spawning from south to north along the Bulgarian and Romanian coasts (Banarascu 1964; Ivanov 1985). They enter the Danube through the southern St Gheorghe branch and to a lesser extent the Sulina branch (Fig. 35.1). The adult fish living in the north west region of the Black Sea, enter the river through the northern Chilia branch, and are generally fished by the Ukraine (Pavlov 1953).

Spawning takes place between Braila (180 km upstream) and about 500 km upstream of the mouth, although some fish do migrate as far as the Iron Gates dam (900 km) (Leonte 1957; I. Navodaru, personal observation). The eggs are pelagic and

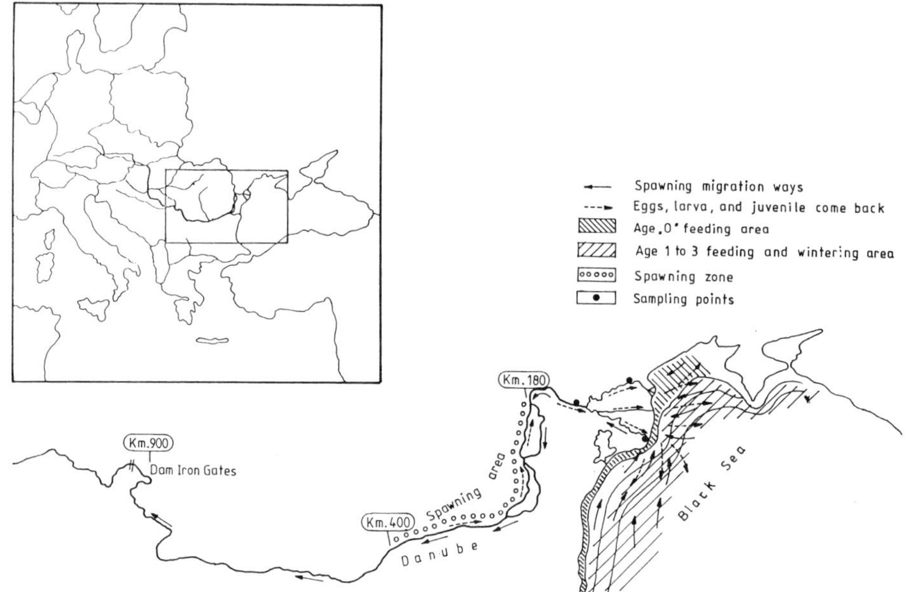


**Fig. 35.1** Migration pattern of Danube shad.

448

float near the surface. Larvae and fingerlings are carried by the water current to the Black Sea where they complete the life cycle. The shad stocks can thus be broken down into two components: the Black Sea for growth and the Danube for spawning stock.

Management of the stocks is difficult because little is known about the dynamics of the species and the impact of exploitation or other factors. The main mortality appears to be associated with fishing of the migratory spawning stock and high natural mortality of the juveniles (Fig. 35.2).

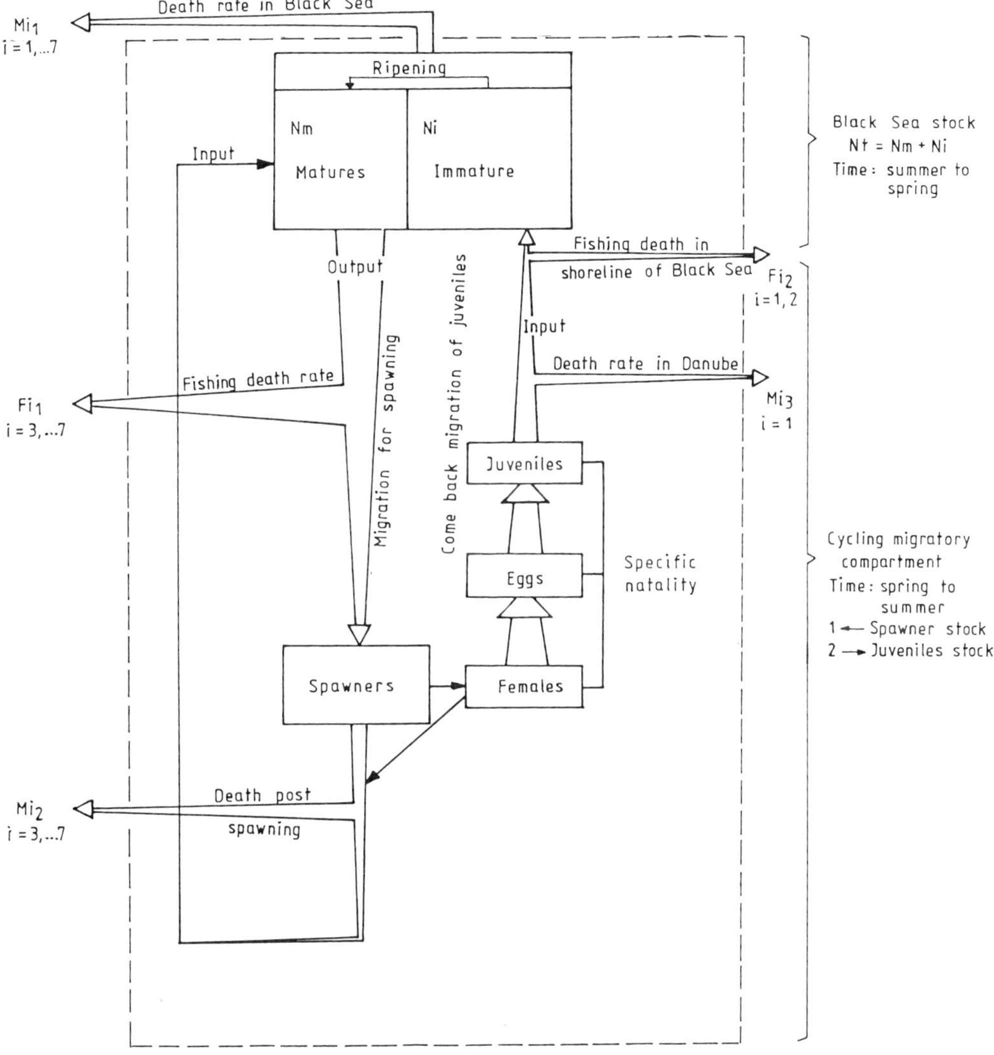


**Fig. 35.2** Diagrammatic representation of the mortality schedule of Danube shad.

This chapter examines further some of the characteristics of the spawning stock of the species as they pass through the Danube Delta in the St Gheorghe channel.

## 35.2     Age structure

The age of the migratory stocks between 1987 and 1993 varied between 2 and 7 years old (Table 35.1). The 3-year-old group dominated although the contribution by 4-year-olds was uncharacteristically high in 1992. This was probably due to the change in the length/weight of this age group: lower in 1992 compared to other years (Table 35.2).

**Table 35.1**   Age structure (% contribution) of migratory spawning stock of *Alosa pontica* between 1987 and 1993.

| Year | Number | Age class | | | | | |
|---|---|---|---|---|---|---|---|
| | | 2 | 3 | 4 | 5 | 6 | 7 |
| 1987 | 645 | 6.3 | 69.5 | 22.4 | 1.8 | | |
| 1988 | 936 | 3.3 | 50.3 | 32.2 | 4.1 | 0.1 | |
| 1989 | 931 | 8.1 | 62.9 | 28.4 | 0.5 | 0.1 | |
| 1990 | 2169 | 4.1 | 55.4 | 38.3 | 2.2 | | |
| 1991 | 2663 | 5.4 | 51.3 | 34.8 | 7.4 | 0.9 | 0.2 |
| 1992 | 2803 | 1.5 | 36.1 | 55.8 | 6.4 | 0.5 | 0.7 |
| 1993 | 734 | 4.7 | 50.0 | 37.3 | 6.9 | 1.1 | |

**Table 35.2**   Size structure (total length cm; weight g) of the migratory spawning stock of *Alosa pontica* between 1987 and 1993.

| Year | Catch (t) | Size | Age class | | | | | |
|---|---|---|---|---|---|---|---|---|
| | | | 2 | 3 | 4 | 5 | 6 | 7 |
| 1987 | 416 | Length | 28.0 | 29.3 | 33.3 | 30.1 | | |
| | | Weight | 193 | 244 | 246 | 242 | | |
| 1988 | 751 | Length | 27.5 | 29.1 | 30.2 | 31.1 | | |
| | | Weight | 196 | 232 | 262 | 291 | | |
| 1989 | 196 | Length | 27.0 | 29.8 | 30.8 | 34.9 | | |
| | | Weight | 183 | 228 | 277 | 398 | | |
| 1990 | 211 | Length | 25.9 | 28.4 | 30.8 | 32.9 | | |
| | | Weight | 179 | 241 | 304 | 359 | | |
| 1991 | 362 | Length | 25.5 | 29.5 | 31.5 | 33.9 | 35.6 | 38.3 |
| | | Weight | 157 | 234 | 290 | 353 | 388 | 487 |
| 1992 | 814 | Length | 27.1 | 28.2 | 30.1 | 32.9 | 35.3 | 37.9 |
| | | Weight | 177 | 202 | 237 | 290 | 349 | 457 |
| 1993 | 414 | Length | 24.1 | 29.6 | 31.8 | 34.5 | 37.1 | |
| | | Weight | 119 | 213 | 263 | 331 | 430 | |

### 35.3     Exploitation

Von Bertalanffy growth parameters for the 1991 migratory stock were $L_\infty$ = 48.1 cm, $k$ = 0.2 and $t_0$ = −1.58. The mortality of the stock was high ($Z$ = 1.68) with exploitation ($F$ = 1.32) and stress due to spawning being the main contributory factors. Indeed, it was estimated that total mortality was so high that it was unlikely that any adults survive to return to the Black Sea for future spawning (I. Navodaru personal observation). Consequently, exploitation each year concentrates on a new component of the overall stock, but the year-class-strength may be variable.

Yield per recruitment isopleths show that the current rate of exploitation ($P_c$) is well above the eumetric point E and thus it is suggested that the length at fish capture ($L_c$) should be increased. Shad are fished with trammel nets with an inside net mesh size of 28–30 cm knot to knot and an outside mesh of 250 cm. The length-frequency distribution/selectivity curve for the stock gave a value for $L_c$ of about 30 cm (I. Navodaru, personal observation), thus to increase the mesh size would considerably reduce the proportion of the stock which would be available for exploitation (see Table 35.2), making the fishing potentially non-viable.

### 35.4     Factors causing variation in catch

The necessity to manage the stocks according to a minimum size at first capture is largely dependent on constant recruitment to the catch cohorts. Evidence suggests that the catch is linked to the flow in the Danube three years previously (Fig. 35.3), i.e. the spawning and juvenile growth conditions encountered by the majority of the fish (3-year-olds) in the catch. If the river flow/water level is low, spawning conditions appear unsatisfactory and catches three to four years later are poor. This problem is

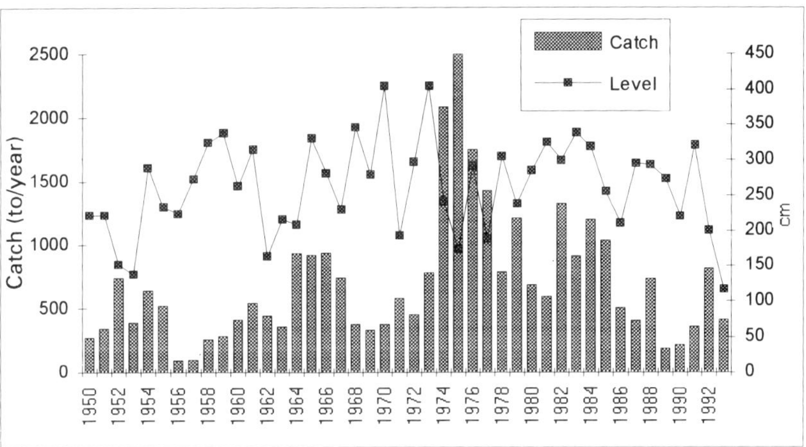



**Fig. 35.3**  Cyclical variation in the catches of Danube shad and the relationship with water level three years previously when the dominant year class in the catch was born.

further exacerbated by an apparent direct relationship between water level in spring (flow) and catch rate (Fig. 35.4).

To manage the fishing on a sustainable basis it is therefore tentatively suggested that fishing effort is decreased in years when poor adult migratory stocks are predicted to ensure an adequate spawning escapement. However, the data relating to these stocks are still poor and much further research and monitoring are required before firm management decisions can be made.

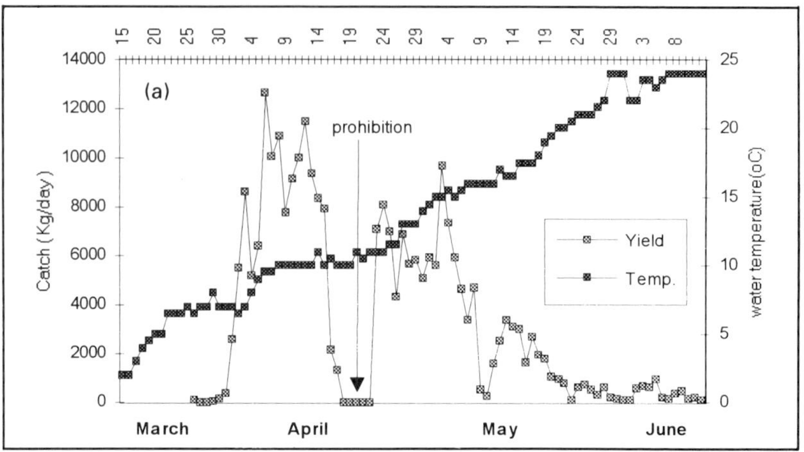


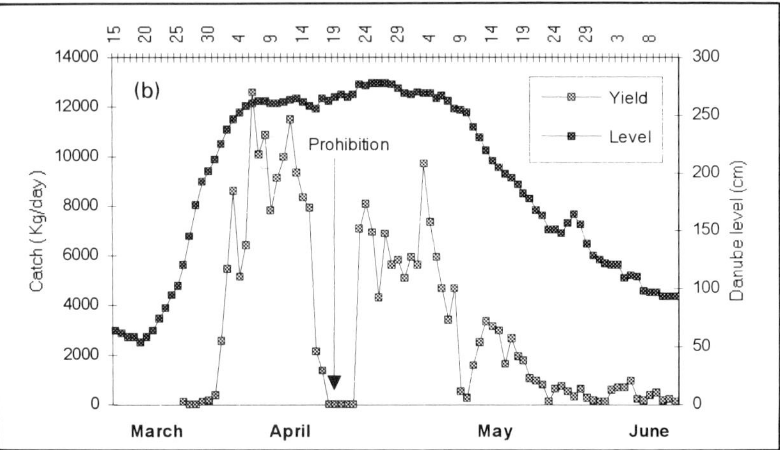


**Fig. 35.4**   Relationship between catches of Danube shad and the water level in the river Danube in 1993.

## References


Banarascu, P. (1964) *Fauna of the Romanian People's Republic* Pisces – Osteichthyes, Vol XIII, 229–232; 239–244.

Ivanov, L. (1985) The fisheries resources of the Mediterranean. Studies and Reviews No 60 Rome: FAO, pp 45–122.

Leonte T.R. (1957) Citeva date asupra reproducerii scrumbiei de Daunare *Bulgarian I.C.P.* **XVI**, 37–46.

Pavlov, P.I. (1953) Biologhiceskaia I promislovaia harakteristika nerestovogo stada dunaiskoi celdi. In: Dunaiskaia seldi u biologhiceskie osnovi ee promisla. Acad. Nauk, Ukrainkoi SSR, Kiev, pp 118–174.

# Chapter 36
# Age and growth of *Petenia splendida* Günther in the San Pedro River (Southern Mexico)

J.-L. NOISET and J.-Cl. MICHA *UNECED/FUNDP Rue de Bruxelles, 61 B-5000 Namur, Belgium*

**Abstract**   The age and growth characteristics of *Petenia splendida* Günther in the San Pedro River in Southern Mexico were determined. Three complementary methods of increasing precision were used: a length-frequency method (ELEFAN), the scale reading, and otolimetry. All methods gave comparable values for the parameters of the von Bertalanffy equation: $L_\infty$ between 435 and 452 mm, $K$ between 0.30 and 0.38 per year and a $t_o$ of –0.18 year. The ELEFAN method and the scale examination highlighted a seasonal growth pattern marked by a reduction or stoppage of growth in the dry-winter season and renewal in the rainy-flood season. The heterogeneity of sizes was large within each age group.

KEYWORDS: Age, back-calculation, ELEFAN, growth, *Petenia spendida*, otoliths

## 36.1   Introduction

The Cichlidae represent an important part of the fishery resources of the water courses and floodplains of the State of Tabasco in Southern Mexico (Tello Dominguez 1988; Chavez & Garrido 1988). However, the biology of the Central American Cichlidae is still relatively unknown (Miller 1976, 1982, 1986; Kullander 1983), and the absence of fishing legislation in Mexico (Almada-Villela 1990) leaves worries about overexploitation of stocks and the disruption of the ecological equilibrium of the river systems.

Consequently a joint project between INIREB (Instituto Necional de Inverstigación sobre los Recursos Bióticos, Mexico) and ADRAI (Association pour le Développement Rural et Artisanal Intégré, Belgium) was established to determine the population dynamics of the Cichlidae, *Petenia splendida* Günther, which is the most abundant and exploited species in the region.

## 36.2   Study area

The study took place between 1988 and 1990 in the tropical region of Southern Mexico in the State of Tabsco (Fig. 36.1). The sources of the San Pedro River are

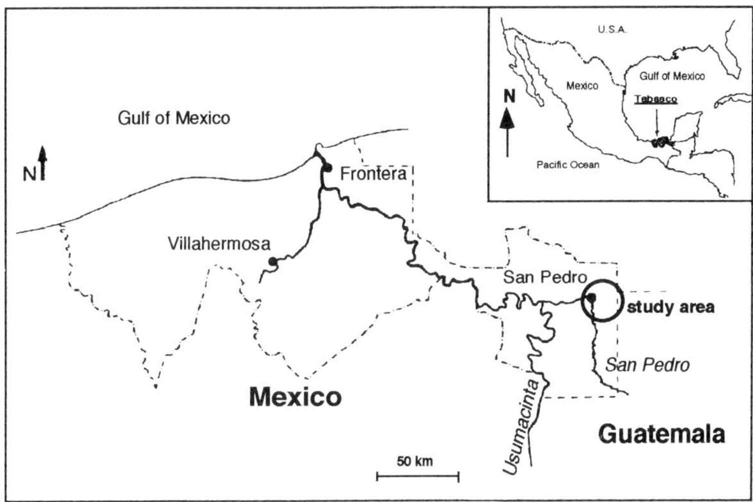


**Fig. 36.1**   Map of the San Pedro River and study area in the State of Tabasco, Southern Mexico.

located in The Petén region in Guatemala. It flows into the biggest river in Central America, the Usumacinta. The floodplain of San Pedro, near the village of the same name, spreads over a maximum of 74 km$^2$ on the right bank. The floodplain is drained by three strongly branched man-made channels (Fig. 36.2)

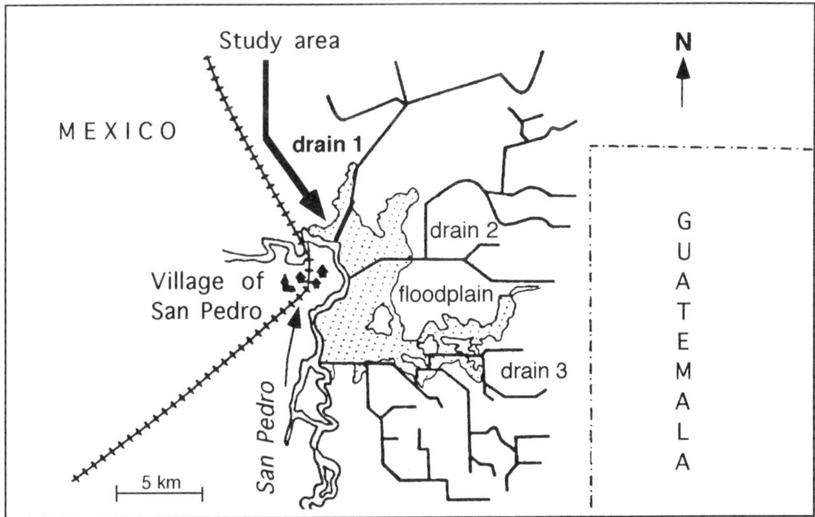


**Fig. 36.2**   Map of the San Pedro River, floodplain and study area near the village of San Pedro (Tabasco, Mexico).

## 36.3    Material and methods

### 36.3.1    *Sampling*

Sampling took place between 1988 and 1990 in the first drain (Fig. 36.2) using three techniques. A 70 m long seine net, 7 m high and 12.7 mm stretched mesh size was used to catch fish approximately every two months, from February to December 1989. The second technique was electrofishing using a peak current at 120 A, 500 V and 80 Hz. As a third method a set of 8 monofilament gillnets (stretched mesh size: 44–89 mm) were deployed transversely in the drain. The fish captured were measured (nearest mm), sexed and scales and sagittae otoliths were removed.

### 36.3.2    Determination of age and growth

Three methods for age and growth determination were used.

The length-frequency method (ELEFAN, Gayanilo *et al.* 1989) and the von Bertalanffy growth curve with seasonal oscillations proposed by Pauly & Gaschutz (1979, in Pauly 1987) and modified by Somers (1988) were used.

Traditional scale reading was used. In total 835 fish were examined. They were separated into seven groups caught over the two years to detect when the annual check was set: i.e. February 1988, 1989 and 1990 (beginning of the dry season); July-August 1988 and 1989 (beginning and during the rainy season); December 1988 and 1989 (end of the rainy season, return migration). Scales were taken from all the individuals caught on the 6 July 1988 to characterize the age structure of the population. Two independent examinations were undertaken and the mean of the two scale measurements was used. The relation between the total length of fish ($L_t$) and the total radius of scale ($S$) was linear but not directly proportional. The Fraser-Lee back-calculation formula was applied (Francis 1990; Ricker 1992): $L_i = (a + [L_t - a]) (s_i / S)$, where $L_i$ is the total length at age $i$; $s_i$ is the scale radius at age $i$, and $a$ is the Y-axis intercept in the relation between $L_t$ and $S$.

The otolimetric method was applied using two complementary techniques (Secor *et al.* 1991): the optical microscopy (OM) was used for small specimens and the scanning electronic microscopy (SEM) for larger ones. For the OM, the 'glue and polish' method was chosen (Secor *et al.* 1992): the sagittae otolith were fixed, in the sagittal plane, to a glass slide by the Crystalbond 509 adhesive of Aremco products. A double polishing procedure (on the two sides) was applied using Carborandum paper No 1200 and a metalographic polishing cloth with an alumina slurry (0.3 μm micropolish II Alumina of Buehler Ltd.). Microstructural examination was made with a Leitz Diaplan microscope (binocular eyepiece 12.5 ×; objective lenses with nominal magnification of 10, 25 and 40 ×).

For SEM, the otoliths were embedded in Epon and sectioned transversely in the anterior part near the core. The posterior part was polished on Carborandum paper no. 800, 1000 and 1200. The preparations were etched for 90 s (small otoliths) to 120 s (larger ones) with 2% HCl, coated with gold (200 A) and examined with a Philipps XL 20 SEM at 20 kV and magnifications running from 200 to 1600.

The von Bertalanffy growth curve was fitted to the points (scalimetric and otolimetric method) and the three growth parameters values ($K$, $L_\infty$ and $t_o$) were determined by the iterative NLIN procedure (DUD method) of the SAS program (v. 6.07) (SAS Institute 1990).

## 36.4    Results

### 36.4.1    *Length-frequency method. Seasonal catches*

The best samples were obtained after regrouping of data in 2 cm classes. Due to the small number of fish, the catches of November and December 1989 were grouped into one sample. The original distributions showed that only the first mode could be followed during the year (Fig. 36.3).

The best combination which permitted the curve to pass amongst the first real peak (Fig. 36.3) gave $L_\infty$ = 43.5 cm (Table 36.1). The $K$ was relatively weak, 0.33 per year. The curve with the maximal amplitude of variation, $C$ = 1, revealed a seasonal cycle of growth. The beginning of the oscillation was fixed at 0.2 per year, which set the growth cessation at mid-March, in the middle of the dry season.

### 36.4.2    *Scale reading*

Examination of the scales showed a reduction in growth during the dry season. A large area of regular circuli (rainy seasons, flooding) gave way to an irregular zone where two marks were generally detectable. The first was situated at the start of the dry season and coincided with the period of lowest temperature (January-February). The second appeared at the end of the dry season and the beginning of the wet season. They were visible at the scale margins and much closer together. During the growth of the fish, they had a progressive tendency to merge and become more noticeable. The first of these two marks was used to represent the annulus.

The length-frequency distribution by age group for the sample of July 1989 in drain 1 showed that practically all the individuals were one-year-olds (87% of sample), but they could achieve 4 years of age (Fig. 36.4). Fishing pressure probably explains the absence of individuals in the older age groups.

The general relationship between the total length of the fish ($L_t$) and the scale radius ($S$) was linear but not directly proportional: $L_t$ = 38 + 3.52.$S$ ($r^2$ = 0.96).

The results of the back-calculation (Table 36.2) were used to fit a von Bertalanffy growth curve (Fig. 36.5). The equation gives comparable values to those obtained by the ELEFAN method: $L_\infty$ = 452 mm; $K$ = 0.30 per year and $t_o$ = –0.18 year ($N$ = 279 individuals).

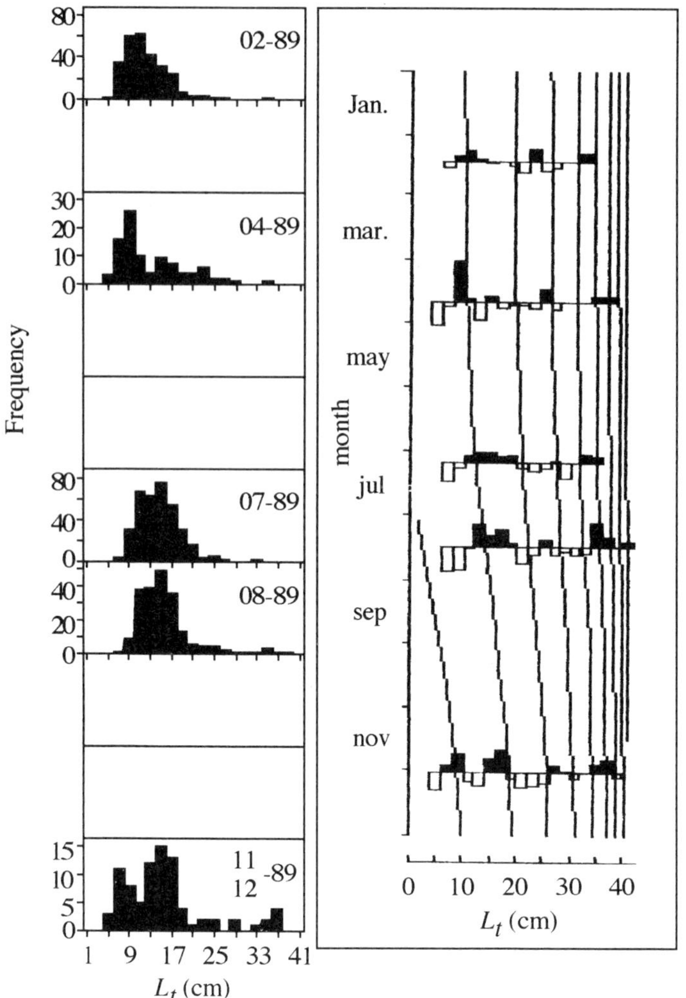



**Fig. 36.3** Growth of *Petenia splendida* in the San Pedro River system, based on length-frequency data (seine nets catches in drain 1) and ELEFAN 1 program (von Bertalanffy growth curve with seasonal oscillations).

**Table 36.1** Parameter estimates for the seasonally oscillating version of the von Bertalanffy growth formula for *Petenia splendida* captured by seine net in drain 1 from February to December 1989.

|  | parameters | value |
|---|---|---|
| $L_\infty$ | Asymptotic length (cm) | 43.50 |
| $K$ | Growth coefficient (per year) | 0.330 |
| $C$ | Amplitude of seasonal growth oscillations | 1.000 |
| $WP$ | Winter point (year) | 0.200 |

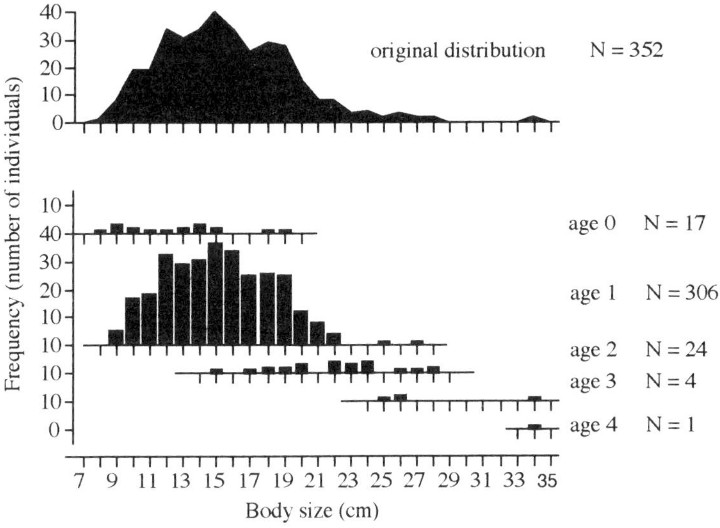



**Fig. 36.4**  Length-frequencies of all *Petenia splendida* caught by a seine net in drain 1 (San Pedro River system) on the 6 July 1989 and classified into age groups according to the number of annuli on their scales.

**Table 36.2**  Back-calculated lengths of *Petenia splendida* based on scale readings ($L_t$, total length; $N_t$, total number).

| Age group | Fish number | $L_t$ (mm) | Mean back-calculated length (mm) at the end of the year | | | | |
|---|---|---|---|---|---|---|---|
| | | | 1 | 2 | 3 | 4 | 5 |
| 0 + | 34 | 156 | | | | | |
| 1 + | 48 | 193 | 132 | | | | |
| 2 + | 24 | 256 | 138 | 222 | | | |
| 3 + | 29 | 304 | 138 | 223 | 281 | | |
| 4 + | 14 | 331 | 138 | 213 | 275 | 312 | |
| 5 + | 4 | 399 | 142 | 220 | 277 | 338 | 380 |
| Number of fish ($N_t = 279$) | | | 129 | 81 | 47 | 18 | 4 |
| Mean back-calculated length | | | 136 | 221 | 279 | 318 | 380 |
| Standard error | | | 3 | 4 | 6 | 11 | 7 |
| min.-max | | | 70–225 | 132–351 | 179–382 | 179–375 | 360–391 |

### 36.4.3  *Otolith microstructure*

The results relate to a total of 68 fish (30 immature, 25 females and 13 males) with a range of lengths between 30 and 265 mm. Individuals were chosen over the two years of sampling to provide an assessment of growth in this period.

A great divergence in total length was found with age (Fig. 36.6). The number of ridges observed on an otolith ranged from 64 (immature of 32 mm) to 585 ($L_t$ = 167 mm: a female). The age of 150 mm fish varied from 60 to 500 days. The length of a 1-

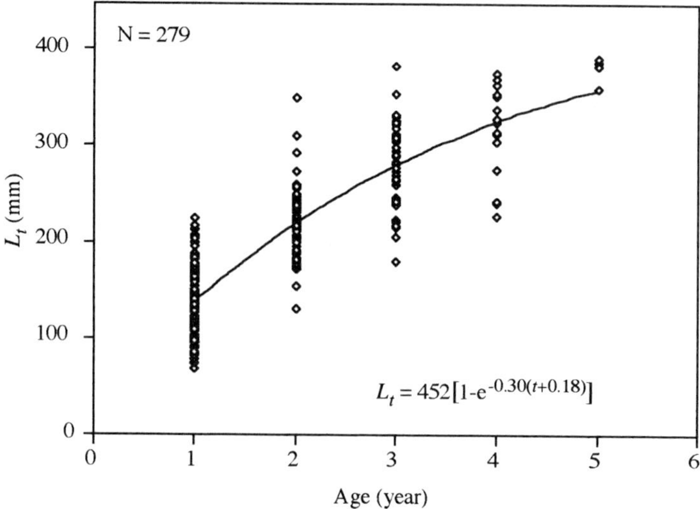



**Fig. 36.5**   Growth of *Petenia splendida* in the San Pedro River system, based on scale readings (von Bertalanffy growth curve).

year-old fish varied between 110 mm and 200 mm. The individual variability was therefore important and confirmed the previous observations made on scales.

A von Bertalanffy growth curve was fitted to the data (Fig. 36.6), but it was impossible to determine $L_\infty$ because only the fish phase of growth was available. The asymptotic length was therefore fixed at the average value obtained by scale reading: 452 mm. The results (Table 36.3) gave a $K$ value similar to that derived by the other methods: 0.38 per year.

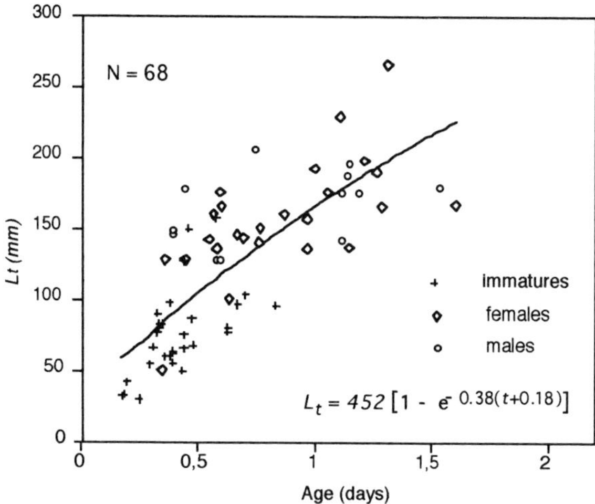



**Fig. 36.6**   Growth of *Petenia splendida* in the San Pedro river system, based on otoliths (von Bertalanffy growth curve model).

**Table 36.3** Parameter estimates for the von Bertalanffy growth formula for *Petenia splendida*, based on otoliths examination ($L_\infty$ fixed from scale readings at 452 mm).

| Parameter | Estimate | Asymptotic Std Error |
|---|---|---|
| $L_\infty$ (mm) | 452 | — |
| $K$ (per year) | 0.389 | 0.040 |
| $t_0$ (year) | −0.179 | 0.085 |

## 36.5 Discussion

The ELEFAN method discriminated growth seasonality in *Petenia splendida* and it is a useful tool for floodplain fisheries where the amplitude and intensity of floods play an essential role in growth (Welcomme 1979). In this study a sample cohort, probably born at the end of the dry season, was sufficient to identify this phenomenon. The ELEFAN method shows a cessation of growth during the dry season, an observation which was less apparent on the scales.

The formation of the annual marks on the scales could also be linked to the flood regimes, fall of temperature, high density of fish in a limited area, and depletion of the food source (Welcomme 1975, 1985; de Merona 1988).

Observations on the biology of fish and determination of the birth dates by the otolimetric method suggested that the reproduction of *Petenia splendida* occurred throughout the year. However, the problems of interannual growth variability in floodplains associated with tropical rivers, may also be an explanation (Welcomme & Hagborg 1977; Welcomme 1979, 1985; Bayley 1988). Consequently, it is probably unrealistic to predict the annual growth of the *Petenia splendida* in the San Pedro River on the basis of environmental characteristics such as: flooding from the previous year and/or two previous years; conductivity; temperature, as proposed by other authors (Welcomme & Hagborg 1977; Welcomme 1985; Maceina 1992).

### Acknowledgements

This research was funded by the 'Instituto Nacional de Investigaciones sobre Recursos Bióticos' (INIREB, Mexico), the 'Association pour le Développement et la Rescherche et de l'Action Intégrée (ADRAI, Belgium) and the DG8 of the European Community Commission (project ONG\2 03\84\B).

### References

Almada-Villela P.C. (1990) Status of threatened Mexican fishes. *Journal of Fish Biology* **37**, 197–199.
Bayley P.B. (1988) Factors affecting growth rates of young tropical floodplain fishes: seasonality and density-dependence. *Environmental Biology of Fishes* **21**, 127–142.
Chavez M.O.L. & Garrido F.M. (1988) Importancia de las pesquerías en los humedales. *In* Ecología y conservación del Delta de los ríos Usumacinta y Grijalva (Memorias del Simposio Internacional sobre

la Ecología y Conservación del Delta de los Ríos Usumacinta y Grijalva. Villahermosa, Tab., México; 2–7 de febrero de 1987) (Regional-Tabasco, I.-D. & Tabasco, G.d.E.d., eda.). INIREB-Division Regional-Tabasco & Gobierno del Estado de Tabasco, Villahermosa, Tab., México: pp 31–38.

de Merona B. (1988) Croissance des poissons d'eau douce africains. In: C. Leveque, M.N. Bruton & G.W. Ssentongo (eds). *Biology and ecology of African freshwater fishes* Paris: Orstom, pp 191–219.

Francis R.I.C.C. (1990) Back-calculation of fish length: a critical review. *Journal of Fish Biology* **36**, 883–902.

Gayanilo F.C.J., Soriano M. & Pauly D. (1989) A draft guide to the complete ELEFAN. ICLARM Software, Manila, Philippines: International Center for Living Aquatic Resources Management, 70 pp.

Kullander S.O. (1983) A revision of the South American genus *Cichlasoma* (Teleostei: Cichlidae). The Swedish Museum of Natural History, Research Department, Stockholm, Sweden, 296 pp.

Maceina M.J. (1992) A simple regression model to assess environmental effects on fish growth. *Journal of Fish Biology* **41**, 557–565.

Miller R.R. (1976) Geographical distribution of Central American freshwater fishes. Reprinted from *Copeia* 1966, No. 4, with addendum. In: T.B. Thorson (ed.) *Investigations of the Ichthyofauna of Nicaraguan lakes* School Life Sci., Univ. Nebraska, Lincoln, pp 125–156.

Miller R.R. (1982) Pisces. In: S.H. Hurlbert (ed.) *Aquatic biota of Mexico, Central America and the West Indies*. San Diego State University, San Diego, California, pp 486-501.

Miller R.R. (1986) Composition and derivation of the freshwater fish fauna of México. *An. Esc. nac. Cienc. biol.* **Mex. 30**, 121–153.

Pauly D. (1987) A review of the ELEFAN system for analysis of length-frequency data in fish and aquatic invertebrates. In: D. Pauly & G.R. Morgan (eds) *Length-based methods in fisheries research*. Iclarm Conference Proceedings **13**, 468 pp, International Center for Living Aquatic Resources Management, Manila, Philippines and Kuwait Institute for Scientific Research, Safat, Kuwait, pp 7–34.

Ricker W.E. (1992) Back-calculation of fish lengths based on proportionality between scale and length increments. *Canadian Journal of Fisheries and Aquatic Sciences* **49**, 1018–1026.

SAS (1990). SAS Language: Reference, Version 6, First Edition. Cary, NC, USA: SAS Institute Inc., 1042 pp.

Secor D.H., Dean J.M. & Laban E.H. (eds) (1991) Manual for otolith removal and preparation for microstructural examination. Electric Power Research Institute & Belle W. Baruch Institute for Marine Biology and Coastal Research, 85 pp.

Secor D.H., Dean J.M. & Laban E.H. (1992) Otolith removal and preparation for microstructural examination. In: D.K. Stevenson & S.E. Campana (eds) Otolith microstructure examination and analysis. *Canadian Special Publication of Fisheries and Aquatic Sciences* **117**, 19–57.

Somers I.F. (1988) On a seasonally oscillating growth function. *Fishbyte* **6**, 8–11.

Tello Dominguez M. (ed.) (1988) Las institutiones y la pesca en Tabsco. Villahermosa, Tabasco, México, 77 pp.

Welcomme R.L. & Hagborg D. (1977) Towards a model of a floodplain fish population and its fishery. *Environmental Biology of Fishes* **2**, 7–24.

Welcomme R.L. (1975) The fisheries ecology of African floodplain. *FAO, CIFA Technical Paper* **3**, 51 pp.

Welcomme R.L. (1979) Fishery management in large rivers. *FAO Fisheries Technical Paper* **194**, 60 pp.

Welcomme R.L. (1985) River fisheries. *FAO Fisheries Technical Paper* **262**, 330 pp.

# VII  FISHERIES MANAGEMENT

# Chapter 37
# Maximization of yields from African lakes

G.F. TURNER *Department of Zoology, University of Aberdeen, Aberdeen AB9 2TN, UK &*

*SOAFD Marine Laboratory, PO Box 101, Victoria Road, Aberdeen AB9 8DB, UK*

**Abstract**   The transfer efficiency from primary production to fish yield is compared for 20 large lakes or areas of lakes in Africa. Lakes containing Nile perch (*Lates* spp.) support fisheries which are less stable and less efficient than those of other lakes. The exception is Lake Tanganyika which has a unique group of endemic *Lates* species and where populations of large predators were severely reduced by fishing in the 1960s. Yields of Nile perch declined rapidly under heavy fishing. However, Nile perch has a very high fecundity and may not be susceptible to recruitment overfishing. It is not clear if overexploitation of Nile perch led to an increase in the transfer efficiency of the fishery as a whole. Thus, introductions of Nile perch into other African lakes cannot be justified.

There was no evidence that the presence of clupeids enhanced the transfer efficiency of lakes. However, clupeids may enhance the productivity of fisheries in deep lakes where no truly pelagic fish are present, particularly in man-made lakes.

While fisheries based on large predators may be more profitable, at least in the short term, fisheries based on small species gave higher yields. The decision to try to manage African lakes in a way which maximizes yields or profit is essentially political, but it is possible that management for larger species is impractical at present. It is suggested that maximization of yields is best achieved by abolishing technical measures such as mesh size restrictions, which are rarely enforced. Licensing schemes and extension campaigns should be put in place in anticipation of the eventual need to restrict access to fisheries. Where fishing poses a threat to biodiversity, closed areas should be established.

KEYWORDS: Africa, cichlids, clupeids, fisheries, lakes, Nile perch

## 37.1   Introduction

Inland fisheries are of great importance in Africa. In 1987, the total fishery yield in Africa (excluding South Africa) was approximately 4.3 million tonnes, of which marine fisheries provided 2.6 million tonnes (60%), aquaculture 70 000 tonnes (1.6%) and inland capture fisheries 1.6 million tonnes (38%) (Vanden Bossche & Bernacsek 1989, 1990, 1991). These data do not reflect the widespread importance of inland capture fisheries: 40% of all marine fish was landed by Namibia and Morocco, while Egypt accounted for 85% of all aquaculture production. Excluding these three countries, inland capture fisheries provided roughly the same yield as marine fisheries and more than 160 times as much as aquaculture. Furthermore, much of the marine fish production was landed by foreign vessels for luxury urban or overseas markets and thus does not contribute to the nutrition of the often impoverished rural population for whom freshwater fish often represents the major source of cheap animal protein. Thus the maximization of inland fishery yields is of paramount importance.

The larger natural lakes and hydroelectric reservoirs support the most important of Africa's freshwater fisheries. For more than 40 years fisheries scientists and administrators have speculated over whether the yields of these lakes could be enhanced by introducing alien fish species. Many introductions have already been carried out.

Often an introduction was accompanied by profound changes in the lake ecosystem, including extinction of indigenous species and alterations to the basic productive system itself (Barel *et al.* 1991; Bootsma & Hecky 1993; Kaufman 1992; Marshall 1991). Causal relationships are often difficult to determine and in many cases the final outcome of these perturbations cannot be confidently predicted.

Wherever an alien species has established breeding populations, fishery yields have increased, often dramatically (Marshall 1993; Pitcher 1995). However, so too have the yields of most other African lakes (Vanden Bossche & Bernacsek 1989, 1990, 1991) and here too the causal link remains unclear. Fishery yields from African lakes are influenced by the differing size and primary productivity of the lakes as well as the social and economic conditions of the lakeshore peoples.

The aim of this chapter is to investigate whether the species of fish found in a lake influences the fishery yields, taking into consideration lake morphology and productivity. Particular consideration is given to the two species whose introduction has produced the most striking effects: the kapenta *Limnothrissa miodon* (Boulenger) and the Nile perch, *Lates niloticus* (L.) and its relatives. The former is a small pelagic zooplankton feeding clupeid, the latter a large piscivorous centropomid.

### 37.2　Classification of lakes by fish taxa present

Lakes are classified into three groups.

***Lates* lakes**　In Lakes Victoria and Kyoga, introduction of *Lates* appears to have had a marked effect on the populations of many other fish species. Therefore, even where *Lates* is not the most important fishery species, its effect on yields of other species might be considerable. Large benthic *Lates* species are indigenous to Lakes Albert, Chad and Turkana, all of which appear to be relatively young lakes and contain few endemic species. The man-made Lakes Kainji, Nasser, and Volta have been colonized by species indigenous to the river before impoundment, including *Lates niloticus*. Lake Tanganyika contains a unique flock of endemic *Lates* species specialized for lacustrine existence. These are classed in the subgenus *Luciolates* (Coulter 1991) and have recently been shown to be more closely related to each other than to any other species (I. Kornfield pers. comm.). For this reason, and because of the dominance of the fisheries by clupeids, comparisons are made where Lake Tanganyika is considered first as a *Lates* lake and then as a clupeid lake.

**Clupeid lakes**    *Limnothrissa miodon* was introduced from Lake Tanganyika to Lakes Kivu and Kariba, where it now forms the basis of the major fishery. In Lake Tanganyika itself, another clupeid *Stolothrissa tanganyikae* Regan is much more important to the fishery. Although indigenous riverine clupeids are also present in Lakes Kainji and Volta, they were not seriously exploited at the time when the most recent available data were collected and so these are not considered as clupeid lakes.

**Cichlid lakes**    Nile perch and clupeids are absent from the majority of African lakes. In Lakes Bangweulu, Edward, George and Mweru the most important fisheries are based on tilapiine cichlids. Haplochromine cichlids dominate the fish catches of Lake Malawi and Lake Malombe, while small cyprinids also support large fisheries on Lake Malawi (*Engraulicypris sardella* (Günther)) and Lake Chilwa (*Barbus paludinosus* Peters), as well on as the *Lates*-dominated Lake Victoria (*Rastrineobola argentea*). In the rest of this chapter, all lakes lacking clupeids and Nile perch are referred to as 'cichlid lakes', although it should be remembered that cichlids often make up the majority of the catch in other lakes.

## 37.3    Data sources

All data were obtained from existing literature. Where different sources gave contradictory information, papers giving unreferenced information (e.g. Craig 1992) were regarded as less reliable than those citing original sources. Otherwise the most recent data were used. The analysis was confined to large (surface area $> 200$ km$^2$) freshwater lakes where at least 10 years' catch data were available and where primary productivity could be estimated.

Previous maximum sustainable yield estimates were generally produced by equilibrium surplus yield models (Schaefer 1954) or morphoedaphic indices (Ryder 1965). The former are now considered to be highly unreliable, especially in cases where the fishery is underexploited (Hilborn & Walters 1992). The morphoedaphic index was criticized as a trivial reflection of lake area (Youngs & Heimbuch 1982). It is also dependent on calibration from yields from existing fisheries, which themselves may be unreliable if they represent underexploited fisheries, or unsustainable overexploited fisheries.

Therefore maximum recorded yields were used, except in cases where fisheries were clearly operated at an unsustainable level.

### 37.4    Are fisheries more stable when certain kinds of fish are present?

Coulter (1991) observed that pelagic fish populations in Lake Tanganyika appeared to have become less stable since the 1960s. This coincided with a decline in the populations of the larger predatory *Lates* species. Coulter suggested that with the reduction of predators, prey populations became resource limited and tended to move into a phase of cyclic population boom and crash. Is there any evidence that fishery yields are more stable in the presence of large efficient predators, such as *Lates*?

The maximum yields and most recent yields for 18 large African lakes can be divided into three groups, according to the behaviour of their fisheries (Table 37.1).

**Table 37.1**   Eighteen large African lakes and reservoirs ranked according to the percentage decline in yields from the recorded maximum. L = Lakes containing *Lates* spp., L* = lakes containing endemic specialized lacustrine *Lates* spp., K = lakes with large kepenta (clupeid) fisheries, K* = lakes with underexploited clupeid stocks.

| Lake | Peak year | Peak yield (t) | Most recent year | Most recent yield (t) | % decline | |
|---|---|---|---|---|---|---|
| Chad | 1974 | 220000 | 1985 | 31334 | 85.8 | L |
| Kainji | 1970 | 28639 | 1977 | 4500 | 84.3 | L, K* |
| Kyoga | 1977 | 167000 | 1988 | 57000 | 65.9 | L |
| Turkana | 1976 | 17044 | 1986 | 7324 | 57.0 | L |
| Albert | 1978 | 303211 | 1987 | 15000 | 50.5 | L |
| Chilwa | 1979 | 25800 | 1982 | 15500 | 39.9 | |
| Edward | 1970 | 16082 | 1976 | 9700 | 38.7 | |
| Volta | 1969 | 61783 | 1979 | 39368 | 36.3 | L, K* |
| Bangweulu | 1973 | 15123 | 1982 | 11006 | 27.2 | |
| Mweru | 1976 | 14123 | 1982 | 10907 | 22.8 | |
| Tanganyika | 1975 | 103563 | 1988 | 85000 | 17.9 | L*, K |
| Malombe | 1982 | 12100 | 1991 | 10300 | 14.9 | |
| Malawi | 1972 | 57000 | 1986 | 51400 | 9.8 | |
| George | 1957 | 5021 | 1988 | 5000 | 0.4 | |
| Kivu | 1987 | 3825 | 1987 | 3825 | 0.0 | K |
| Victoria | 1990 | 500000 | 1990 | 500000 | 0.0 | L |
| Kariba | 1992 | 35000 | 1992 | 35000 | 0.0 | K |
| Nasser | 1982 | 34558 | 1982 | 34558 | 0.0 | L |

Data: Vanden Bossche & Bernacsek (1989; 1990; 1991), except latest yields from Lake Victoria and Kariba (Pitcher 1995).

**Declining fisheries**   In Lakes Chad, Kainji, Kyoga, Turkana and Albert, fishery yields peaked in the 1970s and have subsequently declined to less than half of the maximum level. All of these lakes contain indigenous populations of Nile perch.

**Stable fisheries**   In nine of the lakes, yields have declined from the maximum, but by less than 40%. In most cases, there was no progressive directional decline, but substantial year-to-year variation which was probably well within the range of short-term environmental fluctuations and vagaries of the data collection systems. Of these lakes, only two contain *Lates*: Lake Volta and Lake Tanganyika.

**Expanding fisheries**   In Lakes Kariba, Kivu, Nasser and Victoria, the most recent recorded yields represented the peak to date in a series of progressive increases. In all four of these lakes, there were major ecological perturbations: Kariba and Nasser are man-made lakes, Kariba and Kivu have *Limnothrissa* introduced, and Victoria has *Lates* introduced. It seems possible that the ecology of these lakes is still in an unstable state and thus the expanding fisheries may not be sustainable.

Overall, lakes containing *Lates* showed a greater mean decline in yields than other lakes (mean decline in *Lates* lakes = 44.2%, others = 17.2%, t-test: $t = 2.36$, 9 df, $P = 0.049$). If possibly unstable group 3 lakes are excluded from the analysis and Lake Tanganyika is considered as a non-*Lates* lake, the difference is even more marked (mean decline in *Lates* lakes = 63.3%, mean of others = 21.6%, t-test: $t = 4.48$, 8 df, $P = 0.002$). Although the small sample size precluded statistical comparison, the presence of clupeids had no discernible influence on declines in yields.

### 37.5   Which lakes have the most efficient fisheries?

Lakes differ in size and primary productivity. To compare the efficiencies of fisheries in different lakes, yields were expressed as transfer efficiencies, i.e. the ratio of fish catch to primary productivity, both expressed in grams of carbon per unit surface area per annum (Hecky *et al.* 1981; Oglesby 1977). Fresh fish yields were converted to grams of carbon using Gulland's (1970) constant of 0.1. Where productivities were expressed as grams of oxygen, a photosynthetic productivity quotient of 1.2 was used in the conversion (Melack 1976). The primary productivity of Lake Malombe was estimated from the mean annual chlorophyll-a concentration (Mwanyama 1993) using an empirically derived formula (Lemoalle *et al.* 1981). The south-eastern and south-western arms of Lake Malawi are shallow and support highly productive fisheries based on rather sedentary demersal fish stocks (Turner 1995). The north of the lake is deep, rocky and supports a low intensity of fishing. The two southern arms of the lake were therefore considered separately.

Maximum recorded yields for stable or expanding fisheries (types 2 & 3) were used (Table 37.2). Where yields declined (type 1), the sustainable yields were estimated as the mean of the yields of the last three years or the last two years in the case of Lake Kyoga, to avoid including part of the transient peak.

Transfer efficiencies of lakes containing Nile perch are clearly lower than those of other lakes (Fig. 37.1, Table 37.3a). It could be argued that the use of recent rather than maximum yields in lakes with unstable fisheries tends to underestimate the productivity of fisheries based on *Lates*, which tend to occur in type 1 lakes. Even when the maximum unsustainable yields were used, provided Lake Tanganyika was not included as a *Lates* lake, fisheries were still significantly less efficient in lakes where Nile perch is present when compared to other lakes (Table 37.3b).

**Table 37.2**    Morphometry, yield, productivity and transfer efficiency of 20 lakes or areas of lakes in Africa.

| Lake | Area km$^2$ | Mean depth (m) | S | Max. yield t yr$^{-1}$ | S | Primary productivity g C m$^{-2}$ yr$^{-1}$ | S | Yield/area kg ha$^{-1}$ yr$^{-1}$ | Transfer efficiency % | Main fishery |
|---|---|---|---|---|---|---|---|---|---|---|
| Albert | 5270 | 25 | 1 | 12624 | 1 | 850 | 2 | 24 | 0.03 | Lates |
| Bangweulu | 1721 | 4 | 1 | 15123 | 1 | 207 | 3 | 88 | 0.42 | Tilapia |
| Chad | 10000 | 12 | 4 | 34133 | 4 | 986 | 3 | 34 | 0.03 | Lates |
| Chilwa | 750 | 2 | 1 | 25800 | 1 | 365 | 3 | 344 | 0.94 | Barbus |
| Edward | 2300 | 34 | 1 | 16082 | 1 | 1570 | 2 | 70 | 0.04 | Tilapia |
| George | 250 | 2 | 1 | 5021 | 1 | 1980 | 1 | 201 | 0.10 | Tilapia |
| Kainji | 1280 | 12 | 4 | 4500 | 4 | 890 | 5 | 35 | 0.04 | Tilapia |
| Kariba | 5634 | 29 | 1 | 35000 | 6 | 620 | 5 | 62 | 0.10 | kapenta |
| Kivu | 2370 | 240 | 1 | 3825 | 1 | 303 | 1 | 16 | 0.05 | kapenta |
| Kyoga | 2700 | 2 | 1 | 46000 | 1 | 1204 | 1 | 170 | 0.14 | Tilapia |
| Malawi | 30800 | 426 | 1 | 57000 | 1 | 252 | 2 | 19 | 0.07 | Cichlids |
| Malawi (SE arm) | 1820 | 40 | 1 | 13800 | 7 | 252 | 2 | 76 | 0.30 | Cichlids |
| Malawi (SW arm) | 1210 | 35 | 1 | 7300 | 7 | 252 | 2 | 60 | 0.24 | Cichlids |
| Malombe | 390 | 4 | 1 | 12100 | 1 | 367 | 8 | 310 | 0.85 | Cichlids |
| Mweru | 4650 | 10 | 1 | 14123 | 1 | 124 | 1 | 30 | 0.24 | Tilapia |
| Nasser | 6216 | 25 | 9 | 34558 | 9 | 1934 | 10 | 56 | 0.03 | Tilapia |
| Tanganyika | 33000 | 570 | 1 | 103563 | 1 | 292 | 2 | 31 | 0.11 | kapenta |
| Turkana | 7570 | 30 | 1 | 7744 | 1 | 1177 | 11 | 10 | 0.01 | Tilapia |
| Victoria | 68000 | 20 | 1 | 500000 | 11 | 950 | 12 | 74 | 0.08 | Lates |
| Volta | 8270 | 19 | 4 | 38297 | 4 | 930 | 2 | 46 | 0.05 | Tilapia |

S = Sources: (1) Vanden Bossche & Bernacsek 1989; (2) Hecky & Kling 1987; (3) Burgis & Symoens 1987; (4) Vanden Bossche & Bernacsek 1990; (5) Craig 1992; (6) Marshall 1993; (7) Turner 1995; (8) Mwanyama 1993; (9) Vanden Bossche & Bernacsek 1991; (10) Abdel-Latif 1984; (11) Beadle 1981; (12) Pitcher 1995.

Overall, there is little or no difference in fishery transfer efficiencies between cichlid and clupeid lakes.

### 37.6    What might yields of *Lates* lakes be if the predator were absent?

The presence of Nile perch is associated with unstable, inefficient fisheries. This pattern is robust to a variety of analyses. Assuming that the presence of *Lates* is causing these low yields, it is possible to obtain a very rough approximation to the yields which could be expected from *Lates* lakes if the predator were not present, by fitting a model to the data from lakes where *Lates* is absent (but including Lake Tanganyika).

There is a statistically significant logarithmic relationship between transfer efficiency and mean lake depth ($r = 0.659$, 10 df, $P < 0.05$). Using the resulting regression equation it is possible to predict the yield of each of the 20 lakes or lake areas. However, the model is far too approximate to set reliable maximum sustainable yield estimates for particular lakes, particularly in the absence of confidence intervals. In some lakes, such as Chad, Chilwa and Turkana, dramatic fluctuations in water levels strongly influence yields. Irrespective, the model predicted that fishery yields in lakes

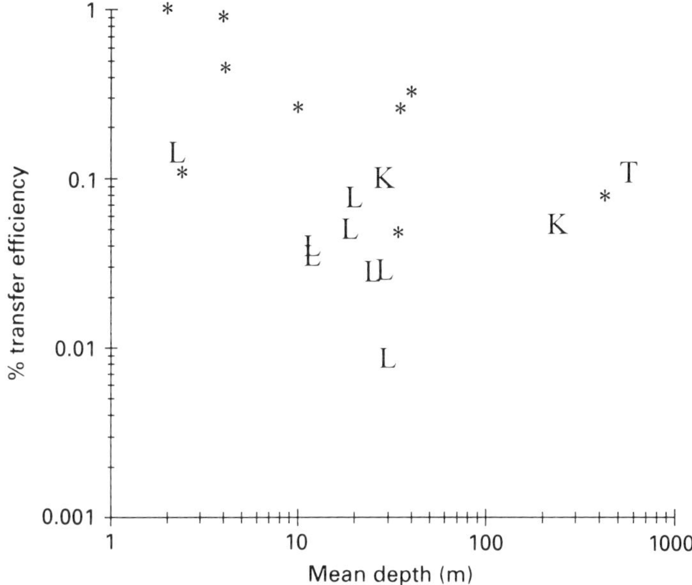

**Fig. 37.1**  Logarithmic plot of transfer efficiency against mean lake depth for 20 African lakes or areas within lakes. See Table 37.2 for sources. * = cichlid lakes; K = clupeid lakes; L = *Lates* lakes; T = Lake Tanganyika.

containing *Lates* (excluding Tanganyika) could be increased to between 2.5 and 20 times the current sustainable levels. For the eight *Lates* lakes (Albert, Chad, Kainji, Kyoga, Nasser, Turkana, Victoria, Volta) combined, this would represent a total increase in sustainable annual yields from 690 000 to 2.2 x $10^6$ t. This is around twice the combined total of the maximum unsustainable yields recorded – 1.06 x $10^6$ t. If it could be realised, this increase would double the total inland capture fishery yields for Africa.

### 37.7  Can the reduction of *Lates* populations by overfishing boost fishery yields?

In many lakes where *Lates* is present, yields of fisheries dominated by the predator have declined. However, details of species abundances are available in few cases and it is difficult to know if declines in *Lates* populations were accompanied by increases in the biomass of its prey species. In some cases there is also likely to be a lag period as available fishing technology is not appropriate for the efficient exploitation of prey species, or local demand for small, low-value species is inadequate to support the capital investment needed to develop a large-scale fishery.

**Lake Tanganyika**  Coulter (1991: p 119) illustrated how local declines in *Lates* populations in Lake Tanganyika led to a rapid increase in catch per unit effort of fisheries for pelagic clupeids. Large benthic species resembling the Nile perch found in

**Table 37.3** Analysis of covariance of transfer efficiencies of African lakes classified according to fish species composition. Mean lake depth is the covariate in all cases. All data logarithmically transformed. In all comparisons, the category on the left has the highest efficiency. *Lates* = Nile perch present; Clupeids = main fishery based on clupeids; cichlids = other lakes, fisheries generally dominated by cichlids. Lake Tanganyika is anomalous is anomalous in that it could be classed as either a clupeid lake or a *Lates* lake.

| Comparison | F value | $P$ | Outliers |
|---|---|---|---|
| a) Recent average yields for lakes with declining fisheries | | | |
| Cichlids v *Lates* | $F_{1,15} = 12.22$ | <0.01 | Lake Tanganyika |
| Cichlids v *Lates* (excluding Tanganyika) | $F_{1,14} = 20.56$ | <0.001 | Lake George |
| Cichlids v Clupeids (Tanganyika = Clupeids) | $F_{1,9} = 0.13$ | 0.727 | |
| Clupeids v *Lates* (Tanganyika = Clupeids) | $F_{1,8} = 5.72$ | <0.05 | |
| Rest v *Lates* (Tanganyika = *Lates*) | $F_{1,17} = 11.87$ | <0.01 | Lake Tanganyika |
| Rest v *Lates* (Tanganyika = Rest) | $F_{1,17} = 24.15$ | <0.001 | Lake George |
| b) Maximum yields from lakes with declining fisheries | | | |
| Cichlids v *Lates* | $F_{1,15} = 2.93$ | = 0.107 | |
| Cichlids v *Lates* (excluding Tanganyika) | $F_{1,14} = 4.58$ | <0.05 | Lake George |
| Clupeids v *Lates* (Tanganyika = Clupeids) | $F_{1,8} = 2.59$ | = 0.146 | |
| Rest v *Lates* (Tanganyika = *Lates*) | $F_{1,17} = 2.70$ | = 0.119 | |
| Rest v *Lates* (Tanganyika = Rest) | $F_{1,17} = 5.68$ | <0.05 | Lake George |

Lake Victoria, and elsewhere, suffered rapid population declines, while the small (45 cm) elongate pelagic *Lates stappersi* was better able to withstand heavy exploitation (Coulter 1991: p.126). Large benthic *Lates* are important predators of cichlids and local reductions in their populations in Zambia were also accompanied by increases in gillnet catches of small demersal cichlids (Coulter 1991: p.198), although these are not seriously exploited as yet.

Coulter (1991: p 135) reported that heavy exploitation in Lake Tanganyika rapidly led to the reduction in populations of large *Lates* species in the early 1960s in Burundi and the mid- to late-1960s in other parts of the lake. The total fishery yield from the Burundian part of Lake Tanganyika was 7100 t in 1962 (Vanden Bossche & Bernacsek 1989). In 1966, Tanzanian and Zambian fisheries landed 15 000 and 6500 t respectively. The only estimate for that period for the Zaïrian shore was 7700 t in 1970. This gives a total estimated yield of 36 300 t for the fishery before the reduction of predator populations, compared to 85 000–100 000 t since. The depletion of *Lates* populations in the 1960s, much earlier than occurred in other lakes, may, in part, explain why the transfer efficiency of Lake Tanganyika is an outlier when it is considered as a *Lates* lake, although the considerable evidence that Lake Tanganyika has an unusually productive pelagic food chain should not be overlooked (Coulter 1991).

**Lake Nasser**     After the closure of the Aswan High Dam, fish yields increased steadily from 1968 to 1982 (Fig. 37.2a). Yields of *Lates niloticus* reached a plateau in 1970 (Fig. 37.2b). However, tilapia yields continued to increase throughout the period, reaching 30 000 tonnes by 1982. The steadily increasing catch per fisher suggests that the fishery is becoming more efficient, perhaps through use of smaller meshes. Thus

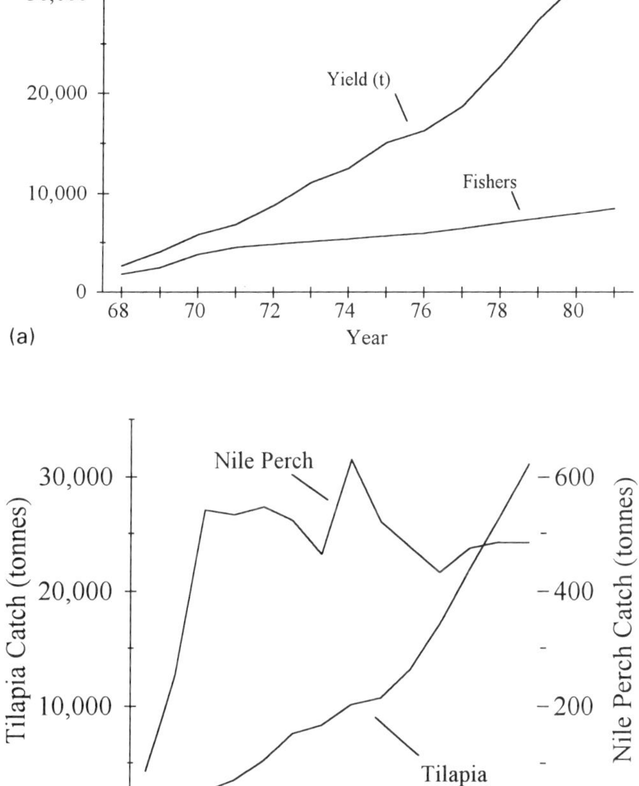


**Fig. 37.2**   (a) Fish yield on Lake Nasser has expanded more rapidly than the population of fishers (data from Vanden Bossche & Bernacsek 1991, with some interpolation of numbers of fishers). (b) The expansion of the fishery has been mainly based on increasing yields of tilapia. Nile perch yields have remained roughly constant since 1970 (data: percentage species composition from Abdel-Latif 1984, Table 17, with some interpolation; yields from Vanden Bossche & Bernacsek 1991).

the roughly constant *Lates* catch may mask a decline in biomass, which may in turn lead to an increase in tilapia populations. Recent data for this fishery are not available.

**Lake Kainji**    Like many other reservoirs, productivity in Lake Kainji was highest in the first few years after impoundment (Kapetsky & Petr 1984). Unusually, a large fisher population was available to exploit the resulting boom in riverine fish populations (Fig. 37.3a). Thereafter, yields rapidly declined and the number of fishers stabilized. Populations of anadromous fishes (cyprinids, characoids) collapsed in the early years of exploitation (Ita 1984), along with *Lates* populations (Fig. 37.3b). Between 1975 and the end of the available data series in 1977, tilapia stocks increased rapidly, perhaps as a result of a reduction in predation by *Lates*. More recent catch data are not available, but a recent survey of fishers (Du Feu 1993a) showed that the

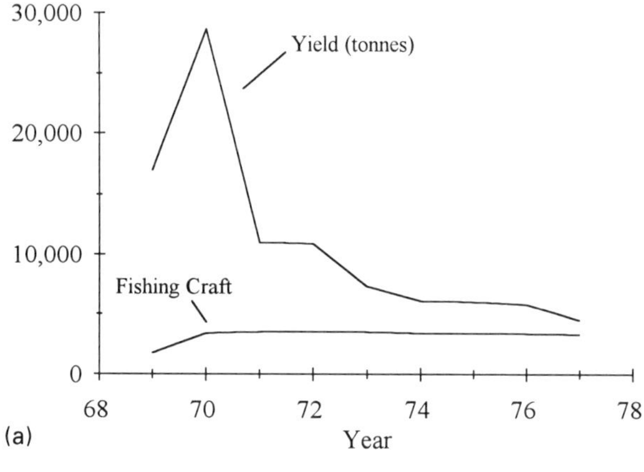


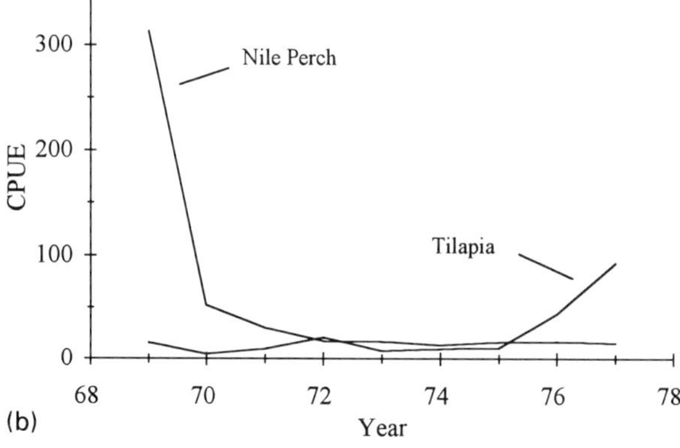


**Fig. 37.3**    (a) Fishery yields on Lake Kainji declined rapidly after an initial post-impoundment boom, while the fishing fleet remained roughly constant (data from Vanden Bossche & Bernacsek 1990). (b) Populations of Nile perch declined rapidly, while those of tilapias increased in the mid-1970s (data: CPUE in weight from experimental fishing, units not given, from Ita (1984), Table 20).

size of the fishing industry was almost three times that in 1977, and most gears employed very small meshes (Du Feu 1993b). It is not known if this expansion of effort led to an increase in catch.

**Lake Kyoga**  Fishery yields in Lake Kyoga declined rapidly in the late 1980s, although this may be more apparent than real, because it appears that the data collection system changed during this period and that the high yields of the late 1980s were greatly overestimated (Vanden Bossche & Bernacsek 1989). *Lates* were previously the main species caught, but by 1985, tilapia catches increased to 79% of the total compared to 16.7% for *Lates* (Ogutu-Ohwayo 1990).

**Lake Victoria**  Lake Victoria supports rapidly increasing yields of Nile perch (Ogutu-Ohwayo 1990; Pitcher 1995). It is presently the only African lake where this is occurring. The most likely explanations for this are: (1) the unstable nature of the ecosystem caused by the elimination of 95% of the indigenous fish species and the rapid eutrophication of the lake; (2) the accumulation of biomass in the long-lived Nile perch as a result of its never-to-be-repeated consumption of the haplochromine cichlid populations in the 1980s; (3) the underexploitation of the fishery: the logarithmic regression model presented above suggests a maximum yield of $1.25 \times 10^6$ t for Lake Victoria in the absence of Nile perch.

It has long been accepted that the tilapia fisheries of Lake Victoria were seriously overexploited in the 1950s. Variations on the same graph of declining tilapia CPUE have appeared in numerous publications (e.g. Fryer & Iles 1972; Lowe-McConnell 1987; Ogutu-Ohwayo 1990). It is possible that Lake Victoria's unique tilapia species were uniquely vulnerable to overfishing, perhaps because they were confined to restricted inshore areas. However, from the data (Vanden Bossche & Bernacsek 1989), it is clear that total yields of all fish species showed a steady increase in all three lakeshore countries throughout the period of the reported decline in tilapia populations: Tanzania: 26 000–48 000 t (1958–1970); Uganda: 10 500–34 800 t (1955–1970); Kenya: 12 000–16 400 t (1964–1970). Even if tilapia CPUE was declining, fishery yields were expanding well before *Lates* became an important part of the catch.

*Lates* is not the only species introduced into Lake Victoria. *Oreochromis niloticus* (L.) is a tilapia which often lives sympatrically with Nile perch and has overtaken the predator as the mainstay of the fisheries in Lakes Kainji, Kyoga and Nasser, as well as in the River Nile (Abdel-Latif 1984). It is possible that the same will eventually happen in Lake Victoria.

Underexploited Nile perch populations, like those of other large species, will include a great many large, old fish which will contribute to high initial yields which cannot be sustained in the long term. Declines in yields of 50–80%, as reported here, are well within the expected range for such species (R.J.H. Beverton, pers. comm.). Although clearly susceptible to growth overfishing, it seems unlikely that Nile perch will be vulnerable to recruitment overfishing (R. Ogutu-Ohwayo pers. comm.). *Lates* has a very high fecundity in comparison to most other freshwater species. If smaller specimens contributed greatly to the reduction in transfer efficiencies, it is possible that some of the species' detrimental effects cannot be reversed. While it appears that yields of other species can be increased, it is not clear if this is due to a reduction in predation, or simply a re-direction of effort to smaller species. From available data, it is uncertain whether overfishing of Nile perch can lead to an increase in the transfer efficiency of the fisheries. Until this question is answered, it remains possible that Nile perch may irreversibly reduce fishery yields in any water body to which it is introduced.

## 37.8    Management goal: profit or protein?

When large predatory *Lates* are present in a lake, the fisheries tend to be unstable and inefficient. Yet the expansion of Nile perch populations in Lake Victoria was accompanied by a five-fold increase in yields and the fish is locally known as the 'saviour'. Initial reports (e.g. Barel *et al.* 1985) that Nile perch was unpopular with fishers and difficult to market proved to be unfounded. Nile perch is economically valuable and is in high demand for export from lakeshore regions for consumption by citizens of developed countries or by the wealthier sector of the local urban population. Thus artisanal fishermen have access to a resource with a high cash value and are better able to invest in expansion of their fishing activities. For this reason, export-driven fisheries will often expand more rapidly than fisheries which meet the demands of a rural population with little access to cash.

Essentially Nile perch can be considered as a cash crop, like tobacco, cotton and cut flowers, while smaller fishes represent a protein source for the poorer sector of the population. Large predators often require capital-intensive filleting and freezing plants and may generate considerable added profit to processors, transporters, and retailers. This may bring valuable foreign exchange to a hard-pressed government.

Like large fish, smaller species may also bring cash to lakeshore regions, but only if there is an adequate market in inland areas, as there is in Malawi (Mdaihli *et al.* 1992). In most African nations the human populations are growing at 2–3% per annum, i.e. doubling roughly every 25–35 years. At present there is still consumer resistance to small fish in some countries, but this is unlikely to last long, as the rapidly growing populations will increase the demand for protein in any form. Small fish are easily and cheaply processed by sun-drying in the lakeshore areas, and transported in small loads via public transport. Thus, they bring little added profit to heavily capitalized organizations. Markets for small species are usually within the country of origin or fish are smuggled over borders. This generates little foreign exchange.

In the long term, the choice is between maximizing profits or protein yields. This is essentially a political decision. Fishery biologists can only point out the implications of alternative strategies and advise on how best to implement them.

### 37.9    Management for protein yields

If African lake fisheries are to be managed to maximize yield rather than profit, this will require an alternative approach to fishery management. At present, many African lake fisheries are nominally managed through technical measures such as restrictions on mesh size, gear headline length, fishing season, and minimum size of fish to be landed. These regulations are usually directed towards preserving stocks of large predators or anadromous fishes which support the least productive fisheries. In practice these regulations are often not enforced. Surveys carried out on Lakes Malawi and Malombe in 1990 and on Lake Kainji in 1993 showed that almost all fishing gears in use were technically illegal (Mdaihli & Donda 1991; Du Feu 1993b). The huge coastlines of these lakes, underfunding of enforcement agencies and nocturnal nature of most artisanal fishing means that these regulations are probably impossible to enforce. Furthermore, it is only by breaking the regulations that fishers are able to exploit the most productive fisheries, i.e. those for small species.

Fishery regulations are often so complex that they are not understood by enforcement staff, let alone fishers: Malawi has 10 different combinations of restrictions of headline length, mesh size, net depth, permitted fishing season and time of day for seine nets alone, while the Sokoto State of Nigeria has minimum landing sizes for 10 different fish taxa.

Attempts to enforce such a bewildering variety of regulations often lead to hostility between fisheries departments and fishing communities, or to apathy or acceptance of bribes by enforcement staff.

In practice, when left alone, fisheries in African lakes tend to follow a path of increasing effort, decreasing mesh sizes and a progressive switch from fisheries based on anadromous species and predators, through tilapias, to small pelagic and benthic species. This results in progressively increasing yields, although it may cause a loss in profits.

In some cases, particular fishery resources may remain underexploited for technological, economic or social reasons. It cannot be assumed that full exploitation of all stocks will come about through the unaided expansion of the artisanal fishery.

Attempts to manage African inland fisheries to maximize profits through sustainable exploitation of large predatory species are not only dubious in their intention, but almost certainly impossible in practice.

### 37.10    The problem of diverse faunas

Lakes Malawi, Tanganyika and Victoria each contain hundreds of endemic species, most of the family Cichlidae. These represent a substantial proportion of all extant vertebrate species and their study may be critical to our understanding of the mechanism of evolution (Turner 1994a). These fishes are not seriously exploited in Lake Tanganyika, although they may be in the future. In Lake Victoria, the introduced Nile perch eliminated most of the commercially exploitable species: those remaining are principally inhabitants of shallow rocky areas where they are largely immune from both *Lates* and fishing gear (Witte *et al.* 1992). By contrast, there are very productive fisheries for small cichlids on Lakes Malawi and Malombe. Although yields from these fisheries, to date, are very stable (Tweddle & Magasa 1989), there were striking changes in the communities (Turner *et al.* 1995). The local distributions and low fecundities, coupled with the economics of the fishing industry make it possible that larger slow-maturing haplochromine cichlid species could be driven to extinction (Turner 1994b). Elimination of predators may lead to increased fishery yields, but in these complex communities there are many species specialized in feeding on resources which might not be accessible to smaller species, such as hard-shelled molluscs or deep-burrowing insect larvae. While maintenance of the full virgin stocks of these species is not likely to be compatible with an efficient fishery, perhaps the preservation of small populations could help boost yields as well as maintain biological diversity. This is probably best achieved, not through attempts to limit the whole fishery through unenforceable technical measures, but through the establishment of areas closed to the more destructive fishing practices such as trawling and seining (Turner 1994b). Staff and resources could be efficiently concentrated on a limited area and the regulations would be very simple: no fishing, except perhaps with hooks and baskets. Similar principles could be applied in other lakes to conserve stocks of anadromous species or large macrophyte or invertebrate feeders.

### 37.11    Preparing for overexploitation

It has been argued above that to maximize total yields of fisheries on African lakes, legislation aimed at preserving fisheries on larger species should be abolished. This does not mean that overexploitation is not possible.

Marine fisheries based on small short-lived pelagic species are vulnerable to overfishing, especially when it is combined with environmentally-induced recruitment failures (Beverton 1990). Simplification of lacustrine pelagic fish communities through heavy exploitation may lead to a similar instability.

Small demersal cichlids generally have low fecundity and produce few large young. This life history strategy is likely to reduce the variability in recruitment and thus lead to more stable populations. However, they are probably also vulnerable to recruitment overfishing. At present, the major exploited species on Lake Malombe are the smallest cichlid species in the community, and these are largely being exploited before they reach sexual maturity (Tweddle *et al.* 1995). It is possible that a reduction in the populations of algivorous fish could lead to eutrophication, deoxygenation and large scale fish kills (Mwanyama 1993).

With fishing activities targeting a wide variety of species in many African lakes, fluctuations in populations of any one species are not of great significance, as fishers can continue to exploit other species. If other species are eliminated, then the socio-economic consequences will be more severe.

It is very difficult to determine the maximum sustainable yield of a fishery until it is overexploited (Hilborn & Walters 1992). When that point is reached, there is inevitably overcapacity in the fishing industry. It will then be essential to limit overall fishing effort, either through buying out existing fishers, or through a reduction in the effort of each fisher. In either case, open access to the fishing industry must be discontinued. In preparation for this, African fisheries departments should license artisanal fishermen. In many cases, licence-holding is a legal requirement, but is rarely enforced. It should not be viewed as a tax on fishers, but should provide fishers with access to positive incentives, in the form of loans, advice, access to landing and processing facilities. The removal of unnecessary technical measures should help establish the goodwill between the fishing communities and fisheries departments which will be necessary for the eventual overall limitation of the fishery. These measures should be introduced at the earliest stage in the development of the fisheries.

### 37.12  Species introductions

Further introductions of Nile perch cannot be justified at present. Although it may boost profits and foreign exchange earnings for a period, it is not clear if this is sustainable for more than a few years. The introduction may lead to an irreversible reduction in the potential yield from the lake. It seems hard to justify this in the face of Africa's rising demand for protein.

In lakes without Nile perch there is no appreciable difference in yields between lakes with or without clupeids. However, it seems unwise to conclude that clupeid introductions are unjustifiable. Prior to the introduction of clupeids, there were no small pelagic fish species in Lakes Cahora Bassa, Kariba and Kivu. Cahora Bassa and Kariba are man-made, while Lake Kivu is less than 20 000 years old (Marshall 1993, Chapter 26). It is likely that clupeid introductions have increased the potential yields of all of these lakes. However, there seems no reason to suppose that introduction of clupeids could be beneficial in any of the 'cichlid lakes' considered in this chapter. Most of these are relatively shallow and benthic species are probably able to fully exploit the primary productivity of such lakes. The exception is Lake Malawi, which has an old and exceptionally diverse fauna, including a number of specialized pelagic cichlids (Turner 1994c).

The introduction of clupeids into other young deep lakes, especially reservoirs, may increase potential yields. It is not yet clear what effect Nile perch has on the productivity of clupeid fisheries, nor is it known if small pelagic cyprinids can support fisheries as efficient as those based on clupeids. Future developments of fisheries in Lakes Kainji, Volta and Victoria may provide answers to these questions.

### References

Abdel-Latif A.-F. (1984) Lake Nasser (Egypt). In: J.M. Kapetsky & T. Petr (eds) *Status of African Reservoir Fisheries*. Rome: *CIFA Technical Paper* **10**, 193–246.
Barel C.D.N., Dorit R., Greenwood P.H., Fryer G., Hughes N., Jackson P.B.N., Kawanabe H., Lowe-

McConnell R.H., Ribbink A.J., Trewavas E., Witte F. & Yamaoka K. (1985) Destruction of fisheries in Africa's Lakes. *Nature* **315**, 19–20.

Barel C.D.N., Ligtvoet W., Goldschmidt T., Witte F. & Goudswaard P.C. (1991) The haplochromine cichlids in Lake Victoria: an assessment of biological and fisheries interests. In: M.H.A. Keenleyside (ed.). *Cichlid Fishes: Behaviour, Ecology and Evolution.* London: Chapman & Hall, pp 258–279.

Beadle L.C. (1981) *The Inland Waters of Tropical Africa, 2nd. Edn.* London: Longman, 475 pp.

Beverton R.J.H. (1990) Small marine pelagic fish and the threat of fishing: are they endangered? *Journal of Fish Biology* **37 (Supplement A)**, 5–16.

Bootsma H.A. & Hecky R.E. (1993) Conservation of the African Great Lakes: a limnological perspective. *Conservation Biology* **7**, 644–656.

Burgis M.J. & Symoens J.J. (eds) (1987) *African Wetlands and Shallow Water Bodies.* Paris, ORSTOM, 650 pp.

Coulter G.W. (ed.) (1991) *Lake Tanganyika and Its Life.* London: Oxford University Press, 354 pp.

Craig J.F. (1992) Human-induced changes in the composition of fish communities in the African Great Lakes. *Reviews in Fish Biology and Fisheries* **2**, 93–124.

Du Feu T. (1993a) *Frame Survey of Kainji Lake.* New Bussa: Nigerian German (GTZ) Kainji Lake Fisheries Promotion Project, 36 pp.

Du Feu T. (1993b) *Fishing Gear Survey of Kainji Lake.* New Bussa: Nigerian German (GTZ) Kainji Lake Fisheries Promotion Project, 25 pp.

Fryer G. & Iles T.D. (1972) *The Cichlid Fishes of the Great Lakes of Africa.* Edinburgh: Oliver & Boyd.

Gulland J.A. (1970) Food chain studies and some problems in world fisheries. In: J.H. Steele (ed.) Marine Food Chains. Berkeley: University of California Press, pp 296–315.

Hecky R.E. & Kling H.J. (1987) Phytoplankton ecology of the great lakes in the rift valleys of Central Africa. *Archiv für Hydrobiologie* **25**, 197–228.

Hecky R.E., Fee E.J., Kling H.J. & Rudd J.M.W. (1981) Relationship between primary production and fish production in Lake Tanganyika. *Transactions of the American Fisheries Society* **110**, 336–345.

Hilborn R. & Walters C.J. (1992) *Quantitative Fisheries Stock Assessment: Choice Dynamics and Uncertainty.* New York: Chapman & Hall, 570 pp.

Ita E.O. (1984) Kainji (Nigeria). In: J.M. Kapetsky & T. Petr (eds.) *Status of African Reservoir Fisheries. CIFA Technical Paper* **10**, 43–103.

Kapetsky J.M. & Petr T. (eds) (1984) *Status of African Reservoir Fisheries. CIFA Technical Paper* **10**, 326 pp.

Kaufman L. (1992) Catastrophic change in species-rich freshwater ecosystems. The lessons of Lake Victoria. *BioScience* **42**, 846–858.

Lemoalle J., Adenji A., Compère P., Ganf G.G., Melack J. & Talling J.F. (1981) Phytoplankton. In: J.J. Symoens, M. Burgis & J.L. Gaudet (eds) *The Ecology and Utilisation of African Inland Waters.* Nairobi: United Nations Environment Programme, pp 37–50.

Lowe-McConnell R.H. (1987) *Ecological Studies in Tropical Fish Communities.* Cambridge: Cambridge University Press, 382 pp.

Marshall B.E. (1991) The impact of the introduced sardine *Limnothrissa miodon* on the ecology of Lake Kariba. *Biological Conservation* **55**, 151–165.

Marshall B.E. (1993) Biology of the African clupeid *Limnothrissa miodon* with reference to its small size in artificial lakes. *Reviews in Fish Biology and Fisheries* **3**, 17–38.

Mdaihli M. & Donda S. (1991) Fishermen entrepreneurs – a baseline survey. *FI:DP/MLW/86/013, Field Document 11, FAO, Rome*, 65 pp.

Mdaihli M., Hara M.M. & Banda M.C. (1992) Fish marketing in Lake Malombe, the Upper Shire River and the South East Arm of Lake Malawi. *FI:DP/MLW/86/013, Field Document 16, FAO, Rome*, 48 pp.

Melack J.M. (1976) Primary productivity and fish yields in tropical lakes. *Transactions of the American Fisheries Society* **105**, 575–580.

Mwanyama N.C. (1993) Relations entre les ressources alimentaires, l'alimentation et la reproduction des Chambos (poissons tilapias du genre *Oreochromis*) dans les lacs Malawi et Malombe (Afrique centrale). PhD thesis, Université de Provence-Aix-Marseille, 158 pp.

Oglesby R.T. (1977) Relationships of fish yield to lake phytoplankton standing crop, production and morphoedaphic factors. *Journal of the Fisheries Research Board of Canada* **34**, 2271–2279.

Ogutu-Ohwayo R. (1990) The decline of the native fishes of Lakes Victoria and Kyoga (East Africa) and the impact of introduced species, especially the Nile perch, *Lates niloticus* and the Nile Tilapia, *Oreochromis niloticus*. *Environmental Biology of Fishes* **27**, 81–96.

Pitcher T.J. (1995) Thinking the unthinkable: a candidate model for predicting sustainable yields of introduced species in African lakes. In: T.J. Pitcher & P.J.B. Hart (eds). *The Impact of Species Changes in African Lakes*. London: Chapman & Hall.

Ryder R.A. (1965) A method for estimating the potential fish production of north-temperate lakes. *Transactions of the American Fisheries Society*, **94**, 214–218.

Schaefer M. (1954) Some aspects of the dynamics of populations important to the management of commercial marine fisheries. *Bulletin of the Inter-American Tropical Tuna Commission* **1**, 27–56.

Turner G.F. (1994a) Speciation mechanisms in Lake Malawi cichlids: a critical review. *Archiv für Hydrobiologie: Advances in Limnology* **44**, 139–160.

Turner G.F. (1994b) Fishing and the conservation of the endemic fishes of Lake Malawi. *Archiv für Hydrobiologie: Advances in Limnology* **44**, 483–496.

Turner G.F. (1994c) A description of a commercially important new pelagic species of the genus *Diplotaxodon* (Pisces: Cichlidae) from Lake Malawi, Africa. *Journal of Fish Biology* **44**, 799–807.

Turner G.F. (1995) Management and conservation of exploited fish stocks in Lake Malawi. In: T.J. Pitcher & P.J.B. Hart (eds) *The Impact of Species Changes in African Lakes*. London: Chapman & Hall.

Turner G.F., Tweddle D. & Makwinja R.D. (1995): Species composition changes of demersal cichlid communities as a result of trawling in southern Lake Malawi. In: T.J. Pitcher & P.J.B. Hart (eds). *The Impact of Species Changes in African Lakes*. London: Chapman & Hall.

Tweddle D. & Magasa J.H. (1989). Assessment of multispecies cichlid fisheries of the Southeast Arm of Lake Malawi, Africa. *Journal du Conseil* **45**, 209–222.

Tweddle D., Turner G.F. & Seisay M.B.D. (1995) Changes in species composition and abundance as a consequence of fishing pressure in Lake Malombe. In: T.J. Pitcher & P.J.B. Hart (eds.) *The Impact of Species Changes in African Lakes*. London: Chapman & Hall.

Vanden Bossche J.-P. & Bernacsek G.M. (1989) Source book for the inland fishery resources of Africa. Volume 1, *CIFA Technical Paper* 18/1 Rome: FAO, 411 pp.

Vanden Bossche J.-P. & Bernacsek G.M. (1990) Source book for the inland fishery resources of Africa. Volume 2. *CIFA Technical Paper* 18/2 Rome: FAO, 240 pp.

Vanden Bossche J.-P. & Bernacsek G.M. (1991) Source book for the inland fishery resources of Africa, Volume 3. *CIFA Technical Paper* 18/3, Rome: FAO, 219 pp.

Witte F., Goldschmidt T., Wanink J., van Oijen M., Goudswaard K., Witte-Maas E. & Bouton N. (1992) The destruction of an endemic species flock: quantitative data on the decline of the haplochromine cichlids of Lake Victoria. *Environmental Biology of. Fishes* **34**, 1–28.

Youngs W.D. & Heimbuch D.G. (1982) Another consideration of the morphoedaphic index. *Transactions of the American Fisheries Society* **111**, 151–153.

# Chapter 38
# Community-based management in inland fisheries: case studies from two Malaysian fishing communities

A.B. ALI *School of Biological Sciences, Universiti Sains Malaysia, 11800 Minden, Penang, Malaysia*

**Abstract**   Artisanal fisheries of Malaysia are characterized by their small-scale, closed-community and limited entry. Fishing is a secondary occupation and catches supplement income, rather than being the main money earner. The fisheries of two fishing communities, one from Chenderoh Reservoir and another from Kerian River, are described. The reservoir fisheries of Chenderoh are more established and better structured compared to the Kerian River fisheries. Almost all fishers use gillnets (30.8 to 93.8% catch; stretched-meshed – 5.7, 10.2 and 11.4 cm), with a few individuals specializing in carnivorous species such as *Channa micropeltis* and *Notopterus chitala* using long lines (30 to 40 m long). Individual fishers use wooden boats (4 to 6 m long; 1.0 to 1.5 m wide) fitted with a 5 hp long-tailed gasoline engine and either utilize 15 to 20 sets of gillnets each 3 m deep and 30–40 m long or 1 to 2 sets of long lines. Entry into the fishery is limited to local villagers and is enforced by setting-up specific landing sites. A middleman controls and monopolizes a landing site. He extends small credit to the fishers for engine or boat repairs when needed. The middleman also determines the fish size to be landed and enforces this by refusing to buy small fish or brooders.

The Kerian River fishery is unlimited entry and less structured compared to Chenderoh Reservoir. The fishing community is more loosely connected and, as such, co-operation among fishers is weaker. The equipment consists of wooden boats (4–5 m long and 1 m wide) operated manually or fitted with outboard engines (2–5 hp). Fish are caught primarily by using 'chandik', a hut built in the middle of the river and fitted with a conical nylon-net bag to trap fish. There is more than one middleman per landing site, hence there is open bidding for fish. However, regulation concerning size of fish landed cannot be enforced since middlemen compete among themselves for limited catch and will accept small-sized fish.

Fishers in both communities have a good understanding of ecological principles. They recognize the importance of protecting brood stocks, hence the taboo against catching gravid fish, especially among fishers in Kerian River. In both areas, riparian and littoral vegetation is well maintained to provide protection and nursery areas for breeding adults, fry and fingerlings.

KEYWORDS: Community-based management, fishing communities, fisheries management, Malaysia

## 38.1    Introduction

Fish is an important protein source for rural communities in Malaysia. Inland settlements established along river banks for easy transportation and accessibility, also fish the rivers for their livelihood.

The need for sustainability in exploiting these resources has resulted in cultures, customs, taboos and beliefs that adapt to, rather than destroy, the environment and deplete the resources. Co-operation among members of such communities has become an important management tool of the commonly-owned resources. This chapter describes the evolution of community-based management practised by two different fishing communities faced with different situations and problems.

## 38.2    Materials and methods

Data were collected from creel survey and unstructured interviews with fishers and middlemen (Johannes 1993). Since there were no previous data available, reliance was placed on interviews with older generations of fishers for historical information.

The Chenderoh study was conducted from 1988 to 1989. Interviews were conducted between 07.00 and 08.00 h at landing sites as the fishers brought in their catches. Since there are only three landing points all 30 fishers and the three middlemen were interviewed either individually or in groups. These interviews were conducted monthly.

For the Kerian River study, two visits were made to the main landing point in 1993 and again in 1994 to interview the fishers. Since no middlemen were present during the visits none was interviewed. The Kerian River fisheries have multiple entry and exit points, therefore the survey trips were taken in boats along the Kerian River and its main tributary (Samagagah branch) during the visits to meet as many fishers as possible. Visits to the homes of older fishers were also made to get historical information on the fisheries.

## 38.3    Description of the study areas

Chenderoh Reservoir is located on the Perak River and was constructed in 1930 for flood control and power generation (Fig. 38.1). The damming resulted in a reservoir with an area of 21 km$^2$. It is the first in the series of four dams built in the upper reaches of the Perak River. The reservoir consists of two major parts; the mainstream channel and the coves. The water in the channel is deep and fast flowing and the

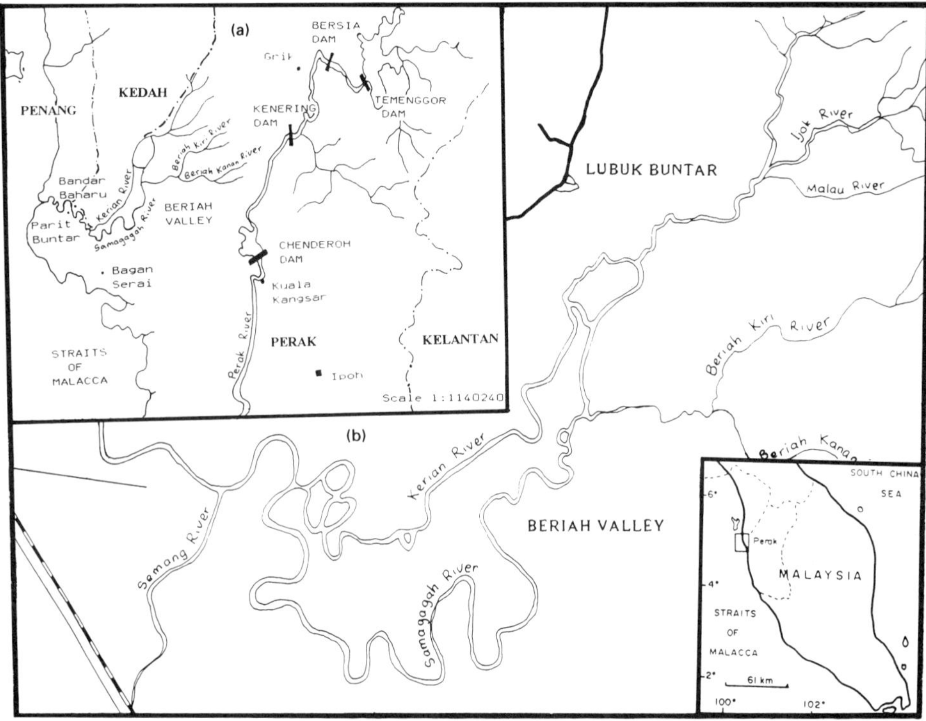


**Fig. 38.1** Locations of the study areas; (A) Chenderoh Reservoir on the Perak River and (B) the Kerian River, Perak, Malaysia.

velocity can increase substantially during power generation (Lee 1989). The reservoir is mesotrophic, receiving effluent from the surrounding littoral villages (Khan 1987; Lee 1989). A summary of some water quality parameters is shown in Table 38.1. Fish populations are affected by the fluctuating water levels, especially during rainy seasons when species such as the cyprinid *Thynnichthys thynnoides* (Bleeker) start breeding (Bat Kamaruzzaman 1992).

**Table 38.1** Selected water quality parameters for Chenderoh Reservoir, Perak, Malaysia (Modified from Lee 1989).

| Parameters | Unit | Coves | Mainstream channel |
|---|---|---|---|
| Temperature | °C | 2.6 | 29.5 |
| Water current | m s$^{-1}$ | 0.3 | not detected |
| Visibility (Secchi disk) | cm | 55 | 95 |
| Depth | m | 3.72 | 4.14 |
| D.O. | mg l$^{-1}$ | 7.9 | 6.9 |
| pH | | 7.1 | 6.7 |
| Conductivity | µS cm$^{-1}$ | 319 | 277 |

Kerian River is located to the north of Perak River and serves as the border between Kedah/Penang and Perak States (Fig. 38.1). The fisheries are not concentrated but spread along the river banks with many access points. The fisheries studied were limited to the portion of the Kerian River around the town of Lubok Buntar. Fisheries in Samagagah River, the main tributary of the Kerian River, were also studied. In general the river bed is shallow, ranging from 3 to just 0.5 m in some areas due to siltation caused by land clearing for agriculture. Tables 38.2 and 38.3 show some water quality parameters for selected stations along the river. River flow is not regulated, thus water levels fluctuate greatly (up to 1.5 m). In the areas studied, the Kerian River and its tributaries drain into swamps. During the rainy season the swamps and riparian areas are flooded to about 1 to 1.5 m of water and fish spawning probably occurs during this period in the riparian zones as well as the swamps, when food supply is plentiful. The swamps also maintain and regulate water flow to the lower parts of the Kerian River during the dry season.

**Table 38.2**  Water quality parameters for Kerian, Samagagah and Beriah rivers, Perak-Kedah, Malaysia in 1993.

| Station | Time (h) | D.O. (mg $l^{-1}$) | °C | Conductivity ($\mu Scm^{-1}$) | TDS (mg $l^{-1}$) | pH |
|---|---|---|---|---|---|---|
| Kerian | | | | | | |
| A | 1042 | 6.2 | 25.5 | 23.1 | 10.1 | 5.7 |
| B | 1100 | 6.2 | 25.5 | 26.5 | 13.2 | 5.7 |
| C | 1120 | 3.9 | 27.0 | 123.7 | 61.9 | 4.2 |
| Samagagah | | | | | | |
| D | 1145 | 3.7 | 27.0 | 117.2 | 58.2 | 4.1 |
| E1 | 1215 | 4.2 | 27.0 | 134.6 | 67.8 | 4.4 |
| E2 | 1225 | 4.8 | 27.0 | 50.9 | 18.7 | 4.5 |
| Beriah | | | | | | |
| F1 | 1255 | 2.0 | 27.0 | 57.5 | 28.7 | 4.1 |

**Table 38.3**  Water quality parameters for the Kerian River, Perak-Kedah, Malaysia in February 1994.

| Station | Time (h) | D.O. (mg $l^{-1}$) | °C | Conductivity ($\mu S\ cm^{-1}$) | pH | $NO_3^--N$ (mg $l^{-1}$) | $PO_4^--P$ (mg $l^{-1}$) |
|---|---|---|---|---|---|---|---|
| 1 | 1122 | 2.8 | 26.0 | 43.4 | 6.0 | 0.036 | 0.29 |
| 2 | 1145 | 1.6 | 26.5 | 66.0 | 6.0 | 0.031 | 0.31 |
| 3 | 1200 | 6.5 | 26.0 | 36.2 | 6.2 | 0.058 | 0.25 |
| 4 | 1236 | 6.2 | 26.0 | 28.0 | 6.2 | 0.028 | 0.48 |
| 5 | 1257 | 7.5 | 27.0 | 41.7 | 6.1 | — | — |

## 38.4     Description of the fisheries

### 38.4.1     *Chenderoh Reservoir*

The Perak River supports many riverine species including *Thynnichthys thynnoides* (Bleeker), *Hampala macrolepidota* van Hasselt, *Labiobarbus festivus* (Heckel), *Puntius schwanenfeldii* (Bleeker), *Puntius bulu* (Bleeker), *Channa striata* Bloch and *Notopterus chitala* (Hamilton) (Table 38.4). Studies on the tailwaters of the Chenderoh Reservoir indicated the presence of these species despite the fisheries presently being adversely affected by the low flow regimes (Ali & Kathergany 1987; Lee and Ali 1989). Thus the fisheries of the Perak River were probably well developed prior to the dam's construction.

With the creation of the reservoir, the lotic environment changed with the concomitant change of fish species to lacustrine types. Species such as T. *thynnoides*, *P. bulu* and *P. schwanenfeldii* became dominant in the catches. Creel studies (Lee 1989; Ali & Lee 1995) indicated that fishers were using long lines and cast nets, and later gillnets (Table 38.5). Wooden rowing boats were used originally; however, in the late 1970s fishers began to use long-tailed petrol engines for propulsion. Outboard motors were not preferred due to the heavy infestation of *Hydrilla verticillata* Royle which tends to entangle these types of motors more than the long-tailed types. With improvements in fishing technology fishers began to land more fish and the problem of overfishing started.

Chenderoh Reservoir fisheries were initially open-access with sport and non-resident fishers competing with local fishers for fish. Open-access, coupled with lack of regulation and concrete management plans, resulted in overfishing. Although there were no previous data available for the 1970s, interviews with older fishers indicate the shift to small-sized fish and overall drop in catches. Studies in the 1980s confirmed a drastic drop in landings from 167 t (Ajan 1983) to 31 t per annum (Lee 1989). As the catches continued to decline local fishers began to assert their authority. Outsiders were strongly discouraged from fishing in the reservoir and the local fishers from different villages organized themselves into fishing groups with specific landing points to be serviced by specific middlemen.

### 38.4.2     *Kerian River*

Fishing in the Kerian River is carried out with 'chandik', which is a wooden hut built over the water in the middle of the river. During high water periods, a conical nylon net is set underneath the hut to catch fish that move downstream with the current. After setting the net for a certain period of time, usually overnight, the net is lifted onto the floor of the hut and the fish sorted. Common riverine species such as *Mystus* sp., *Labiobarbus festivus*, *Puntius schwanenfeldii*, *Hampala macrolepidota* and *Channa striata* are caught (Table 38.4). Usually a catch of between 30 and 40 kg is obtained with one set.

**Table 38.4**  Commonly caught fish species of Chenderoh Reservoir, Perak and Kerian River, Perak-Kedah, Malaysia.

| Species | Chenderoh Reservoir[1] | Kerian River |
| --- | --- | --- |
| Cyprinidae | | |
| *Cyclocheilichthys apogon* (C&V) | + | + |
| *Chela anomalura* (Van Hasselt) | + | + |
| *Hampala macrolepidota* (Van Hasselt) | + | + |
| *Labiobarbus* spp. | + | + |
| *Mystacoleucas marginatus* (C&V) | + | + |
| *Osteochilus hasselti* (C&V) | + | + |
| *Puntius bulu* (Bleeker) | + | + |
| *Puntius schwanenfeldii* (Bleeker) | + | + |
| *Rasbora sumatrana* (Bleeker) | + | + |
| *Thynnichthys thynnoides* (Bleeker) | + | + |
| Clariidae | | |
| *Clarias macrocephalus* (L.) | + | + |
| Siluridae | | |
| *Ompok bimaculatus* (Bloch) | + | + |
| *Pangasius micronemus* (Bleeker) | + | – |
| Bagridae | | |
| *Mystus baramensis* (Regan) | + | + |
| *Mystus nemurus* (C&V) | + | + |
| *Mystus negriceps* (C&V) | + | + |
| Anabantidae | | |
| *Helostoma temminckii* (C&V) | + | + |
| *Osphronemus goramy* (Lacepede) | + | + |
| *Pristolepis fasciatus* (Bleeker) | + | + |
| *Trichogaster pectoralis* (Regan) | + | – |
| Channidae | | |
| *Channa striata* (Bloch) | + | + |
| *C. micropeltis* (C&V) | + | + |
| Notopteridae | | |
| *Notopterus chitala* (Hamilton) | + | – |
| *N. notopterus* (Pallas) | + | + |
| Eleotridae | | |
| *Oxyleotris marmoratus* (Bleeker) | + | + |
| Osteoglossidae | | |
| *Scleropages formosus* (Muller & Schlegel) | + | + |

[1] Modified from Lee (1989).

Other methods used to catch fish are long lines (for *Mystus* sp., *Oxyleotris marmoratus* (Bleeker) and *C. striata*) and gillnets (for *Labiobarbus* sp., *P. schwanenfeldii* and *O. marmoratus*) (Table 38.5). Cast nets are also used but they are not popular because they tend to get snagged during use. Freshwater prawn (*Macrobrachium rosenbergii*) is also harvested using hook and line.

**Table 38.5**  Selected fisheries parameters for Chenderoh Reservoir and Kerian River artisanal fisheries, Perak, Malaysia.

| Parameters | Chenderoh Reservoir[a] | Kerian River[b] |
|---|---|---|
| Estimated number of fishers | 20–30 | 10–20 |
| Age range | 18–55 | 18–45 |
| Highest educational level | High school | High school |
| Dominant gears | Wooden boats (4–6 m long; 1.0–1.5 m wide) 5–7 hp long-tailed petrol engine (1) Gillnets (5.7, 10.2 & 11.4 cm stretched-mesh) (2) Long lines (30–40 m long; 30–40 hooks) (3) Cast nets | Wooden & fibreglass boats (4–6 m long; 1.0–1.5 m wide) 2, 5–15 hp outboard motors (1) 'Chandik' (fish hut) (2) Gillnets (variable mesh) (3) Long lines 2–3 m long; (5–10 hooks) |
| Landing points | 3 major points | Multiple points |
| Middlemen | 3 | Variable |
| Estimated daily catch rate per fisher | 2.7–12.8 kg[c] | 2.33 kg[d] |

[a] Modified from Lee (1989)
[b] Modified from creel surveys
[c] Caught mainly with gillnets which are set about 6 days a week – Modified from Ali & Lee (1995)
[d] Caught mainly with 'chandik' which are set about twice monthly during fishing seasons (March to April and August to September).

Fishing activities on the Kerian River are highly seasonal. Fishers operate actively only during rainy seasons when water levels are high, usually in March-April and August-September. The fishing communities of the Kerian River are more diffused compared to Chenderoh Reservoir. There is a good opportunity for outsiders to enter the fisheries at many access points along the river. The river is divided into sections and each section is loosely controlled by a group of fishers from one village and interactions between different groups seldom occur. The Kerian River fisheries are not controlled by middlemen as in the Chenderoh Reservoir. Fish landings can be sold to any middlemen at landing points or can be taken by the fishers to the nearby towns and villages for marketing.

The Kerian River is also exploited for *Scleropages formosus* (Muller & Schlegel) (Family Osteoglossidae), also known as 'kelisa', a very expensive aquarium fish popular locally as well as in other Southeast Asian countries. Because of its high monetary value, exploitation of this species is highly regulated and managed by the fishers to ensure its sustainability. This species is sold through middlemen and it is in the process of conserving the breeding stocks that management of the Kerian River fisheries evolved.

## 38.5     Fishers' perception and management of the fisheries

### 38.5.1     *Chenderoh Reservoir*

The local fishers depend on the reservoir for their secondary incomes. Fish provides not only protein but also cash for everyday expenses which is about US$110.00 per fisher per month (Lee 1989). Due to different price tiers (Lee 1989) certain fish species, such as *Mystus* sp., *P. bulu* and *T. thynnoides*, are exploited more than the others. Thus there is a need for the fishers to manage and regulate catches to maintain sustainability. Regulation by the fishers began in 1983; prior to that the fisheries were not regulated.

Regulation was started by limiting access to the fisheries (Fig. 38.2). Outsiders were not allowed to enter the fisheries and enforcement was made through strong discouragement and even intimidation. Fishing gears were standardized (Ali & Lee 1995) with each fisher using a wooden boat (4 to 6 m long), a long-tailed petrol engine (5 hp) and 15 to 20 sets ('kepala') of gillnets of certain mesh sizes (5.7, 10.2 and 11.4 cm stretched-mesh and 3 m deep). Although gillnets are the dominant gear used (catching

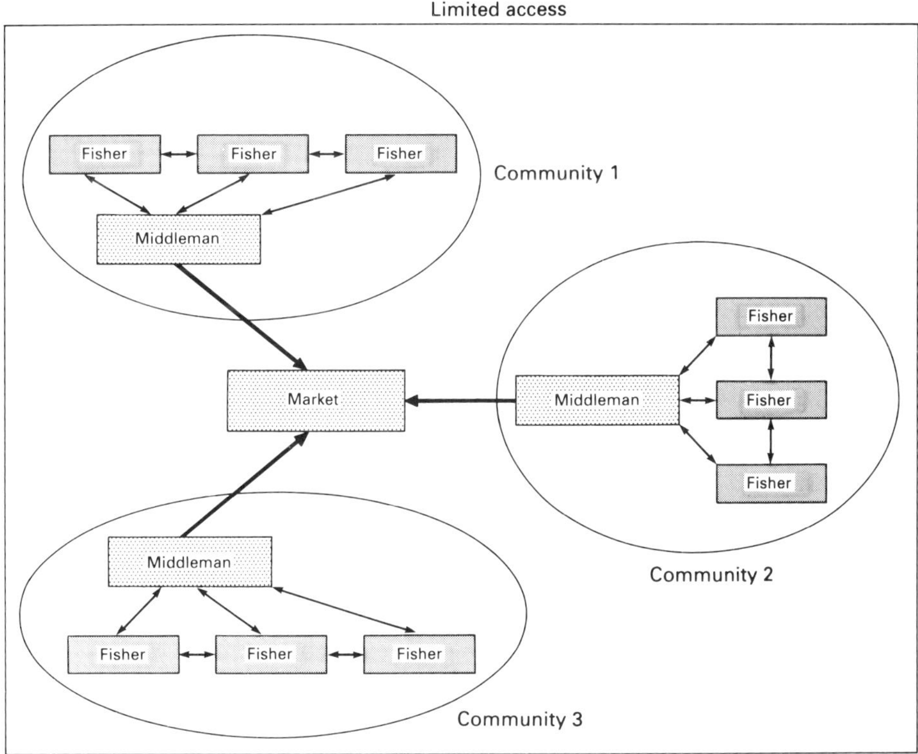


**Fig. 38.2**  Schematic representation of the Chenderoh Reservoir fisheries.

30.8 to 93.8% of the catch) other gears such as long lines (30 to 40 m long and 1 to 2 sets per fisher) are also used (Ali & Lee 1995).

Gillnets are set at around 1600 h and lifted at 0630 h the next morning. The catches are sold to the middleman at around 0730 to 0800 h. Since each fishing community is closely knit, it is difficult for any one fisher to break the rules without the knowledge of others (Fig. 38.2).

Enforcement of the fisheries falls on the middleman in charge of each fishing community (Fig. 38.2). There are presently only three active landing points, managed by three middlemen and serving three different fishing communities. The middleman can refuse to buy certain catches if the sizes are too small. Since the relationship between the fishers and the middleman is monopolistic, the fishers have no choice but to conform to the set rules and regulations. This type of relationship between the fishers and middleman is quite common among artisanal fisheries of Southeast Asia. Pomeroy (1992) discussed the role of the middlemen (the 'suki') in the Philippines small-scale fisheries. In Chenderoh Reservoir, the middleman sets the ex-landing prices for different fish species as well as different fish sizes. These prices are lower than that which the fishers can obtain if they sell their catches at the local markets. However, the fishers still prefer to deal with the middleman because he also provides credit to the fishers, loaning small sums of money (between US$4.00 to US$8.00) to fishers for engine or boat repairs. The fishers can never expect to get this kind of loan from banks or any other credit agencies. Thus although the relationship between the middlemen and the fishers is exploitative in nature the fishers and the middlemen in these small-scale artisanal fisheries depend on each other for their livelihood.

### 38.5.2    *Kerian River*

The Kerian River fisheries are much more loosely organized than those of Chenderoh Reservoir (Fig. 38.3). There are no distinct fishing communities because fishers that utilized the river can come from different villages. Furthermore the entry points to the river can be at any location along the river banks.

There is no set regulation concerning the fisheries. Any fisher can build a 'chandik' and set long lines or gillnets to catch fish (Fig. 38.3). In the early 1980s fish were also caught using derris root poison. Catch is sold to any middleman available. There is no standardization of gears. Boats can be fibreglass or wooden, and measure between 4 and 5 m long and 1 and 1.5 m wide, and are equipped with 2–5 hp outboard motors. In the survey conducted in February 1994, fibreglass boats were becoming more dominant and some fishers were using 10–15 hp outboard motors. There are thus extreme variations in gears and equipment which translate into different fishing efforts. Since the fishers are from different villages it is difficult for them to agree to any set rules and regulations concerning the fisheries. In terms of marketing the catches, currently the fishers themselves transport their catches to the nearest towns or villages, since middlemen have stopped buying fish at the landing points.

There is good regulation of the 'kelisa' fishery in the Kerian River. The high prices

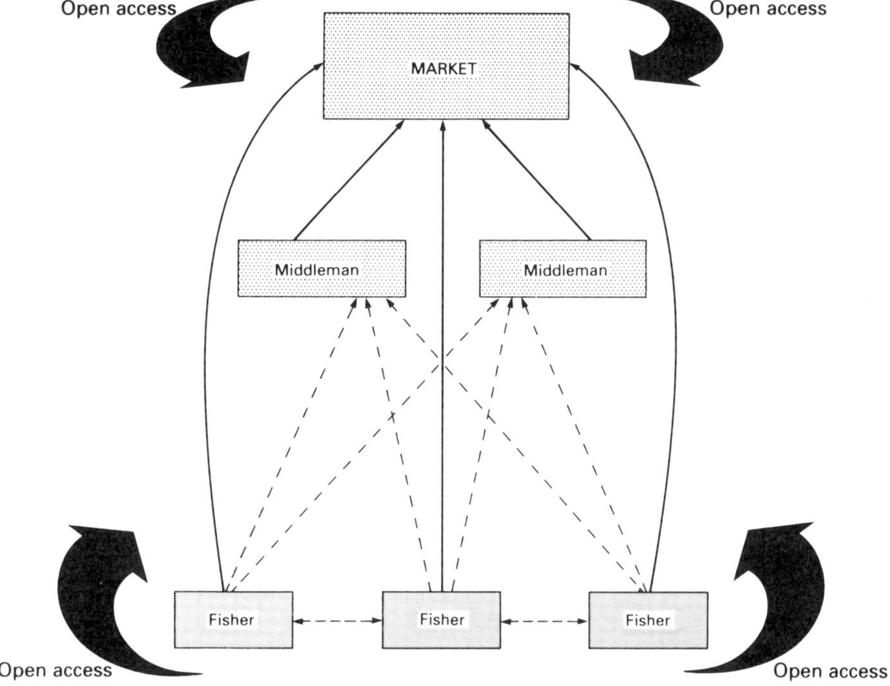


**Fig. 38.3**   Schematic representation of the Kerian River Fisheries.

for the species make it sensible for the fishers to co-operate and ensure that it remains sustainable. The most rigorous unwritten rule among the fishers is against catching adult fish; those who do so will be ostracized. Fishers are also expected not to catch fingerlings larger than 8 cm to ensure recruitment into the breeding populations. Fishing effort is standardized by allowing only the use of dip nets and headlamps to catch the fry at night. To ensure no conflicts arise between various fishing groups, each group can only fish a certain portion of the river and this rule is rigorously enforced. If adults are inadvertently caught either during or outside of their breeding season, the fish must be released immediately.

To ensure the survival of strong year classes, the fishers also agree not to cut down or disturb riparian vegetation along the river banks. This submerged vegetation, especially the rush *Coix* spp. is used as breeding, feeding and nursing habitat by the 'kelisa'. Thus, maintaining their stands along the river banks ensures that the 'kelisa' populations have suitable breeding sites and high chances of producing strong year classes.


### 38.6     Conclusions

The small-scale inland fisheries of Malaysia are similar to other small pelagic fisheries in Southeast Asia; they are open-access with little regulation, multispecies and multigears and large numbers of small-scale fishers (Dalzell *et al.* 1990).

One of the important management pre-requisites for resource sustainability is the creation of property rights empowering 'owners' to enforce measures that would result in optimum benefits (Cruz-Trinidad 1993; Monbiot 1994). Once the property rights of the community are established and recognized, access to the fishery can be controlled (Pomeroy 1993). Thus, in this context, common property rights, as defined by Hardin (1968) are unacceptable. Well defined property rights, in which each individual fisher has equal co-ownership of the resource, are needed to control access to the fishery (Ciriacy-Wantrup & Bishop 1975). Since these equal co-owners of the resource set the rules and regulations that will sustain the fisheries themselves, common property here represents private property for the group of co-owners (Pomeroy 1993). Recognition must also be given to communities that, given the opportunity, are able to manage their own resources wisely, as has been shown by the coastal communities of Thailand (Chansnoh 1993).

In the current study the two fishing communities took two different approaches to managing their respective common property resources. The ability of the Chenderoh fishing communities to work together is due to the fishers all coming from same fishing villages. The problem of determining property rights to the resource did not arise since each community assumed those rights as a matter of course. In closely knit communities each member is expected to adhere and conform to any norms, values or taboos set by the communities. Failure to conform results in 'loss of face' to the other members of the community.

The role of the middleman is that of an enforcer of rules and regulations set up by the community. Pomeroy (1992) described similar arrangements among small-scale fishers in the Philippines, although in that arrangement the middlemen, or 'suki', did not act as enforcers of rules and regulations. The middleman, by exercising his right not to buy certain types of catches, exerts a tremendous influence on the fishing community. The monopolistic nature of the relationship may be unprofitable to the fishers, however, as observed by Pomeroy (1992); in small fishing communities where everyone knows each other the middlemen's exploitative behaviour is inhibited by social and family ties.

In Chenderoh Reservoir fisheries the middleman ensured that all catches are bought. By extending credit (albeit at very small amounts) the middleman enables the fishers to survive in a situation where banks or co-operatives would not be able to help. This paternalistic co-operation between the middleman and the fishers helps to maintain the present fisheries of Chenderoh Reservoir because, in the long run, it is to the advantage of both groups to develop sustainable fisheries.

This community-based management is unique only to Chenderoh Reservoir. It is not practised in other reservoirs or inland fisheries of Malaysia. The other major reservoirs such as the Kenyir and Temenggor do not have established fishing villages, thus access to the fisheries is open to the general public.

In the Kerian River fisheries the loose social and family connections between fishers do not allow community-based management to develop. The highly regulated 'kelisa' fishery evolved due to its high economic returns which in the 1993/1994 season were as high as US$150.00 per fingerling. Co-operation towards developing a sustainable fishery in this case is highly desirable for the fishers. Although community-based management for the general fisheries is not well developed, there is an emergence of new thinking among fishers towards regarding the river as an important ecological and economic resource. The emergence of these 'naturalist-rivermen', who spend most of their free time in the river, bodes well for Kerian River conservation. These rivermen spend their free time cleaning up the river and protecting riverine vegetation for use by birds, small mammals and other organisms. The appearance of this group of fishers will probably result in more appropriate management of the river in the future.

## Acknowledgements

Fund for this study was provided by the Malaysian Government Research & Development (R&D) Grant. I thank Mashhor Mansor, Abu Hassan Ahmad, Zulfigar Yassin, Abd. Wahab Abd. Rahman, Nazlan Najimudin, Mohd. Hifni, Md. Yaakob Md. Yusof, Khalid Putih, Mohd. Yusof Omar and Ganesh for their help, support and encouragement. Special thanks are also due to all the fishers from Chenderoh Reservoir and Kerian River for their help and co-operation.

## References

Ajan J.K. (1983) Assessment of Tasek Chenderoh fisheries based on gill net. BSc Thesis, Universiti Sains Malaysia, Penang, Malaysia, 69 pp.

Ali A.B. & Kathergany M.S. (1987) Preliminary investigation on standing stocks, habitat preference and effects of water level on riverine fish population in a tropical river. *Tropical Ecology* **28**, 264–273.

Ali A.B. & Lee K.Y. (1995) Chenderoh Reservoir, Malaysia: A characterization of a small-scale, multigear and multispecies artisanal fishery in the tropics. *Fisheries Research* **23**, 267–81.

Bat Kamaruzzaman B.A.K. (1992) The study on the biology (reproduction and food habits) of ikan lomah, *Thynnichthys thynnoides* (Bleeker) at Tasik Chenderoh. BSc Thesis, Universiti Sains Malaysia, Minden, Penang, Malaysia, 77 pp.

Chansnoh P. (1993) Community management of coastal resources, Southern Thailand. *NAGA – The ICLARM Quarterly* **16**, 10–12.

Ciriacy-Wantrup S.V. & Bishop R.C. (1975) Common property as a concept in natural resource policy. *Natural Resources Journal* **15**, 713–727.

Cruz-Trinidad A. (1993) Economic exploitation in the Philippines small pelagic fishery and implication for management. *NAGA – The ICLARM Quarterly* **16**, 13–15.

Dalzell P., Corpuz P., Arce F. & Ganaden R. (1990) Philippines small pelagic fisheries and their management. *Aquaculture & Fisheries Management* **21**, 77–94.

Hardin G. (1968) The tragedy of the commons. *Science* **162**, 1243–1248.

Johannes R.E. (1993) The plight of the Osfish, or why quantitative sophistication is no substitute for asking the right questions. *NAGA – The ICLARM Quarterly* **16**, 4–5.

Khan M.S. (1987) Some aspects of the biology of ikan baung, *Mystus nemurus*, C&V with reference to Chenderoh Reservoir. MSc Thesis, Universiti Pertanian Malaysia, Selangor, Malaysia, 196 pp.

Lee K.Y. (1989) Preliminary study on fish populations of Chenderoh Reservoir, Perak. BSc Thesis, Universiti Sains Malaysia, Penang, Malaysia, 109 pp.

Lee K.Y. & Ali A.B. (1989) The status of reservoir fisheries in Tasik Chenderuh, Perak: A case study. Proceedings of the 12th Annual Seminar Malaysian Society of Marine Sciences, **12**, 231–239.

Monbiot G. (1994) The tragedy of enclosure. *Scientific American* **270**, 140.

Pomeroy R.S. (1992) Fish marketing in the Philippines: Is the 'suki' symbiotic or parasitic. *NAGA – The ICLARM Quarterly* **15**, 13–14.
Pomeroy R.S. (1993) Clearing up some misconception: Open access vs. common property. *NAGA – The ICLARM Quarterly* **16**, 40–41.

# Chapter 39
# The integration of fish stock assessment into fisheries management

I.G. COWX *International Fisheries Institute, University of Hull, HU6 7RX, UK*

**Abstract**   The current approach to managing inland fisheries relies heavily on having some measure of stock abundance or associated population parameter(s) on which to formulate the decision-making process or evaluate the impact of a particular management activity. For example, stock assessment is usually deemed necessary to gauge the effects of activities such as the impact of overfishing, a pollution event, land drainage improvement works, river regulation, habitat restoration, stocking, introduction of a new species, or merely to respond to requests for advice on the management of waters. However, in many cases, particularly in large water bodies, the results obtained are disappointing because of the problems of estimating absolute abundance of the fish populations by conventional scientific methods. Often the root cause is poor fish sampling efficiency and subsequent failure to meet the assumptions underlying the assessment methodology/models. Notwithstanding, many practitioners, having knowingly violated the assumptions to some (unmeasured) extent, proceed to interpret the results as if no possibility of error existed. The long-term consequences of such an approach are obvious.

A strategic approach to stock assessment which meets the demands of both the manager and the scientist is considered. The objectives of fishery management activities are reviewed and the level of precision required from stock assessment procedures evaluated. Finally, the advantages and disadvantages of the different sampling strategies and techniques available are assessed and how they can be best used to satisfy the management objectives. Particular attention is paid to the use of relative, cost-effective methods for providing information on the status of fish stocks.

KEYWORDS: Data precision, management, population assessment, sampling strategies

## 39.1   Introduction

The majority of inland fisheries tend to be heavily exploited or utilized for a variety of reasons, and this generally leads to the need for management of the resources. The objectives are usually associated with the types of use (e.g. commercial or recreational fishing), level of exploitation (for supply of fish protein, broodstock or ornamentals), as well as with socio-economic factors connected with the associated community (e.g. maintenance of employment or conservation value).

In terms of management, these objectives would normally demand action aimed at the maintenance of the fish stock. The current approach to managing inland fisheries therefore relies heavily on having some measure of stock abundance or associated population parameter(s) on which to formulate the decision-making process or evaluate the impact of a particular management activity. For example, stock assessment is usually deemed necessary to gauge the effects of activities such as the impact of overfishing, a pollution event, land drainage improvement works, river regulation, habitat restoration, stocking, introduction of a new species, or merely to respond to requests for advice on the management of waters. Unfortunately, in many cases the information available on the fish stocks is either poor or does not meet the demands of management. In these cases the data are often inaccurate or lack precision because they were collected using methods of assessment which were totally unsuitable for the prevailing conditions. Often the root cause is poor fish sampling efficiency and subsequent failure to meet the assumptions underlying the assessment methodology/models. Notwithstanding, many practitioners, having knowingly violated the assumptions to some (unmeasured) extent, proceed to interpret the results as if no possibility of error existed. In some cases, this practice arises because the manager demands a definitive appraisal of the fishery to formulate policy and is unaware of the restrictions of the scientist.

It should be recognized from an early stage that management of inland fisheries is an inexact science and that it is often not possible to obtain precise information on fish stocks. Nonetheless, many of the limitations and restrictions can be overcome or minimized by strategic planning of the stock assessment exercise. It involves consultation and collaboration at all levels of activity from the fishermen through the scientist to the manager, and *vice versa*. This chapter describes some of the issues and problems faced by the scientists and managers and considers a strategic approach to stock assessment which meets the demands of all groups.

## 39.2    Relationship between fisheries stock assessment and management

To manage fisheries successfully there is a prerequisite for 'adequate' information on the status of the fish stocks. The mechanism by which this is achieved is simplified in Figure 39.1. In essence fishery survey data are modelled to provide managers and policy makers with advice on exploitation/utilization of aquatic resources. The stock assessment procedure (data collection and handling) is not an isolated activity but is an integrated component in the overall fisheries management plan. Failure to recognize this can lead to poor stock assessment procedures and inadequate advice. Unfortunately many scientists are faced with increased pressure from managers for more accurate and extensive information. This is paralleled by a reduction in avail-

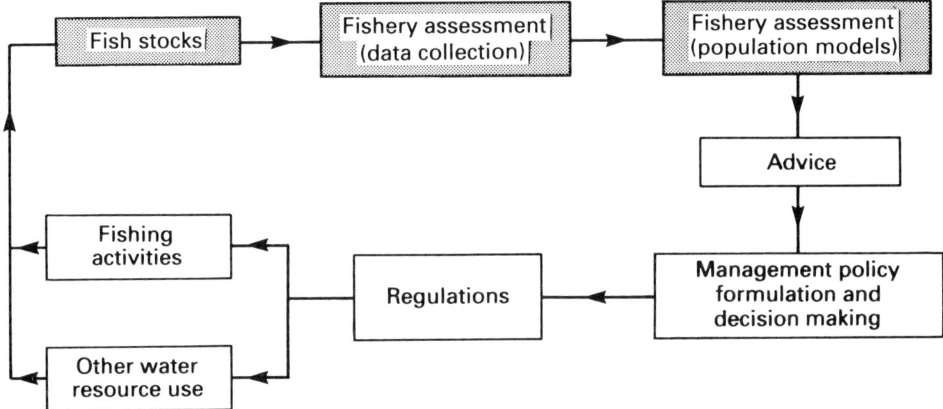

**Fig. 39.1** Relationship between fish stock assessment and management (shaded area represents the activities usually carried out by the fishery scientist).

able resources and often a poor understanding of the practicalities of achieving the desired output.

To ensure that the best possible advice is provided by the scientist a number of steps should be considered (Fig. 39.2). This mechanism is akin to the project cycle developed by international donor agencies to formulate and implement projects (Anon. 1982). It can be broken down into five phases: identification, preparation, appraisal, implementation and post evaluation (Fig. 39.2). This step-wise approach minimizes the potential for introducing gross inaccuracies and biases from the stock assessment procedure.

### 39.3    Objectives of stock assessment

The information that needs to be collected during a stock assessment exercise is, in part at least, dependent on the management objectives for the fishery. In this respect the objectives of fisheries management tend to fall into a number of categories (Cowx 1995; Harvey & Cowx, Chapter 2):

- evaluation of the status of the fish stocks for conservation and enhancement purposes;
- monitoring long-term population changes as a result of natural or anthropogenic activities;
- evaluation of response of management activities directly targeted at the river system or its fisheries, e.g. restocking or introductions, habitat improvements, water quality improvements, flow regulation;
- assessment of environmental damage, e.g. post-impact appraisal of a pollution incident, fish kill or natural catastrophe such as a drought or flood;
- prediction of the impact of development activities on the fisheries (i.e. environmental impact assessment).

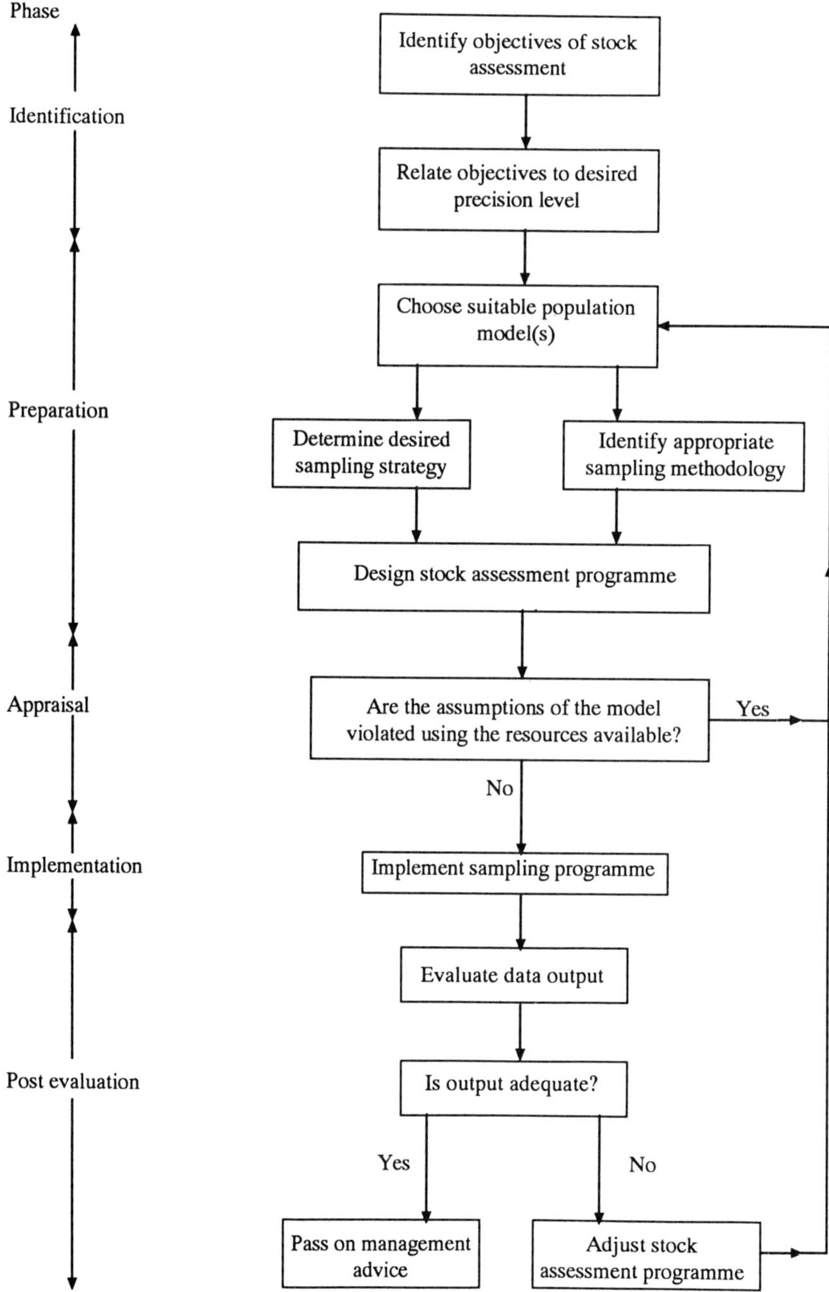


**Fig. 39.2**  Schematic strategy for planning a fish stock assessment programme.

The data requirements for each of these objectives vary according to the precision deemed necessary to support the management decision-making process (see Section 39.4). Sometimes the aim is to assess total population size, e.g. when comparing populations with respect to yield, or population density (number per unit area), e.g. when studying population regulation as a function of stocking. Both outputs are absolute population parameters and require considerable resources to be estimated. Often, however, the aim is to assess temporal or spatial changes and trends, such as in environmental monitoring or the impact of management regulatory measures. In these cases it may be sufficient to make an estimation of relative population parameters (presence/absence, proportional population change or catch per unit effort (CPUE)), thus allowing comparisons between sampling times or locations. Relative estimation of population parameters is usually less costly than absolute estimation, and in many cases may be more than adequate. Finally, the stock assessment exercise may be orientated towards individual fish parameters such as length, weight, growth or fecundity. Again these data are simpler and cheaper to collect because only a representative sample of fish is required.

When defining the objectives it is essential that the type of data output (absolute or relative) is identified and targeted in the planning of the stock assessment exercise. Integral within this procedure is the identification of a mechanism for verifying that the outputs are acceptable and meet (to all parties) the requirements of the exercise. It is also valuable to identify the potential risks of not achieving the desired output. One mechanism for establishing these criteria is the use of the logical project framework (Anon. 1982). This is a simple procedure which is useful in setting out the design of a project in a clear and logical way so that any weaknesses or problems that exist can be brought to the attention of scientist and manager (Crean 1994; Cowx 1995). The identified problems and deficiencies may then be remedied at an early stage, or, if insuperable, the project can be redesigned or, in the worst case scenario, aborted.

## 39.4     Precision level for stock assessment

Before identifying possible mechanisms for undertaking studies with different objectives, it is important to consider the desired information with respect to individual fish or populations, and the accuracy and precision that must be achieved.

In this context, accuracy is associated with the type of error or bias in the data. Poor accuracy tends to lead to assessments that considerably, but consistently, over- or under-estimate. Precision is associated with the 'noise' (usually expressed as the variance or coefficient of variation, CV, of the estimate) generated by the sampling procedure, and is usually reduced by larger sample sizes or repetitive surveys (Southwood 1978). A highly reliable estimate will have a low coefficient of variation. The precision of the stock estimate dictates the change in stock parameters that can be detected

If population parameters are being determined, the required precision of the estimated abundance or magnitude of change (spatial or temporal) that needs to be detected, must be determined in relation to the objectives. This minimizes the risk of obtaining a precision too low or high for the purpose. As the choice of precision level will strongly affect the resource input, it is worth considering this question in relation to the objectives at the planning stage.

Bohlin *et al.* (1989) and Bohlin & Cowx (1990) suggested a rough guide for establishing precision levels for fisheries surveys. This is based on three categories.

- *Class 1:* Studies in this class require a high level of precision; a population change, in time or space by a factor as small as 1.2 (e.g. $83 \leftarrow 100 \rightarrow 120$) has to be detected with about 80% probability when using a 5% significance level. In the case of an independent estimation, this level of precision corresponds approximately to a coefficient of variation not larger than about 0.05.
- *Class 2:* Studies in this class require a high level of precision; a population change, in time or space by a factor as small as 1.5 (e.g. $67 \leftarrow 100 \rightarrow 150$) has to be detected with about 80% probability when using a 5% significance level. In the case of an independent estimation, this level of precision corresponds approximately to a coefficient of variation not larger than about 0.10
- *Class 3:* Studies in this class require a high level of precision; a population change, in time or space by a factor as small as 2.0 (e.g. $50 \leftarrow 100 \rightarrow 200$) has to be detected with about 80% probability when using a 5% significance level. In the case of an independent estimation, this level of precision corresponds approximately to a coefficient of variation not larger than about 0.16.

Although the choice of precision level is ultimately the decision of the manager or scientist there are certain standards that need to achieved to meet the objective of the project/survey. The minimum acceptable precision level for each of the objectives of stock assessment outlined earlier is given in Table 39.1. This illustrates that absolute abundance is not a prerequisite of all stock assessment exercises, and this should be taken into consideration when planning surveys.

In addition to collecting data of the desired precision to formulate the management response, there are two other reasons for considering precision level.

(1)   Data collection exercises tend to be expensive and high precision has to be paid for in time, manpower and financial resources.

**Table 39.1**   Minimum desirable stock assessment data for various management activities.

| Activity | Abundance estimate | Precision level |
| --- | --- | --- |
| Evaluation of the status of the fish stocks | Relative/absolute | 2–3 |
| Monitoring long-term population change | Relative | 3 |
| Evaluating response to management activities | Absolute/relative | 2–3 |
| Assessment of environmental damage | Absolute/relative | 1–2 |
| Environmental impact assessment | Absolute/relative | 1–2 |

(2)  It is often difficult to obtain precise information on the status of the fish populations because the sampling technology is not available.


### 39.5  Methods of sampling fish populations

Numerous methods are available for sampling fish populations; however, not all are appropriate for all situations. For example, seine netting cannot be used on large fast-flowing rivers with many obstacles or electric fishing in the offshore region of a large lake. There is thus a need to marry the appropriate methodology and sampling strategy to optimize on the data collection exercise and ensure the information is collected in the most cost-effective manner. This is a two-stage process where the selection procedure is based on meeting the assumptions of the assessment models (Fig. 39.3), and choosing the most cost-effective sampling method. The mechanism for choosing the most suitable survey design has been considered in detail by Cowx (1995). In essence the planning exercise must identify whether the proposed sampling strategy is feasible with the methodology available. If the assumptions upon which the models are based are violated an alternative must be sought. This is frequently the case when absolute estimates of population size are required, because it is extremely difficult to fulfil the assumptions on which the models are based, except in small streams and ponds. Even then violations are frequently incurred, such as the declining catchability of fish by electric fishing during successive catches for depletion estimates (Bohlin & Cowx 1990). Consequently, it is highly probable that stock assessments on large water bodies will be relative evaluations.

Once the assessment mechanism is determined, consideration should be given to the most cost-effective and efficient sampling method. Generally this is based on the experience of the scientist undertaking the assessment procedure, and scientists tend to have personal preferences. To minimize these selection problems a linear programming exercise can be carried out. This essentially categorizes all the attributes and disadvantages of the various methods and weights (raising factor) them according to a predetermined level of importance for each criterion. A simplified exercise is shown in Table 39.2. Here electric fishing, seine netting and angler catches are compared for a relative assessment procedure. It should be noted that although angler catches appear to be the best data collection mechanism this will not provide a quick response and may take several years to provide a thorough evaluation (Cowx 1990). Also due consideration should be given to the biological characteristics (migration patterns, behaviour, etc.) of the target species as this can affect the sampling success.

A better approach, where resources permit, would be to adopt a multiple gear strategy, i.e. using the combined output of several methods. This will reduce selectivity biases and probably overcome some of the sampling limitations of a single gear approach.

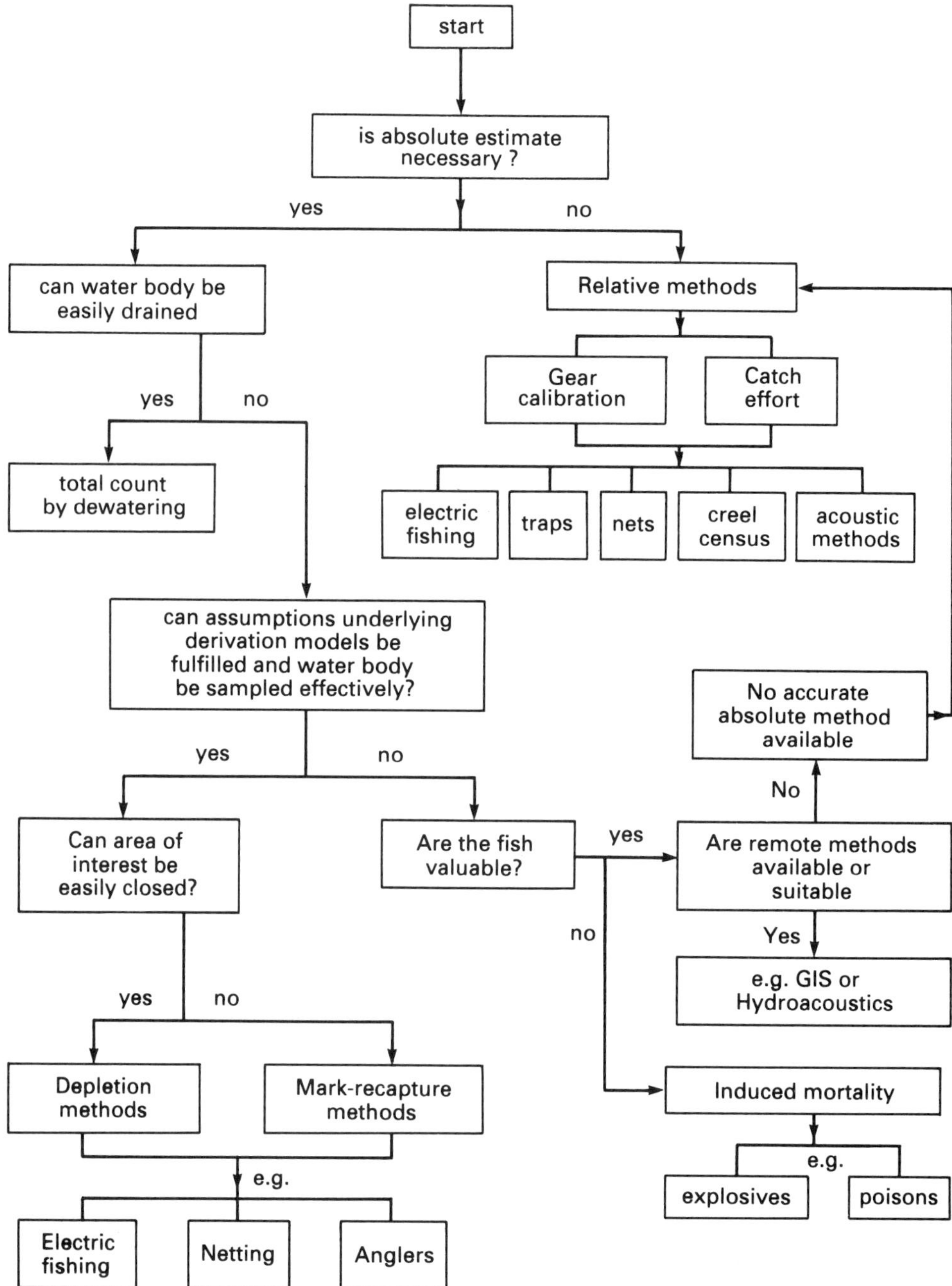


**Fig. 39.3**  Schematic approach to selecting the most appropriate gear and stock abundance methodology (modified from Cowx 1995).

**Table 39.2** Linear programming for stock assessment in a medium flowing stream, conductivity 500 μS cm$^{-1}$, temperature 10°C, many obstructions.

|  | Obstruction | Flow | Conductivity | Temperature | Efficiency | Cost | Sum |
|---|---|---|---|---|---|---|---|
| *Rising factor* | 5 | 5 | 5 | 5 | 20 | 30 | |
| Electric fishing | 2 | 2 | 2 | 2 | 3 | 2 | 160 |
| Seine netting | 1 | 1 | 3 | 2 | 2 | 1 | 105 |
| Anglers | 3 | 3 | 3 | 2 | 1 | 3 | 165 |

## 39.6     Stock assessment models

A wide diversity of stock assessment models are currently in use. These range from detailed, long-term assessments of single species such as trout (*Salmo trutta* L.) in temperate streams (e.g. Kristiansen, Chapter 23) to rapid appraisal of multispecies (e.g. Okada, Agostinho & Petrere, Chapter 12), multigear floodplain fisheries in the tropics (e.g. Hoggarth & Kirkwood, Chapter 20; and from models on the optimal stocking of reservoir fish (e.g. Allison, Chapter 24) to broader simulations of the interactions between fisheries and other components of the aquatic system (e.g. Cryer, Chapter 28). The approaches used are generally well known (see, for example, Cowx 1995) and are summarized in Fig. 39.4.

The four strategies indicated provide output data at different levels of precision and each requires very different levels of input. Experimental and monitoring strategies are typically those used for determining absolute population parameters and use intense, efficient, data collection methods to provide input for the models. It is often this approach which is misused to provide absolute population parameters because the data collected are inadequate to fit the assumptions of the model (particularly mark-recapture and depletion models). The strategy can be, and frequently is, used to provide relative population parameters, which is probably a more appropriate mechanism in large rivers and lakes.

Fisheries statistical methods are the main methods used where the fisheries are commercially or recreationally exploited. They provide an assessment of the status of the existing fishery but the quality of the output is very dependent on the input information which can be highly variable depending on its source, e.g. scientifically collected or provided by fishermen.

Simulation of exploitation strategies (e.g. Lorenzen 1995) and environmental correlation methods (e.g. Cryer, Chapter 28) are less precise in their output but do provide a good basis for initiating policy development. They offer a starting point for the development of the management process but need to be supported by more accurate methods if management is to be formulated on a sound scientific basis. They are particularly useful in fisheries where no fishery data exist, but their application should be treated with caution. This was clearly illustrated by Cryer (Chapter 28),

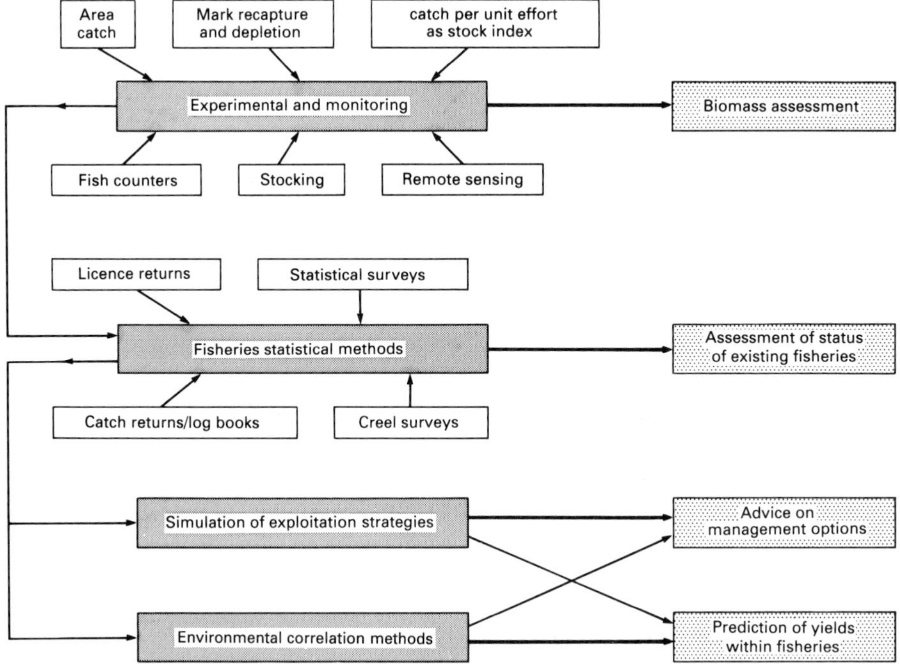


**Fig. 39.4** Suggested combination of methods and models for fish stock assessment and yield predictions. Dark shaded area represents the assessment models and light shaded area the typical output.

who showed the potential for gross misrepresentation of the status of the fisheries when only environmental correlation methods were used.

The model selected for a given situation will be determined by the information required, the data available or obtainable, the assumptions which can be made, and the funding, resources and experience available. The choice of model will always be a compromise between these various factors. For example, a simple, cheap model will be easy to apply, but will require many assumptions with unknown risks attached. By contrast, a more detailed model may provide useful insights into the dynamics of the system, but only at the cost of expensive data collection programmes and staff expertise. It should be noted, however, that more complicated models are not necessarily more precise, although it may be easier to remove some of the uncertainties from the assessment.

### 39.7    Design of stock assessment programme

Once the objectives and desired output criteria of the stock assessment programme have been identified, effort should focus on planning its implementation. As already indicated, the planning phase should consider:

(1)   the limitations of the various gears in different sampling situations;

(2)   limitations of the various stock assessment models;
(3)   the biology and behaviour of the target species and populations;
(4)   the availability of resources.

It is extremely doubtful whether sufficient detailed information is available on fish species in large water bodies. Thus, before embarking on a stock assessment programme a critical appraisal of the procedures should be carried out. The construction of flow charts, such as those shown in Figs 39.2 and 39.3, can be drawn up to guide the operator towards the most appropriate strategy. Goals should be set and activities should be orientated towards meeting the aims. The appraisal should also draw out the limitations of the exercise in relation to the models' assumptions and availability of resources, and the risks of not achieving the objectives. Any weaknesses that exist in the proposed assessment programme can be brought to the attention of the planners or managers at this stage. The identified deficiencies may then be remedied at an early stage, thus avoiding the problem of undesirable data output. The best mechanism of achieving this is the use of the logical project framework (Crean 1994; Cowx 1995).

## 39.8     Implementation of the stock assessment programme

Once the stock assessment strategy has been ratified, detailed work plans and financial/resource arrangements can be drawn up. The stock assessment activity programme is then carried out. During implementation it is important to constantly evaluate the outputs in relation to the established indicators. If the desired standards and outputs are not achieved the strategy may have to be adjusted or reconsidered. It is imperative that the operators do not allude themselves that they are achieving the desired results when it is quite clear they are not.

Finally, the overall stock assessment programme should be evaluated in relation to the objectives defined and verified against indicators established during project identification. This feedback is imperative to improve the methods and strategies for stock assessment. Both successes and failures, problems encountered and the way they were tackled, and limitations (relating to manpower, gear, methods of estimating abundance and finance) of the procedures should all be reported.

## 39.9     Conclusions

The importance of biological assessment in the management of fish stocks is beyond question, but the methods and limitations of the procedures frequently employed continue to raise concern. The planning strategy described above will help to ameliorate some of the potential conflicts that arise from stock assessment programmes, but will not overcome the need for more precise information. It should ensure the most appropriate data are collected to meet the demands of the assessment models or, conversely, the most appropriate model is selected to fit the data available. Decisions on the choice of stock assessment procedures should be made jointly between managers and scientists, depending on the most recent information from the fishery. In certain circumstances, 'adaptive management' strategies (Hilborn & Walters 1992) may produce the most useful information if continual monitoring is undertaken, but only if the risks involved are compatible with the socio-economic status of the fishery. Long-term studies are invaluable in this respect, particularly where environmental fluctuations are pronounced, and greater consideration should be given to such activities. The monitoring of index rivers or lakes is a useful, cost-effective mechanism for achieving this desirable activity.

At present most stock assessment programmes are based on data collected by one sampling method. This can lead to inherent difficulties or inaccuracies. Consideration should be given to linking assessment programmes using different sampling technology, for example the combination of data from electric fishing and hydroacoustic surveys. These data frequently complement each other and allow a better evaluation of the status of the fish stocks.


## Acknowledgements

The author wishes to thank all those participants at the stock assessment workshops and in particular Willem Dekker who provided useful dialogue on stock assessment models.


## References


Anon. (1982) The logical framework approach (LFA). Oslo: Norwegian Agency for Development Corporation, 42 pp.

Bohlin T., Hamrin S., Heggberget T.G., Rasmussen G. & Saltveit S.J. (1989) Electrofishing – Theory and practice with special emphasis on salmonids. *Hydrobiologia* **173**, 9–43.

Bohlin T., Heggberget T.G. & Strange C. (1990) Electric fishing for sampling and stock assessment. In: I.G. Cowx & P. Lamarque (eds) *Fishing with electricity*. Oxford: Fishing News Books, pp 112–139.

Bohlin T. & Cowx I.G (1990) Implications of unequal probability of capture by electric fishing on the estimation of population size. In: I.G Cowx (ed.) *Developments in electric fishing*. Oxford: Fishing News Books, pp 145–155.

Cowx I.G. (1990) Application of creel census data for the management of fish stocks in large rivers in the United Kingdom. In: W.L.T. van Densen, B. Steinmetz & R.H. Hughes (eds) *Management of freshwater fisheries*. Wageningen: Pudoc, pp 526–534.

Cowx I.G. (1995) Fish stock assessment – A biological basis fo sound ecological management. In: D.M. Harper & A.J.D. Ferguson (eds) *The ecological basis for river management*. Chichester: J Wiley & Sons, pp 375–388.

Crean K. (1994) Planning and development of inland fisheries. In: I.G. Cowx (ed.) *Rehabilitation of freshwater fisheries*. Oxford: Fishing News Books, pp 21–33.

Hilborn R. & Walters C.J. (1992) *Quantitative fish stock assessment: choices, dynamics and uncertainty.* London: Chapman & Hall, 570 pp.

Lorenzen K. (1995) Population dynamics and management of culture-based fisheries. *Fisheries Management and Ecology* **2**, 61–73.

Southwood T.R.E. (1978) *Ecological methods*. London: Chapman and Hall.

# Index